The Mycota

Edited by
K. Esser

The Mycota

I	*Growth, Differentiation and Sexuality* 1st edition ed. by J.G.H. Wessels and F. Meinhardt 2nd edition ed. by U. Kües and R. Fischer
II	*Genetics and Biotechnology* Ed. by U. Kück
III	*Biochemistry and Molecular Biology* Ed. by R. Brambl and G. Marzluf
IV	*Environmental and Microbial Relationships* 1st edition ed. by D. Wicklow and B. Söderström 2nd edition ed. by C.P. Kubicek and I.S. Druzhinina
V	*Plant Relationships* 1st edition ed. by G. Carroll and P. Tudzynski 2nd edition ed. by H.B. Deising
VI	*Human and Animal Relationships* 1st edition ed. by D.H. Howard and J.D. Miller 2nd edition ed. by A.A. Brakhage and P.F. Zipfel
VII	*Systematics and Evolution* Ed. by D.J. McLaughlin, E.G. McLaughlin, and P.A. Lemke†
VIII	*Biology of the Fungal Cell* Ed. by R.J. Howard and N.A.R. Gow
IX	*Fungal Associations* Ed. by B. Hock
X	*Industrial Applications* Ed. by H.D. Osiewacz
XI	*Agricultural Applications* Ed. by F. Kempken
XII	*Human Fungal Pathogens* Ed. by J.E. Domer and G.S. Kobayashi
XIII	*Fungal Genomics* Ed. by A.J.P. Brown
XIV	*Evolution of Fungi and Fungal-like Organisms* Ed. by J. Wöstemeyer
XV	*Physiology and Genetics: Selected Basic and Applied Aspects* Ed. by T. Anke and D. Weber

The Mycota

A Comprehensive Treatise
on Fungi as Experimental Systems
for Basic and Applied Research

Edited by K. Esser

VI — *Human and Animal Relationships*
2nd Edition

Volume Editors:
A.A. Brakhage • P.F. Zipfel

Series Editor

Professor Dr. Dr. h.c. mult. Karl Esser
Allgemeine Botanik
Ruhr-Universität
44780 Bochum, Germany

Tel.: +49 (234)32-22211
Fax.: +49 (234)32-14211
e-mail: Karl.Esser@rub.de

Volume Editors

Professor Dr. Axel A. Brakhage

Tel.: +49-3641-5321000
Fax.: +49-3641-656600
e-mail: axel.brakhage@hki-jena.de

Professor Dr. Peter F. Zipfel

Tel.: +49-3641-656900
Fax.: +49-3641-656902
e-mail: peter.zipfel@hki-jena.de

Leibniz Institut für Naturstoff-Forschung und Infektionsbiologie e.V.
Hans-Knöll-Institut
Friedrich Schiller Universität
Beutenbergstr. 11a
00745 Jena, Germany

Library of Congress Control Number: 2008927360

ISBN 978-3-540-79306-9   e-ISBN 978-3-540-79307-6
ISBN 3-540-58007-7 1st ed.

This work is subject to copyright. All rights are reserved, whether the whole or part of the material is concerned, specifically the rights of translation, reprinting, reuse of illustrations, recitation, broadcasting, reproduction on microfilm or in any other way, and storage in data banks. Duplication of this publication or parts thereof is permitted only under the provisions of the German Copyright Law of September 9, 1965, in its current version, and permissions for use must always be obtained fromSpringer-Verlag. Violations are liable for prosecution under the German Copyright Law.

springer.com
© Springer-Verlag Berlin Heidelberg 1996, 2008

The use of general descriptive names, registered names, trademarks, etc. in this publication does not imply, even in the absence of a specific statement, that such names are exempt from the relevant protective laws and regulations and therefore free for general use.

Cover design: Erich Kirchner and WMXDesign GmbH, Heidelberg, Germany

Printed on acid-free paper   5 4 3 2 1 0

Karl Esser

(born 1924) is retired Professor of General Botany and Director of the Botanical Garden at the Ruhr-Universität Bochum (Germany). His scientific work focused on basic research in classical and molecular genetics in relation to practical application. His studies were carried out mostly on fungi. Together with his collaborators he was the first to detect plasmids in higher fungi. This has led to the integration of fungal genetics in biotechnology. His scientific work was distinguished by many national and international honors, especially three honorary doctoral degrees.

Axel A. Brakhage

earned a Master's degree in Biology with the major subject Microbiology at the University of Münster. During his PhD he spent a year at the Institut de Biologie Physico-Chimique in Paris. After obtaining a PhD on the regulation of the phenylalanyl-tRNA synthetase in *Bacillus subtilis*, he joined the Biotechnology Department of BASF as a staff scientist. He then started work on the molecular biology of filamentous fungi as a postdoctoral student at the University of Sheffield (UK). After two years, he became an Assistant Professor at the Institute of Genetics and Microbiology at the University of Munich where he earned his habilitation in Microbiology. In 1998 he became Associate Professor at the Darmstadt University of Technology. During 2001–2004 he was Full Professor of Microbiology at the University of Hanover. From 2004 he has held a chair of Microbiology and Molecular Biology at the Friedrich Schiller University of Jena. In 2005 he became head of the Department of Molecular and Applied Microbiology at the Leibniz Institute of Natural Product Research and Infection Biology – Hans-Knoell-Institute (HKI) – in Jena and, simultaneously, he is the Director of the Institute. He is a member of several scientific boards including a DFG panel on Microbiology and Infection Biology. He is Vice-President of the Association of General and Applied Microbiology (VAAM), coordinator of a DFG excellence graduate school and serves on the editorial boards of several Journals. He has received the Seeliger Award for Bacteriology and Mycology (2006). His research addresses two main areas: (1) pathobiology of the most important airborne fungal pathogen *Aspergillus fumigatus*. (2) molecular regulation of the biosynthesis of fungal secondary metabolites with emphasis on the role of transcription factors.

Peter F. Zipfel

studied Biology and Biochemistry with specialization in Cellular Biology. In 1984 he graduated from the University in Bremen. He performed his postdoctoral studies at the Laboratory of Immunoregulation at the NIAID, National Institutes of Health in Bethesda (Md., USA). Then he moved to the Bernhard Nocht Institute for Tropical Medicine in Hamburg where he headed a junior research group. He received his Habilitation and venia legendi from the University of Hamburg in 1993 in the areas of Immunology and Genetics. In 2000 he was appointed University Professor for Infection Biology at the Friedrich Schiller University in Jena and at the same time chairman of the Department of Infection Biology at the Leibniz Institute for Natural Product Research and Infection Biology – Hans-Knoell-Institute (HKI) – in Jena. He is author of over 200 research papers, book chapters and review articles, several patents and has edited a book on the role of the complement system in human renal diseases. He has received the Research Price for Basic Science from the State of Thueringen (2004) and the "Heinz-Spitzbart Preis" of the European Society for Infectious Diseases in Obstetrics and Gynaecology (ESIDOG; 2007). His scientific interests include: (1) understanding the role of the complement system as a part of the innate immune system in health and disease, (2) how human pathogenic fungi such as *Candida albicans* and *Aspergillus fumigatus* and numerous other pathogens utilize host complement components for immune evasion.

Series Preface

Mycology, the study of fungi, originated as a subdiscipline of botany and was a descriptive discipline, largely neglected as an experimental science until the early years of this century. A seminal paper by Blakeslee in 1904 provided evidence for selfincompatibility, termed "heterothallism", and stimulated interest in studies related to the control of sexual reproduction in fungi by mating-type specificities. Soon to follow was the demonstration that sexually reproducing fungi exhibit Mendelian inheritance and that it was possible to conduct formal genetic analysis with fungi. The names Burgeff, Kniep and Lindegren are all associated with this early period of fungal genetics research.

These studies and the discovery of penicillin by Fleming, who shared a Nobel Prize in 1945, provided further impetus for experimental research with fungi. Thus began a period of interest in mutation induction and analysis of mutants for biochemical traits. Such fundamental research, conducted largely with *Neurospora crassa*, led to the one gene: one enzyme hypothesis and to a secondNobel Prize for fungal research awarded to Beadle and Tatum in 1958. Fundamental research in biochemical genetics was extended to other fungi, especially to *Saccharomyces cerevisiae*, and by the mid-1960s fungal systems were much favored for studies in eukaryotic molecular biology and were soon able to compete with bacterial systems in the molecular arena.

The experimental achievements in research on the genetics andmolecular biology of fungi have benefited more generally studies in the related fields of fungal biochemistry, plant pathology,medicalmycology, and systematics. Today, there ismuch interest in the geneticmanipulation of fungi for applied research. This current interest in biotechnical genetics has been augmented by the development of DNA-mediated transformation systems in fungi and by an understanding of gene expression and regulation at the molecular level. Applied research initiatives involving fungi extend broadly to areas of interest not only to industry but to agricultural and environmental sciences as well.

It is this burgeoning interest in fungi as experimental systems for applied as well as basic research that has prompted publication of this series of books under the title *The Mycota*. This title knowingly relegates fungi into a separate realm, distinct from that of either plants, animals, or protozoa. For consistency throughout this Series of Volumes the names adopted for major groups of fungi (representative genera in parentheses) are as follows:

Pseudomycota

Division:	Oomycota (*Achlya, Phytophthora, Pythium*)
Division:	Hyphochytriomycota

Eumycota

Division:	Chytridiomycota (*Allomyces*)
Division:	Zygomycota (*Mucor, Phycomyces, Blakeslea*)
Division:	Dikaryomycota

Subdivision: Ascomycotina

    Class: Saccharomycetes (*Saccharomyces, Schizosaccharomyces*)
    Class: Ascomycetes (*Neurospora, Podospora, Aspergillus*)

Subdivision: Basidiomycotina
    Class: Heterobasidiomycetes (*Ustilago, Tremella*)
    Class: Homobasidiomycetes (*Schizophyllum, Coprinus*)

We have made the decision to exclude from *The Mycota* the slime molds which, although they have traditional and strong ties to mycology, truly represent nonfungal forms insofar as they ingest nutrients by phagocytosis, lack a cell wall during the assimilative phase, and clearly show affinities with certain protozoan taxa.

The Series throughout will address three basic questions: what are the fungi, what do they do, and what is their relevance to human affairs? Such a focused and comprehensive treatment of the fungi is long overdue in the opinion of the editors.

A volume devoted to systematics would ordinarily have been the first to appear in this Series. However, the scope of such a volume, coupled with the need to give serious and sustained consideration to any reclassification of major fungal groups, has delayed early publication. We wish, however, to provide a preamble on the nature of fungi, to acquaint readers who are unfamiliar with fungi with certain characteristics that are representative of these organisms and which make them attractive subjects for experimentation.

The fungi represent a heterogeneous assemblage of eukaryotic microorganisms. Fungal metabolism is characteristically heterotrophic or assimilative for organic carbon and some nonelemental source of nitrogen. Fungal cells characteristically imbibe or absorb, rather than ingest, nutrients and they have rigid cell walls. The vast majority of fungi are haploid organisms reproducing either sexually or asexually through spores. The spore forms and details on their method of production have been used to delineate most fungal taxa. Although there is a multitude of spore forms, fungal spores are basically only of two types: (i) asexual spores are formed following mitosis (mitospores) and culminate vegetative growth, and (ii) sexual spores are formed following meiosis (meiospores) and are borne in or upon specialized generative structures, the latter frequently clustered in a fruit body. The vegetative forms of fungi are either unicellular, yeasts are an example, or hyphal; the latter may be branched to form an extensive mycelium.

Regardless of these details, it is the accessibility of spores, especially the direct recovery of meiospores coupled with extended vegetative haploidy, that have made fungi especially attractive as objects for experimental research.

The ability of fungi, especially the saprobic fungi, to absorb and grow on rather simple and defined substrates and to convert these substances, not only into essential metabolites but into important secondary metabolites, is also noteworthy. The metabolic capacities of fungi have attracted much interest in natural products chemistry and in the production of antibiotics and other bioactive compounds. Fungi, especially yeasts, are important in fermentation processes. Other fungi are important in the production of enzymes, citric acid and other organic compounds as well as in the fermentation of foods.

Fungi have invaded every conceivable ecological niche. Saprobic forms abound, especially in the decay of organic debris. Pathogenic forms exist with both plant and animal hosts. Fungi even grow on other fungi. They are found in aquatic as well as soil environments, and their spores may pollute the air. Some are edible; others are poisonous. Many are variously associated with plants as copartners in the formation of lichens and mycorrhizae, as symbiotic endophytes or as overt pathogens. Association with animal systems varies; examples include the predaceous fungi that trap nematodes, the microfungi that grow in the anaerobic environment of the rumen, the many insectassociated fungi and themedically important pathogens afflicting humans. Yes, fungi are ubiquitous and important.

There are many fungi, conservative estimates are in the order of 100,000 species, and there are many ways to study them, from descriptive accounts of organisms found in nature to laboratory experimentation at the cellular and molecular level. All such studies expand our knowledge of fungi and of fungal processes and improve our ability to utilize and to control fungi for the benefit of humankind.

We have invited leading research specialists in the field of mycology to contribute to this Series. We are especially indebted and grateful for the initiative and leadership shown by theVolumeEditors in selecting topics and assembling the experts.We have all been a bit ambitious in producing these Volumes on a timely basis and therein lies the possibility of mistakes and oversights in this first edition.We encourage the readership to draw our attention to any error, omission or inconsistency in this Series in order that improvements can be made in any subsequent edition.

Finally, we wish to acknowledge the willingness of Springer-Verlag to host this project, which is envisioned to require more than 5 years of effort and the publication of at least nine Volumes.

Bochum, Germany
Auburn, AL, USA
April 1994

KARL ESSER
PAUL A. LEMKE
Series Editors

Addendum to the Series Preface

In early 1989, encouraged by Dieter Czeschlik, Springer-Verlag, Paul A. Lemke and I began to plan The Mycota. The first volume was released in 1994, 12 volumes followed in the subsequent years, and two more volumes (Volumes XIV and XV) will be published within the next few years. Unfortunately, after a long and serious illness, Paul A. Lemke died in November 1995. Thus, it wasmy responsibility to proceed with the continuation of this series, which was supported by JoanW. Bennett for Volumes X–XII.

The series was evidently accepted by the scientific community, because several volumes are out of print. Therefore, Springer-Verlag has decided to publish completely revised and updated new editions of Volumes I, II, III, IV, V, VI, VIII, and X. I am glad that most of the volume editors and authors have agreed to join our project again. I would like to take this opportunity to thank Dieter Czeschlik, his colleague, Andrea Schlitzberger, and Springer-Verlag for their help in realizing this enterprise and for their excellent cooperation for many years

Bochum, Germany
May 2008

Karl Esser

Volume Preface to the Second Edition

Pathogenic fungi are widely distributed and can infect many organisms, particularly humans but also other vertebrates and insects. The interaction of these pathogens with their host is of considerable interest for basic research, but also for many applications in medicine and plant protection. Understanding the pathogenesis of fungal infections and the interaction of fungi with their hosts is an expanding field of current research. The intention of the current volume VI of the series Mycota was not to give an overview about human and zoopathogenic fungi. Rather, it was our intention to point to some important and specific interactions of fungal pathogens with their respective hosts, including the description of the application of techniques of functional genomics.

The understanding of the biochemical and physiological reactions mediated by fungi is a prerequisite to elucidate infection mechanisms and to understand infection processes that lead to disease. In particular, the switch of a fungus from being a harmless commensal or saprophyte to becoming an aggressive pathogen represents a key question of modern fungal research. This emerging area of research is fostered by the increasing medical relevance of fungal infections. The characterization of the interaction of pathogenic fungi with their animal hosts includes the identification of virulence factors. Furthermore, the study of virulence mechanisms will help identify the novel and unique escape strategies of pathogens.

This volume VI of the Mycota series covers three major and important areas of the field human and animal relationship. The first part summarizes the current understanding of pathogenic fungi and reports important physiological reactions that are relevant for pathogen–host interaction, ultimately leading to a description of the pathophysiological reactions. The characterization of strategies utilized by human pathogenic and entomopathogenic fungi reveals common principles which are utilized by fungi to survive in competent hosts. The second part covers modern, state of the art technology that is currently used to study fungus–host interactions. Transcriptomics and proteomics represent highly relevant techniques that are used for the simultaneous analysis of both gene expression and protein formation patterns to define specific life-stages of the pathogen. These data contribute to a molecular understanding of the communication between pathogen and host. The third area of this volume addresses the characterization of the host response towards pathogenic fungi. The host utilizes a powerful and highly efficient immune system to combat invading pathogens. The innate immune response forms a major barrier against fungi and pathogens. This defence system responds immediately to infection and the sequential response includes activation of the complement system and recognition of the pathogen by toll-like receptors. These aspects represent a developing and highly interesting area of research. Clinical aspects are covered in one chapter and demonstrate how pathogenic fungi, specifically dermatophytes, colonize different niches and hosts, and how they undermine the host immune response. The state of the art reviews written by experts in the field provide the reader with a comprehensive overview about different research topics. We would be glad if the reader shared with us the excitement to follow this interesting area of research. We are in particular grateful for the excellent contributions made by the different authors of the chapters who joined us in preparing this volume.

Jena, Germany
May 2008

AXEL A. BRAKHAGE
PETER F. ZIPFEL
Volume Editors

Volume Preface to the First Edition

The eight volumes of The Mycota represent the first comprehensive treatment of the fungi in 30 years. Volume VI of the series presents a series of individual chapters on the relationship of fungi with humans and other animals. The intention was not to provide a comprehensive coverage of the zoopathogenic fungi; such an approach is already well represented by some fine textbooks, several monographic treatments of certain aspects of host-parasite interactions or of individual mycoses, and a large number of books on the identification of pathogenic fungi. Rather, it was our intention to emphasize biochemical interactions of the fungi with their hosts.

To that end, the topic of the pathogenesis of the mycoses, both general factors and specific enzymatic reactions, opens the volume (Chaps. 1 and 2) on human associations of the fungi with their hosts. This section is followed by a consideration of the host response to invasion by zoopathogens (Chaps. 3 and 4). The host response is often played out nowadays in an immunocompromised host and this topic is covered in three subsequent chapters (Chaps. 5, 6 and 7).

From the outset, our intention was to consider both invasive fungi and those that initiate illness by means of metabolites. The next seetion of the book covers the effects of inhalation to fungal spores, encompassing allergic reactions and organic dust toxic syndrome (Chaps. 8–11). Finally, there is a review of mushroom intoxications including psychoactive substances (Chap. 12).

A consideration of the interactions of fungi with animals other than humans is treated subsequently. An introductory chapter on veterinary mycology (Chap. 13) is followed by separate considerations of anaerobic fungi (Chap. 14), fungal diseases of fish and shellfish (Chap. 15) and arthropods (Chap. 16). Chapters on the entomopathogenic fungi (Chap. 17) and the mutualism between fungi and insects (Chap. 18) conclude V olume VI.

We are grateful to the group of splendid authors who joined us in preparing this volume.

Los Angeles, CA, USA D.H. Howard
Ottawa, Canada J.D. Miller
November 1995 *Volume Editors*

Contents

Pathogens

1 Trichomycetes and the Arthropod Gut
 Robert W. Lichtwardt 3

2 Opportunistic Mold Infections
 Ronald G. Washburn 21

3 Entomopathogenic Fungi: Biochemistry and Molecular Biology
 George G. Khachatourians, Sohail S. Qazi 33

4 Physiology and Metabolic Requirements of Pathogenic Fungi
 Matthias Brock 63

5 CO_2 Sensing and Virulence of *Candida albicans*
 Estelle Mogensen, Fritz A. Mühlschlegel 83

6 Hyphal Growth and Virulence in *Candida albicans*
 Andrea Walther, Jürgen Wendland 95

7 Pathogenicity of *Malassezia* Yeasts
 Peter A. Mayser, Sarah K. Lang, Wiebke Hort 115

Techniques

8 Proteomics and its Application to the Human-Pathogenic Fungi
 Aspergillus fumigatus and *Candida albicans*
 Olaf Kniemeyer, Axel A. Brakhage 155

9 Transcriptomics of the Fungal Pathogens, Focusing on *Candida albicans*
 Steffen Rupp 187

Host

10 Yeast Infections in Immunocompromised Hosts
 Emmanuel Rollides, Thomas J. Walsh 225

11 The Host Innate Immune Response to Pathogenic *Candida albicans*
 and Other Fungal Pathogens
 Peter F. Zipfel, Katharina Gropp, Michael Reuter,
 Susan Schindler, Christine Skerka 233

12 Toll-Like Receptors and Fungal Recognition
 Frank Ebel, Jürgen Heesemann 243

13 Clinical Aspects of Dermatophyte Infections
    Jochen Brasch, Uta-Christina Hipler.................................... 263

Biosystematic Index 287

Subject Index 291

List of Contributors

AXEL A. BRAKHAGE
(e-mail: axel.brakhage@hki-jena.de, Tel.: +49 3641/5321000, Fax: +49 3641/65-6600)
Department of Molecular and Applied Microbiology, Leibniz Institute for Natural Product Research and Infection Biology–Hans Knöll Institute, Beutenbergstrasse 11a, 07745 Jena; and Friedrich Schiller University, Jena, Germany

JOCHEN BRASCH
(e-mail: jbrasch@dermatology.uni-kiel.de , Tel.: +49 431/597-1507, Fax: +49 431/597-1509)
Abteilung Dermatologie, Universität Kiel, Schittenhelmstraße 7, 24105 Kiel, Germany

MATTHIAS BROCK
(e-mail: Matthias.brock@hki-jena.de , Tel.: +49 3641/65-6865, Fax: +49 3641/65-6860)
Microbial Biochemistry and Physiology, Leibniz Institute for Natural Product Research and Infection Biology e.V., Hans Knöll Institute, Beutenbergstrasse 11a, 07745 Jena, Germany

FRANK EBEL
(e-mail: ebel@mvp.uni-muenchen.de , Tel.: +49 89/5160-5263, Fax: +49 89/5160-5223)
Max-von-Pettenkofer-Institut, LMU München, Pettenkoferstrasse 9a, 80336 München, Germany

KATHARINA GROPP
(e-mail: katharina.gropp@hki-jena.de , Tel.: +49 3641/65-6842, Fax: +49 3641/65-6902)
Department of Infection Biology, Leibniz Institute for Natural Product Research and Infection Biology, Hans Knöll Institute, Beutenbergstrasse 11a, 07745 Jena, Germany

JÜRGEN HEESEMANN
(e-mail: heesemann@m3401.mpk.med.uni-muenchen.de , Tel: +49 89 5160-5201, Fax: +49 89 5160-5202)
Max-von-Pettenkofer-Institut, LMU München, Pettenkoferstrasse 9a, 80336 München, Germany

UTA-CHRISTINA HIPLER
(e-mail: chip@derma.uni-jena.de, Tel.: +49 3641-937355, Fax:+49 3641-937437)
 Abteilung Dermatologie, Universität Jena, Erfurter Straße 35, 07743 Jena, Germany

WIEBKE HORT
(e-mail: Wiebke.Hort@derma.med.uni-giessen.de , Tel.: +49 641-99-43284, Fax:+49 641-99-43259)
Zentrum für Dermatologie und Andrologie, Justus Liebig Universität, Gaffkystrasse 14, 35385 Giessen, Germany

GEORGE G. KHACHATOURIANS
(e-mail: george.khachatourians@usask.ca , Tel.: +1 306-966-5032, Fax: +1 306 966-8898)
Bioinsecticide Research Laboratories, College of Agriculture and Bioresources, University of Saskatchewan, Saskatoon, S7N 5A8, Saskatchewan, Canada

OLAF KNIEMEYER
(e-mail: olaf.kniemeyer@hki-jena.de, Tel.: +49 3641/65-6815, Fax: +49 3641/65-6925)
Department of Molecular and Applied Microbiology, Leibniz Institute for Natural Product Research and Infection Biology–Hans Knoell Institute, Beutenbergstrasse 11a, 07745 Jena; and Friedrich Schiller University, Jena, Germany

SARAH K. LANG
(e-mail: sarah.lang@derma.med.uni-giessen.de, Tel.: +49 641-99-43211, Fax:+49 641-99-43259)
Zentrum für Dermatologie und Andrologie, Justus Liebig Universität, Gaffkystrasse 14, 35385 Giessen, Germany

ROBERT W. LICHTWARDT
(e-mail: licht@ku.edu, Tel.: +1 785 864-3740, Fax: +1 785 864-5321)
Department of Ecology & Evolutionary Biology, University of Kansas, Lawrence, Kansas 66045-7534, USA

PETER A. MAYSER
(e-mail: Peter.Mayser@derma.med.uni-giessen.de , Tel.: +49 641-99-43220, Fax:+49 641-99-43209)
Zentrum für Dermatologie und Andrologie, Justus Liebig Universität, Gaffkystrasse 14, 35385 Giessen, Germany

ESTELLE MOGENSEN
Biomedical Science Group, Department of Biosciences, University of Kent, Canterbury, Kent, CT2 7NJ, United Kingdom

FRITZ A. MÜHLSCHLEGEL
(e-mail: F.A.Muhlschlegel@kent.ac.uk, Tel.: +44 1227 764 000, Fax: +44 1227 763912)
Biomedical Science Group, Department of Biosciences, University of Kent, Canterbury, Kent, CT2 7NJ, United Kingdom

SOHAIL S. QAZI
Bioinsecticide Research Laboratories, College of Agriculture and Bioresources, University of Saskatchewan, Saskatoon, S7N 5A8, Saskatchewan, Canada

MICHAEL REUTER
(e-mail: michael.reuter@hki-jena.de, Tel.: +49 3641/65-6842, Fax: +49 3641/65-6902)
Department of Infection Biology, Leibniz Institute for Natural Product Research and Infection Biology, Hans Knöll Institute, Beutenbergstrasse 11a, 07745 Jena, Germany

EMMANUEL ROLLIDES
(e-mail: xxxt@mail.nih.gov, Tel.: +1 301 402-0023, Fax: +1 301 480-2308)
Immunocompromised Host Section, Pediatric Oncology Branch, National Cancer Institute, Building 10, CRC 1-5750, Bethesda, MD 20892, USA

STEFFEN RUPP
(e-mail: rupp@igb.fhg.de, Tel.: +49 711-970-4045, Fax: +49 711-970-4200)
Fraunhofer IGB, Nobelstrasse 12, 70569 Stuttgart, Germany

SUSAN SCHINDLER
(e-mail: susann.schindler@hki-jena.de, Tel.: +49 3641/65-6669, Fax: +49 3641/65-6902)
Department of Infection Biology, Leibniz Institute for Natural Product Research and Infection Biology, Hans Knöll Institute, Beutenbergstrasse 11a, 07745 Jena, Germany

CHRISTINE SKERKA
(e-mail: christine.skerka@hki-jena.de, Tel.: +49 3641/65-6848, Fax: +49 3641/65-6902)
Department of Infection Biology, Leibniz Institute for Natural Product Research and
Infection Biology, Hans Knöll Institute, Beutenbergstrasse 11a, 07745 Jena, Germany

THOMAS J. WALSH
(e-mail: walsht@mail.nih.gov, Tel.: +1 301 402-0023, Fax: +1 301 480-2308)
Immunocompromised Host Section, Pediatric Oncology Branch, National Cancer
Institute, Building 10, CRC 1-5750, Bethesda, MD 20892, USA

ANDREA WALTHER
(e-mail: anwa@crc.dk)
Carlsberg Laboratory, Yeast Biology, Gamle Carlsberg Vej 10, 2500 Copenhagen Valby,
Denmark

RONALD G. WASHBURN
(e-mail: ronald.washburn@med.va.gov)
Overton Brooks Veterans Administration Medical Center and Louisiana State
University Health Sciences Center, 510 East Stoner Avenue, Shreveport, LA 71101, USA

JÜRGEN WENDLAND
(e-mail: juergen.wendland@crc.dk, Tel.: +44 3327 5230, Fax: + 44 3327 4708)
Carlsberg Laboratory, Yeast Biology, Gamle Carlsberg Vej 10, 2500 Copenhagen Valby,
Denmark

PETER F. ZIPFEL
(e-mail: peter.zipfel@hki-jena.de, Tel.: +49 3641/65-6900, Fax: +49 3641/65-6902)
Friedrich Schiller University, Jena, Germany; and Department of Infection Biology,
Leibniz Institute for Natural Product Research and Infection Biology,
Hans Knöll Institute, Beutenbergstrasse 11a, 07745 Jena, Germany

Pathogens

1 Trichomycetes and the Arthropod Gut

Robert W. Lichtwardt[1]

CONTENTS

I. Introduction	3
A. General Characteristics	3
B. Arthropod Host Types and Their Habitats	5
C. Host Specificity	6
D. Ecological Adaptations	6
E. Geographic Distribution	7
F. Culturability	7
II. Effects of Gut Fungi on Arthropods	8
A. Commensalism	8
B. Lethal Species	8
C. Ovarian Cysts	9
D. Possible Benefit to Hosts	10
III. Effects of Arthropods on Gut Fungi	12
A. Ecdysis and Fungal Morphogenesis	12
B. Dispersal	13
C. Spore Extrusion and Host Selection	13
IV. Physiology of Trichomycetes In Vitro	15
A. Nutrition	15
B. Rates of Growth	16
C. Reproduction	16
D. Sterol and Lipid Production	16
V. Conclusions	17
References	18

I. Introduction

The obligate association of trichomycetes with arthropods, whose guts they inhabit, has led to a number of unique morphological and physiological features that distinguish them from other organisms. These include specialized reproductive structures that help to ensure success of transmission from host to host within a population, mechanisms for dispersal among host populations, recognition of suitable types of hosts, the ability of thalli and spores to tolerate the digestive processes within the gut, and a degree of synchrony between fungal and host development.

The Class Trichomycetes is related to the fungal Zygomycota (Fig. 1.1A). Molecular systematics has now revealed that two of the four orders that were formerly included in the Trichomycetes, Amoebidiales and Eccrinales, are not fungi, but rather are protozoans. The relationship of Amoebidiales to the Kingdom Fungi had been suspect for some time (Lichtwardt 1986), and that order now has been demonstrated to belong to the protozoan Mesomycetozoea by several investigators using DNA sequence analysis (Benny and O'Donnell 2000; Ustinova et al. 2000; Mendoza et al. 2002; Cafaro 2005; White et al. 2006a). Those protozoans are a basal group in the tree of life where it is believed fungi and animals may have diverged. Cafaro (2005; Fig. 1.1B) demonstrated that Eccrinales also fall within the Mesomycetozoea, and that Eccrinales and Amoebidiales are sister clades. Thus, only Harpellales and the questionable Asellariales are regarded to be Fungi (O'Donnell et al. 1998; Gottlieb and Lichtwardt 2001; Cafaro 2005; White et al. 2006a). Because of their similarities, in this chapter trichomycetes (lower case *t*) will be used for all of these ecologically similar gut organisms, just as the word "fungus" is sometimes used for fungus-like organisms that are now known to be phylogenetically unrelated.

A. General Characteristics

Most trichomycetes live within the hindgut of their arthropod hosts. Some Harpellales, however, attach to the peritrophic matrix in dipteran midguts; and thalli of a few Eccrinales live in the stomach (foregut) of certain decapods. Thalli of trichomycetes attach firmly to the gut cuticle by means of a secreted holdfast structure or, in relatively few species, by a specialized basal cell that cements the thallus to the chitinous substrate. Only thalli of *Amoebidium* species (Amoebidiales) are attached externally to the exoskeleton of larval insects or small crustaceans. Penetration of the cuticle is known to occur in a few species of trichomycetes, but without damaging the subcuticular tissues

[1] Department of Ecology & Evolutionary Biology, University of Kansas, Lawrence, Kansas 66045-7534, USA; e-mail: licht@ku.edu

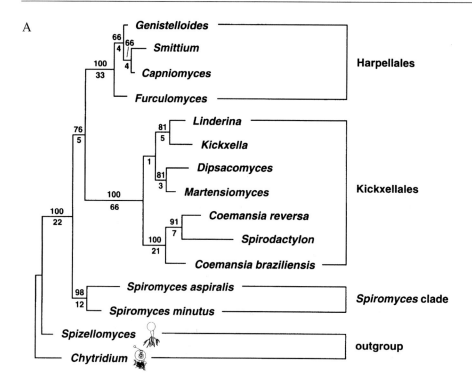

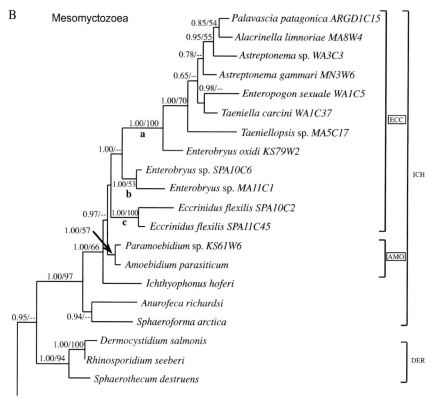

Fig. 1.1. A One of the first molecular trees to indicate that Harpellales (Trichomycetes) and Kickxellales (Zygomycetes) share a common fungal ancestry, based on a maximum parsimony analysis of 18S DNA data (O'Donnell et al. 1998; modified, with permission from The New York Botanical Garden). **B** Part of a phylogram indicating that Eccrinales (*ECC*) and Amoebidiales (*AMO*), formerly considered to be Trichomycetes, are sister clades in the protozoan class Mesomycetozoea, based on a Bayesian analysis of an 18S DNA dataset (Cafaro 2005; modified, with permission from Elsevier)

(Lichtwardt 1986). Fungal nutrients consist of the materials passing through the gut lumen, and conceivably some intestinal secretions.

The basic reproductive structure of trichomycetes is the sporangiospore, as it is in other Zygomycota, but the spores that have evolved in each of the orders are distinctly different, no doubt in response to different host types and habitats (Table 1.1). Harpellales produce series of basipetal monosporous sporangia, known as trichospores, that project laterally from generative cells. Trichospores normally have one to many fine basal appendages upon release, with the number of appendages being a generic character. In Asellariales (at least in species of *Asellaria*), the sporangia disarticulate from thallial branches in a manner resembling arthrospores. The monosporous sporangia in most genera of Eccrinales are produced apically in a basipetal series on the unbranched thalli, and each spore emerges from a pore or tear in the sporangial wall. Amoebidiales have thalli that function as a single sporangium for the release of sporangiospores with rigid walls, or amoeboid cells just prior to molting of the host (Lichtwardt et al. 2001a).

Harpellales possess unique biconical zygospores quite unlike the spherical sexual spores produced by other Zygomycota. Zygospores have not been seen in some common and frequently collected species, and possibly some Harpellales have lost the ability to reproduce sexually, as is the case with some other fungi. In Eccrinales, conjugations between thalli of *Enteropogon sexuale* Hibbits have been described (Hibbits 1978), and somewhat similar conjugations were reported by Lichtwardt et al. (1987) in an *Enteromyces*-like eccrinid living in another marine anomurid. What may have been zygotes formed in both of these cases, but it is not certain whether this type of development represented a normal stage in their life cycles or was merely an aborted attempt at sexuality.

B. Arthropod Host Types and Their Habitats

The known host range of trichomycetes is listed in Table 1.1. Hosts of Harpellales are larval stages of primitive orders of aquatic insects, namely lower Diptera (Nematocera), Ephemeroptera, Plecoptera, Trichoptera, and, in one case, an aquatic beetle larva. Nematoceran hosts of Harpellales include Chironomidae, Simuliidae, Culicidae, Ceratopogonidae, Tipulidae, Dixidae, and Psychodidae. Trichomycetes in other orders live predominantly in adult arthropods, though infestation of immature stages is also known. Species of *Asellaria* (Asellariales) inhabit only Isopoda, and these may be terrestrial, aquatic, or marine. Species of the other asellarid genus *Orchesellaria* live in species of the very primitive insect order Collembola (springtails) sometimes classified in the Hexapoda. The greatest range of host types and habitats is found in the Eccrinales. Many families of millipedes consistently contain species of Eccrinales. Other terrestrial hosts of Eccrinales include various kinds of crustaceans and a few families of beetles. The crustacean hosts are found in terrestrial, freshwater, or intertidal habitats. Deeper marine levels have been explored only minimally, but one eccrinid, *Arundinula abyssicola* Van Dover & Lichtwardt, was found in galatheid squat lobsters living around six Pacific hydrothermal vents at depths of about 2600 m (Van Dover & Lichtwardt 1986). *Amoebidium* species (Amoebidiales) do not seem to have particular host preferences, having been found attached externally to a variety of small crustaceans such as *Daphnia* and various larval stages of freshwater insects. In contrast,

Table 1.1. Trichomycete host types and habitats of the fungal orders Harpellales and Asellariales and protozoan orders Eccrinales and Amoebidiales

Order	Arthropod hosts	Habitats
Harpellales	Larval stages of lower Diptera, Ephemeroptera, Plecoptera, Trichoptera, Coleoptera	Streams and lentic freshwaters
Asellariales	Isopoda, Collembola	Streams, terrestrial, marine (intertidal)
Eccrinales	Crustacea (Isopoda, Amphipoda, Decapoda), Diplopoda, Coleoptera	Terrestrial, freshwater, marine
Amoebidiales	Larval Insecta and small Crustacea	Freshwater (lentic)

Paramoebidium species (also Amoebidiales) live within the hindgut of larval stoneflies, mayflies and black flies, and may have host preferences. In general, hosts of trichomycetes are omnivores, detritivores, shredders, or algivores, whereas arthropods that are leaf-eating herbivores, carnivores, or are predacious apparently are not suitable hosts for trichomycetes. It is not known whether it is the host diet that has deterred adaptation to such arthropods, or other factors such as the structure of the gut and accompanying gut microorganisms.

C. Host Specificity

Data on host specificity are only as reliable as the extent to which particular gut fungi have been sought or potential host groups have been studied. Many species of trichomycetes have been reported as living in a single arthropod species, perhaps in some instances owing to the limited number of collections, though it is possible that some trichomycetes are indeed species-specific. More commonly the host range of trichomycetes is a genus or a family of arthropods. For example, the marine fungus *Asellaria ligiae* Tuzet & Manier ex Manier has been collected in the Mediterranean, east and west coasts of the contiguous USA, Puerto Rico, Hawaii, and Japan in different species of the isopod genus *Ligia* (Lichtwardt et al. 2001a). In the case of the many genera and species of fungi in the guts of Simuliidae (black fly) larvae, it appears to be the family, rather than lower taxa, the defines the host range. A few geographically widespread harpellid species of *Smittium* (*S. culisetae* Lichtw., *S. culicis* Manier, *S. simulii* Lichtw.) are known to inhabit larvae of four to six families of Diptera. Most other species of this large genus, *Smittium*, appear to have narrower host ranges. At the other extreme, *Enteromyces callianassae* Lichtw. and *Taeniella carcini* Léger & Duboscq, two widespread marine eccrinids, live in both true crabs (Brachyura) and shrimplike crabs (Anomura).

D. Ecological Adaptations

Spores of many fungi undergo a germination process that may take hours or even days before germ-tube formation is complete. Trichomycetes cannot afford such a temporal luxury, because of the brief time it takes for an ingested spore to reach the specific region of the gut (foregut, midgut, or hindgut), where it must attach and begin development. In some feeding aquatic dipterans, such as mosquito larvae, the time span for passage of food through the gut can be less than 30 min (Clements 1963). Trichospores of the larva-inhabiting Harpellales are well adapted to such conditions. These elongated, deciduous sporangia rapidly extrude their sporangiospores upon receiving cues from the appropriate host, and the expanded spores immediately produce an adhesive substance that anchors them to the cuticle. The extrusion process is described in Section IIIC of this chapter.

Zygospores of Harpellales are biconical structures, in contrast to the spherical zygospores of other Zygomycota, and this shape may also be related to the germination process in the gut. Zygospores have not been studied in vitro, as have trichospores, but on a few occasions they have been observed under the microscope to expel their inner, walled cell from one end of the biconical zygospores in a manner similar to sporangiospore extrusion in trichospores.

Trichospores, upon release from the thallus, may bear one or more long, fine appendages. Many larval hosts of harpellids select stream substrates where the water is turbulent. Appendages undoubtedly function to entangle and attach some of the defecated trichospores to the substrates where the larvae cling, thus increasing the probability of trichospores being ingested before they wash downstream. Transfer of trichospores between aquatic larvae can be remarkably efficient in some cases. Populations of lotic midges can be well infected with certain species of *Stachylina* whose unbranched thalli in some species normally produce an average of only two to four trichospores. Few extant fungi of any kind have such low reproductive rates and be so obviously successful.

Some gut fungi, such as species of *Orphella* (in stoneflies) and *Pteromaktron* and *Zygopolaris* (in mayflies), have sporulating thallial tips that extend beyond the host's anus. Nymphs bearing these fungi often congregate on rocks, leaves, or other stream substrates, and it is presumed that such exposed trichospores might be grazed directly from the projecting thalli and therefore do not rely on the assistance of appendages in such instances.

Terrestrial trichomycetes do not produce spore appendages, but some genera of Eccrinales, whose hosts live in freshwater or marine habitats

(*Arundinula, Astreptonema, Taeniella*), have one or two stout appendages at the end of their oval, thick-walled sporangiospores. Such spore appendages may also have a function in their transmission within populations. However, the ontogeny of spore and appendage formation appears to be quite different in Eccrinales and Harpellales, and probably originated through different evolutionary pathways. Since most Eccrinales are terrestrial, one hypothesis is that aquatic and marine habitats of eccrinid species are derived, and that appendages evolved de novo within those Eccrinales in response to a shift to an aquatic habitat.

In most species of Harpellales and Asellariales, the spores produced by a thallus must pass out of the gut and be ingested by another individual before germination occurs. Thus, each thallus in the gut originates from one external propagule. In Eccrinales (*Palavascia* excluded), a predominant type of sporangiospore has evolved that attaches to the cuticle immediately upon emergence from the sporangium and then germinates in situ. By this means it is possible for a single source of infestation eventually to build up an aggregate of many thalli in one gut. There are four genera of Harpellales, though capable of producing normal trichospores, that have evolved different means of producing new thalli endogenously within the gut. This is accomplished through: (a) the development of specialized auxiliary cells in *Graminella* (Lichtwardt and Moss 1981), (b) a modified form of trichospore in *Allantomyces* (Williams and Lichtwardt 1993), (c) the production of trichospores that occasionally attach to the cuticle and germinate to produce a small sporulating thallus in *Ephemerellomyces* (White and Lichtwardt 2004), or (d) in *Ejectosporus*, the extrusion and development of lunate cells from trichospore-like structures that remain attached to the thallus (Strongman 2005).

E. Geographic Distribution

The geographic distribution of trichomycetes as a group is limited only by the habitats and distribution of suitable host types. These gut fungi have been found on all continents except Antarctica, and on many large and small islands as well. Many regions of the world have yet to be surveyed, but it is quite evident when new geographic areas are explored that regardless of climate or elevation, if suitable arthropods are present, many are likely to harbor trichomycete symbionts, often including new taxa. This worldwide distribution may be an expression of the antiquity of these gut organisms, rather than the dispersal ability of individual species. There are a few species with very wide distributions. Examples include some species of *Smittium* and *Stachylina*, and *Harpella melusinae* Léger & Duboscq in dipteran larvae, and some marine species in crustaceans (*Enteromyces callianassae* Lichtw., *Taeniella carcini* Léger & Duboscq, *Palavascia sphaeromae* Tuzet & Manier). But the majority of trichomycetes appear to have more limited geographic ranges. Unfortunately, insufficient field data can provide a false appearance of restricted distribution. There is evidence, nonetheless, that many species of Harpellales and some Amoebidiales, which have been studied on a broader geographic scale than species of other orders, are restricted to certain continental regions, or to islands such as New Zealand (Williams and Lichtwardt 1990; Lichtwardt and Williams 1992d). In Australia, the endemic aquatic insect fauna was found to contain many new species of harpellids, and each of these was found in a different biogeographic region of that country (Lichtwardt and Williams 1990, 1992a–c, e; Williams and Lichtwardt 1993).

These examples, among others that could be cited (see Lichtwardt et al. 2001a), document that some species of trichomycetes may be geographically restricted. There are several possible explanations, among which are host specificity, poor long-distance dispersal of the hosts, and young evolutionary age of some trichomycete species. Consideration of geographic distribution of some trichomycetes must also take into account certain types of arthropod hosts that no doubt have been spread with their gut fungi by human activity, such as the greenhouse millipede *Oxidus gracilis* (Koch) and other Diplopoda, various genera of springtails (Collembola), mosquito larvae (Culicidae), and several species of pill bugs (Armadillidiidae).

F. Culturability

Except for *Amoebidium parasiticum* Cienkowski, first cultured by Whisler (1960), and another species, *A. appalachense* Siri, M.M. White & Lichtw. (Amoebidiales), the only other trichomycetes isolated axenically are a relatively small percentage of branched Harpellales wherein culturability varies among genera and species. The author and several other investigators have tried unsuccessfully to isolate species of most genera of Harpellales, *Asellaria ligiae* (Asellariales), and a number of species of Eccrinales. The trichomycete culture collection at the University of Kansas contains over 300 axenic isolates representing eight genera (*Amoebidium, Barbatospora, Capniomyces, Furculomyces, Genistelloides, Simuliomyces, Smittium, Trichozygospora*) consisting of some 35 named species, as well as many currently unnamed species of *Smittium*. The most common isolation media have been a dilute brain–heart infusion, a tryptone-glucose-salts medium, or a mixture of both. The formula for these media, as well as the vitamins that are usually added, antibiotics used for primary isolation, and general techniques for removing and culturing the fungi have been published in some detail (Lichtwardt 1986; Lichtwardt et al. 2001a). All species in culture are aquatic and most require a water overlayer if a medium with agar is used. Freezing in liquid nitrogen is the preferred method for long-term storage.

The vigor with which different isolates grow in vitro varies considerably. A few isolates have been maintained for up to 40 years in refrigerator storage with transfers several times a year. Isolates of other species have died out in refrigerator storage after a year or so, even when transferred frequently. Although all of the current isolates were, of course, species that in nature grew obligately within their larval hosts (except for *Amoebidium* spp. that grow externally), the cultured species do not seem to require unusual nutrients or culture conditions in vitro, based upon laboratory studies of selected genera and species. These studies are discussed in Section IV of this chapter. Because of the relative ease with which some harpellids can be cultured, it is not understood at present why other species that live as coinhabitants in the same insect gut cannot be isolated axenically. For example, black fly larvae (Simuliidae) are hosts to nine genera of trichomycetes (*Barbatospora, Genistellospora, Graminelloides, Harpella, Paramoebidium, Pennella, Simuliomyces, Smittium, Stipella*). It is not uncommon to find species of two or three of these genera in one larval hindgut, yet repeated culture attempts over the years have resulted in isolating only some six species of *Smittium* and *Barbatospora*.

II. Effects of Gut Fungi on Arthropods

A. Commensalism

All trichomycetes are dependent on their hosts for survival, but the need by arthropods for these gut organisms has not been demonstrated. The relationship is often described as commensalistic, though exceptions exist (see Sections that follow). In such a relationship, in which the gut organisms are neither overtly detrimental nor beneficial, several conditions must be met in order to permit investigating the more subtle interactions of host and commensal. Most arthropods with trichomycetes also harbor various kinds of bacteria and sometimes other microorganisms such as protozoans and nematodes. As a consequence, to study the less overt effects of trichomycetes on their hosts, it is necessary to meet two criteria: (a) to be able to raise a host axenically in the laboratory (without the presence of any other microorganisms), and (b) to have a pure culture of the gut fungus capable of infecting that host. These requirements have placed limitations on the kinds of interorganismal systems available for study. Mosquitoes satisfy the first criterion, being easily raised in the laboratory under aseptic conditions, but the experimental fungi are essentially limited to a few species of *Smittium* that are symbionts of mosquito larvae or other lower Diptera.

B. Lethal Species

Mortality in laboratory-raised mosquito larvae was reported by Coluzzi (1966) in Italy, and Dubitskii (1978) in Russia. The deaths were attributed to *Smittium culisetae*. More likely, based on the symptoms, the pathogenic fungus was *S. morbosum* Sweeney, a species that Sweeney (1981) reported to cause 50–95% mortality in *Anopheles hilli* Woodhill & Lee larvae in his rearing tanks in Australia. It apparently had been introduced into his tanks from a field collection of *A. annulipes* Walker. *Smittium morbosum* resembles *S. culisetae*, which he also found growing innocuously in his mosquito larvae, and he was able to culture and compare both species. Larvae infected with *S. morbosum* had an externally visible black spot in the abdomen, as Dubitskii had reported. Sweeney attributed this to a melanization reaction around the hyphae of the fungus which penetrated slightly into the hemocoel of the posterior midgut.

This penetration apparently anchored the fungus in such a way that it prevented successful ecdysis. The fungus grew to some extent into the cells of the Malpighian tubules, and may have blocked the gut as well. Sweeney also reported finding pupae and adults with sporulating thalli.

Natural field infections by *S. morbosum* were reported for the first time by Sato et al. (1989) in *Aedes albopictus* (Skuse) and *Culex pipiens* L. in Japan. López Lastra (1990) found what she described as *S. morbosum* var. *rioplatensis* in Argentina infecting five species belonging to several genera of Culicidae. W. Reeves discovered and C. Beard isolated (personal communication) what appears to be *S. morbosum* infecting mosquito larvae in South Carolina (USA). Thus, the known distribution of *S. morbosum* is geographically scattered. There are no experiments that estimate the potential value of this trichomycete as a biological agent for controlling mosquitoes.

C. Ovarian Cysts

Reports of fungal bodies in the ovaries of Simuliidae, such as those by Garms (1975) in black flies from Guatemala and Liberia, and Undeen and Nolan (1977) and Yeboah et al. (1984) in Newfoundland (Canada), made no reference to trichomycetes. They were thought to be "phycomycetes" until Moss and Descals (1986) demonstrated convincingly that the cysts they discovered, which they called chlamydospores, in association with some oviposited black fly eggs in southern England were a stage of *Harpella melusinae*. Subsequently, other forms of ovarian cysts were linked to several genera of harpellids that inhabit simuliid and ephemeropteran guts and *Smittium* associated with chironomids (Taylor and Moss, unpublished data). Cysts of *H. melusinae* and *Genistellospora homothallica* Lichtw. were later found at Carter Pond outlet near Greenwich (New York, USA) by the present author and Daniel P. Molloy of the New York State Museum (Figs. 1.2, 1.3). Cysts have also been found in adult black fly ovaries in Newfoundland, but attempts to germinate the cysts to obtain trichospores for identifying them were not successful (Lichtwardt et al. 2001b). However, molecular methods allowed some cysts to be identified by using 18S and 28S rDNA sequences to compare cysts with identified fungi taken from simuliid guts (White et al. 2006b).

The infected female host is capable of flying and "ovipositing" the cysts at new sites. The significance of harpellid cysts include: (a) the cysts carried by the flying adults provide an explanation of how simuliid fungi can be found consistently in larvae located in the headwaters of some streams, and how the fungi can disperse to new drainage systems or from one disjunct lentic habitat to another, (b) the fungi completely or partly sterilize the females, resulting in a decrease in the overall number of black fly eggs deposited in given sites, and (c) this parasitic stage of harpellids demonstrates that these "commensals" are not as benign as was previously thought.

The cysts are much smaller than black fly eggs and can be produced in much greater numbers per individual than eggs. It is not yet understood how the fungi invade the ovaries from their normal location in the gut lumen. Yeboah et al. (1984) found histological evidence of fungi in the fat body and ovaries of about 10% of both larvae and pupae of *Stegopterna mutate* (Malloch), so infection of tissue apparently can take place early in the development of preimaginal stages.

The cysts of *H. melusinae* (Fig. 1.2), when exposed to the aquatic environment, germinate at both ends to produce two germ tubes that functions as generative cells (Moss and Descals 1986). A coiled trichospore is produced at the tip of each generative cell, and four or more short basal appendages can be seen upon their release (Fig. 1.2E). The trichospores originating from cysts are smaller and the appendages are shorter than those originating from thalli attached to the peritrophic matrix, perhaps because of the limited resources in the cyst. *Genistellospora homothallica* cysts (Fig. 1.3) germinate at one end, and one (occasionally two) ovoid trichospore develops from the generative cell(s). Released trichospores have about six very fine appendages.

The number of female black flies bearing cysts seems to vary considerably, depending on the site, time of year, and no doubt other factors such as host species and environmental conditions. Moss and Descals (1986) found cysts in their southern England site only in October and November during several years of collecting. Rizzo and Pang (2005) illustrated *H. melusinae*-like zygospores – the first report of a sexual state developing from cysts – attached to oviposited Simuliidae eggs collected in October, also in southern England. Labeyrie et al. (1996) found considerable variation

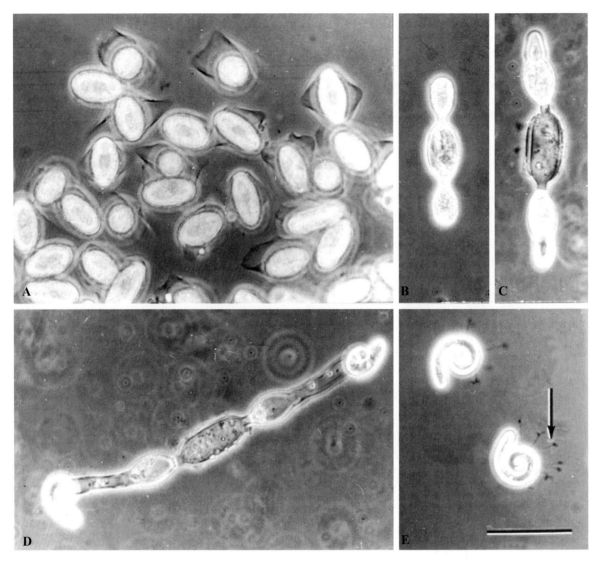

Fig. 1.2. Cysts of *Harpella melusinae* removed from the ovaries of an adult blackfly (Simuliidae) and their in vitro developmental stages. Collected by the author and Daniel P. Molloy, New York State Museum. **A** Fresh cysts, each surrounded by a membranous structure. **B** Germination from both poles. **C** Later stage with empty cyst. **D** Formation of terminal trichospores. **E** Released trichospore bearing multiple short appendages with split ends (*arrow*). *Bar* 40 µm

in the infection rate among females captured from late April through July at six sites near Cambridge (New York, USA). The infection rate at Thurber Pond outlet, for example, ranged from 2% to 80% during one year. Infections in more than 50% of females were reported in Newfoundland (Yeboah et al. 1984), but were generally lower.

The broader question in harpellid–insect relationships is whether many or all species of Harpellales are capable of dispersing by means of ovarian cyst production in their respective hosts, which consist of mayflies, stoneflies, caddisflies, and many families of nematocerans. The latter include important disease vectors or pests such as Simuliidae (black flies), Culicidae (mosquitoes), Ceratopogonidae (biting midges), and Chironomidae (nonbiting midges).

D. Possible Benefit to Hosts

Trichomycete infection of arthropods, in general, appears to be most abundant in vigorous populations of hosts. This suggests that even if the gut

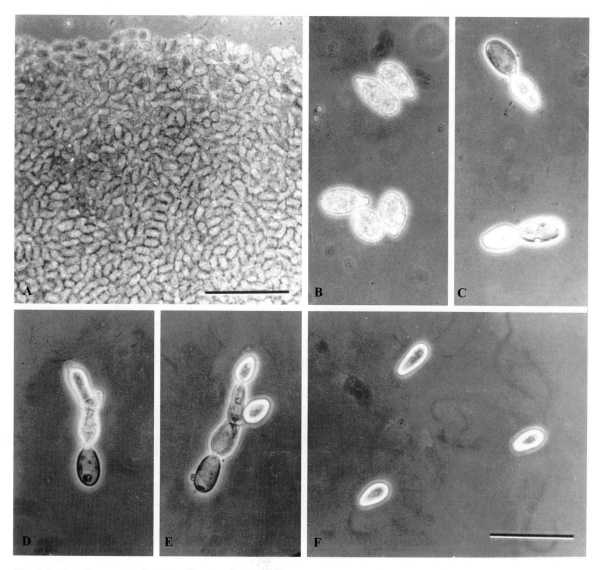

Fig. 1.3. Ovarian cysts of *Genistellospora homothallica* from an adult blackfly. Collected by the author and Daniel P. Molloy, New York State Museum. **A** Part of a large mass of cysts removed from an ovary. **B** Ungerminated cysts. **C** Germination from one end of the cysts. **D** Initial stages of trichospore formation. **E** Two maturing trichospores. **F** Released trichospores with extremely fine and long appendages. *Bars* 100 μm for **A**, 40 μm for **B–F**

fungi do not benefit their hosts, most species do not appear to harm them. Obtaining measurable evidence of possible benefits to the host is not simple, as stated earlier, unless axenically reared hosts and suitable fungal cultures are available to compare the fungus-infected hosts with those bearing no gut microorganisms under aseptic laboratory conditions. The only study meeting these criteria involved use of the mosquito *Aedes aegypti* (L.) larvae and an isolate of *Smittium culisetae* (Horn and Lichtwardt 1981).

Larvae of *A. aegypti* were reared singly on a semidefined medium that included B vitamins and sterols. These essential ingredients were individually excluded from the medium in a series of tests. Establishment of infection by *S. culisetae* was accomplished by feeding trichospores to sets of larvae and comparing them with uninfected sets on complete and deficient media, using an additional undefined medium as a control. Although the results were not unequivocal, larvae with gut fungi developed through more instars and did so at a faster rate than uninfected larvae. It was determined that desmosterol, the major sterol produced by *Smittium* species (Starr et al. 1979), satisfied the sterol requirements of the mosquito larvae, as did dead whole mycelium of *S. culisetae*.

Under natural conditions, one would not expect a complete absence of individual vitamins or a sterol source in a mosquito habitat. More likely, resources might be initially suboptimal, or dense populations of mosquito larvae might reduce particular essential nutrients to a suboptimal level. In such cases the fungi might adequately supplement those nutrients so as to accelerate larval development and permit mosquito eclosion. It is conceivable, therefore, that populations of larvae infected with *Smittium* could have a survival advantage over uninfected populations, and under particular conditions of nutritional stress the relationship of the symbionts might be considered mutualistic.

III. Effects of Arthropods on Gut Fungi

A. Ecdysis and Fungal Morphogenesis

Though sheltered and provided with food by the foraging host, gut fungi nevertheless have had to adapt to a potentially precarious life style, namely the molting stages in arthropod development. Trichomycetes attach to the chitinous linings of the gut which are shed with the exoskeleton in each molting cycle. Two basic forms of adaptation were necessary. One was to ensure some reproductive success before expulsion from the host, and the other was the capacity to survive in the external environment and be able to reinfest another host.

Eccrinales primarily infest adult crustaceans, millipedes, or beetles, though juveniles may also become infested. In some cases spores ingested by an immature arthropod produce a thallus, but there may be insufficient time for the thallus to reach reproductive maturity before the next molting event (Lichtwardt 1961). In a few instances, however, species of eccrinids normally grow and sporulate in larval stages, such as *Lajasiella aphodii* Tuzet & Manier ex Manier in scarabiid beetle larvae (Tuzet and Manier 1950). Even after reaching sexual maturity, some crustaceans and millipedes continue to molt occasionally as they grow in size. Eccrinales generally produce thin-walled sporangiospores during the intermolt period, resulting in endogenous replication of thalli within the gut.

In many eccrinid genera (*Arundinula, Astreptonema, Eccrinidus, Eccrinoides, Taeniella*) special thick-walled sporangiospores begin to develop just prior to host molting. These thick-walled spores are presumed to be resistant to desiccation and other stressful conditions in the external environment. Exuviae are often the best source for locating such spores. Some crustaceans and millipedes eat parts of their exuvia that may contain shed spores, and this may be one way whereby an individual could reinfest itself quickly.

The larval hosts of Harpellales have some of the shortest intermolt periods, thus requiring thalli in the gut to grow and sporulate rapidly. For instance, under optimal conditions a first instar of *Aedes aegypti* may begin to molt in 23–30h (Christophers 1960), and it has been demonstrated that the hindgut fungus *Smittium culisetae* can grow and sporulate in about 22h (Williams and Lichtwardt 1972a). In some harpellids (*Allantomyces, Glotzia, Legeriomyces, Simuliomyces, Zygopolaris*) there may be a shift from trichospore to zygospore production just prior to molting, whereas other harpellids may produce both spore types simultaneously.

Another example of how molting influences morphogenesis of trichomycetes is found in Amoebidiales. *Amoebidium parasiticum*, growing externally on various aquatic insects and small crustaceans, produces allantoid to lunate sporangiospores that break through the thallial (sporangial) wall from where they can be picked up by other aquatic arthropods upon contact. As a host begins to molt, maturing thalli of *Amoebidium* release amoeboid cells instead of sporangiospores, and these amoebae migrate and appear to seek a suitable substrate where they can encyst and later produce cystospores to resume the life cycle. Injury to the host can also result in amoebagenesis. Although it is not known what natural cues in the molting process elicit this morphogenetic shift, Whisler (1968) was able to induce amoebagenesis in cultured *Amoebidium* using fractions of a dried daphnid concentrate that included calcium, glucose, and several amino acids.

Paramoebidium species develop hidden within the larval hindgut of lotic stoneflies, mayflies, caddisflies, and certain dipterans, and lack the sporangiospore stage of development. Thalli of *Paramoebidium* remain unreproductive in the hindgut until the larvae molt or are injured by dissection, whereupon they produce copious amoebae within minutes or a few hours. As in *Amoebidium*, the amoebae of *Paramoebidium* produce cysts and, later, cystospores.

B. Dispersal

The distribution of species of Eccrinales and Asellariales is determined by the dispersal abilities of their adult hosts. Some marine species of these two symbiont orders have considerable geographic distribution, perhaps due in part to the occasional rafting of the crustaceans in which they live. *Enteromyces callianassae* and *Taeniella carcini*, mentioned earlier and which live in an unusually wide range of decapod hosts, are especially widespread in their marine habitats. Human activities can also distribute arthropods together with their gut fungi over great distances. The author (unpublished data) found many Eccrinales on the island of Oahu (Hawaii, USA) but all of them were in introduced species of millipedes and isopods.

Accounting for the dispersal of Harpellales in their nonflying larval hosts is more problematic. There is no evidence that these aquatic fungi can disseminate as airborne propagules. Many harpellids appear to be more or less restricted to particular geographic regions of the world, but even in such instances reliable local dispersal upstream or from one lentic habitat to another is essential for species survival. The most likely dispersal mechanism of harpellids is by means of the ovarian cysts as described earlier for simuliids (Section IIC), but such cysts have yet to be demonstrated in the ovaries of other types of insects. Phoretic dispersal of infested larvae by birds or other animals is possible, but this has not been substantiated, and dependence upon this type of chance dispersal is not likely to assure survival of harpellid species.

C. Spore Extrusion and Host Selection

Streams with abundant and varied insect faunas almost always contain an assortment of gut fungi. For instance, a 20-m stretch of one high-elevation second-order stream in the Colorado Rocky Mountains revealed nine genera and 18 species of Harpellales plus several species of *Paramoebidium* (Lichtwardt and Williams 1988). It can be assumed that the insect larvae in that stream, several species of which were often found clinging to the same benthic substrate, were exposed to a smorgasbord of different trichospores. The question is, why do trichospores that are ingested by a compatible host extrude their inner spore, which then attaches to and develops in the gut, whereas alien trichospores pass through the gut unaffected? This question was addressed by Horn (1989a, b, 1990) through in vitro and in vivo studies using *Aedes aegypti* larvae and two species of *Smittium* (*S. culisetae*, *S. culicis*).

The sporangiospore of *S. culisetae* extrudes from the sporangium (trichospore) by bursting through the apex of the sporangium under pressure, and the extruding sporangiospore approximately doubles its original length (Figs. 1.4B, 1.5A). The extrusion process usually takes 2–10 s (Williams 1983; Horn 1989a). A preliminary holdfast is secreted through a field of canals in the tip of the extruded sporangiospore wall. The origin of this adhesive substance is a series of membrane-bound and usually spherical "apical spore bodies" within the tip of the trichospore (Fig. 1.6; Moss and Lichtwardt 1976; Williams 1983; Horn 1989a, b).

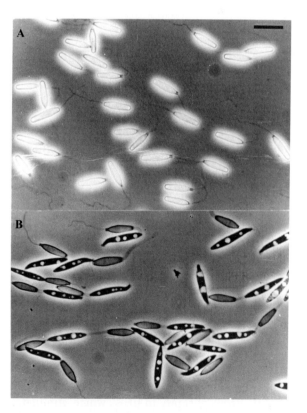

Fig. 1.4. Trichospores of *Smittium culisetae* in vitro, showing: **A** untreated trichospores, and **B** extruded sporangiospores following Phase I and Phase II treatment of trichospores (Horn 1989a). *Bar* 20 µm

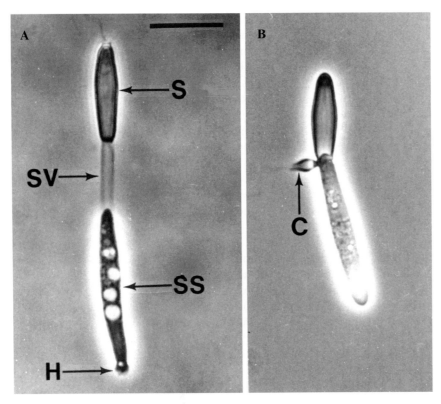

Fig. 1.5. Two forms of in vitro extrusion by trichospores of *Smittium* spp. from mosquito larvae (Horn 1989a). **A** *Smittium culisetae* with apical extrusion following Phase II treatment. **B** *Smittium culicis*

Horn (1989a) obtained up to 98% spore extrusion in vitro with *S. culisetae* by subjecting cultured trichospores to two sequential treatment phases. In Phase I, trichospores were exposed to a minimum of 20 mM potassium chloride at pH 10 for about 15 min, followed by Phase II, which consisted of shifting the buffer to pH 7 (6–8). Phase II led to *S. culisetae* spore extrusion within several minutes. The extrusion process, once started, took 10 s or less. These stimuli mimic the physiological conditions an ingested trichospore would encounter in the mosquito gut. Some potassium, excreted from the Malpighian tubules in primary urine (Bradley 1985), flows forward and is present in the midgut lumen (Ramsay 1950; Jones 1960) where the alkalinity is maintained at approximately pH 10. Upon reaching the pyloric chamber and hindgut region, the pH drops abruptly to pH 7, which stimulates spore extrusion and holdfast formation.

Extrusion in *S. culicis* was slightly different, for it occurred during Phase I (or in the midgut of larvae), but holdfast formation followed Phase II treatment (hindgut conditions). Furthermore, although spore extrusion was sometimes apical, as in *S. culisetae*, extrusion in *S. culicis* was often subbasal or basal (Fig. 1.5B). There were no visible ultrastructural changes in *S. culisetae* trichospores during Phase I treatment. However, an interwall layer may play a part in aiding sporangiospore extrusion, for it swells in Phase II (Horn 1990), and in *S. culisetae* (only) it forms a sleeve-like structure (Fig. 1.5A) connecting the trichospore wall and the expanded and extruded sporangiospore (Horn 1989b).

Pressure within the trichospore of *S. culisetae* increased by 1.0 MPa during Phase I, though this increase was not sufficient to extrude the sporangiospore and could not be attributed to potassium uptake (Horn 1990). It was possible to inhibit extrusion in Phase II by lowering the osmotic potential of the solution to −2.7 MPa. Horn hypothesized that potassium might stimulate the breakdown of certain organic solutes in trichospores which leads to the initial pressure increase during Phase I. Sporangiospore extrusion then occurs during Phase II due to rupture of the sporangial wall

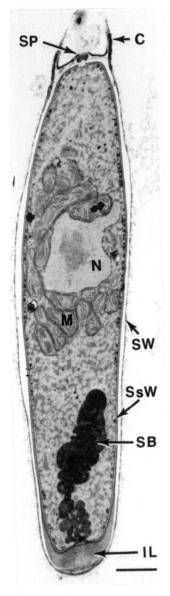

Fig. 1.6. Ultrastructure of *Smittium culisetae* trichospore at Phase I, with spore bodies (*SB*) aligned near the tip of the spore (Horn 1989b). *C* Collar, *IL* interwall layer, *M* mitochondrion, *N* nucleus, *SP* septal plug, *SsW* sporangiospore wall, *SW* sporangial wall. *Bar* 1 μm

caused by additional pressure from swelling of the interwall layer.

A small percentage of the more than 150 axenic isolates of *Smittium* available are known to extrude spontaneously in culture, perhaps a result of their having been transferred repeatedly and the ability to extrude having been a selective factor under such artificial cultural conditions. (Extrusion outside of the host in nature would be counterselective.) Horn (1989a) used Phase I and Phase II treatments on 29 harpellids belonging to four genera and ten species, comparing extrusion in treated and untreated trichospores. Among these harpellids were three species of *Smittium* that are notorious for their wide dipteran host ranges: *S. culisetae* and *S. culicis* (usually in Culicidae), and *S. simulii* (usually in Chironomidae or Simuliidae, but occasionally in Culicidae). The extrusion response to treatment of 22 isolates, with a few exceptions, was generally fair to excellent. The other seven species (belonging to four genera) showed no response to this specific trichospore treatment. These studies indicate that trichospores are attuned to chemical stimuli and may serve as a first level of host recognition by species of harpellids. It is not known specifically what elicits sporangiospore extrusion in other families of aquatic insects.

IV. Physiology of Trichomycetes In Vitro

A. Nutrition

The nutritional requirements of only four of the more than 35 species of trichomycetes that have been isolated axenically have been studied: *Amoebidium parasiticum*, *Smittium culisetae*, *Furculomyces boomerangus* (M.C. Williams & Lichtw.) M.C. Williams & Lichtwardt, and *Austrosmittium biforme* M.C. Williams & Lichtw. Whisler (1962), using a tryptone-glucose-salts medium in his nutritional studies of *A. parasiticum*, obtained a sixfold increase in dry weight with the addition of thiamine. He devised a chemically defined medium by substituting methionine for tryptone, but this resulted in much lower dry weight yields. Glucose, mannose, and fructose were satisfactory carbon sources, but apparently nitrate was not utilized as a nitrogen source. *Amoebidium parasiticum* attaches to the exoskeleton of its hosts, and its nutrition in nature is dependent upon solutes in the aquatic environment. In the author's experience, thalli of *A. parasiticum* are more abundant in waters with at least some obvious organic pollution.

Smittium culisetae, in contrast, is dependent on the nutrients within the gut lumen, whether they are exogenous or provided by secretions or excretions of the host. A medium consisting of 2% tryptone with 0.5% glucose and inorganic salts provided good growth of *S. culisetae* in liquid shake culture (Williams and Lichtwardt 1972b), but a completely defined medium, without tryptone, could not be devised. Difco tryptone was superior to other protein digests. The addition of thiamine, even in concentrations as low as 10 µg/l, was stimulatory. Glucose was the preferred carbon source, though glycerine, mannose, and fructose provided more than 50% as much dry weight. Trehalose, a blood sugar in insects, was not assimilated, nor was soluble starch. Ammonium compounds and urea, which are commonly excreted into insect guts (Chapman 1969), were preferred to other nitrogen sources. A tenfold dilution of Difco brain–heart infusion with added thiamine and biotin (1/10 BHIv) has proven to be satisfactory for growth of the majority of cultured trichomycetes (Lichtwardt 1986; Lichtwardt et al. 2001a). This medium, in general, is superior to tryptone–glucose media (TGv; for formulae, see Lichtwardt et al. 2001a), or various other formulations that have been attempted for primary isolation and storage of trichomycetes. A 1:1 mixture of 1/10 BHIv and TGv (BHIGTv) may be superior to either medium alone. These nutritional data, though basic and important for many kinds of in vitro studies, beg the major question, which is why do most trichomycetes remain unculturable?

B. Rates of Growth

Extensive growth of *Smittium* species in culture belies the limited growth of thalli in the hindgut of their small larval hosts, where excessive growth could be detrimental if not lethal. Considerably greater amounts of growth were obtained in shaken than in stationary flasks (Williams and Lichtwardt 1972b; El-Buni and Lichtwardt 1976). Growth rates as high as 220 mg/day (dry weight) were achieved with *S. culisetae*. Maximum dry weight yields in 4 days were >3 mg/ml medium. As expected, growth dynamics varied among species and strains, and were dependent on several culture parameters. Farr and Lichtwardt (1967) grew *S. culisetae* in stationary cultures at 15 temperatures, and obtained good growth between 10 °C and 32 °C. In general, lower temperatures produced lower growth rates but higher dry weight yields (Williams and Lichtwardt 1972b; El-Buni and Lichtwardt 1976). *Smittium culisetae* is also tolerant of a wide range of pH values. Growth in stationary cultures was satisfactory to good at initial pH values of 6.0 to 9.0, with greatest growth occurring in slightly alkaline media (Farr and Lichtwardt 1967), but in shaken cultures an initial pH of 5.5 produced the best growth (El-Buni and Lichtwardt 1976). The same strain of *S. culisetae* (COL-18-3) grown in tryptone–glucose medium with an initial pH just below neutrality exhibited a considerable drop in pH in stationary cultures after 6 days, whereas there was a rise in pH when cultures were shaken. Misra (2000) studied the growth and sporulation parameters of *Furculomyces boomerangus* and *Austrosmittium biforme*, which had been isolated from Australian Chironomidae, and obtained results similar to studies on *S. culisetae*, though biomass production, rates of growth, and trichospore production were generally less in those two species.

C. Reproduction

To the fungus living in the hindgut of its poikilothermic host where the pH is more or less constant, the amount of growth produced is probably less important to survival than successful sporulation. Cultures of *Smittium culisetae* in aerated liquid medium produced a dramatic number of trichospores (1.4×10^6/ml) after 4 days, in contrast to unshaken flasks in which the numbers of spores produced remained negligible (Williams and Lichtwardt 1972b). Among other factors strongly influencing trichospore production in vitro were several components in the media (El-Buni and Lichtwardt 1976). Maximum sporulation occurred with 2% tryptone, the numbers of spores dropping off considerably with higher or lower percentages of tryptone. Glucose concentrations in excess of 0.5% had a deleterious effect on trichospore production. Tryptone-glucose medium is free of sterols (Starr et al. 1979). The addition of sitosterol acetate and β-sitosterol to the medium (20 mg/l) more than doubled trichospore production, whereas ergosterol and cholesterol were inhibitory (El-Buni and Lichtwardt 1976).

D. Sterol and Lipid Production

Starr et al. (1979) studied sterol production in 14 isolates of *Smittium* species grown on a sterol-free medium and found that the primary sterol synthesized in most isolates was desmosterol, which apparently is rare in fungi. Three isolates produced ergosterol and older cultures of *S. culisetae* (isolate COL-18-3) also produced cholesterol. Several isolates produced one or two unidentified sterols. The amount of sterol produced by *S. culisetae* (isolate COL-18-3) varied quantitatively as well as qualitatively when measured at 3, 7, and 14 days of growth, with maximum production around day 7. The total sterol content per gram of dry weight of mycelium varied from 0.125 mg/g to 3.38 mg/g when harvested at 7 days in each of the 14 isolates, but this may not reflect the maximum potential of these isolates due to their different growth rates.

An analysis of lipid components of *S. culisetae* (isolate HAW-13-2) grown on dilute brain-heart infusion was done by Patrick et al. (1973), a medium which Starr et al. (1979) found to contain 0.715 mg of cholesterol per gram of dry medium. Patrick et al. (1973) reported that lipids comprised 9.9% of the mycelial dry weight, with a ratio of approximately 3:1 neutral to polar lipids. Triglycerides constituted 26.6% of the lipid classes. Of the total fatty acids, palmitoleic was unusually high (38.7%), followed quantitatively by palmitic (34.3%) and oleic acids (16.3%). Steroids made up 12.9% of the total combined lipids, and steroid ester and hydrocarbons 9.2%. The possibility that *S. culisetae* sterols may supplement the requirements of mosquito larvae has been discussed in a previous section of this chapter (Section IID).

V. Conclusions

It is now known that two of the four orders of arthropod gut organisms (Amoebidiales and Eccrinales) are protozoans rather than fungi. Several parallelisms between those orders and the fungal orders (Harpellales and the questionable Asellariales) include some morphological and host similarities and intimate developmental relationships with their respective hosts. Therefore, studying them as an ecological group, trichomycetes, seems justified.

The considerable success of trichomycetes in establishing associations with arthropods can be attributed to numerous adaptations. The morphological and physiological features that they have evolved make trichomycetes distinctively different in several respects from phylogenetically related organisms. These differences include, among others, modified sporangia, zygospores, and other propagules that promote transmission of the fungus from host to host in aquatic or terrestrial environments, and spores that have the ability to recognize suitable hosts when ingested, leading to attachment to the gut cuticle and immediate establishment of a thallus. The production of resistant sporangiospores and zygospores can be influenced in some species by the molting cycle of the host.

Trichomycete associations with arthropods vary from benign to lethal or pathogenic. In particular circumstances, fungal biosynthesis of sterols and B vitamins may provide essential nutrients to the host, thus leading to a more mutualistic relationship. Recent studies on some Harpellales, whose species infect the guts of larval stages of aquatic insects, are providing evidence that fungal dispersal from one body of water to another may be accomplished by flight of adults whose ovaries have become infected with cystlike cells of the fungus. These cells are "oviposited" by the female, even through she has become completely or partially sterilized by the fungus.

Most trichomycetes currently remain unculturable. Among those that can be grown in vitro, there appear to be no major differences in nutritional requirements when compared to many other fungi, and rates of trichomycete growth can be equal to that of free-living saprotrophs. Aeration, among other culture parameters, is especially important in achieving maximum growth and sporulation in vitro. In the restricted and less aerobic environment of the arthropod gut, however, adequate maturation and reproduction prior to ecdysis of the host and successful transmission of spores within populations of arthropods are more important for survival than achievement of maximum growth.

This review makes it clear that many aspects of the symbiotic relationships that have evolved between trichomycetes and their various arthropod hosts are still to be elucidated, thus providing fertile fields to be investigated. It is also evident that theories on the co-evolution of trichomycetes and arthropods, as reflected in current phylogenetic hypotheses, are still in a state of flux, and can only be resolved through the study of additional extant taxa that undoubtedly remain to be discovered.

Acknowledgments. Much of the original information in this chapter, including some unpublished data, was obtained with support of several National Science Foundation awards. Any opinions, findings, and conclusions expressed in this article are those of the author and do not necessarily reflect the views of the National Science Foundation. The author is grateful to the many collaborators and other investigators cited herein who have contributed materially over the years to our foundation of knowledge about trichomycetes that has made this review possible. The author is grateful to Merlin M. White, Bruce W. Horn, and Matías J. Cafaro for reading this manuscript and providing useful suggestions.

References

Benny GL, O'Donnell (2000) *Amoebidium parasiticum* is a protozoan, not a Trichomycete. Mycologia 92:1133–1137

Bradley TJ (1985) The excretory system: structure and physiology. In: Kerkut GA Gilbert LI (eds) Comprehensive insect physiology, biochemistry and pharmacology, vol IV. Pergamon, Oxford, pp 421–465

Cafaro M (2005) Eccrinales (Trichomycetes) are not fungi, but a clade of protists at the early divergence of animals and fungi. Mol Phylogen Evol 35:21–34

Chapman RF (1969) The insects (structure and function). American Elsevier, New York

Christophers SR (1960) *Aedes aegypti* (L.). The yellow fever mosquito. Its life history, bionomics and structure. Cambridge University Press, Cambridge

Clements AN (1963) The physiology of mosquitoes. (Inter Ser Monogr Pure Appl Biol, vol 17) Pergamon, New York

Coluzzi M (1966) Experimental infections with *Rubetella* fungi in *Anopheles gambiae* and other mosquitoes. Proc Int Congr Parasitol 1(vol 1):592–593

Dubitskii AM (1978) Biological control of blood sucking Diptera in the USSR. Inst Zool Kazakhstan Acad Sci, Alma Ata, pp 92–93

El-Buni AM, Lichtwardt RW (1976) Asexual sporulation and mycelial growth in axenic cultures of *Smittium* spp. (Trichomycetes). Mycologia 68:559–572

Farr DF, Lichtwardt RW (1967) Some cultural and ultrastructural aspects of *Smittium culisetae* (Trichomycetes) from mosquito larvae. Mycologia 59:172–182

Garms R (1975) Observations on filarial infections and parous rates of anthropophilic blackflies in Guatemala, with reference to the transmission of *Onchocerca volvulus*. Tropenmed Parasit 26:169–182

Gottlieb AM, Lichtwardt RW (2001) Molecular variation within and among species of Harpellales. Mycologia 93:66–81

Hibbits J (1978) Marine Eccrinales (Trichomycetes) found in crustaceans of the San Juan Archipelago, Washington. Syesis 11:213–261

Horn BW (1989a) Requirement for potassium and pH shift in host-mediated sporangiospore extrusion from trichospores of *Smittium culisetae* and other *Smittium* species. Mycol Res 93:303–313

Horn BW (1989b) Ultrastructural changes in trichospores of *Smittium culisetae* and *S. culicis* during in vitro sporangiospore extrusion and holdfast formation. Mycologia 81:742–753

Horn BW (1990) Physiological changes associated with sporangiospore extrusion from trichospores of *Smittium culisetae*. Exp Mycol 14:113–123

Horn BW, Lichtwardt RW (1981) Studies on the nutritional relationship of larval *Aedes aegypti* (Diptera: Culicidae) with *Smittium culisetae* (Trichomycetes). Mycologia 73:724–740

Jones JC (1960) The anatomy and rhythmical activities of the alimentary canal of *Anopheles* larvae. Ann Entomol Soc Am 53:459–474

Labeyrie ES, Molloy DP, Lichtwardt RW (1996) An investigation of Harpellales (Trichomycetes) in New York State blackflies (Diptera: Simuliidae). J Invert Pathol 68:293–298

Lichtwardt RW (1961) A *Palavascia* (Eccrinales) from the marine isopod *Sphaeroma quadridentatum* Say. J Elisha Mitchell Sci Soc 77:242–249

Lichtwardt RW (1986) The Trichomycetes: fungal associates of arthropods. Springer, Heidelberg

Lichtwardt RW, Moss ST (1981) Vegetative propagation in the new species of Harpellales, *Graminella microspora*. Trans Br Mycol Soc 76:311–316

Lichtwardt RW, Williams MC (1988) Distribution and species diversity of trichomycete gut fungi in aquatic insect larvae in two Rocky Mountain streams. Can J Bot 66:1259–1263

Lichtwardt RW, Williams MC (1990) Trichomycete gut fungi in Australian aquatic insect larvae. Can J Bot 68:1057–1074

Lichtwardt RW, Williams MC (1992a) Two new Australasian species of Amoebidiales associated with aquatic insect larvae and comments on their biogeography. Mycologia 84:376–383

Lichtwardt RW, Williams MC (1992b) Tasmanian trichomycete gut fungi in aquatic insect larvae. Mycologia 84:384–391

Lichtwardt RW, Williams MC (1992c) Western Australian species of *Smittium* and other Trichomycetes in aquatic insect larvae. Mycologia 84:392–398

Lichtwardt RW, Williams MC (1992d) *Smittium bullatum* from a New Zealand midge larva and new records of other trichomycete gut fungi. Can J Bot 70:1193–1195

Lichtwardt RW, Williams MC (1992e) *Furculomyces*, a new homothallic genus of Harpellales (Trichomycetes) from Australian midge larvae. Can J Bot 70:1196–1198

Lichtwardt RW, Kobayasi Y, Indoh H (1987) Trichomycetes of Japan. Trans Mycol Soc Japan 28:359–412

Lichtwardt RW, Cafaro MJ, White MM (2001a) The Trichomycetes, fungal associates of arthropods, rev edn. http:www.nhm.ku.edu/~ fungi

Lichtwardt RW, White MM, Colbo MH (2001b) Harpellales in Newfoundland aquatic insect larvae. Mycologia 93:764–773

López Lastra CC (1990) Primera cita de *Smittium morbosum* var. *rioplatensis* var. nov. (Trichomycetes: Harpellales) patógeno de 5 especies de mosquitos (Diptera: Culicidae) en la República Argentina. Rev Argentina Mycol 13:14–18

Mendoza L, Taylor JW, Ajello L (2002) The Class Mesomycetozoea: a heterogeneous group of microorganisms at the animal-fungal boundary. Ann Rev Microbiol 56:315–344

Misra JK (2000) Growth, sporulation and pH tolerance of *Furculomyces boomerangus* and *Austrosmittium biforme* in axenic culture. Mycologia 92:1051–1056

Moss ST, Descals E (1986) A previously undescribed stage in the life cycle of Harpellales (Trichomycetes). Mycologia 78:213-222

Moss ST, Lichtwardt RW (1976) Development of trichospores and their appendages in *Genistellospora homothallica* and other Harpellales and fine-structural evidence for the sporangial nature of trichospores. Can J Bot 54:2346-2364

O'Donnell K, Cigelnick E, Benny GL (1998) Phylogenetic relationships among the Harpellales and Kickxellales. Mycologia 90:624-639

Patrick MA, Sangar VK, Dugan PR (1973) Lipids of *Smittium culisetae*. Mycologia 65:122-127

Ramsay JA (1950) Osmotic regulation in mosquito larvae. J Exp Biol 27:145-157

Sato H, Shimada N, Aoki J (1989) Light and electron microscopy of *Smittium morbosum* (Trichomycetes), newly recorded from Japan. Trans Mycol Soc Jpn 30:51-59

Starr AM, Lichtwardt RW, McChesney JD, Baer TA (1979) Sterols synthesized by cultured Trichomycetes. Arch Microbiol 120:185-189

Strongman DB (2005) Synonymy of *Ejectosporus magnus* and *Simuliomyces spice*, and a new species, *Ejectosporus trisporus*, from winter-emerging stoneflies. Mycologia 97:552-561

Sweeney AW (1981) An undescribed species of *Smittium* (Trichomycetes) pathogenic to mosquito larvae in Australia. Trans Br Mycol Soc 77:55-60

Tuzet O, Manier J-F (1950) *Lajassiella aphodii*, n.g., n.sp. Palavascide parasite d'une larve d'*Aphodius* (Coléoptère Scarabaeidae). Ann Sci Nat Zool Ser 11 12:465-470

Undeen AH, Nolan RA (1977) Ovarian infection and fungal spore oviposition in the blackfly *Prosimulium mixtum*. J Invert Pathol 30:97-98

Ustinova I, Krienitz L, Huss VAR (2000) *Hyaloraphidium curvatum* is not a green alga, but a lower fungus; *Amoebidium parasiticum* is not a fungus, but a member of the DRIPS. Protist 151:253-262

Van Dover CL, Lichtwardt RW (1986) A new trichomycete commensal with a galatheid squat lobster from deep-sea hydrothermal vents. Biol Bull 171:461-468

Whisler HC (1960) Pure culture of the trichomycete, *Amoebidium parasiticum*. Nature 186:732-733

Whisler HC (1962) Culture and nutrition of *Amoebidium parasiticum*. Am J Bot 49:193-199

Whisler HC (1968) Developmental control of *Amoebidium parasiticum*. Dev Biol 17:562-570

White MM, Lichtwardt RW (2004) Fungal symbionts (Harpellales) in Norwegian aquatic insect larvae. Mycologia 96:891-910

White MM, James TY, O'Donnell K, Cafaro MJ, Tanabe Y, Sugiyama J (2006a) Phylogeny of the Zygomycota based on nuclear ribosomal sequence data. Mycologia 98:872-884

White MM, Lichtwardt RW, Colbo MH (2006b) Confirmation and identification of parasitic stages of obligate endobionts (Harpellales) in blackflies (Simuliidae) by means of rRNA sequence data. Mycol Res 110:1070-1079

Williams MC (1983) Zygospores in *Smittium culisetae* (Trichomycetes) and observations on trichospore germination. Mycologia 75:251-256

Williams MC, Lichtwardt RW (1972a) Infection of *Aedes aegypti* larvae by axenic cultures of the fungal genus *Smittium* (Trichomycetes). Am J Bot 59:189-193

Williams MC, Lichtwardt RW (1972b) Physiological studies on the cultured trichomycete, *Smittium culisetae*. Mycologia 64:806-815

Williams MC, Lichtwardt RW (1990) Trichomycete gut fungi in New Zealand aquatic insect larvae. Can J Bot 68:1045-1056

Williams MC, Lichtwardt RW (1993) A new monotypic fungal genus, *Allantomyces*, and a new species of *Legeriomyces* (Trichomycetes, Harpellales) in the hindgut of a Western Australian mayfly nymph (*Tasmanocoenis* sp.). Can J Bot 71:1109-1113

Yeboah DO, Undeen AH, Colbo MH (1984) Phycomycetes parasitizing the ovaries of blackflies (Simuliidae). J Invert Pathol 43:363-373

2 Opportunistic Mold Infections

Ronald G. Washburn[1]

CONTENTS

I.	Introduction	21
II.	Aspergillosis	21
	A. Mycology	21
	B. Epidemiology	22
	C. Clinical Manifestations	23
	1. Invasive Pulmonary Aspergillosis and Disseminated Infection	23
	2. Invasive *Aspergillus* Sinusitis	23
	3. Noninvasive Mycelial Mass of the Lung	23
	4. Noninvasive Mycelial Mass of the Paranasal Sinuses	24
	5. Bronchopulmonary Aspergillosis	24
	6. Allergic *Aspergillus* Sinusitis	24
	D. Diagnosis	24
	E. Treatment	25
III.	Mucormycosis	25
	A. Mycology	26
	B. Epidemiology	26
	C. Clinical Manifestations	26
	1. Rhinocerebral Mucormycosis	26
	2. Invasive Pulmonary Mucormycosis and Disseminated Infection	26
	3. Gastrointestinal Mucormycosis	26
	4. Skin and Wound Mucormycosis	27
	D. Diagnosis	27
	E. Treatment	27
IV.	Infections Caused by *Fusarium* spp.	27
V.	Infections Caused by *Pseudallescheria boydii*	27
VI.	Dematiaceous Molds	28
VII.	Rare Agents	28
VIII.	Conclusions	28
	References	29

I. Introduction

This chapter provides an overview of opportunistic mold infections, including aspergillosis, mucormycosis, infections caused by *Fusarium* spp. and *Pseudallescheria boydii*, and infections caused by the dematiaceous molds (Table 2.1). Although each of these agents is also capable of causing disease in normal individuals, special emphasis has been placed on infections in compromised hosts.

II. Aspergillosis

Species of *Aspergillus* are responsible for a wide variety of different clinical syndromes, including saprophytic colonization of the pulmonary airspaces in patients with chronic lung disease, chronic noninvasive mycelial masses of the lungs or paranasal sinuses, and rapidly-progressive life-threatening invasive infections of the lungs or paranasal sinuses in neutropenic hosts. In addition, *Aspergillus* spp. produce a number of different allergic syndromes, including extrinsic allergic alveolitis, bronchopulmonary aspergillosis, and allergic *Aspergillus* sinusitis.

A. Mycology

Aspergillus spp. are ascomycetes that are ubiquitous in nature, and most species can be recovered from soil. One hundred and thirty-two species are listed by Raper and Fennell (1965); the species most commonly responsible for human disease are *A. fumigatus*, *A. flavus* (Miloshev et al. 1967; Green et al. 1969; Young et al. 1972), and *A. niger*, with fewer cases caused by relatively nonpathogenic species such as *A. avenaceus* (Washburn et al. 1988), *A. nidulans* (Redmond et al. 1965; Bujak et al. 1974; White et al. 1988), *A. oryzae* (Ziskind et al. 1958), and *A. terreus* (Moore et al. 1988; Hara et al. 1989). The organism grows at temperatures ranging over 4–55 °C, with most abundant growth at 30–37 °C.

Aspergillus grows in air conditioning conduits, fire-proofing material, and decaying vegetation; since it is thermophilic, the organism grows in almost pure culture in compost heaps. *Aspergillus*

---

[1] Overton Brooks Veterans Administration Medical Center and Louisiana State University Health Sciences Center, 510 East Stoner Avenue, Shreveport, LA 71101, USA; e-mail: ronald.washburn@med.va.gov

Table 2.1. Clinical syndromes caused by opportunistic molds

Aspergillus spp.	
Invasive disease	Invasive pulmonary aspergillosis ± disseminated infection
	Invasive *Aspergillus* sinusitis
	Skin and wound invasive aspergillosis
	Keratomycosis
Noninvasive disease	Noninvasive mycelial mass of lung or paranasal sinus
	Bronchopulmonary aspergillosis
	Extrinsic allergic alveolitis
	Allergic *Aspergillus* sinusitis
Agents of mucormycosis	
	Rhinocerebral mucormycosis
	Invasive pulmonary mucormycosis ± disseminated infection
	Gastrointestinal mucormycosis
	Skin and wound mucormycosis
Fusarium spp.	
	Disseminated infection
	Keratomycosis
	Mycetoma
	Onychomycosis
Pseudallescheria boydii	
	Pneumonia
	Brain abscess
	Invasive sinusitis
	Noninvasive mycelial mass of lung or paranasal sinus
	Mycetoma
Dematiaceous molds	
Invasive disease	Brain abscess
	Localized subcutaneous infection
	Disseminated infection
	Invasive sinusitis
Noninvasive disease	Allergic bronchopulmonary disease
	Allergic sinusitis

spp. possess a wide array of different degradative and biosynthetic enzymes, and therefore the organisms are able to grow on many different substrates, including not only rich media (e.g. Sabouraud's agar), but also chemically defined media comprised solely of low molecular weight components (e.g. Czapek-Dox agar).

The infectious particles are airborne conidia (2–5 µm diameter) produced by phialides that are borne on conidiophores. When these tiny spheres are inhaled, they are capable of reaching the pulmonary alveoli. In normal hosts, the organisms are effectively cleared by mucociliary mechanisms and phagocytic killing (Waldorf et al. 1984a; Levitz and Diamond 1985; Washburn et al. 1987); however, in susceptible hosts, severe invasive pneumonia may ensue (discussed below). The elongated hyphae that invade tissue are narrow (2–5 µm), regular, septate, and dichotomously branching (ca. 45-degree angles). In invasive disease, hyphae invade blood vessels, producing distal tissue infarction.

B. Epidemiology

Conidia are continuously inhaled, but typically cause infection only in immunocompromised individuals; e.g. those with prolonged neutropenia (Young et al. 1970), supraphysiologic glucocorticoid therapy, chronic granulomatous disease (Bujak et al. 1974), or AIDS (Pursell et al. 1990; Woods and Goldsmith 1990; Denning et al. 1991). Concentrations of airborne conidia increase during construction and there is some evidence to support a link between hospital construction and nosocomial outbreaks of invasive pulmonary aspergillosis. Thus, the recommendation has been made that susceptible hosts (e.g. bone marrow transplant recipients) should be housed in rooms with high-efficiency particulate air (HEPA) filters (Sherertz et al. 1987).

Less commonly, infection can result from direct percutaneous inoculation in susceptible hosts; e.g. by intravenous catheters or contaminated tape

(Prystowsky et al. 1976; Carlile et al. 1978; McCarty et al. 1986; Googe et al. 1989). Invasive disease of the eye (Cameron et al. 1991) or deep structures (e.g. vertebrae or endocardium) can follow surgical procedures even in previously normal individuals, presumably due to direct inoculation (Corrall et al. 1982; Mawk et al. 1983; Weber and Washburn 1990). Rarely, *Aspergillus* endocarditis may be encountered in intravenous drug abusers. There is no documented human-to-human transmission.

C. Clinical Manifestations

1. Invasive Pulmonary Aspergillosis and Disseminated Infection

The most lethal form of aspergillosis is invasive disease in the neutropenic host with leukemia or lymphoma. The disease is characterized by fever, dyspnea, pleuritic chest pain, cough, and rapidly progressive pulmonary infiltrates after ca. 2–4 weeks of neutropenia. Invasive pulmonary aspergillosis usually progresses rapidly during several weeks. Even in the face of appropriate antifungal therapy the infection is almost uniformly fatal unless bone marrow function returns. Vascular invasion and pulmonary infarction may lead to hemoptysis. In approximately one-third of cases the infection disseminates to distant organs; e.g. brain (Boon et al. 1990), kidneys, liver, and spleen. Gastrointestinal (GI) lesions may produce bleeding and perforation (Meyer et al. 1973b).

In patients with abnormal phagocyte function (e.g. those with chronic granulomatous disease; CGD) or those taking daily high-dose glucocorticoids (e.g. bone marrow transplant patients with graft versus host disease) the infection is more slowly progressive than that seen in patients with prolonged neutropenia. In CGD, invasive pulmonary aspergillosis typically presents with multiple nodules and the infection often invades contiguous structures such as mediastinum, ribs, vertebrae, or clavicle (Altman 1977). Hyphae are scarce, but the granulomatous inflammation is exuberant.

Several cases of invasive pulmonary aspergillosis have been reported in small children following inhalation of large inoculae of conidia or in alcoholics with chronic liver disease (Zellner et al. 1969; Blum et al. 1978; Brown et al. 1980). Spontaneous resolution is the rule in the childhood form of the disease, whereas invasive aspergillosis in alcoholics carries a poor prognosis.

2. Invasive *Aspergillus* Sinusitis

Invasive sinusitis can be either acute (e.g. in the neutropenic host) (Young et al. 1970; Swerdlow and Deresinski 1984) or chronic (e.g. in immunologically normal individuals) (Washburn et al. 1988; Washburn 1998). In general, the acute form of the disease resembles rhinocerebral mucormycosis (McGill et al. 1980), but acute *Aspergillus* sinusitis is most commonly encountered in the setting of prolonged neutropenia, and rhinocerebral mucormycosis usually occurs in poorly controlled diabetics.

Acute *Aspergillus* spp. sinusitis typically begins in the maxillary or ethmoid sinuses and rapidly invades blood vessels, producing tissue necrosis and violating anatomic barriers (Peterson and Schimpff 1989; Dyken et al. 1990; Talbot et al. 1991). Thus, the disease may be recognized because of a black eschar on the hard palate due to inferior extension through the floor of the maxillary sinus. Unilateral proptosis and chemosis may occur because of superior erosion through the roof of the maxillary sinus or lateral extension through the ethmoid sinus into the lamina papyracea. Focal neurological findings and altered mental status are signs that the disease has already extended into the brain and such involvement in neutropenic individuals usually predicts a fatal outcome. However, patients with chronic granulomatous disease have more indolent brain involvement that may respond to antifungal therapy.

Aspergillus may produce a more chronic form of invasive sinusitis in apparently normal hosts (Washburn 1998). Patients typically present with painless unilateral proptosis, and they are not systemically ill. The infection is endemic in the arid portion of the Sudan where *A. flavus* is the predominant causative organism presumably because of high ambient concentrations of airborne *A. flavus* conidia. It has been postulated that chronic upper respiratory mucosal fissuring due to the arid climate facilitates passage of the organism into submucosal tissue. In this disease erosion through anatomic barriers requires months or years and produces granulomatous inflammation.

3. Noninvasive Mycelial Mass of the Lung

Aspergillus spp. may grow as noninvasive mycelial masses in preexisting lung cavities caused by emphysema, tuberculosis, cancer, or sarcoidosis (Varkey and Rose 1976; Jewkes et al. 1983).

The fungus ball is usually mobile and not attached to the wall of the cavity. Radiographically, a crescent of air can often be seen above the mass. *Aspergillus* spp. may be recovered repeatedly from expectorated sputum, and when *Aspergillus niger* is the causative organism calcium oxalate crystals may be observed (Wilson and Wilson 1961). Patients with noninvasive pulmonary mycelial masses may develop life-threatening hemoptysis (Jewkes et al. 1983). Repeated severe hemoptysis is an indication for surgery (discussed below).

4. Noninvasive Mycelial Mass of the Paranasal Sinuses

Patients with chronic bacterial sinusitis and impaired sinus drainage (e.g. due to polypoid obstruction) may accumulate inspissated mucus that supports noninvasive mycelial growth. Intermittent unilateral maxillary sinus pain, vertex headaches due to sphenoid sinus pressure, and cacosmia are common complaints (Washburn 1998). In this disease the mucosa and submucosa may be chronically thickened and infiltrated by lymphocytes and plasmacytes but hyphae do not invade tissue.

5. Bronchopulmonary Aspergillosis

Bronchopulmonary aspergillosis represents an allergic response that is usually seen in patients with preexisting asthma. The syndrome is characterized by bronchospasm, fleeting pulmonary infiltrates, bronchiectasis, mucus plugging, repeated recovery of *Aspergillus* spp. from sputum, and eosinophils and Charcot–Leyden crystals in sputum (Washburn 1996). Additional features include immediate cutaneous hypersensitivity to *Aspergillus* antigens, peripheral eosinophilia, elevations of total IgE, and specific anti-*Aspergillus* precipitins, including IgE and IgG.

6. Allergic *Aspergillus* Sinusitis

Allergic *Aspergillus* sinusitis shares many features with bronchopulmonary aspergillosis, except that the inflammation is located in the paranasal sinuses instead of the lungs (Katzenstein et al. 1983; Waxman et al. 1987). These individuals have chronic sinusitis with positive cultures and visible hyphae in direct smears of mucus. The characteristic mucus is termed "allergic mucin" because it contains eosinophils and Charcot–Leyden crystals.

There may be mucosal and submucosal thickening, with eosinophilic inflammation. Eventually, patients may have bony rearrangement, probably due to sustained increases in intrasinus pressure. By definition, the infection does not cross anatomic barriers and thus, for example, extension into the brain has not been described in carefully defined cases of allergic *Aspergillus* sinusitis. Similar to patients with bronchopulmonary aspergillosis, those with allergic *Aspergillus* sinusitis exhibit peripheral eosinophilia, immediate-type hypersensitivity to *Aspergillus* antigens, elevated total IgE, and specific anti-*Aspergillus* precipitins, including IgE and IgG.

D. Diagnosis

Diagnosis of invasive pulmonary aspergillosis is usually made at autopsy. To firmly establish the diagnosis antemortem, biopsies must demonstrate invasive hyphae; cultures are confirmatory. Diagnostic yield of transbronchial biopsy is low, so open-lung biopsy is the gold standard. Unfortunately, thrombocytopenia or coagulopathy often render biopsy inadvisable. In that setting a low attenuation halo surrounding a nodule visualized by computerized tomography (CT) is suggestive of invasive fungal disease. In nonsmokers a presumptive diagnosis can be established by repeated recovery of *Aspergillus* from expectorated sputum, bronchoalveolar lavage fluid, or nasopharyngeal cultures. In contrast, smokers are so frequently colonized that recovery of *Aspergillus* spp. from these same sources carries less diagnostic weight (Yu et al. 1986).

New cerebral mass lesions in the setting of neutropenia and progressive pulmonary infiltrates raise the specter of central nervous system dissemination; and GI hemorrhage raises suspicions for GI dissemination. Unfortunately, blood cultures are virtually always negative, but galactomannan detection by enzyme immunoassay is now gaining practical diagnostic value.

Invasive *Aspergillus* spp. sinusitis is diagnosed by visualizing characteristic hyphae in mucosa and deeper structures, and cultures are confirmatory. In the neutropenic host with thrombocytopenia, the ability to safely obtain deep biopsy specimens may be limited. However, in immunologically intact hosts biopsies are helpful for defining the invasive nature of the infection and delineating the associated inflammatory response.

Noninvasive mycelial mass of the lung is diagnosed by the appearance of a fungus ball on chest roentgenograms, combined with repeated recovery of *Aspergillus* spp. from expectorated sputum and/or bronchoalveolar lavage specimens; more invasive procedures would rarely be warranted. Many patients have circulating anti-*Aspergillus* precipitins, usually IgG (Kurup and Fink 1978).

Noninvasive mycelial mass of the paranasal sinus is diagnosed by sinoscopic visualization of a greasy or friable mass that shows entangled hyphae limited to the airspace, and thickened submucosa that may contain lymphocytic inflammation. Occasionally, conidiophores may be seen within the mycelial mass, providing a clue about the causative species; cultures are confirmatory.

Diagnostic criteria for bronchopulmonary aspergillosis and allergic *Aspergillus* sinusitis are outlined in the respective sections concerning clinical manifestations (discussed above).

E. Treatment

Therapy for invasive pulmonary aspergillosis with or without dissemination is vorizonazole 4 mg/kg every 12 h (Steinbach and Stevens 2003) or amphotericin B deoxycholate to a total dose of at least 1–2 g. Liposomal amphotericin B is approved as salvage therapy and has the advantage of being less nephrotoxic than amphotericin B deoxycholate. There is some evidence to support adjunctive use of 5-fluorocytosine or rifampin with amphotericin B-based regimens in especially difficult cases (Kitahara et al. 1976; Arroyo et al. 1977; Hughes et al. 1984). Additionally, caspofungin is approved for invasive aspergillosis in patients who are refractory or intolerant of other therapy (Patterson 2005). Some authorities advocate surgical resection of infected lung tissue for patients in whom hemostasis can be achieved (Weiland et al. 1983). Unfortunately the disease remains highly lethal unless bone marrow function returns and immunosuppression is reduced. For those patients in whom the infection comes under control, a course of oral voriconazole or itraconazole 400 mg/day could be considered for follow-up therapy (Denning et al 1994; Patterson 2005). Posaconazole also has useful clinical activity against invasive aspergillosis (Walsh et al. 2007).

The same antifungal options are available for invasive *Aspergillus* sinusitis. Aggressive surgical extirpation of necrotic tissue is recommended when hemostasis can be achieved, and follow-up therapy with oral voriconazole, itraconazole, or posaconazole would be an option for patients who can be discharged from the hospital.

Noninvasive mycelial masses of the lung do not require therapy except when severe hemoptysis ensues, and in those cases surgical excision can be life-saving (Jewkes et al. 1983). Noninvasive mycelial masses of the paranasal sinuses can be cured with complete surgical removal of the fungus ball (Washburn et al. 1988). No antifungal chemotherapy is required for noninvasive masses of lung or sinuses.

For patients with bronchopulmonary aspergillosis, bronchodilation with cromolyn sodium facilitates clearance of infected secretions. Oral steroid therapy may also be required for control of acute exacerbations; dosage recommendations are reviewed elsewhere (Washburn 1996). A few patients become steroid dependent. There is now some evidence to support the use of itraconazole to prevent recurrent episodes of bronchopulmonary aspergillosis.

There is no controlled data to guide therapy of allergic *Aspergillus* sinusitis. However, accumulating evidence suggests that a reasonable approach would be surgical removal of inspissated mucus and inflamed mucosa, combined with postoperative topical steroids, e.g. beclomethasone (Washburn 1998). The roles of systemic antifungal agents and systemic glucocorticoids remain controversial.

III. Mucormycosis

Mucormycosis is caused by zygomycetous fungi of the order Mucorales. There are several different clinical forms of mucormycosis, including acute rhinocerebral disease, invasive pulmonary disease (often accompanied by disseminated infection), GI infection, and locally invasive cutaneous disease arising from direct inoculation. Patients at risk include poorly controlled diabetics and individuals with prolonged neutropenia (e.g. leukemia or lymphoma), chronic renal insufficiency, kidney transplantation, deferoxamine chelation therapy for iron- or aluminum-overload, metabolic acidosis, burns, malnutrition, and intravenous drug abuse.

A. Mycology

Three different genera account for the majority of mucormycosis cases: *Rhizopus* (*R. oryzae* and *R. microsporus* var. *rhizopodiformis*; Bottone et al. 1979), *Cunninghamella bertholletiae* (Boyce et al. 1981; Sands et al. 1985; Rex et al. 1988; Mostaza and Barbado 1989; Zeilender et al. 1990), and *Saksenaea vasiformis* (Pierce et al. 1987; Kaufmann et al. 1988; Goldschmied-Reouven et al. 1989). The infectious particles, tiny sporangiospores measuring 2–3 µm in diameter, are produced within sporangia borne on sporangiophores. Alveolar macrophages bind to the spores and inhibit germination (Waldorf et al. 1984a, b). Coenocytic hyphae are the tissue-invasive form of the organism. Those structures are morphologically distinct from *Aspergillus* hyphae, because they are aseptate or only sparsley septate, broad and irregular (up to 15 µm wide), and they exhibit right-angle branching. The broad hyphae of mucormycosis invade blood vessels and produce extensive tissue infarction similar to invasive aspergillosis. Neutrophils kill *Rhizopus* hyphae (Schaffner et al. 1986). Deferoxamine enhances growth of the fungus (Boelaert et al. 1993, 1994).

B. Epidemiology

The airborne spores are ubiquitous; for example, the mold grows on decaying vegetation, stale bread, and fruit. Infection is acquired by inhalation, or rarely by direct inoculation into skin e.g. by Elastoplast bandages (Gartenberg et al. 1978; Hammond and Winkelmann 1979; Sheldon and Johnson 1979), or by intravenous and peritoneal dialysis catheters (Baker et al. 1962; Branton et al. 1991), or into the eye, e.g. by lens insertion (Orgel and Cohen 1989). In severely malnourished children GI infection is believed to be acquired by oral ingestion (Watson 1957; Isaacson and Levin 1961; Washburn and Bennett 1995). There is no documented human-to-human transmission.

C. Clinical Manifestations

1. Rhinocerebral Mucormycosis

Rhinocerebral mucormycosis is a catastrophic illness, usually seen in patients with poorly controlled diabetes mellitus; it may also be encountered in renal allograft recipients and patients with other forms of severe immunocompromise. The infection represents a true medical and surgical emergency because it can progress rapidly to death. Patients typically present with unilateral nasal discharge, headache, and fever. As the disease progresses, an eschar may form over the hard palate due to erosion through the floor of the maxillary sinus. Unilateral proptosis and chemosis are indicators of superior extension through the roof of the maxillary sinus or lateral extension through the ethmoid. Contiguous areas of involved skin progress rapidly from erythematous to violaceous, and ultimately become necrotic.

Lateral extension from the sphenoid sinus can lead to cavernous sinus thrombosis and cranial nerve palsies (cranial nerves III–VI). Brain invasion presents with focal neurological deficits and altered mental status progressing to coma and death. Central nervous system lesions may be visualized by CT scan or magnetic resonance imaging, but cerebrospinal fluid (CSF) cultures are almost uniformly negative and other CSF parameters may be normal. Intravenous drug abusers may present with brain abscesses, presumably due to infected emboli (Pierce et al. 1982; Smith et al. 1989; Stave et al. 1989; Fong et al. 1990).

2. Invasive Pulmonary Mucormycosis and Disseminated Infection

Invasive pulmonary mucormycosis is a life-threatening infection usually found in patients with prolonged neutropenia (Meyer et al. 1972), diabetes mellitus, or deferoxamine therapy. Clinical presentations of the disease are similar to those of invasive pulmonary aspergillosis; the infection may extend locally into the mediastinum (Connor et al. 1979) or across the diaphragm, and hemoptysis may result from vascular invasion and pulmonary infarction (Murray 1975). Disseminated disease most commonly affects the brain and GI tract.

3. Gastrointestinal Mucormycosis

GI mucormycosis is principally a disease of neutropenic and kidney transplant patients and is usually a manifestation of disseminated infection (Lyon et al. 1979; Washburn and Bennett 1995). The disease typically produces localized tissue infarction and thus presents with abdominal pain and GI bleeding. The stomach is most commonly involved, but infection of the colon (Agha et al.

1985) and esophagus have also been described. Small children with protein-calorie malnutrition may develop gastric mucormycosis (Ismail et al. 1990; Thomson et al. 1991).

4. Skin and Wound Mucormycosis

In susceptible patients (e.g. those with diabetes mellitus or immunosuppression), skin and underlying structures may become infected by inoculation with tape (e.g. Elastoplast), intravenous or intraperitoneal catheters, or minor trauma (e.g. spider bite), leading to extensive localized tissue destruction. Even in previously normal hosts, tissue that becomes devitalized through crush injury or burns can become locally infected by agents of mucormycosis (Pierce et al. 1987; Vainrub et al. 1988). Rarely, disseminated mucormycosis produces skin lesions that resemble echthyma gangrenosa (Meyer et al. 1973a).

D. Diagnosis

Blood and urine cultures are almost uniformly negative and there is no readily available serologic test for invasive mucormycosis, so diagnosis relies on documentation of tissue-invasive hyphae with characteristic morphologic features. Positive tissue cultures are confirmatory but the yield is extremely low. Unfortunately, thrombocytopenia and coagulopathy often render biopsy procedures inadvisable, so most cases of invasive pulmonary mucormycosis evade diagnosis until autopsy. Patients with rhinocerebral or skin disease have more accessible lesions, so biopsy is safer.

E. Treatment

Therapy for all forms of mucormycosis includes amphotericin B deoxycholate or lipid formulations of amphotericin B to reduce nephrotoxicity. Reduction of immunosuppression and control of diabetic ketoacidosis contribute to likelihood of successful outcome. The recommended daily dosage of amphotericin B deoxycholate is at least 1 mg/kg for life-threatening disease until the infection has been brought under control; the daily dose may then be reduced to 0.5 mg/kg during completion of an 8-week course of therapy. Lipid formulations of amphotericin B reduce nephrotoxicity. Posaconazole has clinical activity against mucormycosis, but at present there is less experience to support its use compared with amphotericin B-based regimens (Sugar 2005).

Surgical extirpation of necrotic tissue is helpful when hemostasis can be achieved, and is strongly advised for diabetic patients with acute sinusitis if intracranial extension has not yet occurred. Mortality for all forms of invasive mucormycosis is significant, with the exception of localized skin, wound, and eye disease. With appropriate therapy, mortality from rhinocerebral mucormycosis can be limited to 15–24% (Meyers et al. 1979; Lehrer et al. 1980).

IV. Infections Caused by *Fusarium* spp.

Fusarium spp. are soil saprophytes that are responsible for a number of different clinical syndromes, including disseminated infection in neutropenic hosts or patients with burns (Anaissie et al. 1988; Merz et al. 1988; Richardson et al. 1988; Venditti et al. 1988; Gamis et al. 1991; Robertson et al. 1991) and localized disease in normal hosts; e.g. mycetoma, onychomycosis (DiSalvo and Fickling 1980), and keratomycosis (Zapater and Arrechea 1975). The species most commonly responsible for human disease are *F. solani*, *F. oxysporum*, and *F. moniliforme*.

Hyphae of *Fusarium* spp. appear similar to those of *Aspergillus* spp. in tissue. However, three key features help to distinguish disseminated fusariosis from invasive aspergillosis: (1) unlike aspergillosis, a majority of patients with disseminated fusariosis have painful skin lesions resembling echthyma gangrenosa that can provide diagnostic tissue (60–70%), (2) the majority of patients with fusariosis have positive blood cultures (~60%), and (3) fusariosis is even more refractory to antifungal chemotherapy than is aspergillosis. Because of the highly lethal nature of fusariosis (Anaissie et al. 1988), it is recommended that patients should be treated with high daily doses of amphotericin B (1.0–1.5 mg/kg daily; Merz et al. 1988). Voriconazole is approved as second-line therapy (Hospenthal 2005).

V. Infections Caused by *Pseudallescheria boydii*

Pseudallescheria boydii, a fungus found in soil and water, gives rise to a number of clinical entities, including pneumonia following near-drowning (often complicated by brain abscess; Fisher et al. 1982; Dubeau et al. 1984; Hachimi-Idrissi et al. 1990), invasive sinusitis in normal or immunocompromised hosts (Bryan et al. 1980; Winn et al. 1983; Schiess et al. 1984; Salitan et al. 1990), noninvasive pulmonary mass (Louria et al. 1966; Arnett and Hatch 1975; Rippon and Carmichael 1976), paranasal sinus mass (Washburn et al. 1988), and mycetoma (Green and Adams 1964). The course of *Pseudallescheria* infections in compromised hosts is usually more rapidly progressive than in previously normal individuals (Winston et al. 1977; Gumbart 1983; Shih and Lee 1984; Smith et al. 1985).

In tissue, hyphae of *P. boydii* appear similar to those of *Aspergillus* spp., and cultures are required to make a positive identification. However, in noninvasive mycelial masses *P. boydii* can sometimes be presumptively identified even without culture confirmation, on the basis of fascicle-like structures called coremia (Gluckman et al. 1977), and teardrop-shaped conidia.

Since *P. boydii* is resistant to amphotericin B but susceptible to imidazoles in vitro, and since clinical successes also support the use of imidazoles, most authorities agree that intravenous miconazole is preferable to amphotericin B for treatment of *Pseudallerscheria* infections in patients that require intravenous therapy (Lutwick et al. 1979; Anderson et al. 1984; Collignon et al. 1985; Berenguer et al. 1989). Oral itraconazole (Piper et al. 1990) or ketoconazole (Schiess et al. 1984) have been used successfully to complete therapy. More recently voriconazole (Nesky et al. 2000) and posaconazole have been shown to possess clinical activity and voriconazole has been approved for treatment of patients who are refractory to or intolerant of other antifungal therapy. Surgical debridement of infected tissue improves the chances for a permanent cure.

VI. Dematiaceous Molds

Dematiaceous molds are soil- and plant-fungi that produce brown pigments in vitro and sometimes in vivo. Human infection caused by these organisms is termed phaeohyphomycosis. Genera producing human infections include species of *Alternaria*, *Bipolaris*, *Cladosporium*, *Curvularia*, *Drechslera*, *Exophiala*, *Exserohilum*, and *Phialophora* (Adam et al. 1986; Washburn et al. 1988).

Clinical syndromes that affect compromised hosts include brain abscess (Seaworth et al. 1983; Anandi et al. 1989; Aldape et al. 1991), localized subcutaneous inoculation disease (Fincher et al. 1988), and disseminated disease. Syndromes found principally in previously normal hosts include chronic invasive fungal sinusitis (Washburn 1998), allergic fungal sinusitis with clinical features paralleling those of allergic *Aspergillus* sinusitis (Brummund et al. 1986; MacMillan et al 1987; Bartynski et al. 1990; Gourley et al. 1990; Friedman et al. 1991), and allergic bronchopulmonary disease (Dolan et al. 1970; Halwig et al. 1985). In tissue, the septate branching hyphae of dematiaceous fungi are irregular in diameter, and may contain globose swellings. Therapy for chronic invasive fungal sinusitis includes surgical extirpation of infected tissue combined with a prolonged course of antifungal chemotherapy, usually amphotericin B to a total dose of ca. 2 gm. Therapy for allergic fungal sinusitis includes surgical removal of inspissated mucus and inflamed mucosa, combined with topical steroids; in contrast, roles of systemic steroids and antifungal chemotherapy remain to be defined. Anecdotal literature concerning allergic bronchopulmonary disease suggests that combined therapy with bronchodilators and systemic steroids is effective. Disseminated disease in normal hosts or immunocompromised patients is treated with a prolonged course of amphotericin B.

Itraconazole has been used successfully in nonlife-threatening disease and there is anecdotal evidence for posaconazole as salvage therapy after amphotericin B failure (Hospenthal 2005). The in vitro activity of voriconazole against dematiaceous molds is more potent than that of itraconazole.

VII. Rare Agents

The following molds only rarely cause infection in humans: *Acremonium*, *Paecilomyces*, *Penicillium*, *Schizophyllum*, and *Scopulariopsis*. For a full discussion of those organisms, the reader is referred to Kwon-Chung and Bennett (1992).

VIII. Conclusions

The molds discussed in this chapter are those that take opportunistic advantage of an immunocompromised host.

Aspergillosis is an infection caused by one or another species in the genus *Aspergillus*. These molds produce pigmented colonies in culture. The conidia produced are phialoconidia arranged in a characteristic fashion on the conidiophore (Raper and Fennell 1965). *Aspergillus* spp. appear in the tissues of an infected host as septate hyphae with branches occurring at acute angles. Mycelial fungi such as *Fusarium* spp. and *Pseudallescheria boydii*, among others, resemble *Aspergillus* spp. in tissue.

The disease mucormycosis is caused by zygomycetes of the order Mucorales. Both deep-seated, systemic diseases and inoculation mycoses may be produced. These zygomycetes produce asexual spores called sporangiospores. The mycelium is composed of nonseptate hyphae, and this morphologic form is observed in tissues of an infected host.

There are a large number of fungi other than species of *Aspergillus* and mucoraceous zygomycetes that can cause infections in an immunosuppressed host.

References

Adam RD, Paquin ML, Petersen EA, Saubolle MA, Rinaldi MG, Corcoran JG, Galgiani JN, Sobonya RE (1986) Phaeohyphomycosis caused by the fungal genera *Bipolaris* and *Exserohilum*. Medicine 65:203–217

Agha FP, Lee HH, Boland CR, Bradley SF (1985) Mucormycoma of the colon: early diagnosis and successful management. Am J Roentgenol 145:739–741

Aldape KD, Fox HS, Roberts JP, Ascher NL, Lake JR, Rowley HA (1991) *Cladosporium trichoides* cerebral phaeohyphomycosis in a liver transplant recipient. Am J Clin Pathol 95:499–502

Altman AR (1977) Thoracic wall invasion secondary to pulmonary aspergillosis: a complication of chronic granulomatous disease of childhood. Am J Roentgenol 129:140–142

Anaissie E, Kantarjian H, Ro J, Hopfer R, Rolston K, Fainstein V, Bodey G (1988) The emerging role of *Fusarium* infections in patients with cancer. Medicine 67:77–83

Anandi V, John TJ, Walter A, Shastry JCM, Lalitha MK, Padhye AA, Ajello L, Chandler FW (1989) Cerebral phaeohyphomycosis caused by *Chaetomium globosum* in a renal transplant recipient. J Clin Microbiol 27:2226–2229

Anderson RL, Carroll TF, Harvey JT, Myers MG (1984) *Petriellidium (Allescheria) boydii* orbital and brain abscess treated with intravenous miconazole. Am J Ophthalmol 97:771–775

Arnett JC, Hatch HB (1975) Pulmonary allescheriasis: report of a case and review of the literature. Arch Intern Med 135:1250–1253

Arroyo J, Medoff G, Kobayashi GS (1977) Therapy of murine aspergillosis with amphotericin B in combination with rifampin or 5-fluorocytosine. Antimicrob Agents Chemother 11:21–25

Baker RD, Seabury JH, Schneidau JD (1962) Subcutaneous and cutaneous mucormycosis and subcutaneous phycomycosis. Lab Invest 11:1091–1102

Bartynski JM, McCaffrey TV, Frigas E (1990) Allergic fungal sinusitis secondary to dematiaceous fungi: *Curvularia lunata* and *Alternaria*. Otolaryngol Head Neck Surg 103:32–39

Berenguer J, Diaz-Mediavilla J, Urra D, Munoz P (1989) Central nervous system infection caused by *Pseudallescheria boydii*: case report and review. Rev Infect Dis 11:890–896

Blum J, Reed JC, Pizzo SV, Thompson WM (1978) Miliary aspergillosis associated with alcoholism. Am J Roentgenol 131:707–709

Boelaert JR, deLocht M, Van Cutsem J, Kerrels V (1993) Mucormycosis during deferoxamine therapy is a siderophore-mediated infection. In vitro and in vivo animal studies. J Clin Invest 91:1979–1986

Boelaert JR, Van Cutsem J, deLocht M, Schneider YJ, Crichton RR (1994) Deferoxamine augments growth and pathogenicity of *Rhizopus* while hydroxypyridinone chelators have no effect. Kidney Int 45:667–671

Boon AP, Adams DH, Buckels J, McMaster P (1990) Cerebral aspergillosis in liver transplantation. J Clin Pathol 43:114–118

Bottone EJ, Weitzman L, Hanna BA (1979) *Rhizopus rhizopodiformis*: emerging etiological agent of mucormycosis. J Clin Microbiol 9:530–537

Boyce JM, Lawson LA, Lockwood WR, Hughes JL (1981) *Cunninghamella bertholletiae* wound infection of probable nosocomial origin. South Med J 74:1132–1135

Branton MH, Johnson SC, Brooke JD, Hasbargen JA (1991) Peritonitis due to *Rhizopus* in a patient undergoing continuous ambulatory peritoneal dialysis. Rev Infect Dis 13:19–21

Brown E, Freedman S, Arbeit R, Come S (1980) Invasive pulmonary aspergillosis in an apparently nonimmunocompromised host. Am J Med 69:624–627

Brummund W, Kurup VP, Harris GJ, Duncavage JA, Arkins JA (1986) Allergic sino-orbital mycosis. J Am Med Assoc 256:3249–3253

Bryan CS, DiSalvo AF, Kaufman L, Kaplan W, Brill AH, Abbott DC (1980) *Petriellidium boydii* infection of the sphenoid sinus. Am J Clin Pathol 74:846–851

Bujak JS, Kwon-Chung KJ, Chusid MJ (1974) Osteomyelitis and pneumonia in a boy with chronic granulomatous disease of childhood caused by a mutant strain of *Aspergillus nidulans*. Am J Clin Pathol 61:361–367

Cameron JA, Antonios SR, Cotter JB, Habash NR (1991) Endophthalmitis from contaminated donor corneas following penetrating keratoplasty. Arch Ophthalmol 109:54–59

Carlile JR, Miller RE, Cho CT, Vats TS (1978) Primary cutaneous aspergillosis in a leukemic child. Arch Dermatol 114:78–80

Collignon PJ, Macleod C, Packham DR (1985) Miconazole therapy in *Pseudallescheria boydii* infection. Aust J Dermatol 26:129–132

Connor BA, Anderson RJ, Smith JW (1979) Mucor mediastinitis. Chest 75:524–526

Corrall CJ, Merz WG, Rekedal K, Hughes WT (1982) *Aspergillus* osteomyelitis in an immunocompetent adolescent: a case report and review of the literature. Pediatrics 70:455–461

Denning DW, Follansbee SE, Scolaro M, Norris S, Edelstein H, Stevens DA (1991) Pulmonary aspergillosis in the acquired immunodeficiency syndrome. N Engl J Med 324:654–662

Denning DW et al (1994) NIAID Mycoses Study Group multicenter trial of oral itraconazole therapy for invasive aspergillosis. Am J Med 97:135–144

DiSalvo AF, Fickling AM (1980) A case of nondermatophytic toe onychomycosis caused by *Fusarium oxysporum*. Arch Dermatol 116:699–700

Dolan CT, Weed LA, Dines DE (1970) Bronchopulmonary helminthosporiosis. Am J Clin Pathol 53:235–242

Dubeau F, Roy LE, Allard J, Laverdiere M, Rousseau S, Duplantis F, Boileau J, Lachapelle J (1984) Brain abscess due to *Petriellidium boydii*. Can J Neurol Sci 11:395–398

Dyken ME, Biller J, Yuh WTC, Fincham R, Moore SA, Justin E (1990) Carotid-cavernous sinus thrombosis caused by *Aspergillus fumigatus*: magnetic resonance imaging with pathologic correlation – a case report. Angiology 41:652–657

Fincher RME, Fisher JF, Padhye AA, Ajello L, Steele JC Jr (1988) Subcutaneous phaeohyphomycotic abscess caused by *Phialophora parasitica* in a renal allograft recipient. J Med Vet Mycol 26:311–314

Fisher JF, Shadomy S, Teabeaut R, Woodward J, Michaels GE, Newman KA, White E, Cook P, Seagraves A, Yaghmai F, Rissing JP (1982) Near-drowning complicated by brain abscess due to *Petriellidium boydii*. Arch Neurol 39:511–513

Fong KM, Seneviratne EME, McCormack JG (1990) Mucor cerebral abscess associated with intravenous drug abuse. Aust N Z J Med 20:74–77

Friedman GC, Hartwick WJ, Ro JY, Saleh GY, Tarrand JJ, Ayala AG (1991) Allergic fungal sinusitis. Report of three cases associated with dematiaceous fungi. Am J Clin Pathol 96:368–372

Gamis AS, Gudnason T, Giebink GS, Ramsay NK (1991) Disseminated infection with *Fusarium* in recipients of bone marrow transplants. Rev Infect Dis 13:1077–1088

Gartenberg G, Bottone EJ, Keusch GT, Weitzman I (1978) Hospital acquired mucormycosis (*Rhizopus rhizopodiformis*) of skin and subcutaneous tissue. N Engl J Med 299:1115–1118

Gluckman SJ, Ries K, Abrutyn E (1977) *Allescheria (Petriellidium) boydii* sinusitis in a compromised host. J Clin Micro 5:481–484

Goldschmied-Reouven A, Shvoron A, Topaz M, Black C (1989) *Saksenaea vasiformis* infection in a burn wound. J Med Vet Mycol 27:427–429

Googe PB, DeCoste SD, Herold WH, Mihm MC Jr (1989) Primary cutaneous aspergillosis mimicking dermatophytosis. Arch Lab Med 113:1284–1286

Gourley DS, Whisman BA, Jorgensen NL, Martin ME, Reid MJ (1990) Allergic *Bipolaris* sinusitis: clinical and immunopathologic characteristics. J Allergy Clin Immunol 85:583–591

Green WO, Adams TE (1964) Mycetoma in the United States. Am J Clin Pathol 42:75–91

Green WR, Font RL, Zimmerman LE (1969) Aspergillosis of the orbit: report of ten cases and review of the literature. Arch Ophthalmol 82:302–313

Gumbart CH (1983) *Pseudallescheria boydii* infection after bone marrow transplantation. Ann Intern Med 99:193–194

Hachimi-Idrissi S, Willemsen M, Desprechins B, Naessens A, Goossens A, De Meirleir L (1990) *Pseudallescheria boydii* and brain abscesses. Pediatr Infect Dis J 9:737–741

Halwig JM, Brueske DA, Greenberger PA, Dreisin RB, Sommers HM (1985) Case reports: allergic bronchopulmonary curvulariosis. Am Rev Respir Dis 132:186–188

Hammond DE, Winkelmann RK (1979) Cutaneous phycomycosis: report of three cases with the identification of *Rhizopus*. Arch Dermatol 115:990–992

Hara KS, Ryu JH, Lie JT, Roberts GD (1989) Disseminated *Aspergillus terreus* infection in immunocompromised hosts. Mayo Clin Proc 64:770–775

Hospenthal DR (2005) Uncommon fungi. In: Mandell GL, Bennnett JE, and Dolin R (eds) Principles and practice of infectious diseases, vol 2. Elsevier, Philadelphia, pp 3068–3079

Hughes CE, Harris C, Moody JA, Peterson LR, Gerding DN (1984) In vitro activities of amphotericin B in combination with four antifungal agents and rifampin against *Aspergillus* spp. Antimicrob Agents Chemother 25:560–562

Isaacson C, Levin SE (1961) Gastro-intestinal mucormycosis in infancy. S Afr J Med 35:581–584

Ismail MHA, Hodkinson HJ, Setzen G, Sofianos C, Hole MJ (1990) Gastric mucormycosis. Trop Gastroenterol 11:103–105

Jewkes J, Kay PH, Paneth M, Citron KM (1983) Pulmonary aspergilloma: analysis of prognosis in relation to haemoptysis and survey of treatment. Thorax 38:572–578

Katzenstein A-LA, Sale SR, Greenberger PA (1983) Allergic *Aspergillus* sinusitis: a newly-recognized form of sinusitis. J Allergy Clin Immunol 72:89–93

Kaufman L, Padhye AA, Parker S (1988) Rhinocerebral zygomycosis caused by *Saksenaea vasiformis* J Med Vet Mycol 26:237–246

Kitahara M, Seth VK, Medoff G, Kobayashi GS (1976) Activity of amphotericin B, 5-fluorocytosine, and rifampin against six clinical isolates of *Aspergillus*. Antimicrob Agents Chemother 9:915–919

Kurup VP, Fink JN (1978) Evaluation of methods to detect antibodies against *Aspergillus fumigatus*. Am J Clin Pathol 69:414–417

Kwon-Chung KJ, Bennett JE (1992) Medical Mycology. Lea and Febiger, Malvern

Lehrer RI, Howard DH, Sypherd PS, Edwards JE, Segal GP, Winston DJ (1980) Mucormycosis. Ann Intern Med 93:93–108

Levitz SM, Diamond RD (1985) Mechanisms of resistance of *Aspergillus fumigatus* conidia to killing by neutrophils in vitro. J Infect Dis 152:33–42

Louria DB, Lieberman PH, Collins HS, Blevins A (1966) Pulmonary mycetoma due to *Allescheria boydii*. Arch Intern Med 117:748–751

Lutwick LI, Rytel MW, Yañez JP, Gaigiani JN, Stevens DA (1979) Deep infections from *Petriellidium boydii* treated with miconazole. J Am Med Assoc 241:272–273

Lyon DT, Schubert TT, Mantia AG, Kaplan MH (1979) Phycomycosis of the gastrointestinal tract. Am J Gastroenterol 72:379–394

MacMillan RH, Cooper PH, Body BA, Mills AS (1987) Allergic fungal sinusitis due to *Curvularia lunata*. Hum Pathol 18:960–964

McCarty JM, Flam MS, Pullen G, Jones R, Kassel SH (1986) Outbreak of primary cutaneous aspergillosis related to intravenous arm boards. J Pediatr 108:721–724

McGill TJ, Simpson G, Healy GB (1980) Fulminant aspergillosis of the nose and paranasal sinuses. Laryngoscope 90:748–754

Mawk JR, Erickson DL, Chan SN, Seljeskog EL (1983) *Aspergillus* infections of the lumbar disc spaces. J Neurosurg 58:270–274

Merz WG, Karp JE, Hoagland M, Jett-Goheen M, Junkins JM, Hood AF (1988) Diagnosis and successful treatment of fusariosis in the compromised host. J Infect Dis 158:1046–1055

Meyer RD, Rosen P, Armstrong D (1972) Phycomycosis complicating leukemia and lymphoma. Ann Intern Med 77:871–879

Meyer RD, Kaplan MH, Ong M, Armstrong D (1973a) Cutaneous lesions in disseminated mucormycosis. J Am Med Assoc 225:737–738

Meyer RD, Young LS, Armstrong D, Yu B (1973b) Aspergillosis complicating neoplastic disease. Am J Med 54:6–15

Meyers BR, Wormser G, Hirschman SZ, Blitzer A (1979) Rhinocerebral mucormycosis: premortem diagnosis and therapy. Arch Intern Med 139:557–560

Miloshev B, Davidson CM, Gentles JC, Sandison AT (1967) Aspergilloma of paranasal sinuses and orbit in northern Sudanese. Sabouraudia 6:57

Moore CK, Hellreich MA, Coblentz CL, Roggli VL (1988) *Aspergillus terreus* as a cause of invasive pulmonary aspergillosis. Chest 94:889–891

Mostaza JM, Barbado FJ (1989) Cutaneoarticular mucormycosis due to *Cunninghamella bertholletiae* in a patient with AIDS. Rev Infect Dis 11:316–318

Murray HW (1975) Pulmonary mucormycosis with massive fatal hemoptysis. Chest 68:65–68

Nesky MA, McDougal EC, Peacock JE Jr (2000) *Pseudallescheria boydii* brain abscess successfully treated with voriconazole and surgical drainage: case report and literature review of central nervous system pseudallescheriasis. Clin Infect Dis 31:673–677

Orgel IK, Cohen KL (1989) Postoperative zygomycetes endophthalmitis. Ophthalm Surg 20:584–587

Patterson TF (2005) *Aspergillus* species. In: Mandell GL, Bennett JE, Dolin R (eds) Principles and practice of infectious diseases, vol 2. Elsevier, Philadelphia, pp 2958–2973

Peterson DE, Schimpff SC (1989) *Aspergillus* sinusitis in neutropenic patients with cancer: a review. Biomed Pharmacother 43:307–312

Pierce PF Jr, Solomon SL, Kaufman L, Garagusi VF, Parker RH, Ajello L (1982) Zygomycetes brain abscesses in narcotic addicts with serological diagnosis. J Am Med Assoc 248:2881–2882

Pierce PF, Wood MB, Roberts GD, Fitzgerald RH Jr, Robertson C, Edson RS (1987) *Saksenaea vasiformis* osteomyelitis. J Clin Microbiol 25:933–935

Piper JP et al (1990) Successful treatment of *Scedosporium apiospermum* suppurative arthritis with itraconazole. Pediatr Infect Dis J 9:674–675

Prystowsky SD, Vogelstein K, Ettinger DS, Merz WG, Kaizer H, Sulica VI, Zinkham WH (1976) Invasive aspergillosis. N Engl J Med 295:655–658

Pursell KJ, Telzak EE, Armstrong D (1990) *Aspergillus* species colonization and invasive disease in patients with AIDS. Clin Infect Dis 14:141–148.

Raper KB, Fennell DI (1965) The genus *Aspergillus*. Williams & Wilkins, Baltimore

Redmond A et al (1965) *Aspergillus* (*Aspergillus nidulans*) involving bone. J Pathol Bacteriol 89:391

Rex JH, Ginsberg AM, Fries LF, Pass HI, Kwon-Chung KJ (1988) *Cunninghamella bertholletiae* infection associated with deferoxamine therapy. Rev Infect Dis 10:1187–1194

Richardson SE, Bannatyne RM, Summerbell RC, Milliken J, Gold R, Weitman SS (1988) Disseminated fusarial infection in the immunocompromised host. Rev Infect Dis 10:1171–1181

Rippon JW, Carmichael JW (1976) Petriellidiosis (allescheriosis): four unusual cases and review of literature. Mycopathologia 58:117–124

Robertson MJ, Socinski MA, Soiffer RJ, Finberg RW, Wilson C, Anderson KC, Bosserman L, Sang DN, Salkin IF, Ritz J (1991) Successful treatment of disseminated *Fusarium* infection after autologous bone marrow transplantation for acute myeloid leukemia. Bone Marrow Transplantation 8:142–145

Salitan MD, Lawson W, Som PM, Bottone EJ, Biller HF (1990) *Pseudallescheria* sinusitis with intracranial extension in an immunocompromised host. Otolaryngol Head Neck Surg 102:745–780

Sands JM, Macher AM, Ley TJ, Nienhuis AW (1985) Disseminated infection caused by *Cunninghamella bertholletiae* in a patient with beta-thalassemia. Ann Intern Med 102:59–63

Schaffner A, Davis CE, Schaffner T, Markert M, Douglas H, Braude AI (1986) In vitro susceptibility of fungi to killing by neutrophil granulocytes discriminates between primary pathogenicity and opportunism. J Clin Invest 78:511–524

Schiess RJ, Coscia MF, McClellan GA (1984) *Petriellidium boydii* pachymeningitis treated with miconazole and ketoconazole. Neurosurgery 14:220–224

Seaworth BJ, Kwon-Chung KJ, Hamilton JD, Perfect JR (1983) Brain abscess caused by a variety of *Cladosporium trichoides*. Am J Clin Pathol 79:747–752

Sheldon DL, Johnson WC (1979) Cutaneous mucormycosis. J Am Med Assoc 241:1032–1034

Sherertz RJ, Belani A, Kramer BS, Elfenbein GJ, Weiner RS, Sullivan ML, Thomas RG, Samsa GP (1987) Impact of air filtration on nosocomial *Aspergillus* infections. Unique risk of bone marrow transplant recipients. Am J Med 83:709–718

Shih L, Lee N (1984) Disseminated petriellidiosis (allescheriasis) in a patient with refractory acute lymphoblastic leukemia. J Clin Pathol 37:78–82

Smith AG, Crain SM, Dejongh C, Thomas GM, Vigorito RD (1985) Systemic pseudallescheriasis in a patient with acute myelocytic leukemia. Mycopathologia 90:85–89

Smith AG, Bustamante CI, Gilmor GD (1989) Zygomycosis (absidiomycosis) in an AIDS patient. Mycopathologia 105:7–10

Stave GM, Heimberger T, Kerkering TM (1989) Zygomycosis of the basal ganglia in intravenous drug users. Am J Med 86:115–117

Steinbach WJ, Stevens DA (2003) Review of newer antifungal and immunomodulatory strategies for invasive aspergillosis. Clin Infect Dis 37[Suppl 3]:S157–S187

Sugar AM (2005) Agents of mucormycosis and related species. In: Mandell GL, Bennett JE, Dolin R (eds) Principles and practice of infectious diseases, vol 2. Elsevier, Philadelphia, pp 2973–2984

Swerdlow B, Deresinski S (1984) Development of *Aspergillus* sinusitis in a patient receiving amphotericin B. Am J Med 76:162–166

Talbot GH, Huang A, Provencher M (1991) Invasive *Aspergillus* rhinosinusitis in patients with acute leukemia. Rev Infect Dis 13:219–232

Thomson SR, Bade PG, Taams M, Chrystal V (1991) Gastrointestinal mucormycosis. Br J Surg 78:952–954

Vainrub B, Macareno A, Mandel S, Musher D (1988) Wound zygomycosis (mucormycosis) in otherwise healthy adults. Am J Med 84:546–548

Varkey B, Rose HD (1976) Pulmonary aspergilloma: a rational approach to treatment. Am J Med 61:626–631

Venditti M, Micozzi A, Gentile G, Polonelli L, Morace G, Bianco P, Avvisati G, Papa G, Martino P (1988) Invasive *Fusarium solani* infections in patients with acute leukemia. Rev Infect Dis 10:653–660

Waldorf AR, Levitz SM, Diamond RD (1984a) In vivo bronchoalveolar macrophage defense against *Rhizopus oryzae* and *Aspergillus fumigatus*. J Infect Dis 150:752–760

Waldorf AR, Ruderman N, Diamond RD (1984b) Specific susceptibility to mucormycosis in murine diabetes and bronchoalveolar macrophage defense against *Rhizopus*. J Clin Invest 74:150–160

Walsh TJ et al (2007) Treatment of invasive aspergillosis with posaconazole in patients who are refractory to or intolerant of conventional therapy: an externally controlled trial. Clin Infect Dis 44:2–12

Washburn RG (1996) Bronchopulmonary aspergillosis. In: Kelley WN et al (eds) Textbook of internal medicine. Lippincott, Philadelphia, pp 2000–2001

Washburn RG (1998) Fungal sinusitis. In: Remington JS, Swartz MN (eds) Curr Clin Top Infect Dis 18:60–74

Washburn RG, Bennett JE (1995) Deep mycoses. In: Blaser MJ et al (eds) Infections of the Gastrointestinal Tract. Raven, New York, pp 957–966

Washburn RG, Gallin JI, Bennett JE (1987) Oxidative killing of *Aspergillus fumigatus* conidia proceeds by parallel myeloperoxidase-dependent and myeloperoxidase-independent pathways. Infect Immun 55:2088–2092

Washburn RG, Kennedy DW, Begley MG, Henderson DK, Bennett JE (1988) Chronic fungal sinusitis in apparently normal hosts. Medicine 67:231–247

Watson KC (1957) Gastric perforation due to the fungus *Mucor* in a child with kwashiorkor. S Afr J Med 31:99–101

Waxman JE, Spector JG, Sale SR, Katzenstein A-LA (1987) Allergic *Aspergillus* sinusitis: concepts in diagnosis and treatment. Laryngoscope 97:261–266

Weber SF, Washburn RG (1990) Invasive *Aspergillus* infections complicating coronary artery bypass grafting. South Med J 83:584–588

Weiland D, Ferguson RM, Peterson PK, Snover DC, Simmons RL, Najarian JS (1983) Aspergillosis in 25 renal transplant patients. Ann Surg 198:622–629

White CJ, Kwon-Chung KJ, Gallin JI (1988) Chronic granulomatous disease of childhood; an unusual case of infection with *Aspergillus nidulans* var. *echinulatus*. Am J Clin Pathol 90:312–316

Wilson BJ, Wilson CH (1961) Oxalate formation in moldy feedstuffs as a possible factor in livestock toxic disease. Am J Vet Res 22:961–969

Winn RE, Ramsey PD, McDonald JC, Dunlop KJ (1983) Maxillary sinusitis from *Pseudallescheria boydii*: efficacy of surgical therapy. Arch Otolaryngol 109:123–125

Winston DJ, Jordan MC, Rhodes J (1977) *Allescheria boydii* in the immunosuppressed host. Am J Med 63:830–835

Woods GL, Goldsmith JC (1990) *Aspergillus* infection of the central nervous system in patients with acquired immunodeficiency syndrome. Arch Neurol 47:181–184

Young RC, Bennett JE, Vogel CL, Carbone PP, DeVita VT (1970) Aspergillosis: The spectrum of the disease in 98 patients. Medicine 49:147–173

Young RC, Jennings A, Bennett JE (1972) Species identification of invasive aspergillosis in man. Am J Clin Pathol 58:554–557

Yu VL, Muder RR, Poorsattar A (1986) Significance of isolation of *Aspergillus* from the respiratory tract in diagnosis of invasive pulmonary aspergillosis. Am J Med 81:249–254

Zapater RC, Arrechea A (1975) Mycotic keratitis by *Fusarium*: a review and report of two cases. Ophthalmologia 170:1–12

Zeilender S, Drenning D, Glauser FL, Bechard D (1990) Fatal *Cunninghamella bertholletiae* infection in an immunocompetent patient. Chest 97:1482–1483

Zellner SR, Selby JB, Loughrin JJ (1969) Aspergillosis: an unusual presentation. Am Rev Respir Dis 100:217–220

Ziskind J, Pizzolato P, Buff EE (1958) Aspergillosis of the brain. Am J Clin Pathol 29:554–559

3 Entomopathogenic Fungi: Biochemistry and Molecular Biology

GEORGE G. KHACHATOURIANS[1], SOHAIL S. QAZI[1]

CONTENTS

I. Introduction.......................... 33
II. Life Cycle and Mass Culturing............ 33
III. Biochemical Aspects of Disease
    Development 35
    A. Key Enzymes in Pathogen–Host–
       Environment Interaction 35
       1. Proteases and Peptidases........... 36
       2. Chitinases...................... 38
       3. Lipases and Lipoxygenases 39
       4. DNA Repair Enzymes.............. 39
    B. Toxins and Pigments 40
       1. Non-peptide Toxins and Pigments... 40
       2. Linear and Cyclic Peptide Toxins 41
    C. Fungal Virulence Factors 42
IV. Physico-Chemical Aspects of Disease
    Development 42
    A. Spore Adhesion 43
    B. Spore Germination–Growth.......... 44
    C. Growth Within Insects 47
V. Genetics and Molecular Biology
    of Disease Development................. 48
    A. Enzymes and Virulence Factors 48
    B. Molecular Genetic Analysis 49
       1. Cloning and Sequencing
          of Chromosomal Genes 49
       2. Cloning and Sequencing
          of Mitochondrial Genes............ 50
    C. Molecular Probing into Pathogenicity
       and Tracking....................... 51
VI. Conclusions and Perspectives 52
    References............................ 53

I. Introduction

Entomopathogenic fungi (EPF) have become a significant force in shaping the larger context of insecticides within contemporary insect pest management schemes (Lord 2005; Roy et al., 2006; Khachatourians 2008). Needless to say, as mycologists we need the perspective and understanding to explain the diversity of EPF and their spatial and temporal distribution within the insect ecosystem. In the past decade, the accelerated focus of research and scholarly studies has generated two perspectives: (a) the molecular biology, genomics and proteomics of EPF, and (b) the practical use of EPF in insect pest management schemes. Their value therefore is two-fold, first in the study of microbial pathogenicity and second in their application to the microbial control of phytophagous insects as much as biting and hematophagous insect pest populations. Altogether some 90 genera and 700 species are involved with entomopathogenicity, only a few members of the Entomophthorales and Hyphomycetes have been well studied. In the past decade, major new developments in the realm of application of the knowledge of EPF to insect pest management have been realized (Khachatourians 1996). New developments in genomic and molecular research and serious interest in commercialization of EPF for pest control have become the new drivers of understanding in the field, challenges that were forecasted to meet the promise of new biotechnology (Khachatourians 1986). With such knowledge, physiological manipulations, isolation of mutants with enhanced virulence, and construction of environmentally safe strains with limited persistence should be possible within the near future. This chapter primarily reviews the literature since 1995 on the biochemistry and molecular biology of EPF and their involvement in the disease of insects. Additional sources of information can be followed from Table 3.1.

II. Life Cycle and Mass Culturing

The life cycle of EPF is composed of the spore–mycelia–spore phases. The germination of spores leads to polar growth, and branching ends with sporulation. The medium ingredients and types of commercially available agars can affect developmental

---

[1] Bioinsecticide Research Laboratories, College of Agriculture and Bioresources, University of Saskatchewan, Saskatoon, S7N 5A8, Saskatchewan, Canada; e-mail: george.khachatourians@usask.ca

Table 3.1. Recent reviews on entomopathogenic fungi

Subject area	References
Ecology and epizootiology	Hu and St Leger (2002)
Environmental release	Hu and St Leger (2002)
Evolution and habitat	Bidochka and Small (2005), Chandler (2005)
Host–pathogen interaction	Hegedus and Khachatourians (1995), Khachatourians (1996), Roy et al. (2006)
Mass production	Deshpande (1999)
Mycopesticides potential	Deshpande (1999), Shah and Pell (2003), Roberts and St Leger (2004)
Phylogenetics/taxonomy	Sugimoto et al. (2001), Khachatourians and Uribe (2004), Bidochka and Small (2005), Rehner (2005)
Physiology and genetics	Hegedus and Khachatourians (1995), Khachatourians and Uribe (2004), Roberts and St Leger (2004)
Toxic metabolites	Vey et al. (2001), Pedras et al. (2002)
Use as insecticides	Deshpande (1999), Chandler et al. (2000), Shah and Pell (2003), Roberts and St Leger (2004), Roy et al. (2006)
World status	Khachatourians and Valencia (1998), Khachatourians et al. (2002), Qazi and Khachatourians (2005)

(Kamp and Bidochka 2002a) and cell densities. Braga et al. (1999) reported oxygen consumption by *Metarhizium anisopliae* during germination and growth on monosaccharides, polysaccharides, amino acids, and proteins. Germination was marked by a significant increase in O_2 consumption, which was drastically reduced after depletion of the exogenous carbon source, when casein, hydrolyzed casein, and N-acetylglucosamine (NAGA) accelerated germination, reduced the lag phase, and increased the growth rate.

A yeast-like phase for *Beauveria bassiana* was described by Alves et al. (2002); and *M. flavoviride* production of oblong blastospore-like propagules (Fargues et al. 2002) is reminiscent of *B. bassiana* GK 2016 placed under nutrient stressed growth, as reported by Thomas et al. (1987). The life cycle of *M. anisopliae* under liquid culture conditions has been described (Uribe and Khachatourians 2008).

The relationship between EPF and their utilization of substrate for growth and development whether in vitro and in vivo has now been established by expressed sequence tag (EST) analysis. Cho et al. (2006a, b) examined EST from cDNA libraries of *B. (Cordyceps) bassiana* aerial conidia, blastospores, and submerged conidia. The authors indicated that the unique and divergent representation of the *B. bassiana* transcriptome from each developmental cell type corresponded to environmental conditions. Wang et al. (2005a) and Freimoser et al. (2005) had earlier reported variation in gene expression patterns in *M. anisopliae* in response to various growth environments or hosts. Differential gene expression through cDNA microarrays were constructed from an EST) collection of 837 genes. During growth in culture cuticle (*Manduca sexta*), *Metarhizium anisopliae* up-regulated 273 genes, e.g., proteases), amino acid/peptide transport, transcription regulation, and many others with unknown function. The 287 down-regulated genes included a large set of ribosomal protein genes which were also distinctive.

The ambient temperatures have important consequences to EPF in their diversity and fitness in particular agricultural and or forestry climates. A grouping of *B. bassiana* was found to be equally associated with habitat and thermal growth preferences (Bidochka et al. 2002). Smits et al. (2003) have provided a model for comparison of non-linear temperature-dependent in vitro growth of EPF. Xavier et al. (1999) showed a shift in growth due to a change from ambient temperature to a higher temperature as a stressor. When several members of EPF were placed under heat shock conditions, changes in the expression of proteins in general and those specific to heat shock were observed (Xavier and Khachatourians 1999; Xavier et al. 1999).

Water stress and solutes are known to have significant implications for conserving the ecophysiological quality of fungal biocontrol propagules during harvesting, storage, and formulation. Hallsworth and Magan (1999) showed a positive role for water and temperature on the growth of EPF. Ypsilos and Magan (2004) showed that germinability of *M. anisopliae* blastospores is affected by water stress (water activity, a_w, of 0.98, 0.97 and 0.96) imposed by either polyethylene glycol (PEG) 200, KCl, or NaCl at 25 °C. The combined effects of water and temperature in relation to field conditions was examined by Devi et al. (2005) for 29 isolates of *B. bassiana* with 8 h high/16 h low temperature (similar to field conditions), low water availability, and a combination of these two stress conditions to show the complexity of outcomes. Clearly, these studies are useful for the production and performance of EPF.

In *Paecilomyces fumosoroseus* phototropism is important in resetting of the circadian rhythm, the induction of carotenogenesis and the development of reproductive structures (Sanchez-Murillo et al. 2004). These attributes are controlled by blue light at 180 µmol m^{-2}. Growth of the fungus in continuous illumination or under a night–day regime resulted in increased conidia.

Mass culturing of EPF continues to be a challenging issue. Solid-state fermentation (SSF) and submerged fermentation using various culture media, whether synthetic or inexpensive agricultural waste products or industrial residues, are pertinent to propagule production. Dalla Santa et al. (2004) reported conidia production of *Beauveria* sp. strain LAG by SSF using blends of agro-industrial residues composed of potato residues and sugar-cane bagasse at a blend of 60% and 40% respectively and a yield of 1.07×10^{10} conidia g^{-1} dry substrate after a 10-day fermentation. Kang et al. (2005) reported conidia production by *B. bassiana* during SSF was 4.9×10^{8} g^{-1} and was affected more by antifoaming agent than forced aeration. Tarocco et al. (2005) reported on optimization of erythritol and glycerol accumulation in conidia of *B. bassiana* by SSF, using response surface methodology. Deshpande (1999) reviewed parameters and indicated the importance of inocula preparation and maintenance for industrial scale production. Ryan et al. (2002) found subculturing of an isolate of *M. anisopliae* led to phenotypic degeneration and Shah and Butt (2005) showed *M. anisopliae* (strains V245, V275) formed sectors which differed in their stability and enzyme profiles when grown on different nutrient media. Interestingly, sectors also produced a lower concentration of destruxins than the parent cultures, independent of the strain. Wang et al. (2005a) suggest sectorization of *M. anisopliae* is a sign of aging.

Fargues et al. (2002) showed that seven liquid culture media (Adamek, Catroux, Jackson, Jenkins-Prior, Goral, Kondryatiev, Paris media) affected morphology, growth, propagule production, and pathogenicity of *M. flavoviride* strains.

III. Biochemical Aspects of Disease Development

In order for the EPF to support their growth and disease development, they must confront their hosts and access the unique trophic niche. Confrontation occurs by utilization of surface layers and/or entry into the host. Within the hemolymph, EPF encounter the defense reactions, grow, multiply, and mummify, to exit as mycelia and spores from the host.

Since the publication of *Mycota V*, new information on the triad host–pathogen–environment and the reactions between host and EPF has accumulated. The cumulative effects of biochemical reactions can manifest as changes in food consumption, host growth, and behavior. For example, stimulation followed by the reduction and stoppage of feeding after host infections with *B. bassiana* and *M. flavoviride*, reduced mobility, behavioral fever response after infection, changed migrational pattern, and vertical climbing of EPF-infected insects represent a few such situations (Tefera and Pringle 2003a, b; Ouedraogo et al. 2004).

A. Key Enzymes in Pathogen–Host–Environment Interaction

It is generally accepted that both mechanical force and enzymatic processes and perhaps certain metabolic acids mediate the initial interaction. Genes responsible for the outcomes of successful interaction with the host and environment are now known (Table 3.2). Additionally, radiation and free radicals can be mitigated by a variety of responses, including those involving DNA-repair enzymes. Together these enzymes play a pivotal role in the adaptation and evolution of EPF.

Cuticular penetration by the germ tubes of *B. bassiana* (Bidochka and Khachatourians 1994a, b) and other EPF show a zone of clearing surrounding the penetration peg before invasion by hyphae. Cuticle-degrading enzymes once absorbed into target structures processively degrade polymers into utilizable monomeric precursors. The obvious candidate catabolic enzymes are those which affect proteins, chitin, waxes and lipids, and other exo-skeletal layers, insect tissues, hemocoel and hemolymph.

Table 3.2. Entomopathogenic fungal protein coding genes isolated and sequenced

Fungus	Gene	Enzyme	References
Metarhizium anisopliae	sod	Superoxide dismutase	Shrank et al. (1993)
	Pr1B	Subtilisin-like protease	Joshi et al. (1997)
	Pr1 (A–K)	Protease	Bagga et al. (2004)
	CRR1	DNA-binding protein	Screen et al. (1997)
	nrr1	Nitrogen response regulator	Screen et al. (1998)
	chit1	Chitinase	Bogo et al. (1998)
		Chitinase	Kang et al. (1998)
		Chitin synthase	Nam et al. (1998)
	chi2	Chitinase	Baratto et al. (2003, 2006), Screen et al. (2001)
	MeCPAA	Zinc carboxypeptidase	Joshi and St Leger (1999)
	ssgA	Hydrophobin	Bidochka et al. (2001)
		Trehalase	Zhao et al. (2006)
		Peptide synthetase	Bailey et al. (1996)
	trp1	Tryptophan synthetase	Staats et al. (2004)
Beauveria bassiana	prt1	Protease	Joshi et al. (1995)
		Bassianin I	Kim et al. 1999
	prt1-like	Serine endoprotease	Fang et al. (2002)
	chit	Chitinase	Fang et al. (2005)
		Endonuclease	Yokoyama et al. (2002)
	buv1	UV repair	Chelico et al. (2006)
B. brongniartii	buv1	UV repair	Chelico et al. (2006)

It is clear EPF adapt to fit their ecological niches (St Leger et al. 1997). Bidochka et al. (1999) found that members of the genus *Verticillium*, which are virulent towards insects and plants, have a shared commonality in extracellular proteases and carbohydrases. Phylogenetic relationships of 18 isolates in the genus *Verticillium*, representing 13 species of diverse eco-nutritional groups (pathogens of insects, plants, mushrooms, nematodes and spiders, and saprobes), suggest that the entomopathogenicity might have evolved independently over time. The evolutionary genetics of both EPF and phytopathogens are intriguing, as these groups of fungi have developed particular relationships with a diverse and large number of species. A model system to answer some initial questions with particular attention to host restriction and evolution towards obligate fungal pathogens has been addressed by Scully and Bidochka (2005, 2006a, b, c).

1. Proteases and Peptidases

Proteolytic enzymes of EPF have the following roles in insect pathogenesis: (1) cuticle degradation, (2) activation of the prophenol oxidase in the hemolymph, and (3) virulence. The protein-degrading enzymes proteases, collagenases, and chymoleastases have been identified and characterized from the following EPF: *A. aleyrodis, B. bassiana, B. brongniartii, E. coronata, Erynia* spp., *Lagenidium giganteum, Nomuraea rileyi, M. anisopliae*, and *V. lecanii* (Charnley and St Leger 1991; Khachatourians 1991, 1996; Sheng et al. 2006). Proteases and peptidases of EPF are also required for saprophytic growth.

Insect cuticle is primarily composed up of chitin and protein, 75% of which holds the chitin fibers together. Joshi et al. (1995) cloned extracellular subtilisin-like serine endoprotease (Pr1) from *B. bassiana*. A cDNA clone of the protease was isolated from mycelial culture of *B. bassiana* grown on cuticle/chitin cultures. Pr1 is synthesized as a large precursor (M_r 37 460) containing a signal peptide for translocation, a propeptide, and the mature protein with a M_r of 26 832. Joshi et al. (1997) identified a differentially expressed subtilisin-like protease (Pr1B), which had 54% sequence homology to Pr1A. However, *Pr1A* and *Pr1B* are located on different chromosomes. *Pr1B* is synthesized as a large precursor of 1158 nucleotides with a signal peptide, a propeptide but the mature protease (283 amino acids with a deduced molecular mass of 28 714 Da). Pr1B possesses the substitution of Thr^{220} by serine as well as Asn^{155} by glycine, which have not been reported in any other known subtilisins. Screen and St Leger (2000) found chymotrypsin (CHY1) from *M. anisopliae* was synthesized as a precursor (374 amino acids; pI/MW: 5.07/38 279) with a 186-amino-acid N-terminal fragment. Subtilisin-like proteases were recently classified by expressed sequence tags (EST). Polymerase chain reaction amplified ten orthologs from *M. anisopliae* sf. *anisopliae*; and seven from *M. anisopliae* sf. *acridum* were found. These subtilisins were placed into four major groups/clusters, based on the sequence similarity and exon–intron structure (Bagga et al. 2004).

Freimoser et al. (2005) measured gene expression responses to diverse insect cuticles by using cDNA microarrays constructed from an EST clone collection of 837 genes. *M. anisopliae*-infected cuticles from different insects hosts indicated 273 genes to have been up-regulated during fungal growth on *M. sexta* cuticle, including cuticle-degradation enzymes. Freimoser and coworkers identified some overlapped gene responses with unique expression patterns in response to cuticles from *Lymantria dispar*, *Blaberus giganteus* and *Popila japonica*.

The Pr1 protease is considered to be a key virulence determinant of EPF, believed to be up-regulated during appressorium formation and during conidiogenesis (Small and Bidochka 2005). They identified the sequence of seven conidiation-associated genes (*cag*) in *M. anisopliae* by using subtractive hybridization. Out of seven identified genes, they found *cag7* vital for cuticle evasion, by encoding an extracellular subtilisin-like proteinase (Pr1). Reverse-transcription polymerase chain reaction (RT-PCR) analysis confirmed that *cag* cDNA is expressed during conidiation under nutrient-starved conditions. RT-PCR analysis was also performed for Pr1 during infection of *G. mellonella*. Data revealed the up-regulation of subtilisin like-Pr1 when the mycelia appear on the surface of cuticle and produce conidia on the surface of the cadaver.

Gene structure and expression for a novel *B. bassiana* protease (bassianin I) has been described (Kim et al. 1999). It is 1137 bp or 379 amino acids long, has high homology with (Pr 1) and proteinase K and has three introns which are 69, 62, and 68 bp. Bidochka and Melzer (2000) reported genetic polymorphisms in three subtilisin-like protease isoforms (Pr1A, Pr1B, Pr1C) from isolates of *M. anisopliae*. RFLP variation was not observed in any Pr1 genes from isolates within the same genetically related group. Comparative analysis between *B. bassiana* (Pr1) and bassianin I indicated 82.1% homology, which shared 78.2% similarity with the Pr1 of *B. bassiana*. An extracellular *B. bassiana* protease, designated BBP, has also been purified and characterized (Urtiz and Rice 2000). In comparison to Pr1, BBP has a lower isoelectric point (pI 7.5) than Pr1 and is 0.5 kDa smaller. Both proteases has equal activity against cuticle, although the timing of expression and substrate specificities differ. Fang et al. (2002) reported the cloning and characterization of a cuticle-degrading protease (CDEP-1) from *B. bassiana*. It contained an 1134-bp ORF and predicting a protein of 377 amino acids (M_r 38,616, pI 8.302). Its amino acid sequence showed 57.9%, 83.3%, and 54.7% identity to *M. anisopliae* Pr1, *B. bassiana* Pr1, and proteinase K, respectively. Southern analysis indicated that CDEP-1 is a single-copy gene.

Insect and plant pathogenic species of the genus *Verticillium* have been evaluated for the production and regulation of hydrolytic enzymes (Bidochka et al. 1999). The facultative plant pathogens, including *V. albo-atrum* and *V. dahliae*, showed a greater production of cellulase and xylanase than the facultative insect pathogen, *V. lecanii*. Conversely, *V. lecanii* produced extracellular subtilisins (Pr1) upon induction in insect cuticle-supplemented medium. No subtilisins were detected in the plant cell wall-supplemented medium for opportunistic plant pathogens such as *V. fungicola* and *V. coccosporum* and saprophytic species like *V. rexianum*. Segers et al. (1999) described the distribution and variation between the subtilisins of entomogenous and phytopathogenic fungi. They reported that multiple isoforms with unique N-terminal sequences in single strains represent gene families. There were neither subtilisin nor homologous genes from plant pathogenic *Verticillium* spp. but Pr1-like enzymes and genes from weak plant-pathogenic species or saprotrophs.

Genetic polymorphism of gene family subtilisin-like protease from *M. anisopliae* strains has been described (Bidochka and Melzer 2000). RFLP variation was not observed from any *Pr1* genes from isolates of the same geographic group. However, in the case of *Pr1 A*, RFLP pattern variation was obtained both from genetically related groups and between isolates of disparate geographical areas. For *Pr1B* and *Pr1C*, a variation was observed at the *Eco*R1 site. Two fungal isolates, *M. anisopliae* var. *majus* strain 473 and *M. flavoviride* were found most dissimilar in RFLP pattern at all *Pr1* genes. Bidochka and Melzer further stated/hypothesized that the *Pr1* genes represent a gene family of subtilisin-like proteases. Moreover, the *Pr1A* gene encodes for the ancestral subtilisin-like protease, which was subsequently duplicated and rearranged within the genome. Wang et al. (2005b) isolated and characterized three spontaneous single-spore *pr1A* and *pr1B* mutants of *M. anisopliae* (V275) by using Nested PCR approach. The RAPD data revealed that an overall similarity between the wild type and the mutant stain was less than 70%. These mutants had low Pr1A and elastase activity and reduced lethality to *Tenebrio molitor*.

Freimoser et al. (2003b) studied the insect pathogen *Conidiobolus coronatus* (Zygomycota) during growth on insect cuticle. Expressed sequence tags cDNA clones were sequenced to analyze gene expression. They found genes that can encode chitinases and multiple subtilisins, trypsin, metalloprotease, and aspartyl protease activities, but in comparison to *M. anisopliae*) fewer hydrolases, antimicrobial agents and secondary metabolites. Possibly EPF such as *C. coronatus* adapted via a modification of the saprophytic ruderal-selected strategy, enabling its rapid growth to overwhelm the host and exploit the cadaver before other fungal competitors overrun it.

The crucial role of proteases in pathogenesis was also tested by the first genetically engineered mycoinsecticide based on an *M. anisopliae* construct, which over-expresses a toxic protease (St Leger et al. 1996). They main attribute of this construct was to hasten the speed of the kill by expressing additional copies of Pr1 from *M. anisopliae* into its genome. The constitutive expression resulted in the overexpression of Pr1 in the hemolymph of *M. sexta*, activating the phenoloxidase system. Larvae challenged with the engineered fungus exhibited 25% reduction in the time of death and 40% reduced food consumption. Moreover, a rapid melanization of the insect, preventing further fungal sporulation, helped to provide biological containment of the engineered strain.

2. Chitinases

Chitinases perform critical functions by their involvement in the growth and degradation of the fungal cell wall and insect cuticle, as chitin is a major component of both. The classification of chitinases indicates the presence of both endo- and exo-chitinases for the cleavage of NAGA polymer into smaller units or monomers. Extracellular chitinases are virulence determinance factors (Khachatourians 1991, 1996; Charnley 1997).

St Leger et al. (1996) reported the presence of chitinolytic enzymes, N-acetyl-β-D-glucosaminidases and endochitinases, in *M. anisopliae*, *M. flavoviride*, and *B. bassiana* during growth in media supplemented with insect cuticle. However, *M. flavoviride* also secreted 1,4-β-chitobiosidases into the cuticle media. The chitinase from *M. anisopliae* comprised acidic (pI 4.8) proteins of 43.5 kDa and 45 kDa. The identified N-terminal sequences of both bands were similar to an endochitinase from *Trichoderma harzianum*.

Valadares-Inglis and Peberdy (1997) located chitinolytic enzymes in enzymatically produced protoplasts and whole cells (mycelia) of *M. anisopliae*. No significant induction was observed from mycelia, yet protoplasts induced these enzymes significantly. The majority of chitinolytic activity was cell-bound in both whole cells and protoplast preparations, where this activity was mainly located in the membrane fraction.

Kang et al. (1998, 1999) reported a chitinase cDNA from *M. anisopliae* grown in a medium containing chitin as the sole carbon source. The molecular mass was approximately 60 kDa and the optimum pH was 5.0, which is different from values of 33.0, 43.5, and 45 kDa for endo- and 110 kDa for exo-chitinases reported previously by St Leger et al. (1996).

Screen et al. (2001) cloned the chitinase gene (*Chit*1), from *M. anisopliae* sf. *acridum* ARSEF strain 324 in *M. anisopliae* sf. *anisopliae* ARSEF strain 2575 (*Chit*1). They used the promoter of *Aspergillus* (*gpd*) for constitutive expression. Screen et al. (2001) did not detect the native, acidic form chitinase on zymograms (isoelectric focusing) from non-inducing medium. Conversely, the 2575-Chit strain produced chitinase earlier then the wild-type strain in chitin containing medium. They proposed that the formation of soluble chitin inducer after hydrolysis of chitin by CHIT1 is necessary for its production. However, the chitinase overproducers did not reveal any altered virulence to *M. sexta*, suggesting that wild-type levels of chitinase are not limiting for cuticle penetration.

Baratto et al. (2003) reported the expression and characterization of the 42-kDa chitinase of *M. anisopliae* in *Escherichia coli* using a bacteriophage T7- based promoter expression vector. Baratto et al. (2006) performed transcriptional analysis of the chitinase *chi2* gene of *M. anisopliae* var. *anisopliae* to show it has 1542 bp encoding for a deduced 419 amino acids. The gene itself is interrupted by two introns and contains a signal peptide of 19 amino acids. It was putatively identified to contain CreA/CreI/Crr1-binding domains aiding the up- and down-regulation of transcription of the *chi2* gene by chitin and glucose respectively. Nahar et al. (2004) found that the extracellular constitutive chitin deacetylase (CDA) produced by *M. anisopliae* converted chitin, a β-1, 4-linked N-acetylglucosamine polymer, into its deacetylated form chitosan, a glucosamine polymer. This CDA was not inhibited by solubilized melanin.

An endochitinase (Bb chit1) was purified to homogeneity from colloidal chitin containing liquid cultures of *B. bassiana* (Fang et al. 2005). Bbchit1 was 33 kDa (pI 5.4) and the encoding gene, *Bbchit1*, and their upstream regulatory sequences were cloned based on N-terminal amino acid sequence. *Bbchit1* contains no introns and is present as a single copy in the *B. bassiana* genome. The regulatory sequence of *Bbchit1* contains putative CreA/CreI-binding elements which regulate fungal carbon metabolism. The amino acid sequence of Bbchit1 is similar to the endochitinase of *Streptomyces avermitilis*, *S. coelicolor*, and *T. harzianum* (Chit36Y) but not to EPF reflecting a novel chitinase. Fang and co-worker (2005) constructed a *B. bassiana* transformant (*gpd-Bbchit1*), which overproduced *Bbchit1* and had enhanced virulence.

3. Lipases and Lipoxygenases

The epicuticule of the insect integument contains lipoproteins, fats, and waxy layers. Without the action of lipases and lipoxygenases, some of these materials would be barriers to EPF for their entry as some have anti-fungal activity, and they would be of no use as substrates for EFP (Khachatourians 1996). The growth of some EPF can be inhibited with various short chain saturated fatty acids. Lord et al. (2002) showed a role for the lipoxygenase pathway through eicosanoid-mediated cellular immune response to the *B. bassiana*. James et al. (2003) showed that cuticular lipids and silverleaf whitefly (*B. argentifolii*) affect conidial germination of *B. bassiana* and *P. fumosoroseus*. Cuticular lipids are toxic or inhibitory on the conidia of *B. bassiana* and *P. fumosoroseus*. The thick coating of long-chain wax esters produced by whitefly nymphs affects spore germination. Lord and Howard (2004) proposed the cuticular fatty amides of *Liposcelis bostrychophila* to have a role in preventing adhesion of dry-conidial preparations of *B. bassiana*, *P. fumosoroseus*, *A. parasiticus* or *M. anisopliae* and hence producing low (16%) mortality. *L. bostrychophila* is the only insect for which fatty acid amides have been identified as cuticular components that do not reduce the germination of *B. bassiana* or *M. anisopliae* conidia or the growth of their mycelia. This evidence indicates that cuticular fatty amides may contribute to insect tolerance for EPF by decreasing hydrophobicity and static charge, thereby reducing conidial adhesion.

House fly (*Musca domestica*) males are highly attracted to dead female flies infected with *E. muscae*. Apparently, male insects orient to the larger abdominal regions due to both visual and chemical cues to fungus-infected flies. Zurek and co-workers (2002) examined effect of the fungus on sex pheromone and other cuticular hydrocarbons. Behavioral assays demonstrated that the attraction is sex-specific and males were attracted more to infected females than to infected males, regardless of cadaver size. The main component of the house fly sex pheromone, (Z)-9-tricosene, n-tricosane, n-pentacosane, (Z)-9-heptacosene, and total hydrocarbons of young (7-day-old) and old (18-day-old) virgin females of young *E. muscae*-infected flies accumulated significantly less sex pheromone and other hydrocarbons on their cuticular surface than uninfected females. The cuticular hydrocarbons of older flies were unaffected by fungus infection.

4. DNA Repair Enzymes

Survival of EPF under exposure of ultraviolet (UV) light from solar radiation has been reported in several papers (Fargues et al. 1997, 2001a; Braga et al. 2001a, b). It is well known that several factors can affect UV tolerance in fungi, e.g., pigmentation in *M. anisopliae*. (Rangel et al. 2005, 2006a, b; Braga et al. 2006), choice of carbon sources, and particulars of the growth environment (Rangel et al. 2005, 2006a, b), and enzymatic DNA repair proficiency. In a series of studies, we (Chelico et al. 2005, 2006; Chelico and Khachatourians 2008) characterized nucleotide excision repair (NER) and photoreactivation (Phr) in six EPF and compared them for *N. crassa* and *A. niger*. *Beauveria bassiana* and *M. anisopliae* are the most UV-tolerant EPF. Molecular evidence of NER was identified by its homology to a conserved region of the *Saccharomyces. cerevisiae rad1* gene. This region was amplified from *B. brongniartii* by degenerate PCR as a 255-bp product. Sequencing showed that the *B. brongniartii* homolog, *buv1*, was most similar to the homolog *mus38* from *N. crassa*. The *rad1* homolog was present in *B. bassiana*, *B. nivea*, *B. cylindrosporum*, *M. anisopliae*, and *P. farinosus*, based on Southern blot analysis. We also isolated the first UV-sensitive mutants from *B. bassiana* and characterized NER and Phr of three mutants (Chelico and Khachatourians 2008). Swollen and germinating conidia and blastospores, which represent transcriptionally active stages, showed differential NER between transcriptionally active and dormant spore stages. Collateral sensitivity was tested with methyl methanosulfonate. UV-sensitive mutants affected in NER permitted testing of DNA damage and repair in EPF.

Tolfo Bittencourt et al. (2004) found *M. anisopliae* contains three superoxide dismutases (SOD), one of which was purified and partially characterized as a CuZn-SOD. The native enzyme is a dimer consisting of two subunits. Polyclonal antiserum were raised against purified CuZn-SOD and used to determine its subcellular localization by immunoelectron microscopy to show its presence in the cell wall.

B. Toxins and Pigments

Fungal toxins and pigments complement enzymes as key determinants of the outcome of EPF interaction with target pests, their survival, morbidity or mortality. Based on structure and function, toxins produced by EPF can be divided into non-peptide and peptide groups (see Table 3.3).

1. Non-Peptide Toxins and Pigments

The non-peptide group of toxins includes the red pigment oosporin, the yellow pigment tenellin, and bassianin. Tenellin and bassianin have a similar mechanism and an alternative pathway for tenellin biosynthesis implicates a role for tyrosine (Caragh Moore et al. 1998). Oosporin inactivates

Table 3.3. Toxins of entomopathogenic fungi

Toxin	Fungus	References
Non-peptide		
Oosporin	*B. tenella*	Michelitsch et al. (2004)
Tenellin	*B. bassiana*	Khachatourians (1991, 1996)
Bassianin		Caragh-Moore et al. (1998)
Toxic metabolite	*H. thompsonii* var. *thompsonii*	Vey et al. (1993)
Oxalic acid	*B. bassiana*	Bidochka and Khachatourians (1993), Alverson (2003), Kirkland et al. (2005)
Paecilomycine	*P. tenuipes*	Kikuchi et al. (2004)
Kojic acid	*T. inflatum*	Alverson (2003)
Tolypin		Khachatourians (1991, 1996)
Tolypocin	*T. geodes*	Jegorov et al. (1993)
Linear peptide toxins		
Leucinostins	*Paecilomyces*	Krasnoff and Gupta (1991)
Efrapeptins	*T. geode, T. cylindrosporum T. niveum*	Bandani et al. (2001), Bandani (2004) Krasnoff and Gupta (1991)
Cicapeptins	*C. heteropoda*	Krasnoff et al. (2005)
Cyclic peptide toxins		
Beauvericin	*B. bassiana, P. fumosoroseus*	Khachatourians (1991), Tang et al. (2005)
Destruxins	*Metarhizium* spp.	Khachatourians (1991), Jegorov et al. (1998), Pedras et al. (2002)
Enniatin complex	*F. avanacium, F. sambucinum*	Khachatourians (1991, 1996)
Hirsutide	*Hirsutella* sp.	Lang et al. (2005)
Protein Toxins		
10-kDa toxin	*B. bassiana*	Khachatourians (1996)
TF-1 and TF-2	*B. bassiana*	Khachatourians (1996)
Hirsutellin	*H. thompsonii* var. *thompsonii B. bassiana*	Liu et al. (1996)
Bassiacridin	*B. bassiana*	Quesada-Moraga and Vey (2004)

proteins through redox reactions of sulfhydryl groups, inhibits ATPase activity, and promotes cell lysis (Jeffs and Khachatourians 1997b). Michelitsch et al. (2004), using differential pulse polarography, determined oosporin in the culture broth of the fungus *B. brongniartii* and is accurate, rapid, and low cost.

Oxalic acid is also an important virulence factor since EPF penetrating through an external barrier are assisted by it (Bidochka and Khachatourians 1991). *B. brongniartii*, an EPF, produces crystals of ammonyl oxalate in liquid cultures. Kojic and oxalic acid have been shown to reduce the overall biological fitness of the plant bug *L. hesperus* (Alverson 2003). Oxalic acid in the lone star tick *Amblyomma americanum* at 50 mM induces >60% mortality after two weeks (Kirkland et al. 2005). Asaff et al. (2005) isolated dipicolinic acid from *P. fumosoroseus* and showed it has insecticidal toxicity to whiteflies *B. tabaci* and *B. argentifolii* nymphs by topical applications. Dipicolinic acid was detected after 24h when the fungus started growing. Non-peptide toxins are produced in rich medium in vitro and to a lesser extent in vivo (Strasser et al. 2000a, b).

2. Linear and Cyclic Peptide Toxins

Linear and cyclic peptide toxins which may have a varying numbers of amino acids are synthesized by many phytopathogenic and EPF. Cultures filtrates of EPF can be toxic upon feeding, topical application or injection into susceptible insects (see, Khachatourians 1991). Hirsutide is a cyclic tetrapeptide from a spider-derived entomopathogenic fungus, *Hirsutella* sp. isolated from an infected spider (Lang et al. 2005). This is a new cyclotetrapeptide cyclo-(L-NMe-Phe-L-Phe-L-NMe-Phe-L-Val), which along with the known cytochalasin Q, show cytotoxicity. The depsipeptides beauvericin and destruxins (DTXs) are most important. The name "destruxin" or "destructor" relates to the fungal species *Oospora destructor* from which these metabolites were first isolated.

Leucinostins and efrapeptins are linear peptide toxins isolated respectively from *Paecilomyces* and *T. geodes* Gams or *T. niveum* Rostrup (Krasnoff and Gupta 1991). Both toxins are ATPase inhibitors, with efrapeptins having been shown to bind strongly to V-ATPases in the brush border membrane of the midgut of *G. mellonella* (Bandani et al. 2001; Bandani 2004). When beauvericin was tested against the greenbug *Schizaphis graminum*, an overall decrease in the number of offspring produced was observed. Histopathological analysis showed an increased level of nucleic acid-specific DAPI staining, suggesting that beauvericin has an effect on the endosymbionts of this insect (Ganassi et al. 2002). In vivo tests using *S. frugiperda* cell line SF-9 showed cytotoxicity and growth inhibition (Calo et al. 2003; Fornelli et al. 2004).

Vilcinskas and co-workers (1999) showed the effects of cyclosporin A by intrahemocoelic injection in *G. mellonella* larvae. Beauverolide L metabolite caused no mortality but activated humoral responses, by inducing a significant release of lysozyme and cecropin-like activity into the hemolymph. Pedras et al. (2002) reviewed the chemical synthesis, biosynthesis, biotransformation, and biological activity of DTX. These cyclic hexadepsipeptides are composed of an α-hydroxy acid and five amino acid residues. There are 36 naturally occurring forms, most from fungi and rapeseed plant, *Brassica napus* (Pedras and Montaut 2003). Individual DTX differ on the hydroxy acid, N-methylation, and R group of the amino acid residues. The mode of action of DTX has been reviewed (Strasser et al. 2000a; Pedras et al. 2002). In general, DTX A and DTX E are the most toxic, with reported effects of blocking H^+-ATPase activity (Muroi et al. 1994) and interfering with cell membrane integrity (Bradfisch and Harmer 1990). In *M. anisopliae*, a spontaneous mutant deficient for the subtilisin isoforms pr1A and pr1B was also found to be unable to produce DTX (Wang et al. 2003) as a result of the loss of a conditionally dispensable 1.05-Mb chromosome. In the absence of destruxin production, the fungus was still equally virulent to *G. mellonella*. Amiri-Besheli et al. (2000) also found that virulence of isolates of *M. anisopliae* to Coleoptera, Orthoptera, and Hemiptera were variable even when the level of DTX production was low.

While DTX A and B are produced by *M. anisopliae* (Liu et al. 1996), limited or lack of production was reported in strains V245 and V275 on rice for up to 10–30 days (Wang et al. 2004). Yet the amount increased with the increasing content of peptone in the medium. When insect homogenate was used as single nutrient source or added to CD medium, no toxins were detected in the culture filtrate. Hsiao and Ko (2001) showed regulation of levels of DTX produced by different strains of *M. anisopliae* and mutant strain F061 that had significantly increased the production of DTXs.

Beauvericin is synthesized in a manner similar to enniatins. The biosynthesis of enniatins involves the enniatins synthetase, a multifunctional enzyme (Panaccione et al. 1992). Calo et al. (2003) found that beauvericin is cytotoxic to the insect cell line SF-9; and Tang et al. (2005) reported beauvericin to activate Ca^{2+}-activated Cl^- currents and induces cell deaths in *Xenopus* oocytes via influx of extracellular Ca^{2+}. Supothina et al. (2004) found enniatin production by *V. hemipterigenum* BCC 1449 grown in yeast extract sucrose medium had the highest levels for seven known enniatins and three new analogs: O1, O2, and O3.

Finally, Krasnoff et al. (2005) showed that cicadapeptins I and II are new α-aminoisobutyric acid-containing peptides from *Cordyceps heteropoda*. Fermentation extracts of this fungus yielded a known antifungal compound, myriocin, and a complex of novel non-ribosomal peptides: cicadapeptins I and II.

C. Fungal Virulence Factors

The ability of EPF to cause mortality in their host insects relates uniquely to the biochemical and mechanical attributes of this group of fungi. Although virulence measurements for EPF are expressed in terms of the LC_{50}, pathogenicity is the quality or state of being pathogenic (Shapiro-Ilan et al. 2004). Pathogenicity is a qualitative term, an "all-or-none" concept, whereas virulence quantifies it. It is in this context that research for the identification of virulence factors is being addressed.

Phenotypes including rapid germination, high sporulation rate, presence of toxins, metabolic acids, and high levels of extracellular tissue-destroying enzymes have been frequently associated with virulence (Valencia 2002). Virulence in terms of pathogen–host interactions and the contribution of virulence factors must depend not only on genetic proficiency [so far as the "virulence" gene(s) product(s) or factor(s) are concerned] but also on their coupling with transmission of the pathogen (Carruthers and Soper 1987). The immunological profile of the host and the molecular mimicry by EPF were equally important in virulence. Because of the multiplicity of determinants, the molecular dissection of virulence will remain challenging.

Fargues and Remaudiere (1977) categorized EPF into four groups: accidental, occasional, facultative, and obligate pathogens. The category facultative pathogens include at least 200 insect species susceptible to *M. anisopliae* and nearly 500 species sensitive to *B. bassiana*. Fargues and Remaudiere (1977) description of two extremes (so far as the pathogen's nutritional requirements, growth rates, and other interactions with the host were concerned) is still the only working scenario. First, there were species that had a wide spectrum of hosts and were capable of growing on simple synthetic media without added vitamins or organic nitrogenous supplements; and second, there were species with an extremely narrow host range, which had complex nutritional requirements. The influence of the environment (i.e., temperature, RH, sunlight) and the behavioral influences of social insects (e.g., ants, bees) should be considered in any discussion of virulence. For example, the Entomophthorales are pathogens of rapidly growing insects or those which have short-lived adult stages, i.e., aphids and dipteran insects. Here, the development of both the host and the pathogen could be affected by temperature where lower temperatures would retard the developmental cycle of the insect more than that of EPF. At higher temperatures through the shortening of the intermolt period, the growth of the fungus could be eliminated or circumvented before their ability to infect. The restricted maximal growth temperature of an EPF can work in favor of non-target insects, e.g., honeybees becoming infected. Additionally, host-specific factors such as developmental stage or age, intactness of the integument, the presence of wounds (Gillespie and Khachatourians 1992; Hayden et al. 1992; Bidochka and Khachatourians 1994a), the structure and composition of the cuticle, the anatomy of the respiratory system, and the mechanisms of resistance and immunity have a significant bearing on the determination of virulence of EPF. Ideas and models such as those reported by Bidochka and Small (2005) and others are urgently needed to determine the key evolutionary genetic divergence for fungal insect pathogens.

IV. Physico-Chemical Aspects of Disease Development

The physico-chemical aspect of disease development requires favorable interaction of the pathogen with the host outer tissues. Thereafter pathogenesis involves: (1) growth of the germ tube on the cuticle with the concomitant production of extracellular hydrolytic enzymes, (2) production of adhesive mucilaginous substance(s), (3) formation of appressoria on the cuticle surface, and (4) penetration of the infectious peg into the epicuticule layer, then the procuticle, and finally the hemocoel and hemolymph. Inside the hemocoel and hemolymph, fungal growth and multiplication occur, leading to the production of hyphal bodies. These bodies distribute themselves throughout the hemocoel, produce toxic metabolites, and interact with and/or evade insect defense mechanisms.

Pendland and Boucias (1998) produced monoclonal antibodies (MAbs) against epitopes on yeast-like hyphal bodies and hyphae of the entomopathogenic hyphomycete, N. riley, and in a subsequent paper, Pendland and Boucias (2000) generated polyclonal antibodies (PAB) in mice against S. exigua (beet armyworm) larval hemolymph and hemocytes and against the cell wall surfaces of hyphal bodies and hyphae of N. rileyi. The PAB exhibited strong activity against their original antigenic substrates and cross-reacted extensively with other substrates, including insect fat body basement membrane extracellular matrix (ECM) and N. rileyi and B. bassiana cell wall surfaces. Thus, these cross-reactivities suggest that the host mimicry expressed by surface components of EPF represents an important pathogenic determinant.

A. Spore Adhesion

The insect cuticle, the major barrier to infection by EPF, is comprised of three layers: the epicuticule, the procuticle, and the epidermis. So far as the spore–cuticle interaction is concerned, the most important mechanism is that of the hydrophobic interactions (Boucias et al. 1988).

The hydrophobicity of EPF spores is in part responsible for their attachment to the host cuticle. Also, the outer hydrophobic layer is known to protect the spore from dehydration in the environment (Boucias et al. 1988). Conidia of B. bassiana, whether produced under aerial or submerged conditions, showed similar hydrophobic characteristics; but blastospores were less hydrophobic, suggesting that conidia can bind to the cuticle much more efficiently (Hegedus et al. 1992). Dunlap et al. (2005) indicated that the physicochemical surface properties of blastospores of P. fumosoroseus are best described as having a basic monopolar surface, and classified as hydrophilic. Blastospores are also negatively charged under neutral conditions, with a pI of 3.4. Once attracted, other specific recognition systems, such as lectins, may strengthen the attachment by mucilaginous substances (Lecuno et al. 1991) and facilitate subsequent germination. The exogenous carbohydrates in general do not influence the types of glycoprotein production on the cell surface (Hegedus et al. 1990, 1992). However, Tartar et al. (2005) have shown differential expression of chitin synthase and glucan synthase genes in B. bassiana impact on the formation of a modified, thinner cell wall in vivo.

The adhesion of fluorescently labeled B. (Cordyceps) bassiana can be used to quantify the kinetics of adhesion of various cell types to surfaces having different hydrophobic or hydrophilic properties (Holder and Keyhani 2005). The adhesion of aerial conidia associated with hydrophobins, which are absent on either blastospores or submerged conidia, can explain the above data.

In general, B. bassiana conidia are hydrophobic (Jeffs and Khachatourians 1997a; Jeffs et al. 1999) and therefore better suited for suspension and dispersion in non- or partially aqueous formulation. Thus, it should be of no surprise that the evaluation by Vandenberg et al. (1998) of B. bassiana (Bb) for the control of diamondback moth on cruciferous plants grown in growth chambers, the greenhouse, and the field showed that an emulsifiable suspension applied at a high rate provided more significant reductions in larval counts than the aqueous formula. Carballo (1998) showed that Bb formulations of 10, 15, or 20% oil and 5×10^8 fungal conidia/ml produced 100% mortality of Cosmopolites sordidus. Masuka and Manjonjo (1996) used M. flavoviride and Bb spores formulated in commercial soya oil for the control of Mecostibus pinivorus. However, some oils can be phytotoxic and therefore unpractical. Akbar et al. (2004, 2005) showed that diatomaceous earth and plant essential oils or in mineral oil and organosilicone carriers increase the efficacy of B. bassiana conidial attachment. Vega et al. (1997) examined the in vitro effects of seven secondary plant compounds (catechol, chlorogenic acid, gallic acid, salicylic acid, saponin, sinigrin, tannic acid) mixed with Noble agar as an additional constraint to the survival on germination of blastospores of P. fumosoroseus.

While the overall spore surface hydrophobicity is not a major determinant of spore association with grasshoppers, Jeffs et al. (1997a) examined the attachment of *B. bassiana* conidia to the migratory grasshopper's body areas and found that hairs and cavities accumulate spores in higher densities and the palps and mouth parts can also be primary infection sites. Sosa-Gomez et al. (1997) examined the conidial attachment of *M. anisopliae* on the southern green stinkbug, *Nezara viridula*, using the exuviae and nymphal stage of the host as a substrate for *M. anisopliae* conidia. The hydrocarbon fraction (nC13, nC21–nC31 hydrocarbon series) served as a binding substrate for *M. anisopliae*, but the conidia did not degrade these hydrocarbons and did not use them as a carbon source.

Bidochka et al. (1995) and Jeffs et al. (1999) have identified hydrophobins of *B. bassiana* as SDS-insoluble and formic-acid-extractable (FAE) proteins in the 12.8 kDa and 14.0 kDa ranges in fractured aerial and submerged conidia but not from blastospores. Oxidation of the FAE sample resulted in a single protein band (15.4 kDa) as judged by SDS-PAGE, termed cell wall protein (cwp1), that occurs primarily next to the rodlet layer. Western blot analysis of two-dimensional gels revealed at least three acidic isoforms (pI 4.0–4.8) of cwp1. The N-terminal sequence of cwp1 showed no similarities with other published sequences.

Kamp and Bidochka (2002b) showed pleomorphic deterioration in *M. anisopliae*, which is a process where a fungal isolate loses the ability to produce conidia during repeated subculturing related to differences in hydrophobins). Ying and Feng (2004) illustrated the relationship between the thermotolerance and the contents of hydrophobin-like FAE proteins in aerial conidia of *B. bassiana* and *P. fumosoroseus* produced on rice-based substrate. Approximately 80% of variability in conidial thermotolerance was attributed to either 15.0-kDa or 17.5-kDa FAE protein or both.

B. Spore Germination–Growth

The first step in fungal spore germination and penetration must be spore hydration, activation, and production of hydrolytic enzymes. The ability of the spores to germinate requires the presence of carbon and energy source, anchorage on a substratum, and the production and utilization of precursors of cellular growth and macromolecular synthesis derived from the insect host. James (2001) examined the effects of exogenous nutrients on conidial germination and virulence of *B. bassiana* and *P. fumosoroseus* against the silver leaf whitefly. When spores were germinated before being applied to third-instar *B. argentifolii*, mortality was as much as 2.45 times greater and occurred more rapidly than that for fresh spores. For ungerminated conidia, the mean time to death from infection was 5.45±0.16 days and 4.74±0.08 days for application rates of 37 and 144 conidia/mm^2, respectively. When conidia were germinated before application, infection times dropped to 4.58±0.16 days and 4.45±0.10 days, respectively. A likely explanation for the greater pathogenicity and virulence of germinated over ungerminated *B. bassiana* conidia is that only a fraction of the spores applied to whitefly nymphs actually germinate on the cuticle. Degradation of monomethyl alkanes by germinating spores in some instances is a prerequisite to germ tube penetration in the host (Lecuona et al. 1991), whereas in the non-host insects, hyphal growth over the cuticle is observed. Finally, there is a relationship between the nitrogen source and lipid/carbohydrate reserves of *B. bassiana* and spore adhesion, germination, and virulence. The relationship between susceptibility of cuticular proteins, glycoproteins, and specific molecular characteristics is a critical step in the spore germination of EPF on insect cuticular layers.

Exoenzymes whether from spores or mycelia play a pivotal role in cuticle evasion. Pr1, chitinase, and esterase release from conidia after treatment, which allows the placement of these in the category of membrane-bound enzymes (See Qazi and Khachatourians 2007). Boucias and Pendland (1991) screened spores of *B. bassiana*, *M. anisopliae*, and *N. rileyi* for the presence of alkaline phosphatase, esterase (C8), lipase (C14), leucine aminopeptidase, acid phosphatase, phosphohydrolase, β-glucosidase, N acetyl-β-glucosaminidase and β-manosidase upon direct application of spores (10^6) to an APIZYM strip. However, these particular studies did not prove that the spores were membrane-bound or extracellular. We have found a novel aspect of pH regulation of the microenvironment by the conidia of *B. bassiana* and subsequent release or synthesis of protease isozymes in a water supernatant (Qazi and Khachatourians 2008). We hypothesize that *B. bassiana* conidia release citric acid and ammonia upon hydration and create a microenvironment favorable for the release of exo-enzymes so that they can function optimally. This facet of EPF has significance in the initiation of the infection cycle and dissolution of at least the epicuticle of insects. Moreover, we report here for the first time that conidia upon hydration and incubation release both chymoelastase (Pr1)- and trypsin (Pr2)-like enzymes from *B. bassiana* and *M. anisopliae* as determined by pNA substrate hydrolysis (Fig. 3.1) and zymopheretograms. These activities are comparable to the enzymes reported from the mycelia of *M. anisopliae*. Various detergents and buffers also affect the subsequent release of those enzymes from the conidia of *B. bassiana* and *M. anisopliae* (Fig. 3.2). These data indicate that both EPF regulate the ambient pH by secreting a varying level of ammonia and hence the activity of conidial proteases.

Hydration and the pH of the aqueous microenvironment in *B. bassiana* and *M. anisopliae* conidia determines the release and functionality of protease isozymes, and the release of citrate and ammonia during an initial conidial treatment in 0.03% Tween and water and upon further incubation in water for 2 days (Qazi and Khachatourians 2007, 2008). Conidia release four and eight protease (gelatinase-like) isozymes, respectively, from the conidia of *B. bassiana* and *M. anisopliae* (Qazi and Khachatourians 2007). These isozymes have similarities (on the basis of pI and substrate specificity) to Pr1, Pr2, and metalloprotease-like

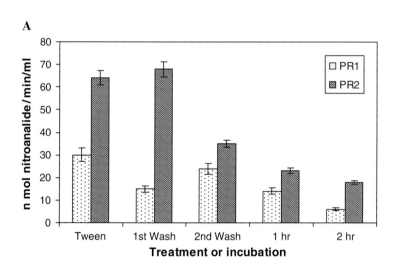

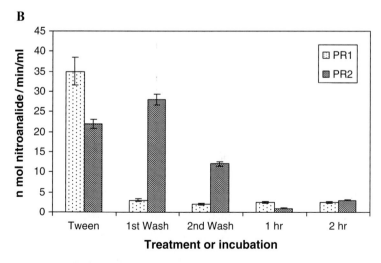

Fig. 3.1. Subtilisin-like (Pr1) and Trypsin-like (Pr2) enzyme activities (±SE) for *M. anisopliae* (**A**) and *B. bassiana* (**B**) were measured by using nitoroanalide substrates. The resultant activities are shown on the y axis. The treatments used for the recovery or incubation periods are given on the x axis

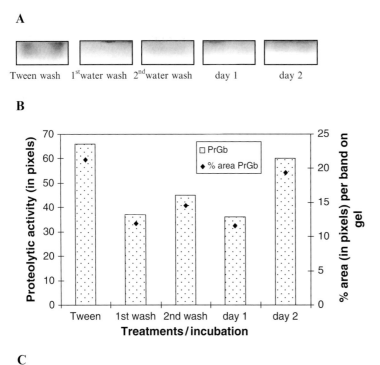

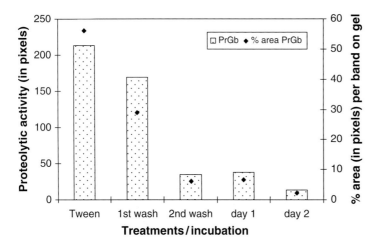

Fig.3.2. Cationic 15% PAGE zymography was performed with copolymerized gelatin (1 mg/ml) as substrate to analyze the basic gelatinase like proteinases (PrGb) released by the conidia of *B. bassiana* (**A**) and *M. anisopliae* (**C**). Two-dimensional spot densitometry was used to determine the proteolytic activity, which are presented in pixels values *B. bassiana* (**B**) and *M. anisopliae* (**D**). Gels were incubated in 50 mM TRIS/HCl, pH 7.4, 37 °C, for 2.5 h. The pixel values of the clearing bands/zones were subtracted from portions of gel at the multiple places which indicated undigested (control) gel, by using Alpha imager ver. 5.5. For quantitative determination of the proteinase activity of the gels, images were photographed, by using a Kodak zoom digital camera, and digital files were saved in TIFF format. The percent area of a given zymography pattern of a band on gel represents the total density that each band has (after background subtraction)

enzyme. Physiologically, upon hydration, chymoelastase (Pr1)- and trypsin (Pr2)-like enzymes are released from conidia.

Cristine et al. (2003) reported that the distribution and dual regulation of chitinase production by GlcNAc included the occurrence of at least two distinct, cell-bound (mycelial), chitinolytic enzymes in *M. anisopliae*. Wang et al. (2005b) by using ESTs and cDNA microarrays, showed adaptive responses of *M. anisopliae* to the varying environmental conditions. During cuticle evasion, the processes and genes involved in carbon metabolism, proteolysis, cell surface properties, and synthesis of metabolites are up-regulated. When *M. anisopliae* enters the hemolymph the expression of proteases is turned off.

C. Growth Within Insects

The infection of insect host hemocoel, hemolymph, and tissues during molting results in disruption of physiological functions, disease, and ultimate death of the insect. In this regard, temperature profiles in conjunction with infectivity assays can be useful in selecting appropriate isolates for particular climates. The relationship between temperature and relative humidity on virulence was demonstrated with V. *lecanii* towards several aphid species: *Rhopalosiphum padi* (Hsiao et al. 1992), *E. neoaphidis*, and *Sitobion avenae*. Schmitz et al. (1993) showed the thermodependancy of virulence and reproductive stages instars of alate and apterous populations to EPF. In studies by Hsiao et al. (1992), the average fecundity in infected and uninfected (control) aphids differed according to the incubation temperature at 15, 21, and 27 °C. Yeo et al. (2003) performed a laboratory evaluation of temperature effects on the germination and growth of EPF and on their pathogenicity to two aphid species. The rate of in vitro conidial germination of all isolates was slower at 10 °C and 15 °C than at 20 °C and 25 °C. Similarly, in vitro growth of most isolates was adversely affected at 10 °C and 15 °C. The greatest reduction in *M. anisopliae* rates of conidial germination and colony growth was at 10 °C.

The susceptibility of insects to fungal infections during molting stages has provided further understanding. In their classic study, Vey and Fargues (1977) showed infection during the ecdysis of Colorado potato beetle by *B. bassiana*. The penetration and development of the fungus occurred in larvae dying before ecdysis. Mycelial elements were present and shed with the ecdysal cuticle. Infection in post-molting larvae having fungal wounds was associated with extensive fungal growth in the area of the wound and was heavily melanized. Larvae which had contact with fungus but did not display symptoms of the disease within 4 or 5 days of shedding their exuvium developed melanized patches which spread slowly over the integument. Melanization was associated with clumps of hemocytes at the contact point and activity of phenoloxidases. Such lesions may be important in the processes of cellular immune systems, such as activation of the prophenoloxidase in insects (Gillespie and Khachatourians, unpublished data). Studies of infection of other insects during molting has demonstrated the production of appressoria-like structures which play a part in the adhesion and penetration of fungal hyphae into the ecdysial cuticle, fungal growth, reproduction including budding blastospores, and injuries of the ecdysial cuticle. Samuels and Paterson (1995) reviewed cuticle-degrading proteases from insect molting fluid and culture filtrates to show the protein in insect cuticle represents a significant barrier to the invading fungus. da Silva et al. (2005) reported that the cuticle-induced endo/exochitinase CHIT30 from *M. anisopliae* is encoded by an ortholog of the chi3 gene which has both endo- and exochitinase activities and is a potential determinant of pathogenicity.

Campos et al. (2005) used scanning electron microscopic analysis of *Boophilus microplus* infected with *Beauveria. amorpha* and *B. bassiana*. They showed appressorium formation during penetration on cattle tick cuticle and subtilisin-like proteases and chitinases in the tick.

In the hemolymph, fungal propagules distribute, produce toxic metabolites, and interact with defense mechanism determinants deterrent to their growth and role as pathogens.

Three mechanisms are responsible for EPF evasion of insect defenses here: (1) changes in the fungal cellular outer layer(s), (2) production fungal immunomodulating substances, and (3) tolerance of insect immunological defense reactions by fungi. The changes in EPF cell surfaces occur (Hegedus et al. 1992; Pendland and Boucias 1992; Pendland et al. 1993) to disguise in vivo surface receptor recognition by opsonins (agglutinins) or hemocytes. Wang and St Leger (2006) found a collagenous protective coat to enable the evasion of *M. anisopliae* in insects. The production of antifungal or immunomodulating substances, which suppress cellular defense systems, is not fully understood; however the presence of such proteins has been observed (Iijima et al. 1993; Khachatourians, unpublished data). Hung et al. (1993) studied the effect of *B. bassiana* and *C. albicans* on the cellular defense response of *S. exigua*. They showed that both fungi were phagocytized by the hemocytes of *S. exigua* although the blastospores of *B. bassiana* were able to suppress the spreading ability of hemocytes, in order to escape and produce germ tubes. Dean et al. (2004) showed changes in the appearance of hemocytes of *Manduca sexta*, associated with the ability of spreading of monolayers.

In vivo gene expression of *B. bassiana* in the hemolymph of *M. sexta* larvae led to the construction of a cDNA library (Tartar and Boucias 2004). Expressed sequence tags were generated which led to the cloning of two protease genes, a tripeptidyl peptidase (Bb TPP) and a dipeptidyl peptidase (Bb DPP). The Bb TPP protease was shown to be up-regulated during infection and its secretion in the host hemolymph. Proteases of EPF can affect cellular immune reactions of insects. Griesh and Vilcinksas (1998) have described the role of proteases from *B. bassiana* and *M. anisopliae* in the suppression of the immune response of isolated *G. mellonella* plasmatocytes. Bidochka and Hajek (1998) showed increased activation of insect prophenoloxidase by *Entomophaga maimaiga* and *E. aulicae* which grow as protoplasts in the hemolymph of permissive insect hosts. They found higher levels of phenoloxidase activity for up to 96 h post-challenge, as well as a prophenoloxidase-activating trypsin activity for *L. dispar* challenged with *E. aulicae* when compared to an *E. maimaiga* challenge. Three isoforms of phenoloxidase (pI 5.0–5.5) and at least six isoforms of trypsin activity (four basic trypsins pI 8–10, two acidic trypsins pI 4–6) with preferences for small amino acid residues were activated in *L. dispar* after challenge. In vitro prophenoloxidase activation showed that treatment of *L. dispar* hemolymph with *E. aulicae* protoplast plasma membranes consistently resulted in higher prophenoloxidase activation than that with *E. maimaiga*. They suggest that differences in the surface glycoproteins, implicated in the activation of zymogenic trypsins, which in turn activate the prophenoloxidase cascade as a non-permissive response.

After the death of an insect infected with EPF, fungal outgrowth from the insect body and, coincidentally, the production and dispersal of spores to new hosts and the environment occurs. Luz and Fargues (1997, 1998) reported on factors affecting conidial production of *B. bassiana* from fungus-killed cadavers of *Rhodnius prolixus*. The conidial production of *B. bassiana* from mummified cadavers required high RH levels of at least 96.5% RH. At 97% RH and 25 °C, 5.3×10^6 to 1.7×10^8 conidia/insect (on first-instar and adult cadavers, respectively) are required depending on their size, the stages of host development, and the conditions of sporulation within 4–5 days after death. At 97% RH conidial production from *Rhodnius* cadavers was unaffected by temperatures of 15–25 °C, but declined at 28–30 °C and was null at 35 °C. Tefera and Pringle (2003a, b) showed that the exposure method to *B. bassiana* and conidia concentration affected mortality, mycosis, and sporulation in cadavers of *Chilo partellus* in the laboratory with the optimum temperatures for mycosis and for sporulation being 20 °C and 15 °C, respectively.

V. Genetics and Molecular Biology of Disease Development

Progress in research during the past 10 years now permits manipulation of the genetics of EPF. As a result the genetics and molecular biology of EPF represent an opportunity for integrated understanding of the triad, the environment, the fungus, and the insect host(s). The general and molecular genetics, genomics and population level analysis of only three genera (*Beauveria*, *Metarhizium*, *Verticillium*) and then only one species in each case (*B. bassiana*, *M. anisopliae*, *V. lecanii*) is well established.

A. Enzymes and Virulence Factors

There are four lines of evidence which deal with virulence determinants in EPF: mutants, somatic hybrids, natural selection, and passage through hosts.

Extracellular protease(s) are virulence factors, as deficient mutants have lowered virulence (Bidochka and Khachatourians 1990) and gene duplication have higher virulence (St Leger et al. 1996). Paradoxically, although oxalic acid is a factor in the cuticle penetration by *B. bassiana* in vitro, yet oxalic acid production mutants did not show increased virulence (Bidochka and Khachatourians 1993). However, Kirkland et al. (2005) report that oxalic acid as a fungal acaricidal virulence factor and cell-free culture supernatants from *B. bassiana* mutants with decreased oxalate production displayed lower acaricide activity.

Viaud et al. (1998) found hypervirulent somatic hybrids of *B. bassiana* and *B. sulfurescens* through protoplast fusion of diauxotrophic mutants. Hybrids appeared to be diploid or aneuploid, with portions of the genome being heterozygous. The pathogenicity and the ploidy of the hybrids remained stable after passage through the host insect.

Ansari et al. (2004) found that two isolates of *M. anisopliae* (CLO 53, CLO 54) caused maximally 90% mortality at 10 weeks post-inoculation of June beetle, *Hoplia philanthus*), an insect which has become a widespread and destructive insect pest of lawns, sport turf, pastures, and horticultural crops in Belgium.

Finally, passage of EPF through a host can affect subsequent virulence. Brownbridge et al. (2001) showed the effects of in vitro passage of *Beauveria bassiana* on its virulence to *Bemisia argentifolii*, as did Vandenberg and Cantone (2004) on three strains of *Paecilomyces fumosoroseus* after passage in vitro or in vivo in *Diuraphis noxia* or *Plutella xylostella*. Vandenberg and Cantone (2004) indicated a combination of positive or no changes in virulence toward host insects after passages in vitro, including lost virulence, reduced speed of germination, reduced mycelial dry weight, and conidiation. Some of these negative changes were restored after passage in vivo. No change in banding pattern was observed for any strain using 14 primers for RAPD-PCR. These results demonstrate the intraspecific variability and phenotypic plasticity of strains of *Paecilomyces fumosoroseus*. Vandenberg and Cantone (2004) conclude that, while stability of virulence after in vitro passage is desirable for mass production, any change in host specificity or productivity in vitro must be monitored and minimized. We have observed RAPD pattern changes in *B. bassiana* isolates under in vitro and in vivo passages as well (Valencia, McCaferry and Khachatourians, unpublished data).

B. Molecular Genetic Analysis

There are many approaches for analyzing host–pathogen relationships. The specific interactions of EPF with hosts and involvement of catabolic enzymes, virulence factors that are subjects of active pursuit. What remains to be investigated is the hypothesis of the relationships between the host and the pathogen in a few model gene-for-gene systems or a combination of determinants.

1. Cloning and Sequencing of Chromosomal Genes

The past decade saw reports on the isolation, characterization, and expression of chromosomal genes of EPF. The first cloning and molecular analysis for the *M. anisopliae* protease structural gene from a cDNA was followed by other explorations of genomic DNA (Hajek and St Leger 2004). Below we chronicle this progress.

Freimoser et al. (2003a) performed EST analysis of genes expressed by *Conidiobolus coronatus* (Zygomycota), which is a facultative saprobe and a pathogen of many insect species. Nearly 2000 EST cDNA clones were sequenced to analyze gene expression during growth on insect cuticle. Of the ESTs that could be clustered, 60% were placed into functional groups ($E \leq 10^{-5}$) and had their best BLAST hits among fungal sequences. These included chitinases and multiple subtilisins, trypsin, metalloprotease, and aspartyl protease activities with the potential to degrade host tissues and disable anti-microbial peptides. Otherwise, compared to the ascomycete *M. anisopliae*, *Con. coronatus* produced many fewer types of hydrolases (e.g., no phospholipases), antimicrobial agents, toxic secondary metabolites, and no ESTs with putative roles in the generation of antibiotics. Instead, *Con. coronatus* produced a much higher proportion of ESTs encoding ribosomal proteins and enzymes of intermediate metabolism that facilitate its rapid growth. These results are consistent with *Con. coronatus* having adapted a modification of the saprophytic ruderal-selected strategy, using rapid growth to overwhelm the host and exploit the cadaver before competitors overrun it. This strategy does not preclude specialization to pathogenicity, as *Con. coronatus* produces the greatest complexity of proteases on insect cuticle, indicating an ability to respond to conditions in the cuticle.

Many EPF secrete subtilisin proteinases to acquire nutrients and breach host barriers. The regulation of protease synthesis was first reported by Bidochka and Khachatourians (1988a, b) and detailed at the molecular level in a study by Screen et al. (1997), showing that synthesis of the pr1 gene product of *M. anisopliae* is subject to both carbon and nitrogen repression. The pr1 promoter region was sequenced revealing the presence of putative CREA- and AREA-binding sites. Using a PCR-based strategy, the *M. anisopliae* crr1 gene was identified; it encodes a putative C2H2-type DNA-binding protein with significant sequence similarity to *A. nidulans* CREA. Complementation experiments with an *A. nidulans* strain-carrying creA204 demonstrated that CRR1 could partially substitute for CREA function. In addition, Screen et al. (1998) identified *M. anisopliae* nrr1 (nitrogen response-regulator gene) using a PCR-based strategy. Bagga et al. (2004) sought a global characterization of the diversity of subtilisins in the insect pathogen *M. anisopliae*. EST expressed 11 subtilisins during growth on insect cuticle and seven from the locust specialist *M. anisopliae* sf. *acridum* (strain 324).

Bogo et al. (1996) showed the complete nucleotide sequence and analysis of a single-copy chromosomal chitinase gene, chit1 of *M. anisopliae* with its three short introns, that encoded for a protein of 423 amino acids with a stretch of 35 amino acid residues as signal peptide. Fang et al. (2005) found an intronless single-copy endochitinase, Bbchit1, a gene whose regulatory sequence contained putative CreA/CreI carbon catabolic repressor-binding domains. At the amino acid level, Bbchit1 showed significant similarity to endochitinases from *S. avermitilis*, *S. coelicolor*, and *T. harzianum*.

The 28s rDNA group I introns of nuclear ribosomal DNA represent a powerful tool for identifying EPF strains (Neuvéglise et al. 1997). Mavridou et al. (2000) identified group I introns at three different positions within the 28S rDNA gene of *M. anisopliae* var. *anisopliae*, using a set of heterologous primers designed from the 3' end of the 28S rRNA gene of *V. dahliae*. Reverse-transcription polymerase chain reactions indicated that all these introns were absent from the mature RNA molecules. DNA sequence alignment of the nuclear 5.8S rRNA gene and internal transcribed spacers (ITS) is now well established in many EPF. Fargues et al. (2002) studied variability in rDNA-ITS in 48 isolates of *P. fumosoroseus* from various geographical and host insect origins, e.g., *Bemisia tabaci-argentifolii*. Of the three distinct groups, only one representing 25 isolates came only from the host *B. tabaci-argentifolii*, whereas those in group 3 were more diffuse and came from various insect host and geographical origins.

Wang et al. (2003a) examined nuclear large subunit rDNA group I intron distribution in a population of *B. bassiana* and showed a strong correlation between specific insertion sites and intron subgroups, fully supported by corresponding clades, suggesting a common ancestry of the site-specific LSU introns. On the above basis, Wang et al. (2003b) showed that the average genetic diversity index of geographical populations of *B. bassiana* to be significantly smaller than that of populations derived from insect host orders. Sugimoto et al. (2003) attempted to clarify relationships among genetic diversity, virulence, and other characteristics of conidia amongst 46 isolates of *V. lecanii* from various hosts and geographical locations. They examined the ITS and IGS regions of rDNA, mt-SrDNA, histone 4, and β-tubulin genes by PCR-RFLP and PCR single-stranded conformational polymorphism (SSCP). There were no relationships among the results of RFLP, SSCP, isolation source, and location. However, the size of IGS did have relationships with conidia size and sporulation.

It appears that the overall genetic diversity in some EPF resulted from the genetic variations within geographical populations. Future research on the use of genomics in population, taxonomic, and phylogenetic studies will be important in the production and use of EPF and their contribution to alternatives in pest management and the history of agriculture and forestry sciences.

2. Cloning and Sequencing of Mitochondrial Genes

In addition to chromosomal genes, the mitochondrial (mt) DNA genes have utility in taxonomy and possibly in the examination of the progression of pathogenesis. In other phytopathogenic fungi, particular mtRFLP patterns have been correlated with pathogenicity, mating type, and host preference. Furthermore the identification of mt markers in EPF can have a significant impact in the study of organelle integrity structure and function. mtDNA can provide probes: (1) to study polymorphisms of mtDNA within natural population structures or taxonomy, and (2) for tracking after the release of EPF into the environment. Khachatourians and Hegedus (2004) and Rodriguez et al. (2004) have published a comprehensive review of the methodological aspects of fungal mtDNA. Khachatourians and Uribe (2004) have reviewed the genomics of the mtDNA of EPF.

mtDNA is a suitable tool for the taxonomic characterization of EPF (Typas et al. 1992; Hegedus and Khachatourians 1993b; Kouvelis et al. 1999; Khachatourians and Hegedus 2004; Khachatourians and Uribe 2004). Mitochondrial genome studies have determined the genome sizes for *B. bassiana* (28.5 kb), *M. anisopliae* (approx. 32 kb), and *V. lecanii* (24.5 kb).

The classic work of Typas et al. (1992) examined the mtDNA of the genus *Verticillium*. This genus contains a heterogeneous group of species that are relevant in agriculture, because they can be both plant pathogens and/or insect pathogens. In that work they were able to differentiate seven species of the genus *Verticillium* and also clearly separated alfalfa pathogenic strains of *V. albo-atrum* from non-pathogenic strains by using mtDNA RFLPs analysis of 29 *Verticillium* isolates. A wider study presented later by the same research group showed 20 different band patterns after the analysis of 54 isolates mainly of *V. lecanii* (51 out of the 54), using mtDNA RFLPs (Kouvelis et al. 1999). Some interesting suggestions such as the subtropical origin of the species was supported by the higher level of divergences found in mt genotypes from strains belonging to this region of the world, in comparison to those found in temperate countries.

By using a similar approach, Mavridou and Typas (1998) found a high degree of polymorphism between 25 isolates of the EPF *M. anisopliae*, obtained from at least 16 different hosts and 15 different countries from five continents. Twenty different groups were suggested for those 25 isolates, 16 of which presented a unique mt RFLP pattern, showing the huge power of this technique to identify DNA polymorphism. Despite the high level of resolution found in the system, it was not possible to identify any correlation between the isolates in terms of either host or geographical origin. Despite clonality, the mt genome may be exposed to individual mutagenic events that can be perpetuated due to the presence of mt incompatibility and even geographic isolation (Hegedus and Khachatourians 1993a).

The polymorphic analysis of *B. bassiana* mtDNA showed only two populations of mtDNA (Hegedus and Khachatourians 1993a), suggesting a very conservative mt genome in comparison to other EPF. However, Uribe and Khachatourians (2004) used whole genomic DNA from 18 *B. bassiana* and one each of *B. amorpha*, *B. cylindrospora*, and *B. nivea*, which came from three continents and four target insect species. Single- and double-restriction enzyme digestion of total genomic or mtDNA with *Eco*RI, *Eco*RI-*Hin*dIII, or *Eco*RI-*Bgl*II, and probed with BbmtE2 showed the predominance of mito-types A–B and an additional nine types (C–K) with *B. bassiana*.

Hegedus et al. (1998) performed characterization and structure of the mitochondrial small rRNA gene of *B. bassiana*. The entire mt srRNA gene from *B. bassiana* was sequenced. Several features were identified that were common only to the Hyphomycetous fungi examined. Phylogenetic analysis indicated that the anamorph *B. bassiana* was more closely related to Pyrenomycete than to Plectomycete ascomycetous fungi. Our compilation reported the phylogenetic analysis of 14 filamentous fungal mtDNA for srRNA sequence, seven of which were EPF (Uribe and Khachatourians 2004). Both *Deuteromycetes* and *Ascomycetes* were considered. These result suggest that EPF came from different ancestral roots of filamentous fungi, including the association of an ancestral mtDNA within the fungal kingdom where, in spite of evolutionary development, the same srRNA is shared between sexual and asexual fungal genera.

Kouvelis et al. (2004) presented the analysis of the complete mitochondrial genome of *Lecanicillium muscarium* (synonym *V. lecanii*), which has a total size of 24 499 bp. It contains the 14 typical genes coding for proteins related to oxidative phosphorylation, the two rRNA genes, one intronic ORF coding for a possible ribosomal protein (*rps*), and a set of 25 tRNA genes which recognize codons for all amino acids, except alanine and cysteine. All genes are transcribed from the same DNA strand. Ghikas et al. (2006) have reported the complete mitochondrial genome of the *M. anisopliae* var. *anisopliae*, including the *trn* gene clusters. Their data reveals a common evolutionary course shared by all Sordariomycetes.

Tymon and Pell (2005) examined ISSR, enterobacterial repetitive intergenic consensus (ERIC), and RAPD PCR-based DNA fingerprint analyses of 30 isolates of *P. neoaphidis* worldwide, together with six closely related species of *Entomophthorales*. Their data showed that, although *P. neoaphidis* isolates were highly polymorphic, they could be separated into a monophyletic group compared with the other *Entomophthorales* tested. ERIC, ISSR, and RAPD analyses allowed the rapid genetic characterization and differentiation of isolates with the generation of potential isolate- and cluster-specific diagnostic DNA markers.

C. Molecular Probing into Pathogenicity and Tracking

Chromosomal DNA probes from EPF could have utility in either dissecting the progression of the infection or tracking of released biocontrol fungi.

The DNA-based technology systems can be ideal for tracking studies. DNA sequences used for probing could include genomic DNA probes or could differentiate between isolates of rDNA and rRNA or mtDNA sequence comparisons, PCR amplification RAPD, RFLPs, SSCP, which can identify the minimum number of base changes (Hegedus and Khachatourians 1996a–d; Urtz and Rice 1997; Sugimoto et al. 2003). These types of probes can be useful for field studies of released EPF.

Castrillo et al. (2003) used strain-specific detection tools for introduced *B. bassiana* in agricultural fields; and Coates et al. (2002) used a nuclear small-subunit rRNA group I intron variation among *Beauveria* for strain identification and evidence of horizontal transfer. Entz et al. (2005) provide an example of a PCR-based diagnostic assay for the specific detection of *M. anisopliae* var. *acridum* using a PCR-based 420-bp sequence for the detection. Other fungal entomopathogens, plant pathogens, mycopathogens, and soil saprophytes were also not detected by the pathogen-specific primers.

The use of contemporary genomics and gene data-mining sources, through EST, has made it easier for the search and identification of the genomes of many organisms. It is not a surprise that EST analysis identified two subspecies of *M. anisopliae* and *M. anisopliae* sf. *acridum*, as reported by Freimoser et al. (2003a). Approximately 1700 5' end sequences from each subspecies were generated from cDNA libraries representing fungi grown under conditions that maximized the secretion of cuticle-degrading enzymes. Both subspecies had ESTs for virtually all pathogenicity-related genes cloned to date from *M. anisopliae*, but many novel genes encoding potential virulence factors were also tagged. Indeed Freimoser et al. (2005), using cDNA microarrays, constructed an EST clone collection of 837 genes. During growth in culture containing caterpillar cuticle (*M. sexta*), *M. anisopliae* up-regulated 273 genes including cuticle-degradation (e.g., proteases), amino acid/peptide transport, and transcription regulation. The 287 down-regulated genes were also distinctive and included a large set of ribosomal protein genes. The response to nutrient deprivation partially overlapped with the response to *M. sexta* cuticle, but unique expression patterns in response to cuticles from another caterpillar (*Lymantria dispar*), a cockroach (*Blaberus giganteus*), and a beetle (*Popilla japonica*) indicated that EPF can respond in special way to specific conditions.

VI. Conclusions and Perspectives

Current progress in the biochemistry and molecular genetics of EPF during recent years provides exciting understanding of the pathogenic mechanims of EFP. Without a firm handle on molecular biology of EPF, exploitation of their biotechnological potential will be lessened. However, it is self-evident that their contribution as control agents in agriculture, forestry, and the world economy has been growing (see Khachatourians 1986, 1991; Ferron et al. 1991; Feng et al. 1994; Khachatourians and Valencia 1999; Qazi and Khachatourians 2006). In the past decade opportunities to explore and exploit the physiology and the genetics of bacterial entomopathogens have led to the invention and production of insecticidal protein toxins of transgenic microorganisms and plants. The same challenge with regard to the EPF, as best appreciated from this chapter, should be recognized and research in applied mycology should be sustained as new challenges illustrated below are confronted.

Traniello et al. (2002) studied the development of immunity in a social insect: to establish group facilitation of disease resistance. The extraordinary diversity and ecological success of the social insects has been attributed to their ability to cope with the rich and often infectious microbial community inhabiting their nests and feeding sites. The mechanisms of disease control used by eusocial species include antibiotic glandular secretions, mutual grooming, the removal of diseased individuals from the nest, and the innate and adaptive immune responses of colony members. Traniello et al. (2002) demonstrated that, after a challenge exposure to *M. anisopliae*, the dampwood termite *Zootermopsis angusticollis* had higher survivorship when individuals developed immunity as group members. Furthermore, termites significantly improve their ability to resist infection when they were placed in contact with previously "immunized" nestmates. This "social transfer" of infection resistance was a previously unrecognized mechanism of disease control in the social insects. To explain how group living may improve the survivorship of colony members (despite the possibly increased risks of accompanying pathogen transmission) could have lessons for the strategic disruption of this behavior and control of an economically significant urban insect pest.

Fungal control of hematophagous insects, some of which are vectors of many debilitating diseases, was described by Miranpuri and Khachatourians (1994). Scholte et al. (2003) showed the infection of *Anopheles gambiae* s.s. and *Culex quinquefasciatus* (vectors of malaria and filariasis) with *M. anisopliae*. Following their successful use of an entomopathogenic fungus against tsetse flies, they investigated the potency of *M. anisopliae* as a biological control agent for adult malarial and filariasis vector mosquitoes. In the laboratory, both sexes of *An. gambiae* and *C. quinquefasciatus* were passively contaminated with dry conidia of *M. anisopliae* under varying conditions. This first attempt for EPF to be used against adult Afrotropical disease vectors shows the potential for novel biocontrol agent-targeted indoor application methods for the control of endophagic host-seeking females. Scholte et al. (2005) described the practical delivery of an entomopathogenic fungus that infected and killed adult *An. gambiae* in rural African village houses in Tanzania. An equally important development was the work of Kirkland et al. (2004), showing the pathogenicity of *B. bassiana* and *M. anisopliae* to the tick species *Dermacentor variabilis*, *Rhipicephalus sanguineus*, and *Ixodes scapularis*.

Developments in the use of EPF have demonstrated that isolates suitable for use in particular regions may not be universally applicable. Adaptations of isolates to forest or agricultural habitats (Bidochka et al. 2002) are a quintessential example of appropriate isolate selection. Further, we need the development of EPF into the pest management system that is economically sustainable for the producer. Isolates must have a low environmental persistence for both environmental considerations and to retain the market need for the produced material. Further, the use of EPF in IPM programs may require new thinking. For example, the application of EPF for management of an insect pest with a rapid intrinsic rate of population growth would require the release of the pathogen earlier than the traditional approach, as the EPF takes longer to reduce the population. In temperate climates, this may mean the selection of isolates capable of cooler activities; and it is likely that a common strain may not be suitable for use throughout an entire crop season as cool-active isolates may be killed at the higher temperatures of the end of the season. Thus the development of an EPF for integration into a pest management scheme may be novel for each case, with success precedents serving as starting points.

While the promise of the application of EPF for sustainable development is the rubric of the day, many most intriguing questions could occupy mycologists and molecular biologists for the next decade. Among the outstanding research questions are the molecular and biochemical aspects of EP, the epigenetic signals governing spore germination, developmental biology, molecular facets of penetration steps into the host, the role of toxins in defense, and the genetic basis of opportunistic fungi in becoming pathogens capable of species evolution. Bidochka's group is using a novel approach to address host–pathogen relationships with, among other questions, particular attention to host restriction and evolution towards obligate fungal pathogens (Scully and Bidochka 2005, 2006a, b, c). Finally, with the release of EPF, whether naturally occurring or genetically engineered, strain identification and tracking by DNA and antibody probes will be required (Fuxa 1991). It is certain that the degree of success with which the EPF suppression of insect pests will be met depends on both incremental as much as serendipitous discoveries. Hopefully the production and use of EPF can be fully developed and their contribution to alternatives in pest management and the history of agriculture and forestry sciences will be acknowledged.

Acknowledgements. The work in our laboratory was made possible through the support of University of Saskatchewan, College of Graduate Studies and Research, and College Agriculture Graduate Scholarships, and Natural Sciences and Engineering Research Council Grant 0493. I am grateful to my many students for their contributions to this project. I thank many colleagues for supplying reprints or communications and helping with the completeness of this review.

References

Akbar W, Lord JC, Nechols JR, Howard RW (2004) Diatomaceous earth increases the efficacy of *Beauveria bassiana* against *Tribolium castaneum* larvae and increases conidia attachment. J Econ Entomol 97:273–280

Akbar W, Lord JC, Nechols JR, Loughin TM (2005) Efficacy of *Beauveria bassiana* for red flour beetle when applied with plant essential oils or in mineral oil and organosilicone carriers. J Econ Entomol 98:683–688

Alverson J (2003) Effects of mycotoxins, kojic acid and oxalic acid, on biological fitness of *Lygus hesperus* (Heteroptera: Miridae). J Invert Pathol 83:60–62

Alves SB, Rossi LS, Lopes RB, Tamai MA, Pereira RM (2002) *Beauveria bassiana* yeast phase on agar medium and its pathogenicity against *Diatraea saccharalis* (Lepidoptera: Crambidae) and *Tetranychus urticae* (Acari: Tetranychidae). J Invert Pathol 81:70–77

Amiri-Besheli B, Khambay B, Cameron S, Deadman ML, Butt TM (2000) Inter- and intra-specific variation in destruxin production by insect pathogenic *Metarhizium* spp., and its significance to pathogenesis. Mycol Res 104:447–452

Ansari MA, Vestergaard S, Tirry L, Moens M (2004) Selection of a highly virulent fungal isolate, *Metarhizium anisopliae* CLO 53, for controlling *Hoplia philanthus*. J Invert Pathol 85:89–96

Asaff A, Cerda-Garcia-Rojas C, Torre M de la (2005) Isolation of dipicolinic acid as an insecticidal toxin from *Paecilomyces fumosoroseus*. Appl Microbiol Biotechnol 68:542–547

Bagga S, Hu G, Screen SE, St Leger RJ (2004) Reconstructing the diversification of subtilisins in the pathogenic fungus *Metarhizium anisopliae*. Gene 324:159–169

Bailey AM, Kershaw MJ, Hunt BA, Paterson IC, Charnley AK, Reynolds SE, Clarkson JM (1996) Cloning and sequence analysis of an intron-containing domain from a peptide synthetase-encoding gene of the entomopathogenic fungus *Metarhizium anisopliae*. Gene 173:195–197

Bandani AR (2004) Effect of entomopathogenic fungus *Tolypocladium* species metabolite efrapeptin on *Galleria mellonella* agglutinin. Commun Agric Appl Biol Sci 69:165–169

Bandani AR, Amiri B, Butt TM, Gordon-Weeks R (2001) Effects of efrapeptin and destruxin, metabolites of entomogenous fungi, on the hydrolytic activity of a vacuolar type ATPase identified on the brush border membrane vesicles of *Galleria mellonella* midgut and on plant membrane bound hydrolytic enzymes. Biochim Biophys Acta 1510:367–377

Baratto CM, Silva MV da, Santi L, Passaglia L, Schrank IS, Vainstein MH, Schrank A (2003) Expression and characterization of the 42 kDa chitinase of the biocontrol fungus *Metarhizium anisopliae* in *Escherichia coli*. Can J Microbiol 49:723–726

Baratto CM, Dutra V, Boldo JT, Leiria LB, Vainstein MH, Schrank A (2006) Isolation, characterization, and transcriptional analysis of the chitinase chi2 gene (DQ011663) from the biocontrol fungus *Metarhizium anisopliae* var. *anisopliae*. Curr Microbiol 53:217–221

Bidochka MJ, Hajek AE (1998) A nonpermissive entomophthoralean fungal infection increases activation of insect prophenoloxidase. J Invert Pathol 72:231–238

Bidochka MJ, Khachatourians GG (1988a) Regulation of extracellular protease in the entomopathogenic fungus *Beauveria bassiana*. Exp Mycol 12:161–168

Bidochka MJ, Khachatourians GG (1988b) N-acetyl-D-glucosamine-mediated regulation of extracellular protease in the entomopathogenic fungus *Beauveria bassiana*. Appl Environ Microbiol 54:2699–2704

Bidochka MJ, Khachatourians GG (1991) The implication of metabolic acids produced by *Beauveria bassiana* in pathogenesis of the migratory grasshopper, *Melanoplus sanguinipes*. J Invert Pathol 58:106–117

Bidochka MJ, Khachatourians GG (1993) Oxalic acid hyperproduction in *Beauveria bassiana* mutants is related to a utilizable carbon source but not to virulence. J Invert Pathol 62:53–57

Bidochka MJ, Khachatourians GG (1994a) Protein hydrolysis in grasshoppers by entomopathogenic fungal proteases. J Invert Pathol 62:7–13

Bidochka MJ, Khachatourians GG (1994b) Effect of cuticular modification on their degradation by entomopathogenic fungal extracellular proteases. J Invert Pathol 64:26–32

Bidochka MJ, Melzer MJ (2000) Genetic polymorphisms in three subtilisin-like protease isoforms (Pr1A, Pr1B, and Pr1C) from *Metarhizium* strains. Can J Microbiol 46:1138–1144

Bidochka MJ, Small CL (2005) Phylogeography of *Metarhizium*, an insect pathogenic fungus. In: Vega FE, Blackwell M (eds) Insect–fungal associations: ecology and evolution. Oxford University Press, Oxford, pp 28–50

Bidochka MJ, Low NH, Khachatourians GG (1990) Storage carbohydrates of the entomopathogenic fungus *Beauveria bassiana*. Appl Environ Microbiol 56:3186–3190

Bidochka MJ, St Leger RJ, Joshi L, Roberts DW (1995) An inner cell wall protein (cwp1) from conidia of the entomopathogenic fungus *Beauveria bassiana*. Microbiology 141:1075–1080

Bidochka MJ, St Leger RJ, Stuart A, Gowanlock K (1999) Nuclear rDNA phylogeny in the fungal genus *Verticillium* and its relationship to insect and plant virulence, extracellular proteases and carbohydrases. Microbiology 145:955–963

Bidochka MJ, Menzies FV, Kamp AM (2002) Genetic groups of the insect-pathogenic fungus *Beauveria bassiana* are associated with habitat and thermal growth preferences. Arch Microbiol 178:531–537

Bogo MR, Vainstein MH, Aragao FJ, Rech E, Schrank A (1996) High frequency gene conversion among benomyl resistant transformants in the entomopathogenic fungus *Metarhizium anisopliae*. FEMS Microbiol Lett 142:123–127

Bogo MR, Rota CA, Pinto H Jr, Ocampos M, Correa CT, Vainstein MH, Schrank A (1998) A chitinase encoding gene (chit1 gene) from the entomopathogen *Metarhizium anisopliae*: isolation and characterization of genomic and full-length cDNA. Curr Microbiol 37:221–225

Boucias DG, Pendland JC (1991) The fungal cell wall and its involvement in the pathogenic process in insect hosts. In: Latge JP, Boucias DG (eds) Fungal cell wall and immune response. Springer, Heidelberg, pp 121–137

Boucias DG, Pendland JC, Latge JP (1988) Nonspecific factors involved in attachment of entomopathogenic Deuteromycetes to host insect cuticle. Appl Environ Microbiol 54:1795–1805

Bradfisch GA, Harmer SL (1990) Omega-conotoxin GVIA and nifedipine inhibit the depolarizing action of the fungal metabolite destruxin B on muscle from the tobacco budworm (*Heliothis virescens*). Toxicon 28:1249–1254

Braga GU, Destefano RH, Messias CL (1999) Oxygen consumption by *Metarhizium anisopliae* during germination and growth on different carbon sources. J Invert Pathol 74:112–119

Braga GU, Flint SD, Messias CL, Anderson AJ, Roberts DW (2001a) Effects of UVB irradiance on conidia and germinants of the entomopathogenic Hyphomycete

Metarhizium anisopliae: a study of reciprocity and recovery. Photochem Photobiol 73:140–146

Braga GU, Flint SD, Miller CD, Anderson AJ, Roberts DW (2001b) Both solar UVA and UVB radiation impair conidial culturability and delay germination in the entomopathogenic fungus *Metarhizium anisopliae*. Photochem Photobiol 74:734–739

Braga GU, Rangel DE, Flint SD, Anderson AJ, Roberts DW (2006) Conidial pigmentation is important to tolerance against solar-simulated radiation in the entomopathogenic fungus *Metarhizium anisopliae*. Photochem Photobiol 82:418–422

Brownbridge M, Costa S, Jaronski ST (2001) Effects of in vitro passage of *Beauveria bassiana* on virulence to *Bemisia argentifolii*. J Invert Pathol 77:280–283

Calo L, Fornelli F, Nenna S, Tursi A, Caiaffa MF, Macchia L (2003) Beauvericin cytotoxicity to the invertebrate cell line SF-9. J Appl Genet 44:515–520

Campos RA, Arruda W, Boldo JT, Silva MV da, Barros NM de, Azevedo JL de, Schrank A, Vainstein MH (2005) *Boophilus microplus* infection by *Beauveria amorpha* and *Beauveria bassiana*: SEM analysis and regulation of subtilisin-like proteases and chitinases. Curr Microbiol 50:257–261

Caragh Moore M, Cox RJ, Duffin GR, O'Hagan D (1998) Synthesis and evaluation of a putative acyl tetramic acid intermediate in tenellin biosynthesis in *Beauveria bassiana*. A new role for tyrosine. Tetrahedron 54:9195–9206

Carballo M (1998) *Cosmopolites sordidus* mortality with certain *Beauveria bassiana* formulations. Manejo Integrad Plagas 48:45–48

Carruthers RI, Soper RS (1987) Fungal disease. In: Fuxa JR, Tanada Y (eds) Epizootiology of insect diseases. Wiley-Interscience, New York, pp 357–416

Chandler D (2005) Understanding the evolution and function of entomopathogenic fungi. http://www2.warwick.ac.uk/fac/sci/hri2/about/staff/dchandler. Accessed 26 September 2005

Chandler D, Davidson G, Pell JK, Ball BV, Shaw KE, Sunderland KD (2000) Fungal biocontrol of Acari. Biocontrol Sci Technol 10:357–384

Charnley AK (1997) Entomopathogenic fungi and their role in pest control. In: Wicklow DT, Soderstrom BE (eds) Environmental and microbial relationships. (Mycota IV) Springer, Heidelberg, pp 185–201

Charnley AK, St Leger RJ (1991) The role of cuticle degrading enzymes in fungal pathogenesis in insects. In: Cole GT, Hoch HC (eds) The fungal spore and disease initiation in plants and animals. Plenum, New York, pp 267–286

Chelico L, Khachatourians GG (2007) Isolation and characterization of nucleotide excision repair deficient mutants of the entomopathogenic fungus, *Beauveria bassiana*. J Invert Pathol (in press)

Chelico L, Haughian JL, Woytowich AL, Khachatourians GG (2005) Quantification of ultraviolet-C irradiation induced cyclobutane pyrimidine dimers and their removal in *Beauveria bassiana* conidiospores. Mycologia 97:621–627

Chelico L, Haughian JL, Khachatourians GG (2006) Nucleotide excision repair and photoreactivation in the entomopathogenic fungi *Beauveria bassiana, B. brongniartii, B. nivea, Metarhizium anisopliae, Paecilomyces farinosus*, and *Verticillium lecanii*. J Appl Microbiol 100:964–972

Cho EM, Liu L, Farmerie W, Keyhani NO (2006a) EST analysis of cDNA libraries from the entomopathogenic fungus *Beauveria (Cordyceps) bassiana*. I. Evidence for stage-specific gene expression in aerial conidia, in vitro blastospores and submerged conidia. Microbiology 152:2843–2854

Cho EM, Boucias D, Keyhani NO (2006b) EST analysis of cDNA libraries from the entomopathogenic fungus *Beauveria (Cordyceps) bassiana*. II. Fungal cells sporulating on chitin and producing oosporein. Microbiology 152:2855–2864

Coates BS, Hellmich RL, Lewis LC (2002) Nuclear small subunit rRNA group I intron variation among *Beauveria* spp provide tools for strain identification and evidence of horizontal transfer. Curr Genet 41:414–424

Cristine CB, Staats C, Schrank A, Vainstein MH (2004) Distribution of chitinases in the entomopathogen *Metarhizium anisopliae* and the effect of N-acetylglucosamine in protein secretion. Curr Microbiol 48:102–107

Dalla Santa HS, Sousa NJ, Brand D, Dalla Santa OR, Pandey A, Sobotka M, Paca J, Soccol CR (2004) Conidia production of *Beauveria* sp. by solid-state fermentation for biocontrol of *Ilex paraguariensis* caterpillars. Folia Microbiol 49:418–22

Dean P, Richards EH, Edwards JP, Reynolds SE, Charnley AK (2004) Microbial infection causes the appearance of hemocytes with extreme spreading ability in monolayers of the tobacco hornworm *Manduca sexta*. Dev Comp Immunol 28:689–700

Deshpande MV (1999) Mycopesticide production by fermentation: potential and challenges. Crit Rev Microbiol 25:229–243

Devi KU, Sridevi V, Mohan CM, Padmavathi J (2005) Effect of high temperature and water stress on in vitro germination and growth in isolates of the entomopathogenic fungus *Beauveria bassiana* (Bals.) Vuillemin. J Invert Pathol 88:181–189

Dunlap CA, Biresaw G, Jackson MA (2005) Hydrophobic and electrostatic cell surface properties of blastospores of the entomopathogenic fungus *Paecilomyces fumosoroseus*. Colloids Surf B Biointerfaces 46:261–266

Entz SC, Johnson DL, Kawchuk LM (2005) Development of a PCR-based diagnostic assay for the specific detection of the entomopathogenic fungus *Metarhizium anisopliae* var. *acridum*. Mycol Res 109:1302–1312

Fang W, Zhang Y, Yang X, Wang Z, Pei Y (2002) Cloning and characterization of cuticle degrading enzyme CDEP-1 from *Beauveria bassiana* (in Chinese). Yi Chuan Xue Bao 29:278–282

Fang W, Leng B, Xiao Y, Jin K, Ma J, Fan Y, Feng J, Yang X, Zhang Y, Pei Y (2005) Cloning of *Beauveria bassiana* chitinase gene Bbchit1 and its application to improve fungal strain virulence. Appl Environ Microbiol 71:363–370

Fargues J, Bon MC (2004) Influence of temperature preferences of two *Paecilomyces fumosoroseus* lineages on their co-infection pattern. J Invert Pathol 87:94–104

Fargues J, Luz C (2000) Effects of fluctuating moisture and temperature regimes on the infection potential of *Beauveria bassiana* for *Rhodnius prolixus*. J Invert Pathol 75:202–211

Fargues J, Remaudiere G (1977) Considerations on the specificity of entomopathogenic fungi. Mycopathologia 62:31–37

Fargues J, Rougier M, Goujet R, Smits N, Coustere C, Itier B (1997) Inactivation of conidia of *Paecilomyces fumosoroseus* by near-ultraviolet (UVB and UVA) and visible radiation. J Invert Pathol 69:70–78

Fargues J, Smiths N, Viial C, Vey A, Vega F, Mercadier G, Quimby P (2002) Effect of liquid culture media on morphology, growth, propagule production, and pathogenic activity of the Hyphomycete, *Metarhizium flavoviride*. Mycopathologia 154:127–138

Feng M-G, Poprawski TJ, Khachatourians GG (1994) Production, formulation and application of the entomopathogenic fungus *Beauveria bassiana* for insect control. Biocontrol Sci Technol 4:3–34

Ferron P, Fargues J, Riba G (1991) Fungi as microbial insecticides against pests. In: Arora DK, Mukerji KG, Drouhet E (eds) Handbook of applied mycology: humans, animals and insects. Dekker, New York, pp 665–706

Fornelli F, Minervini F, Logrieco A (2004) Cytotoxicity of fungal metabolites to lepidopteran (*Spodoptera frugiperda*) cell line (SF-9). J Invert Pathol 85:74–79

Freimoser FM, Screen S, Bagga S, Hu G, St Leger RJ (2003a) Expressed sequence tag (EST) analysis of two subspecies of *Metarhizium anisopliae* reveals a plethora of secreted proteins with potential activity in insect hosts. Microbiology 149:239–247

Freimoser FM, Screen S, Hu G, St Leger R (2003b) EST analysis of genes expressed by the zygomycete pathogen *Conidiobolus coronatus* during growth on insect cuticle. Microbiology 149:1893–1900

Freimoser FM, Hu G, St Leger RJ (2005) Variation in gene expression patterns as the insect pathogen *Metarhizium anisopliae* adapts to different host cuticles or nutrient deprivation in vitro. Microbiology 151:361–371

Fuxa JR (1991) Release and transport of entomopathogenic microorganisms. In: Morris LA, Strauss HS (eds) Risk assessment in genetic engineering. McGraw Hill, New York, pp 83–113

Ganassi S, Moretti A, Bonvicini-Pagliai AM, Logrieco A, Sabatini MA (2002) Effects of beauvericin on *Schizaphis graminum* (Aphididae). J Invert Pathol 80:90–96

Ghikas DV, Kouvelis VN, Typas MA (2006) The complete mitochondrial genome of the entomopathogenic fungus *Metarhizium anisopliae* var. *anisopliae*: gene order and *trn* gene clusters reveal a common evolutionary course for all Sordariomycetes, while intergenic regions show variation. Arch Microbiol 85:393–401

Gillespie JP, Khachatourians GG (1992) Characterization of the *Melanoplus sanguinipes* hemolymph after infection with *Beauveria bassiana* or wounding. Comp Biochem Physiol 103B:455–463

Gillespie JP, Kanost MR, Trenczek T (1997) Biological mediators of insect immunity. Annu Rev Entomol 42:611–643

Griesch J, Vilcinskas A (1998) Proteases released by entomopathogenic fungi impair phagocytic activity, attachment and spreading of plasmatocytes isolated from hemolymph of the greater wax moth *Galleria mellonella*. Biocontrol Sci Technol 8:517–531

Hajek AE, St Leger RJ (1994) Interactions between fungal pathogens and insect hosts. Annu Rev Entomol 39:293–322

Hallsworth JE, Magan N (1999) Water and temperature relations of growth of the entomogenous fungi *Beauveria bassiana*, *Metarhizium anisopliae*, and *Paecilomyces farinosus*. J Invert Pathol 74:261–266

Hayden T, Bidochka MJ, Khachatourians GG (1992) Virulence of several entomopathogenic fungi and host-passage strains of *Paecilomyces farinosus* toward the Blackberry-Cereal Aphid *Sitobion fragariae*. J Econ Entomol 85:58–64

Hegedus DD, Khachatourians GG (1993a) Construction of cloned DNA probes for the specific detection of the entomopathogenic fungus *Beauveria bassiana* in grasshoppers. J Invert Pathol 62:233–240

Hegedus DD, Khachatourians GG (1993b) Identification of molecular variants in mitochondrial DNAs of members of the genera *Beauveria*, *Verticillium*, *Paecilomyces*, *Tolypocladium* and *Metarhizium*. Appl Environ Microbiol 59:4283–4288

Hegedus DD, Khachatourians GG (1995) The impact of biotechnology on hyphomycetous fungal insect biocontrol agents. Biotechnol Adv 13:455–490

Hegedus DD, Khachatourians GG (1996a) Identification and differentiation of the entomopathogenic fungus *Beauveria bassiana* using polymerase chain reaction and single-strand conformation polymorphism analysis. J Invert Pathol 67:289–299

Hegedus DD, Khachatourians GG (1996b) Detection of the entomopathogenic fungus *Beauveria bassiana* within infected migratory grasshoppers (*Melanoplus sanguinipes*) using polymerase chain reaction and DNA probe. J Invert Pathol 67:21–27

Hegedus DD, Khachatourians GG (1996c) Identification and differentiation of the entomopathogenic fungus *Beauveria bassiana* using polymerase chain reaction and single strand conformation polymorphism analysis. J Invert Pathol 67:289–299

Hegedus DD, Khachatourians GG (1996d) The effect of temperature on the pathogenicity of heat sensitive mutants of the entomopathogenic fungus *Beauveria bassiana* toward the migratory grasshopper, *Melanoplus sanguinipes*. J Invert Pathol 68:160–165

Hegedus DD, Bidochka MJ, Khachatourians GG (1990) Chitin and chitin monomers for the production of *Beauveria bassiana* submerged conidia. Appl Microbiol Biotechnol 30:637–642

Hegedus DD, MacPherson JM, Pfeifer TA, Khachatourians GG (1991) DNA sequence of tRNA^{Val}-tRNA^{Ile} tandem genes from the mitochondria of the entomopathogenic fungus *Beauveria bassiana* GK2016. Gene 109:149–154

Hegedus DD, Bidochka MJ, Miranpuri GS, Khachatourians GG (1992) A comparison of the virulence, stability and cell-wall surface characteristics of three spore types produced by the entomopathogenic fungus *Beauveria bassiana*. Appl Microbiol Biotechnol 36:785–789

Hegedus DD, Pfeifer TA, Mulyk DS, Khachatourians GG (1998) Characterization and structure of the mitochondrial small rRNA gene of the entomopathogenic fungus *Beauveria bassiana*. Genome 41:471–476

Holder DJ, Keyhani NO (2005) Adhesion of the entomopathogenic fungus *Beauveria* (*Cordyceps*) *bassiana* to substrata. Appl Environ Microbiol 71:5260–5266

Hsiao WF, Bidochka MJ, Khachatourians GG (1992) Effect of temperature, relative humidity on the virulence of the entomopathogenic fungus *Verticillium lecanii* toward the oat-bird berry aphid *Rhopalosiphum padi*. (Homoptera: Aphididae). J Appl Entomol 114:484–490

Hsiao YM, Ko JL (2001) Determination of destruxins, cyclic peptide toxins, produced by different strains of *Metarhizium anisopliae* and their mutants induced by ethyl methane sulfonate and ultraviolet using HPLC method. Toxicon 39:837–841

Hu G, St Leger RJ (2002) Field studies using a recombinant mycoinsecticide (*Metarhizium anisopliae*) reveal that it is rhizosphere competent. Appl Environ Microbiol 68:6383–6387

Hung S-Y, Boucias DG, Vey AJ (1993) Effect of *Beauveria bassiana* and *Candida albicans* on the cellular defense response of *Spodoptera exigua*. J Invert Pathol 61:179–187

Iijima R, Kurata S, Natori S (1993) Purification, characterization and cDNA cloning of an antifungal protein from the hemolymph of *Sarcophaga peregrina* (flesh fly) larvae. J Biol Chem 268:12055–12061

James RR (2001) Effects of exogenous nutrients on conidial germination and virulence against the silverleaf whitefly for two hyphomycetes. J Invert Pathol 77:99–107

James RR, Buckner JS, Freeman TP (2003) Cuticular lipids and silverleaf whitefly stage affect conidial germination of *Beauveria bassiana* and *Paecilomyces fumosoroseus*. J Invert Pathol 84:67–74

Jeffs LB, Khachatourians GG (1997a) Estimation of spore hydrophobicity for members of the genera *Beauveria, Metarhizium, and Tolypocladium* by salt-mediated aggregation and sedimentation. Can J Microbiol 43:23–28

Jeffs LB, Khachatourians GG (1997b) Toxic properties of *Beauveria* pigments on erythrocyte membranes. Toxicon 35:1351–1356

Jeffs LB, Xavier IJ, Matai RE, Khachatourians GG (1999) Relationships between fungal spore morphologies and surface properties for entomopathogenic members of the genera *Beauveria, Metarhizium, Paecilomyces, Tolypocladium* and *Verticillium*. Can J Microbiol 45:936–948

Jegorov A, Sedmera P, Matha V (1993) Biosynthesis of destruxins. Phytochemistry 33:1403–1405

Joshi L, St Leger RJ (1999) Cloning, expression, and substrate specificity of MeCPA, a zinc carboxypeptidase that is secreted into infected tissues by the fungal entomopathogen *Metarhizium anisopliae*. J Biol Chem 274:9803–9811

Joshi L, St Leger RJ, Bidochka MJ (1995) Cloning of a cuticle-degrading protease from the entomopathogenic fungus, *Beauveria bassiana*. FEMS Microbiol Lett 125:211–217

Joshi L, St Leger RJ, Roberts DW (1997) Isolation of a cDNA encoding a novel subtilisin-like protease (Pr1B) from the entomopathogenic fungus, *Metarhizium anisopliae* using differential display-RT-PCR. Gene 197:1–8

Kamp AM, Bidochka MJ (2002a) Conidium production by insect pathogenic fungi on commercially available agars. Lett Appl Microbiol 35:74–77

Kamp AM, Bidochka MJ (2002b) Protein analysis in a pleomorphically deteriorated strain of the insect-pathogenic fungus *Metarhizium anisopliae*. Can J Microbiol 48:787–792

Kang SC, Park S, Lee DG (1998) Isolation and characterization of a chitinase cDNA from the entomopathogenic fungus, *Metarhizium anisopliae*. FEMS Microbiol Lett 165:267–271

Kang SC, Park S, Lee DG (1999) Purification and characterization of a novel chitinase from the entomopathogenic fungus, *Metarhizium anisopliae*. J Invert Pathol 73:276–281

Kang SW, Lee SH, Yoon CS, Kim SW (2005) Conidia production by *Beauveria bassiana* during solid-state fermentation in a packed-bed bioreactor. Biotechnol Lett 27:135–139

Khachatourians GG (1986) Production and use of biological pest control agents. Trends Biotechnol 4:120–124

Khachatourians GG (1991) Physiology and genetics of entomopathogenic fungi. In: Arora DK, Ajello L, Mukerji KG (eds) Handbook of applied mycology, vol 2: humans, animals, and insects. Dekker, New York, pp 613–661

Khachatourians GG (1996) Biochemistry and molecular biology of entomopathogenic fungi. In: Howard DH, Miller JD (eds) Human and animal relationships. (Mycota VI) Springer, Heidelberg, pp 331–363

Khachatourians GG (2008) Insecticides, microbial. In: Schaecter M, Summers WC et al (eds) Encyclopedia of microbiology, vol 2, 3rd edn. Elsevier, New York

Khachatourians GG, Hegedus DD (2004) Mitochondrial genome. In: Muller GM, Bills GF, Foster MS (eds) Biodiversity of fungi: inventory and monitoring methods. Elsevier, San Diego, pp 94–102

Khachatourians GG, Uribe D (2004) Genomics of entomopathogenic fungi. In: Arora DK, Khachatourians GG (eds) Fungal genomics. (Applied mycology and biotechnology, vol 4) Elsevier, London, pp 353–377

Khachatourians GG, Valencia E (1999) Integrated pest management and entomopathogenic fungal biotechnology in the Latin Americas II. Key research and development prerequisites. Rev Acad Colomb Cienc Exact Fis Nat 23:489–496

Khachatourians GG, Valencia E, Miranpuri GS (2002) *Beauveria bassiana* and other entomopathogenic fungi in the management of insect pests. In: Koul O, Dhaliwal GS (eds) Microbial biopesticides, vol 2. Taylor & Francis, London, pp 239–275

Kikuchi H, Miyagawa Y, Sahashi Y, Inatomi S, Haganuma A, Nakahata N, Oshima Y (2004) Novel trichothecanes, paecilomycine A, B, and C, isolated from entomopathogenic fungus, *Paecilomyces tenuipes*. Tetrahedron Lett 45:6225–6228

Kim H-K, Hoe H-S, Suh DS, Kang SC, Hwang C, Kwon S-T (1999) Gene structure and expression of the gene from *Beauveria bassiana* encoding bassiasin I, an insect cuticle-degrading serine protease. Biotechnol Lett 21:777–783

Kirkland BH, Westwood GS, Keyhani NO (2004) Pathogenicity of entomopathogenic fungi *Beauveria bassiana* and *Metarhizium anisopliae* to Ixodidae tick species *Dermacentor variabilis*, *Rhipicephalus sanguineus*, and *Ixodes scapularis*. J Med Entomol 41:705–711

Kirkland BH, Eisa A, Keyhani NO (2005) Oxalic acid as a fungal acaricidal virulence factor. J Med Entomol 42:346–351

Kouvelis VN, Zare R, Bridge PD, Typas MA (1999) Differentiation of mitochondrial subgroups in the *Verticillium lecanii* species complex. Lett Appl Microbiol 28:263–268

Kouvelis VN, Ghikas DV, Typas MA (2004) The analysis of the complete mitochondrial genome of *Lecanicillium muscarium* (synonym *Verticillium lecanii*) suggests a minimum common gene organization in mtDNAs of Sordariomycetes: phylogenetic implications. Fungal Genet Biol 41:930–940

Krasnoff SB, Gupta S (1991) Identification and biosynthesis of efrapeptins in fungus *Tolypocladium geodes* Gams (Deuteromycotina:Hyphomycetes). J Chem Ecol 17:1953–1960

Krasnoff SB, Reategui RF, Wagenaar MM, Gloer JB, Gibson DM (2005) Cicadapeptins I and II: new Aib-containing peptides from the entomopathogenic fungus *Cordyceps heteropoda*. J Nat Prod 68:50–55

Lang G, Blunt JW, Cummings NJ, Cole AL, Munro MH (2005) Hirsutide, a cyclic tetrapeptide from a spider-derived entomopathogenic fungus, *Hirsutella* sp. J Nat Prod 68:1303–1305

Lecuona R, Riba G, Cassier P, Clement JL (1991) Alterations of insect epicuticular hydrocarbons during infection with *Beauveria bassiana* or *Beauveria brongniartii*. J Invert Pathol 58:10–18

Liu JC, Boucias DG, Pendland JC, Liu WZ, Maruniak J (1996) The mode of action of hirsutellin A on eukaryotic cells. J Invert Pathol 67:224–228

Lord JC (2005) From Metchnikoff to Monsanto and beyond: the path of microbial control. J Invert Pathol 89:19–29

Lord JC, Howard RW (2004) A proposed role for the cuticular fatty amides of *Liposcelis bostrychophila* (Psocoptera: Liposcelidae) in preventing adhesion of entomopathogenic fungi with dry-conidia. Mycopathologia 158:211–217

Lord JC, Anderson S, Stanley DW (2002) Eicosanoids mediate *Manduca sexta* cellular response to the fungal pathogen *Beauveria bassiana*: a role for the lipoxygenase pathway. Arch Insect Biochem Physiol 51:46–54

Luz C, Fargues J (1997) Temperature and moisture requirements for conidial germination of an isolate of *Beauveria bassiana*, pathogenic to *Rhodnius prolixus*. Mycopathologia 138:117–125

Luz C, Fargues J (1998) Factors affecting conidial production of *Beauveria bassiana* from fungus-killed cadavers of *Rhodnius prolixus*. J Invert Pathol 72:97–103

Masuka A, Manjonjo V (1996) Laboratory screening of *Metarhizium flavoviride* and *Beauveria bassiana* for the control of *Mecostibus pinivorus*, a *Pinus patula* defoliator in Zimbabwe. J Appl Sci S Afr 2:91–96

Maurer P, Rejasse A, Capy P, Langin T, Riba G (1997) Isolation of the transposable element hupfer from the entomopathogenic fungus *Beauveria bassiana* by insertion mutagenesis of the nitrate reductase structural gene. Mol Gen Genet 256:195–202

Mavridou A, Typas MA (1998) Intraspecific polymorphism in *Metarhizium anisopliae* var *anisopliae* revealed by analysis of rRNA gene complex and mtDNA RFLPs. Mycol Res 102:1233–1241

Mavridou A, Cannone J, Typas MA (2000) Identification of group-I introns at three different positions within the 28S rDNA gene of the entomopathogenic fungus *Metarhizium anisopliae* var. *anisopliae*. Fungal Genet Biol 31:79–90

Michelitsch A, Ruckert U, Rittmannsberger A, Seger C, Strasser H, Likussar W (2004) Accurate determination of oosporein in fungal culture broth by differential pulse polarography. J Agric Food Chem 52:1423–1426

Miranpuri GS, Khachatourians GG (1994) Bacterial and fungal control of hematophagous vectors: a review. J Insect Sci 6:1–14

Muroi M, Shiragami N, Takatsuki A (1994) Destruxin B, a specific and readily reversible inhibitor of vacuolar type H^+-translocating ATPase. Biochem Biophys Res Commun 205:1358–1365

Nahar P, Ghormade V, Deshpande MV (2004) The extracellular constitutive production of chitin deacetylase in *Metarhizium anisopliae*: possible edge to entomopathogenic fungi in the biological control of insect pests. J Invert Pathol 85:80–88

Nam JS, Lee DH, Lee KH, Park HM, Bae KS (1998) Cloning and phylogenic analysis of chitin synthase genes from the insect pathogenic fungus, *Metarhizium anisopliae* var. *anisopliae*. FEMS Microbiol Lett 159:77–84

Neuvéglise C, Brygoo Y, Riba G (1997) 28s rDNA group-I introns: a powerful tool for identifying strains of *Beauveria brongniartii*. Mol Ecol 6:373–381

Ouedraogo RM, Goettel MS, Brodeur J (2004) Behavioral thermoregulation in the migratory locust: a therapy to overcome fungal infection. Oecologia 138:312–319

Panaccione DG, Scott-Craig JS, Pocard J-A, Walton JD (1992) Acyclic peptide synthetase gene required for pathogenicity of the fungus *Cochiliobolus carbonum* on maize. Proc Natl Acad Sci USA 89:6590–6594

Pedras MS, Montaut S (2003) Probing crucial metabolic pathways in fungal pathogens of crucifers: biotransformation of indole-3-acetaldoxime, 4-hydroxyphenylacetaldoxime, and their metabolites. Bioorg Med Chem 11:3115–3120

Pedras MS, Irina ZL, Ward DE (2002) The destruxins: synthesis, biosynthesis, biotransformation, and biological activity. Phytochemistry 59:579–596

Pendlou JC, Boucias DG (1992) Ultrastructural localization of carbohydrate in cell walls of the entomogenous hyphomycete *Nomuraea rileyi*. Can J Microbiol 38:377–386

Pendland JC, Boucias DG (1998) Characterization of monoclonal antibodies against cell wall epitopes of the insect pathogenic fungus, *Nomuraea rileyi*: differential binding to fungal surfaces and cross-reactivity with host hemocytes and basement membrane components. Eur J Cell Biol 75:118–127

Pendland JC, Boucias DG (2000) Comparative analysis of the binding of antibodies prepared against the insect *Spodoptera exigua* and against the mycopathogen *Nomuraea rileyi*. J Invert Pathol 75:107–116

Pendland JC, Hung S-Y, Boucias DG (1993) Evasion of host defense by in vivo produced protoplast-like cells of the insect mycopathogen *Beauveria bassiana*. J Bacteriol 175:5962–5969

Qazi S, Khachatourians GG (2006) Insect pests of Pakistan and their management practices: prospects for the use of entomopathogenic fungi. Biopest Int 1:13–24

Qazi SS, Khachatourians GG (2007) Hydrated conidia of *Metarhizium anisopliae* release a family of metalloproteases. J Invert Pathol 95:48–59

Qazi SS, Khachatourians GG (2008) Addition of exogenous carbon and nitrogen sources to aphid exuviae modulates synthesis of proteases and chitinase by germinating conidia of *Beauveria bassiana*. Arch Microbiol. doi: 10.1007/w00203-008-0355-9

Quesada-Moraga E, Vey A (2004) Bassiacridin, a protein toxic for locusts secreted by the entomopathogenic fungus *Beauveria bassiana*. Mycol Res 108:441–452

Rangel DE, Braga GU, Anderson AJ, Roberts DW (2005) Influence of growth environment on tolerance to UV-B radiation, germination speed, and morphology *of Metarhizium anisopliae* var. *acridum* conidia. J Invert Pathol 90:55–58

Rangel DE, Anderson AJ, Roberts DW (2006a) Growth of *Metarhizium anisopliae* on non-preferred carbon sources yields conidia with increased UV-B tolerance. J Invertebr Pathol 93:127–134

Rangel DE, Butler MJ, Torabinejad J, Anderson AJ, Braga GU, Day AW, Roberts DW (2006b) Mutants and isolates of *Metarhizium anisopliae* are diverse in their relationships between conidial pigmentation and stress tolerance. J Invertebr Pathol 93:170–182

Rehner SA (2005) Phylogenetics of the insect pathogenic genus *Beauveria*. In: Vega FE, Blackwell M (eds) Insect–fungal associations: ecology and evolution. Oxford University Press, Oxford, pp 3–27

Roberts DW, St Leger RJ (2004) *Metarhizium* spp., cosmopolitan insect-pathogenic fungi: mycological aspects. Adv Appl Microbiol 54:1–70

Rodriguez R, Cullen D, Kurtzman CP, Khachatourians GG, Hegedus DD (2004) Molecular methods for discriminating taxa, monitoring species and assessing fungal diversity. In: Muller GM, Bills GF, Foster MS (eds) Biodiversity of fungi: inventory and monitoring methods. Elsevier, San Diego, pp 77–94

Roy HE, Steinkraus DC, Eilenberg J, Hajek AE, Pell JK (2006) Bizarre interactions and endgames: entomopathogenic fungi and their arthropod hosts. Annu Rev Entomol 51:331–357

Ryan MJ, Bridge PD, Smith D, Jeffries P (2002) Phenotypic degeneration occurs during sector formation in *Metarhizium anisopliae*. J Appl Microbiol 93:163–168

Samuels RI, Paterson IC (1995) Cuticle degrading proteases from insect moulting fluid and culture filtrates of entomopathogenic fungi. Comp Biochem Physiol B Biochem Mol Biol 110:661–669

Sanchez-Murillo RI, Torre-Martinez M de la, Aguirre-Linares J, Herrera-Estrella A (2004) Light-regulated asexual reproduction in *Paecilomyces fumosoroseus*. Microbiology 150:311–319

Schmitz V, Dedryver CA, Pierre JS (1993) Influence of an *Erynia neoaphidis* infection on the relative rate of increase of the cereal aphid *Sitobion avenae*. J Invert Pathol 61:62–68

Scholte EJ, Njiru BN, Smallegange RC, Takken W, Knols BG (2003) Infection of malaria (*Anopheles gambiae* s.s.) and filariasis (*Culex quinquefasciatus*) vectors with the entomopathogenic fungus *Metarhizium anisopliae*. Malaria J 2:29

Scholte EJ, Ng'habi K, Kihonda J, Takken W, Paaijmans K, Abdulla S, Killeen GF, Knols BG (2005) An entomopathogenic fungus for control of adult African malaria mosquitoes. Science 308:1641–1642

Screen SE, St Leger RJ (2000) Cloning, expression, and substrate specificity of a fungal chymotrypsin. Evidence for lateral gene transfer from an actinomycete bacterium. J Biol Chem 275:6689–6694

Screen S, Bailey A, Charnley K, Cooper R, Clarkson J (1997) Carbon regulation of the cuticle-degrading enzyme PR1 from *Metarhizium anisopliae* may involve a trans-acting DNA-binding protein CRR1, a functional equivalent of the *Aspergillus nidulans* CREA protein. Curr Genet 31:511–518

Screen S, Bailey A, Charnley K, Cooper R, Clarkson J (1998) Isolation of a nitrogen response regulator gene (nrr1) from *Metarhizium anisopliae*. Gene 221:17–24

Screen SE, Hu G, St Leger RJ (2001) Transformants of *Metarhizium anisopliae* sf. *anisopliae* overexpressing chitinase from *Metarhizium anisopliae* sf. *acridum* show early induction of native chitinase but are not altered in pathogenicity to *Manduca sexta*. J Invert Pathol 78:260–266

Scully LR, Bidochka MJ (2005) Serial passage of the opportunistic pathogen *Aspergillus flavus* through an insect host yields decreased saprobic capacity. Can J Microbiol 51:185–189

Scully LR, Bidochka MJ (2006a) A cysteine/methionine auxotroph of the opportunistic fungus *Aspergillus flavus* is associated with host-range restriction: a model for emerging diseases. Microbiology 152:223–232

Scully LR, Bidochka MJ (2006b) Developing insect models for the study of current and emerging human pathogens. FEMS Microbiol Lett 263:1–9

Scully LR, Bidochka MJ (2006c) The host acts as a genetic bottleneck during serial infections: an insect-fungal model system. Curr Genet 50:335–345

Segers R, Butt TM, Carder JH,. Keen JN, Kerry BR, Peberdy JF (1999), The subtilisins of fungal pathogens of insects, nematodes and plants: distribution and variation. Mycol Res 103:395–402

Shah FA, Butt TM (2005) Influence of nutrition on the production and physiology of sectors produced by the insect pathogenic fungus *Metarhizium anisopliae*. FEMS Microbiol Lett 250:201–207

Shah PA, Pell JK (2003) Entomopathogenic fungi as biological control agents. Appl Microbiol Biotechnol 61:413–423

Shapiro-Ilan DI, Fuxa JR, Lacey LA, Onstad DW, Kaya HK, (2004) Definitions of pathogenicity and virulence in invertebrate pathology. J Invert Pathol 88:1–7

Sheng J, An K, Deng C, Li W, Bao X, Qiu D (2006) Cloning a cuticle-degrading serine protease gene with biologic control function from *Beauveria brongniartii* and its expression in *Escherichia coli*. Curr Microbiol 53:124–128

Silva MV da, Santi L, Staats CC, Costa AM da, Colodel EM, Driemeier D, Vainstein MH, Schrank A (2005) Cuticle-induced endo/exoacting chitinase CHIT30 from *Metarhizium anisopliae* is encoded by an ortholog of the chi3 gene. Res Microbiol 156:382–392

Small C-LN, Bidochka MJ (2005) Up-regulation of Pr1, a subtilisin-like protease, during conidiation in the insect pathogen *Metarhizium anisopliae*. Mycol Res 109:307–313

Smits N, Briere JF, Fargues J (2003) Comparison of non-linear temperature-dependent development rate models applied to in vitro growth of entomopathogenic fungi. Mycol Res 107:1476–1484

Sosa-Gomez DR, Boucias DG, Nation JL (1997) Attachment of *Metarhizium anisopliae* to the southern green stink bug *Nezara viridula* cuticle and fungistatic effect of cuticular lipids and aldehydes. J Invert Pathol 69:31–39

St Leger R, Joshi L, Bidochka MJ, Roberts DW (1996) Construction of an improved mycoinsecticide overexpressing a toxic protease. Proc Natl Acad Sci USA 93:6349–6354

St Leger RJ, Joshi L, Roberts DW (1997) Adaptation of proteases and carbohydrates of saprophytic, phytopathogenic and entomopathogenic fungi to the requirements of their ecological niches. Microbiology 143:1983–1992

St Leger RJ, Screen SE, Shams-Pirzadeh B (2000) Lack of host specialization in *Aspergillus flavus*. Appl Environ Microbiol 66:320–324

Staats CC, Silva MS, Pinto PM, Vainstein MH, Schrank A (2004) The *Metarhizium anisopliae* trp1 gene: cloning and regulatory analysis. Curr Microbiol 49:66–70

Strasser H, Abendstein D, Stuppner H, Butt TM (2000a) Monitoring the distribution of secondary metabolites produced by the entomogenous fungus *Beauveria brongniartii* with particular reference to oosporein. Mycol Res 104:1227–1233

Strasser H, Vey A, Butt TM (2000b) Are there any risks in using entomopathogenic fungi for pest control, with particular reference to the bioactive metabolites of *Metarhizium*, *Tolypocladium* and *Beauveria* species? Biocontrol Sci Technol 10:717–735

Sugimoto M, Koike M, Hiyama N, Nagao H (2003) Genetic, morphological, and virulence characterization of the entomopathogenic fungus *Verticillium lecanii*. J Invert Pathol 82:176–187

Supothina S, Isaka M, Kirtikara K, Tanticharoen M, Thebtaranonth Y (2004) Enniatin production by the entomopathogenic fungus *Verticillium hemipterigenum* BCC 1449. J Antibiot 57:732–738

Tang CY, Chen YW, Jow GM, Chou CJ, Jeng CJ (2005) Beauvericin activates Ca^{2+}-activated Cl^- currents and induces cell deaths in *Xenopus* oocytes via influx of extracellular Ca^{2+}. Chem Res Toxicol 18:825–833

Tarocco F, Lecuona RE, Couto AS, Arcas JA (2005) Optimization of erythritol and glycerol accumulation in conidia of *Beauveria bassiana* by solid-state fermentation, using response surface methodology. Appl Microbiol Biotechnol 68:481–488

Tartar A, Boucias DG (2004) A pilot-scale expressed sequence tag analysis of *Beauveria bassiana* gene expression reveals a tripeptidyl peptidase that is differentially expressed in vivo. Mycopathologia 158:201–209

Tartar A, Shapiro AM, Scharf DW, Boucias DG (2005) Differential expression of chitin synthase (CHS) and glucan synthase (FKS) genes correlates with the formation of a modified, thinner cell wall in in vivo-produced *Beauveria bassiana* cells. Mycopathologia 160:303–314

Tefera T, Pringle KL (2003a) Effect of exposure method to *Beauveria bassiana* and conidia concentration on mortality, mycosis, and sporulation in cadavers of *Chilo partellus* (Lepidoptera: Pyralidae). J Invert Pathol 84:90–95

Tefera T, Pringle KL (2003b) Food consumption by *Chilo partellus* (Lepidoptera: Pyralidae) larvae infected with *Beauveria bassiana* and *Metarhizium anisopliae* and effects of feeding natural versus artificial diets on mortality and mycosis. J Invert Pathol 84:220–225

Thomas KC, Khachatourians GG, Ingledew WM (1987) Production and properties of *Beauveria bassiana* conidia cultivated in submerged culture. Can J Microbiol 33:12–20

Tolfo Bittencourt SE, Amaral de Castro L, Estrazulas Farias S, Nair Bao S, Schrank A, Henning Vainstein M (2004) Purification and ultrastructural localization of a copper-zinc superoxide dismutase (CuZnSOD) from the entomopathogenic and acaricide fungus *Metarhizium anisopliae*. Res Microbiol 155:681–687

Traniello JF, Rosengaus RB, Savoie K (2002) The development of immunity in a social insect: evidence for the group facilitation of disease resistance. Proc Natl Acad Sci USA 99:6838–6842

Tymon AM, Pell JK (2005) ISSR, ERIC and RAPD techniques to detect genetic diversity in the aphid pathogen *Pandora neoaphidis*. Mycol Res 109:285–293

Typas MA, Griffin AM, Bainbridge BW, Heale JB (1992) Restriction fragment length polymorphism in mitochondrial DNA and ribosomal RNA gene complexes as an aid to the characterization of species and subspecies populations in the genus *Verticillium*. FEMS Microbiol Lett 95:157–162

Uribe D, Khachatourians GG (2004) Restriction fragment length polymorphism of mitochondrial genome of the entomopathogenic fungus *Beauveria bassiana* reveals high intraspecific variation. Mycol Res 108:1070–1078

Uribe D, Khachatourians GG (2008) Identification and characterization of an alternative oxidase in the entomopathogenic fungus *Metarhizium anisopliae*. Can J Microbiol 54:1–9

Urtz BE, Rice WC (1997) RAPD-PCR characterization of *Beauveria bassiana* isolates from the rice water weevil *Lissorhoptrus oryzophilus*. Lett Appl Microbiol 25:405–409

Urtz BE, Rice WC (2000) Purification and characterization of a novel extracellular protease from *Beauveria bassiana*. Mycol Res 104:180–186

Valadares-Inglis MC, Peberdy JF (1997) Location of chitinolytic enzymes in protoplasts and whole cells of the entomopathogenic fungus *Metarhizium anisopliae*. Mycol Res 101:1393–1396

Valencia E (2002) General and molecular characterization of *Beauveria bassiana* and other fungal isolates for the control of *Bemisia tabaci* whiteflies. PhD thesis, University of Saskatchewan, Saskatoon, pp 356

Vandenberg JD, Cantone FA (2004) Effect of serial transfer of three strains of *Paecilomyces fumosoroseus* on growth in vitro, virulence, and host specificity. J Invert Pathol 85:40–45

Vandenberg JD, Ramos M, Altre JA (1998) Dose-response and age- and temperature-related susceptibility of the Diamondback moth (Lepidoptera: Plutellidae) to two isolates of *Beauveria bassiana* (Hyphomycetes: Moniliaceae). Environ Entomol 27:1017–1021

Vega FE, Dowd PF, McGuire MR, Jackson MA, Nelsen TC (1997) In vitro effects of secondary plant compounds on germination of blastospores of the entomopathogenic fungus *Paecilomyces fumosoroseus* (Deuteromycotina: Hyphomycetes) J Invert Pathol 70:209–213

Vey A, Fargues J (1977) Histological and ultrastructural studies of *Beauveria bassiana* infection in *Leptinotarsa decemlineta* larvae during ecdysis. J Invert Pathol 30:207–215

Vey A, Hoagland R, Butt TM (2001) Toxic metabolites of fungal biocontrol agents. In: Butt TM, Jackson CW, Magan N (eds) Fungi as biocontrol agents. CAB International, Wallingford, pp 311–345

Viaud M, Couteaudier Y, Riba G (1998) Molecular analysis of hypervirulent somatic hybrids of the entomopathogenic fungi *Beauveria bassiana* and *Beauveria sulfurescens*. Appl Environ Microbiol 64:88–93

Vilcinskas A, Jegorov A, Landa Z, Gotz P, Matha V (1999) Effects of beauverolide L and cyclosporin A on humoral and cellular immune response of the greater wax moth, *Galleria mellonella*. Comp Biochem Physiol C Pharmacol Toxicol Endocrinol 122:83–92

Wang CS, St Leger RJ (2006) A collagenous protective coat enables *Metarhizium anisopliae* to evade insect immune responses. Proc Natl Acad Sci USA 103:6647–6652

Wang CS, Li Z, Typas MA, Butt TM (2003a) Nuclear large subunit rDNA group I intron distribution in a population of *Beauveria bassiana* strains: phylogenetic implications. Mycol Res 107:1189–2000

Wang CS, Shah FA, Patel N, Li Z, Butt TM (2003b) Molecular investigation on strain genetic relatedness and population structure of *Beauveria bassiana*. Environ Microbiol 5:908–915

Wang CS, Skrobek A, Butt TM (2003c) Concurrence of losing a chromosome and the ability to produce destruxins in a mutant of *Metarhizium anisopliae*. FEMS Microbiol Lett 226:373–378

Wang CS, Skrobek A, Butt TM (2004) Investigations on the destruxin production of the entomopathogenic fungus *Metarhizium anisopliae*. J Invert Pathol 85:168–174

Wang CS, Butt TM, St Leger RJ (2005a) Colony sectorization of *Metarhizium anisopliae* is a sign of ageing. Microbiology 151:3223–3236

Wang CS, Hu G, St Leger R (2005b) Differential gene expression by *Metarhizium anisopliae* growing in root exudate and host (*Manduca sexta*) cuticle or hemolymph reveals mechanisms of physiological adaptation. Fungal Genet Biol 42:704–718

Xavier IJ, Khachatourians GG (1996) Heat-shock response of the entomopathogenic fungus *Beauveria brongniartii*. Can J Microbiol 42:577–585

Xavier IJ, Khachatourians GG, Ovsenek N (1999) Constitutive and heat-inducible heat shock element binding activities of heat shock factor in a group of filamentous fungi. Cell Stress Chaperones 4:211–222

Yeo H, Pell JK, Alderson PG, Clark SJ, Pye BJ (2003) Laboratory evaluation of temperature effects on the germination and growth of entomopathogenic fungi and on their pathogenicity to two aphid species. Pest Manage Sci 59:156–165

Ying SH, Feng MG (2004) Relationship between thermotolerance and hydrophobin-like proteins in aerial conidia of *Beauveria bassiana* and *Paecilomyces fumosoroseus* as fungal biocontrol agents. J Appl Microbiol 97:323–331

Yokoyama E, Yamagishi K, Hara A (2002) Group-I intron containing a putative homing endonuclease gene in the small subunit ribosomal DNA of *Beauveria bassiana* IFO 31676. Mol Biol Evol 19:2022–2025

Ypsilos IK, Magan N (2004) Impact of water-stress and washing treatments on production, synthesis and retention of endogenous sugar alcohols and germinability of *Metarhizium anisopliae* blastospores. Mycol Res 108:1337–1345

Zurek L, Wes WD, Krasnoff SB, Schal C (2002) Effect of the entomopathogenic fungus, *Entomophthora muscae* (Zygomycetes: Entomophthoraceae), on sex pheromone and other cuticular hydrocarbons of the house fly, *Musca domestica*. J Invert Pathol 80:171–177

Zhao H, Charnley AK, Wang Z, Yin Y, Li Z, Li Y, Cao Y, Peng G, Xia Y (2006) Identification of an extracellular acid trehalase and its gene involved in fungal pathogenesis of *Metarizium anisopliae*. J Biochem 140:319–327

4 Physiology and Metabolic Requirements of Pathogenic Fungi

Matthias Brock[1]

CONTENTS

I. Introduction 63
II. Impact of Phospholipases on Membrane
   Destruction 64
   A. Phospholipids 64
   B. Classification of Phospholipases 64
   C. Phospholipases in the Virulence
      of *Candida albicans* 65
   D. Phospholipases in the Virulence
      of *Cryptococcus neoformans* 66
   E. Phospholipases of *Aspergillus
      fumigatus* 67
III. Fatty Acid Metabolism
    and the Glyoxylate Cycle 67
    A. The Glyoxylate Cycle 67
    B. Role of the Glyoxylate Cycle
       in the Pathogenesis of *Can. albicans* . . . 68
    C. Lipid Metabolism and Glyoxylate
       Cycle in *Cry. neoformans* 69
    D. Impact of Lipid Metabolism
       and Glyoxylate Cycle on
       A. fumigatus Pathogenesis 69
IV. Impact of Proteases and Peptidases
    on Virulence 71
    A. Proteins as Nutrients 71
    B. Classification of Proteases 71
       1. Serine Proteases 72
       2. Aspartic Proteases 72
       3. Cysteine Proteases 72
       4. Metalloproteases 72
    C. Impact of Proteases in *Can. albicans*
       Pathogenesis 73
    D. Proteases from *Cry. neoformans* 74
    E. Proteases and Their Role in
       A. fumigatus Pathogenesis 75
V. Role of Carbohydrate
   and Nitrogen Metabolism
   in the Virulence of Pathogenic Fungi 76
VI. Conclusion 78
    References 79

[1] Microbial Biochemistry and Physiology, Leibniz Institute for Natural Product Research and Infection Biology e.V., Hans Knöll Institute, Beutenbergstrasse 11a, 07745 Jena, Germany; e-mail: matthias.brock@hki-jena.de

I. Introduction

This chapter mainly focuses on fungi which can grow, replicate and distribute without the urgent need for a mammalian host but which may, under certain conditions, cause severe invasive infections. A focus is given on *Candida*, *Saccharomyces*, *Aspergillus*, and *Cryptococcus* species. Because they are free-living, these micro-organisms have to be able to adapt rapidly to changing environmental conditions. The enzymatic toolbox of these fungi has to provide the ability for the de novo synthesis of all growth factors, the metabolism of various carbon and nitrogen sources, resistance against heat- and cold-shock, and resistance against osmotic stress.

Fungal infections, defined here as diseases caused by fungi (not only the acquisition or persistence), mostly require a suppressed host immune system. This immunosuppression can either be caused by drugs in transplant medicine or by illnesses like leukaemia, AIDS, and neutropenia. Nevertheless, these fungi have to be able to overcome or hide from the residual immune system. In this respect, *Cryptococcus* and *Candida* species may have adapted more specifically to life within a host compared to *Aspergillus* species. For example, *Can. albicans* is able to escape attack from the complement cascade by binding the complement regulators factor H and FHL-1 (Meri et al. 2002, 2004). In addition, environmental conditions such as contact with serum or a temperature shift to at least 37 °C mediate a switch from yeast to hyphal growth. The hyphae have a much higher capacity to invade tissues and are attributed as a prerequisite for invasive infections. *Cryptococcus neoformans* is also able to form filaments under specific conditions, but filamentous growth is not required for infection rather than for the formation of basidiospores. However, genes involved in mating (e.g. *STEA12*) seem to play a role in capsule formation. This ability to form a polysaccharide capsule seems to create some advantages in growth within infected tissues and dissemination into the brain. In addition, particles from the capsule are released after phagocytosis by macrophages, which accumulate in vesicles and may cause macrophage dysfunction or even cell death (Steenbergen and Casadevall 2003). Although the ability for capsule formation is an important virulence factor, the capsule is not specifically formed during infection but also forms in the natural habitat, the soil, and may protect the fungus from desiccation.

For *Aspergillus* species, specific proteins or cellular structures, which may only be used for invasive growth, have not yet been reported. Although the ability to form pigmented conidia is important for resistance of conidia against the immune response, most, if not all Aspergilli are able to form such pigments (Langfelder et al. 1998; Jahn et al. 2000, 2002).

Nevertheless, only a small number of Aspergilli are able to cause invasive fungal infections, predominantly *A. fumigatus* but also *A. terreus* and *A. flavus*. These fungi also have to possess specific determinants which enable them to grow within infected tissues and which are missing in the closely related species. An obvious advantage of *A. fumigatus*, compared to all other close relatives, is the ability to grow at temperatures up to 55 °C. Consequently, fever caused by the host immune response does not harm the fungus.

Despite the differences in protection and adaptation towards the host immune response, all of these fungi have in common that they are able to acquire nutrients from the host for maintenance and growth. This process is often accompanied by host cell lysis. Since membranes of the mammalian cells are made up of proteins and phospholipids, extracellular-secreted enzymes like proteases and phospholipases may play an important role in virulence. Additionally, the peptides and fatty acids released by this process may serve as important carbon sources, substituting in the growth of the pathogens. However, only a limited number of investigations have focused on the metabolic requirements. Especially, the carbon acquisition of these fungi under infectious conditions is presented. The following sections summarise the current knowledge on the metabolism of carbon compounds during infection and invasive growth. Several enzymes and metabolic pathways involved in the degradation of these carbon sources are discussed in detail.

II. Impact of Phospholipases on Membrane Destruction

A. Phospholipids

Phospholipids are amphiphilic compounds, i.e., they contain a hydrophilic head and a hydrophobic tail. Therefore, they are able to form the bilayer of membranes. In phosphoglycerols the hydrophobic tail is composed of various long chain fatty acids, while the hydrophilic head is made up by a phosphor ester linkage between the glycerol residue and a hydrophilic substance, like serine (phosphatidylserine), ethanolamine (phosphatidylethanolamine), or choline (phosphatidylcholine). Another class of phospholipids is made up of sphingolipids, like sphingomyelin. In sphingomyelin a fatty acid is bound to sphingosin via an amide bond, which together give the hydrophobic tail. The hydrophilic head is formed as in the phosphoglycerols, where a phosphor ester bond combines the C1 hydroxyl group of the sphingosin moiety. This phosphate group can also be linked to serine, ethanolamine, or choline. However, in contrast to phosphoglycerols, sphingolipids are more common to nerve cells.

B. Classification of Phospholipases

Phospholipases are a heterogenous group of enzymes. All of them have the ability to hydrolyse one or more ester linkages in glycerophospholipids. The classification of phospholipases (A_1, A_2, B, C, D) results from their specificity to cleave a specific position within the phosphoglycerolipid substrate (Ghannoum 2000). As shown in Fig. 4.1, phospholipids contain four possible ester cleavage sites. Phospholipase A_1, for example, cleaves the ester bond at position 1 by releasing one fatty acid

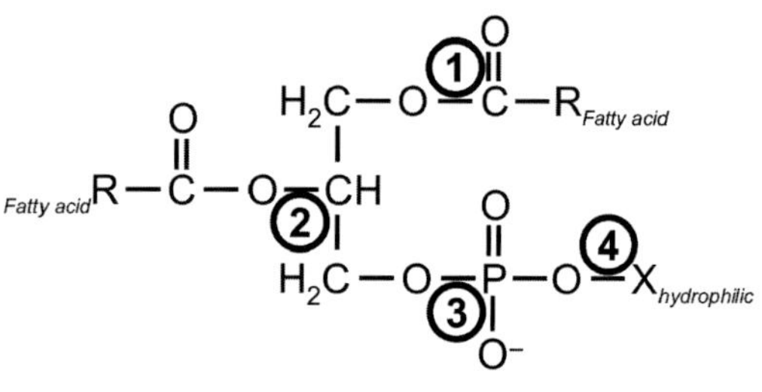

Fig. 4.1. Schematic drawing of a phospholipid. The numbers denote cleavage sites for specific phospholipases. For detailed explanation refer to the text

molecule, whereas phospholipase A_2 cleaves the ester bond at position 2. In contrast to A_1 and A_2, the phospholipase B can cleave at positions 1 and 2, either from lysophospholipids (one fatty acid already removed) or from phospholipids. Additionally, phospholipase B can also add a free fatty acid to lysophospholipids, which results in a phospholipid. Phospholipase C hydrolyses the phosphodiesterbond at position 3, which results in the release of a diacylglycerol and the phosphatidyl moiety, like phosphatidylcholine. Last but not least, phospholipase D cleaves the second phosphodiesterbond (position 4) and yields phosphatidic acid and the hydrophilic compound, like serine, ethanolamine, or choline (denoted by $X_{hydrophilic}$ in Fig. 4.1). The combined action of several phospholipases is supposed to be able to destroy the membrane of tissue cells (Birch et al. 1996) and the accumulation of of lysophospholipids is toxic for the cells, as was shown by the action of several snake venoms, which mainly contain highly active phospholipase A2 and/or B (Shiloah et al. 1973; Takasaki and Tamiya 1982; Bernheimer et al. 1987).

C. Phospholipases in the Virulence of *Candida albicans*

Can. albicans possesses genes coding for all four types of phospholipases (A, B, C, D). Among different *Candida* species, *Can. albicans* seems to produce phospholipases in higher amounts and more frequently than others. However, the amount of phospholipases excreted also varies between different *Can. albicans* isolates. A clear correlation was observed between the source of the isolated strains and enzyme activity. Strains isolated from the blood stream generally produce much higher amounts of extracellular phospholipase than isolates from wounds or the urine. Therefore, a correlation between phospholipase activity and tissue penetration was drawn (Price et al. 1982; Ibrahim et al. 1995). *Can. albicans* contains three genes coding for putative phospholipase C proteins, but none of them seems to contain an N-terminal secretion signal peptide (Kunze et al. 2005). Reduced expression of CaPLC1 resulted in decreased growth on osmotic stress-inducing media and on non-glucose carbon sources, and higher sensitivity to temperature shifts. Therefore, this enzyme seems to be involved in several cellular processes and is not specifically involved in virulence. The role of the other to phospholipase C proteins (CaPLC2, CaPLC3) is unclear. Deletion of neither of these two genes, or the double deletion, showed significant phenotypes (Knechtle et al. 2005; Kunze et al. 2005).

Whether there is any impact of a specific phospholipase A on virulence has yet not been reported. However, a clear impact on virulence was shown for phopholipase B (especially phospholipase B1 CaPLB1) and phospholipase D. Deletion of phospholipase B1 resulted in a 99% reduction in phospholipase B activity and a 80% reduction in lysophospholipase activity when compared to the wild type (Leidich et al. 1998). This indicates that CaPLB1 is the major extracellular phospholipase B (PLB) in *Can. albicans*. Deletion of a single allele reduced the activity by approximately 50%, which shows that both alleles contribute to activity. However, deletion of one or both alleles of *CaPLB1* did not alter the growth phenotype or germination of *Can. albicans* (Leidich et al. 1998). The mutant was also analysed for a possible attenuation in virulence by using different model systems. In a haematogenous dissemination model the mean survival time of infected mice increased from 4.4 ± 2.1 days to 13.3 ± 2.6 days and, furthermore, the PLB-deficient strains were cleared much faster from infected tissues (Leidich et al. 1998). Another model focused on the ability of the mutant to cross the gastrointestinal mucosa with subsequent dissemination to other organs. Although the mutant still had some ability to cause a systemic *Candida* infection dissemination was reduced by approximately 50% when compared to the parental wild-type strain. Therefore, it was concluded that CaPLB1 is important for the dissemination by the gastrointestinal and haematogenous routes (Ghannoum 2000; Mukherjee et al. 2001). In addition to CaPLB1 the impact of phospholipase D1, which also displays an intracellular activity (CaPLD1), was investigated. Both alleles of *pld1* were deleted and the resulting mutants showed no detectable release of phosphatidic acid from phospholipids and strongly reduced amounts of released diacylglycerol, but neither on minimal nor on complex media did the mutant display any growth defects. Also the formation of chlamydospores was not influenced by this mutation. However, the mutant was strongly impaired in hyphal formation on solid Spider medium and the ability to invade agar of solid medium was attenuated. The *pld1* mutant was not attenuated in an oral model of candidosis.

In contrast, oral infection of immunodeficient mice showed that all mice were killed by the parental wild-type strain but the mutant was attenuated in virulence (Hube et al. 2001). Further investigations revealed that a *pld1* mutant was unable to survive in internal organs after intravenous challenge and had a drastically reduced ability to penetrate epithelial monolayers (Dolan et al. 2004). It can be concluded that CaPLD1 plays an important role in hyphal formation when attached to surfaces (solid media and cells) and that this reduced capability in hyphal formation attenuates virulence. Therefore, it can be concluded that phospholipase B1 and phospholipase D1 contribute mostly to the virulence of this pathogen. The release of fatty acids by the action of CaPLB1 can furthermore provide nutrients during the infection process which is discussed in a later section of this chapter.

D. Phospholipases in the Virulence of *Cryptococcus neoformans*

The secretion of phospholipases by various *Cry. neoformans* isolates was shown by different groups. In one study, 23 cryptococcal isolates were tested, 22 of which turned out to be positive for extracellular phospholipase activity (Vidotto et al. 1996). In a second study, 50 independent isolates were tested for extracellular phospholipase activity and 49 of them showed significant, although varying, levels of phospholipase activity (Chen et al. 1997). Purified phospholipase B from a virulent *Cry. neoformans* strain was characterised for its substrate specificity and turned out to combine all phospholipase B, lysophospholipase, and lysophospholipase/transacylase activity in a single protein. The enzyme was not specific for a single substrate and accepted all phospholipids, with the exception of phosphatidic acid, and used dipalmitoyl phosphatidylcholine and dioleoyl phosphatidylcholine as the preferred substrates (Chen et al. 2000). A deletion of the *plb1* gene did not alter growth at 37 °C, capsule formation, laccase activity, and urease activity, but in animal models virulence of the *plb1* mutant was reduced. In addition, a growth defect was observed when incubated with a macrophage-like cell line (Cox et al. 2001). When mice were infected via the respiratory tract, the *plb1* mutant generated a protective immune response that was able to control the infection, whereas infection with the wild type caused a non-protective inflammatory response. This led to the assumption that PLB1 may be involved in the production of eicosanoids. Eicosanoids mainly derive from arachidonic acid and have the ability to down-regulate the immune response. PLB1 is required for the release of arachidonic acid from phospholipids and can be used for fungal eicosanoid production (Noverr et al. 2003). Further studies revealed that macrophage-derived arachidonic acid is incorporated into cryptococcal lipids during the cryptococcus–phagocyte interaction and can be sequestered for the fungal eicosanoid production which, in turn, suppresses the activity of macrophages (Wright et al. 2007). In this context, it was also shown that PLB1 facilitated the entry of cryptococci into the lung and was required for the lymphatic spread from the lung to the lymph nodes and the blood entry. Nevertheless, the use of macrophages as a vehicle for dissemination to the brain was not dependent on PLB1 (Santangelo et al. 2004). Secretion of PLB1 is regulated by a glycosylphosphatidylinositol (GPI) anchor, which targets PLB1 dominantly to the cell membrane (Djordjevic et al. 2005). Within the membrane PLB1 can mainly be found in lipid rafts which consist of the association of sphingolipids and sterols. Within the raft environment the activity of PLB1 is strongly reduced but increases 15-fold upon release to the external environment. This is supposed to help specific virulence determinants to be clustered at the cell surface to allow efficient access to the substrate (Siafakas et al. 2006).

Besides this important role of PLB1 for the infection process, an impact on virulence of the inositol phosphosphingolipid-phospholipase C (ISC1) was recently discovered. ISC1 is involved in the protection of *Cry. neoformans* from acidic, oxidative, and nitrosative stress, conditions which are induced by immune effector cells. Therefore, an ISC1 deficient mutant is not able to survive the phagolysosome of a macrophage and its dissemination to the brain is strongly reduced. In an immunosuppressed mouse, however, such a mutant regains the ability to disseminate (Shea et al. 2006).

In conclusion, phospholipases are important for the virulence of not only *Can. albicans* (see above) but also *Cry. neoformans*. However, the main role of the candidal phospholipases seems to lie in the penetration of epithelial cells, a reduction in systemic infection after intravenous application of yeast cells, and the switch from yeast to hyphal form, whereas the cryptococcal enzymes seem to play a major role in protection against and regulation of the host inflammatory response.

E. Phospholipases of *Aspergillus fumigatus*

Until now, only a limited number of studies on the role of phospholipases from *Aspergillus* species have been performed. A preliminary study on extracellular phospholipase activity during growth on lipid-containing media was carried out in which the hydrolysed products were analysed by fast atom bombardment–mass spectrometry (FAB-MS). From the data of this spectrometric analysis it was concluded that all types of phospholipases are secreted by *A. fumigatus*. Phospholipase activity was detectable within the first 25 h after inoculation of medium; and it steeply declined afterwards (Birch et al. 1996). Additionally, lysophospholipids and glycerophosphoryl compounds were detected, which is indicative, at least, for the action of phospholipase B. However, by this method it is hard to predict whether a lysophospholipid derives from the action of phospholipase A1, A2, or from phospholipase B. Besides these activities the existence of phospholipase C activity was also assumed, but diacylglycerols were hardly represented in the anion spectra. Therefore, the presence of phospholipase C was confirmed by use of the artificial substrate *p*-nitrophenylphosphorylcholine (*p*NPPC). This determination revealed that PLC activity was detectable 30 h after inoculation and reached a maximum at 50 h, which coincided with entering the stationary growth phase of the culture (Birch et al. 1996). The idea that a phospholipase A may contribute to hydrolase activity was abolished because no gene sequence was found in the *A. fumigatus* genome which could possibly code for such an enzyme (Birch et al. 2004). Comparison of 53 clinical isolates with 11 environmental isolates revealed that the mean value for the PLC activity for the clinical isolates was 2.5-fold higher than in the environmental strains, whereas the environmental strains revealed a 1.4-fold higher PLB activity. However, in plate diffusion assays precipitation zones correlated well with PLC activity but not with PLB activity. Due to the importance of phospholipase C in several pathogenic bacteria it was concluded that PLC may have a higher impact on virulence, if at all, than PLB (Birch et al. 2004). However, another study focussed on the identification of phospholipases of the B-type from *A. fumigatus*. Three candidate genes were found in the *A. fumigatus* genome which possibly code for PLBs. Real-time PCR analysis revealed that only two of these genes were responding to lecithin. *Afplb1* was induced by a factor of five and *Afplb3* by a factor >300, whereas *Afplb2* showed no change in the expression pattern. In addition, AfPLB1 and AfPLB3 possessed an N-terminal secretion signal, which was not found in AfPLB2. Therefore, it was concluded that AfPLB1 and AfPLB3 together contribute to *A. fumigatus* phospholipase B activity and that both genes may be involved in the degradation of phospholipids present in the lung during infection (Shen et al. 2004). However, until now it has not been possible to attribute an important role to one of the phospholipases in the pathogenesis of *A. fumigatus* and further studies will be needed, to test the phenotypes of deletion mutants in infection models.

III. Fatty Acid Metabolism and the Glyoxylate Cycle

A. The Glyoxylate Cycle

Various micro organisms are able to grow on lipids, fatty acids, acetate, and ethanol without the need of other carbon sources as complements. All mentioned carbon sources exclusively create acetyl-CoA units during degradation. Acetyl-CoA cannot easily become converted to oxaloacetate, but this conversion is essential for gluconeogenesis and for the formation of building blocks for biosyntheses. Humans are not able to feed solely on these carbon sources, because they do not possess the ability to synthesise oxaloacetate de novo from acetyl-CoA. Micro organisms and plants utilise a specialised cycle for the anaplerotic synthesis of oxaloacetate, the so-called glyoxylate cycle. A scheme of the glyoxylate cycle is shown in Fig. 4.2. Isocitrate lyase circumvents the decarboxylation of isocitrate by cleaving isocitrate into glyoxylate and succinate. Oxaloacetate is regenerated from succinate via fumarate and malate and re-enters the cycle by a condensation with acetyl-CoA. The second reaction, specific for the glyoxylate cycle, is the condensation of the released glyoxylate with acetyl-CoA via malate synthase, which yields malate. The anaplerotic reaction is completed by the conversion of malate to oxaloacetate (Lorenz and Fink 2002).

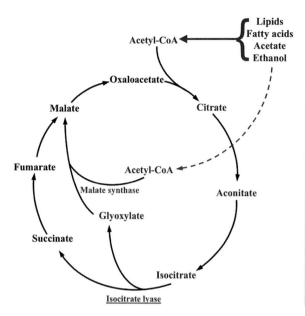
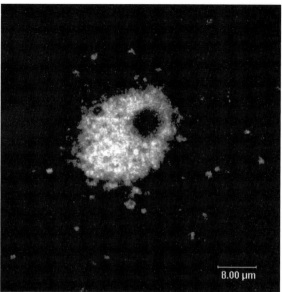

Fig. 4.2. Scheme of the glyoxylate cycle. Acetyl-CoA derives from direct activation of acetate or from oxidative degradation of ethanol, lipids or single fatty acids. The key enzymes of the glyoxylate cycle consist of isocitrate lyase and malate synthase. Isocitrate lyase cleaves isocitrate into succinate and glyoxylate, whereas malate synthase condenses one molecule of glyoxylate with acetyl-CoA leading to the formation of malate. The final oxidation of malate leads to a replenishment of the oxaloacetate pool

Fig. 4.3. Nile red staining of an isolated primary mouse alveolar macrophage. Fluorescence was detected by the lipid-specific yellow-gold fluorescence of Nile red. The cell has accumulated a high amount of lipoproteins (indicated by the *bright white colour*) but is still able to phagocytose pathogens

Lipids and fatty acids are supposed to play a major role in carbon supply for pathogenic microorganisms. Most pathogens become phagocytosed by macrophages in the early phase of infection, which makes necessary the utilisation of carbon sources available within the macrophages. Macrophages contain significant amounts of lipids and cholesterol (Greenspan et al. 1985), which can be found in the form of low-density lipoprotein (LDL) and high-density lipoprotein (HDL). An example of a primary alveolar macrophage with high lipid content is shown in Fig. 4.3. It was therefore assumed that the ability to utilise lipids via the glyoxylate cycle is a prerequisite for survival within lipid-rich environments such as the macrophage.

B. Role of the Glyoxylate Cycle in the Pathogenesis of *Can. albicans*

In order to get insights into the carbon sources used by *Can. albicans* within macrophages a model system involving *Saccharomyces cerevisiae* was employed in which *S. cerevisiae* cells were co-cultivated with macrophages and re-isolated. Micro-array analyses of the yeast cells showed that genes involved in the glyoxylate cycle were highly induced (Lorenz and Fink 2001). Experiments with *Can. albicans* confirmed that isocitrate lyase and malate synthase were highly induced in the presence of macrophages but not when cells were grown in cell culture medium or serum. Further studies confirmed the induction of glyoxylate cycle enzymes during co-incubation with macrophages and other immune effector cells like neutrophils (Fradin et al. 2003; Lorenz et al. 2004; Fradin et al. 2005).

Isocitrate lyase deletion strains were unable to use acetate or ethanol as sole carbon source and, when tested in a systemic mouse infection model, a significant attenuation in virulence of the isocitrate lyase mutant was observed (Lorenz and Fink 2001). This led to the conclusion that isocitrate lyase, and thus the glyoxylate cycle, is essential for virulence of *Can. albicans*. However, recent investigations revealed that the importance of isocitrate lyase in virulence is restricted to specific infection stages (Barelle et al. 2006). The induction of isocitrate lyase under in vitro, ex vivo, and in vivo conditions was followed by single-cell profiling with a fusion of the isocitrate lyase promoter with the enhanced green fluorescent protein (eGFP). Glucose concentrations in the range of 0.01% to 0.1%, as present in the blood stream, were sufficient to suppress the induction of isocitrate lyase. However, cells which were completely phagocytosed were able to induce isocitrate lyase; and also phagocytosis by neutrophils led to a strong fluorescence. In contrast, cells which were able to enter the kidney in a mouse infection model did not always display induction of isocitrate lyase. The picture was quite heterogenous, with a positive immunofluorescence in approximately 50% of the cells. The re-investigation of the fungal burden of an isocitrate lyase mutant, generated independently from the investigation described above, revealed that the number of cells within different organs was reduced. However, cells within these organs were still able to establish lethal infections (Barelle et al. 2006). Isocitrate lyase, therefore, is not essential for a systemic infection but accelerates the process. Therefore, it remains questionable whether a drug specifically targeting isocitrate lyase would be effective when a systemic infection has already been manifested.

C. Lipid Metabolism and Glyoxylate Cycle in *Cry. neoformans*

Cry. neoformans contains up to 10% of its dry weight as lipids, with a large proportion of phospholipids, triacylglycerols (in lipid particles), and diacylglycerols (DAG). Especially the intracellular balance and metabolism of diacylglycerols seems to play an important role in virulence (Shea et al. 2006). The protein kinase C (PKC1) contains a so-called C1 domain, which is known from mammalian DAG-dependent protein kinase C. Deletion of this C1 domain prevents the association of the virulence factor laccase with the cell wall. This laccase is involved in the synthesis of the black pigment melanin. The lack of laccase association to the cell wall leads to cell-wall defects like capsule microfibrils (Heung et al. 2005). In addition, high DAG levels upregulate the expression of the antiphagocytic protein APP1 (Mare et al. 2005). Therefore, intracellular DAG is directly involved in the control of specific virulence-associated genes. Furthermore, triacylglycerols accumulate in lipid particles when *Cry. neoformans* cells reach the stationary growth phase and are assumed to provide an important carbon source under nutrient-limiting conditions such as the phagolysosome of macrophages. Since *Cry. neoformans* is disseminated to the brain within macrophages, an importance of the glyoxylate cycle was assumed. Profiling of the genes up- and down-regulated during infection of rabbits with *Cry. neoformans* revealed that a single gene, coding for isocitrate lyase was significantly up-regulated one week after infection. An isocitrate lyase mutant was tested for virulence in different infection models and, surprisingly, displayed no attenuation in virulence (Rude et al. 2002). Evaluation of the yeast counts from cerebrospinal fluids revealed that the initial number of cells was slightly decreased with the mutant but increased to wild-type levels in later stages of infection, indicating that isocitrate lyase might not provide an antifungal drug target against *Cry. neoformans* infections. Furthermore, it was assumed that lipids do not provide the main carbon sources of *Cry. neoformans* during pathogenesis. However, it was not excluded that lipid metabolism contributes to virulence, since phospholipases, especially phospholipase B1, release arachidonic acid from host phospholipids and this can act as a direct precursor for eicosanoid production. As mentioned above, eicosanoids are supposed to be involved in immune modulation of the host response. In addition, the glyoxylate cycle may be of less importance when lipids are metabolised together with sugars and amino acids, which might provide sufficient amounts of oxaloacetate.

D. Impact of Lipid Metabolism and Glyoxylate Cycle on *A. fumigatus* Pathogenesis

Due to the different impact of isocitrate lyase on virulence of *Can. albicans* and *Cry. neoformans* it was questionable whether the glyoxylate cycle is important for the pathogenesis of *A. fumigatus*. Expression of isocitrate lyase under different growth conditions was checked by use of a strain carrying a *lacZ* fusion with the isocitrate lyase promoter and by Western blot experiments with a monoclonal antibody directed against isocitrate lyase. Results confirmed that isocitrate lyase is only produced when carbon sources generating C_2 units like acetate, ethanol, fatty acids, or lipids are present. Additionally it was observed that isocitrate lyase is present in germinating conidia

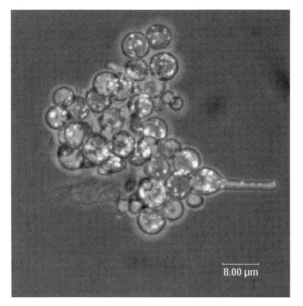

Fig. 4.4. *A. fumigatus* conidia pre-swollen in RPMI medium for 5 h and stained with Nile red. Lipid droplets were visualised by lipid-specific yellow-gold fluorescence. An overlay of the bright field with the fluorescent picture is shown. Lipid droplets are visible as bright spots

but is rapidly lost when conidia are germinated on complete media. This observation coincides with the presence of significant amounts of lipid droplets in conidia (see Fig. 4.4; Ebel et al. 2006); but their role in germination of conidia has not yet been investigated. When the presence of isocitrate lyase after phagocytosis was monitored by immunofluorescence microscopy it was observed that engulfed conidia showed a strong immunofluorescence and that fluorescence was maintained in the germinating conidium. This fluorescence remained visible as long as the hyphae elongated within the macrophage. After destruction of the macrophage cell membrane immunofluorescence was rapidly lost (Ebel et al. 2006), indicating that isocitrate lyase contributes to nutrient acquisition during intracellular growth. However, within a few hours, elongating hyphae overcome the distance which is needed to escape the macrophage and the nutrients like trehalose and lipids, which are stored within the conidium, could provide sufficient energy for this process. To prove this assumption an isocitrate lyase deletion mutant was created and tested in a mouse infection model for invasive aspergillosis. Virulence of the mutant was not attenuated and, additionally, the fungal burden and the index of inflammation revealed no difference in comparison with the wild type (Schöbel et al. 2007). Figure 4.5 shows a comparison of lung thin sections after infection of mice with either the wild-type or the isocitrate lyase mutant. It was concluded that isocitrate lyase and thus the glyoxylate cycle have, if any, only a minor impact on the virulence of *A. fumigatus*. Nevertheless, the metabolism of lipids cannot be

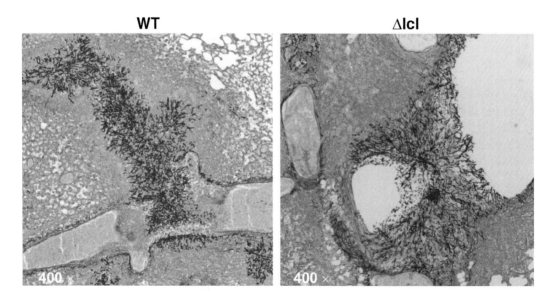

Fig. 4.5. Histopathology of lung tissue sections 5 days after infection. The *left panel* shows a tissue infected with an *A. fumigatus* wild-type strain. The *right panel* shows a tissue section after infection with an isocitrate lyase deletion strain. Both strains show a comparable invasiveness

totally neglected. In a further experiment, a wild-type strain as well as the isocitrate lyase mutant were grown on media containing peptone (as a source of amino acids) and olive oil (resembling lipids). Both strains were able to utilise both carbon sources at the same time withouth a phenotype, although the isocitrate lyase mutant was not able to grow on olive oil as the sole carbon source. Determination of the isocitrate lyase activity of the wild type on peptone/olive oil revealed that isocitrate lyase activity was hardly induced on this mixed carbon source. Therefore, it seems quite likely that the anaplerosis of oxaloacetate derives from amino acids, whereby the acetyl-CoA from β-oxidation of fatty acids can be directly used for energy metabolism (Schöbel et al. 2007).

However, in contrast to *Cry. neoformans*, arachidonic acid and prostaglandin production (one of the eicosanoids) seems to have a negative effect on the virulence of *A. fumigatus*. Three genes, most likely coding for cyclooxygenases were identified from the genome of *A. fumigatus* and deletion of the single genes reduced prostaglandin production on media supplemented with arachidonic acid (Tsitsigiannis et al. 2005). Surprisingly, in a mouse model for invasive aspergillosis the mutants turned out to display a hypervirulent phenotype. Therefore, immune-modulation mediated by these eicosanoids negatively affects survival of *A. fumigatus* in infected tissues. Nevertheless, these results imply that phospholipases may release arachidonic acids from phospholipids and convert them, at least to some extent, to eicosanoids. Consequently, it will be interesting to investigate the virulence of phospholipase mutants. In the case that phospholipases only provide arachidonic acid, a gene deletion will cause a hypervirulent phenotype. If phospholipases are involved in nutrition, an attenuation in virulence may occur.

IV. Impact of Proteases and Peptidases on Virulence

A. Proteins as Nutrients

As mentioned in the previous sections, lipids cannot provide the sole carbon source in pathogenesis. This is unlikely even for *Can. albicans*, which showed the strongest attenuation in virulence, when isocitrate lyase was deleted. During acute tissue infection the glyoxylate cycle is of minor importance for all of the fungi mentioned here. Therefore, it can be assumed that alternative nutrients are utilised. An attractive carbon source, which also supplies nitrogen and sulfur, is afforded by the degradation of proteins. Lysis of cells from the tissues surrounding the pathogen releases all internally stored proteins. Additionally, membranes which may become attacked by phospholipases also contain proteins that may be released by the action of proteases, accompanied with a weakening of the membrane stability. However, all pathogenic fungi seem to possess far more than a single secreted protease and the substrate specificity of several proteases may show some redundancy. Therefore, it is difficult to predict the importance of single proteases in virulence. Nevertheless, at least for dermatophytes proteases seem to play an essential role in virulence, because these fungi invade mainly keratinised tissues like skin, nails, and hair; and all dermatophytes isolated so far secrete proteases. Even more, closely related species of dermatophytes can be distinguished by analysis of the secretion pattern of proteases (Giddey et al. 2007). The following section deals with the classification of different proteases and what is known about the impact of these proteases on pathogenesis of different fungal species.

B. Classification of Proteases

Proteases are also called peptidases, proteolytic enzymes, and peptide hydrolases. All proteases have in common that they catalyse the cleavage of peptide (CO–NH) bonds in proteins, which leads to the release of peptides or free amino acids. Proteases are classified as endoproteases (endopeptidases) which cleave the protein internally, and exoproteases (exopeptidases), which cleave their substrate only at the N- or C-terminal part of the polypeptide chain. N-terminally cleaving exoproteases are called amino-peptidases when they release a single amino acid, dipeptidyl-peptidases when they release a dipeptide, and tripeptidyl-peptidases when they release a tripeptide. Cleavage at the C-terminal side leads to the designation carboxy-peptidase (single amino acid), peptidyl-dipeptidase (dipeptide), or peptidyl-tripeptidase (tripeptide). By contrast, endoproteases generally cleave at a specific amino acid residue within the protein substrate. A classification by their recognition site is not possible. Therefore, these proteases are generally classified by their active site residues.

The combined action of both endo- and exoproteases has a co-operative effect in protein degradation, since the cleavage by endoproteases produces new ends which act as substrate for exoproteases (http://www.chem.qmul.ac.uk/iubmb/enzyme/EC34/).

1. Serine Proteases

Serine proteases are characterised by an enzymatically active serine residue, which is organised in a so-called catalytic triade comprising an additional histidine and an aspartate residue. Members of this class are chymotrypsins and trypsin, the latter being generally used for protein digests in peptide mass determinations. Elastases, which may be important for different pathogenic fungi, prefer small amino acids as substrates, like alanine, glycine, and valine. Secreted serine proteases from fungi generally have a molecular mass of 28–30 kDa and the optimal pH of the reaction is between pH 8 and 9. Therefore, they are also called alkaline proteases (Alp). Due to some serine proteases having a sequence similarity to that of bacterial subtilisins these fungal proteases are subclassified as subtilisin-type (Barrett and Rawlings 1995; Siezen and Leunissen 1997).

2. Aspartic Proteases

A prominent example of aspartic proteases is given by pepsin and most fungal aspartic proteases belong to the pepsin-type proteases. Aspartic proteases cleave their substrate protein without the formation of covalently bound substrate–enzyme intermediates, although a tetrahedral intermediate exists. Aspartic proteases are bilobed, which means that they have two lobes and each contains a catalytically active aspartate residue. In the active enzyme, these amino acids come into close proximity, forming the catalytically active dyad of aspartates. This proximity of the aspartate residues is ensured by four cysteine residues which form two disulfide bridges and are conserved in all proteases of the pepsin type. A water molecule in the active site is essential, because it transfers a proton to one of the aspartate residues, whereby another proton from the second aspartate is transferred to the CO–NH bond of the substrate, leading to cleavage of the peptide bond. Most of the aspartic proteases are active in an acidic environment but some still display activity at neutral pH. Secreted aspartic proteases (Saps) seem to have an impact on the cell adhesion of *Candida* species (Kvaal et al. 1999; Hube and Naglik 2001).

3. Cysteine Proteases

The role of cysteine proteases in fungal virulence has not been studied in detail. A search in the MEROPS database (http://merops.sanger.ac.uk/; Rawlings et al. 2006) reveals that all cysteine proteases from *Can. albicans*, *Cry. neoformans*, and *A. fumigatus* have no functional assignment, which does not exclude a role of these enzymes in pathogenesis. As in serine proteases, a covalently bound intermediate between substrate and enzyme is formed. However, in contrast to the serine proteases, the serine is replaced by a cysteine (http://www.rpi.edu/dept/bcbp/molbiochem/MBWeb/mb2/part1/protease.htm) and a thiolate ion rather than a hydroxyl group attacks the peptide bond. The thiolate ion is stabilised by a neighbouring imidazolium group of a histidine. The attacking group is, therefore, the thiolate-imidazolium pair, which eliminates the need for a water molecule during the reaction.

4. Metalloproteases

Metalloproteases are a quite inhomogenous group of enzymes with respect to structure and primary amino acid sequence. Nevertheless, a great majority of these enzymes contain a zinc, a calcium, or both metals, which are essential for the catalytic activity (http://www.rpi.edu/dept/bcbp/molbiochem/MBWeb/mb2/part1/protease.htm). Many metalloproteases contain the HEXXH motif which binds the metal ion on the histidine residues in close proximity to an acidic amino acid. As in the aspartic proteases, a water molecule is essentially involved in the reaction. In addition, the reaction does not involve a covalently bound intermediate complex between substrate and enzyme, because the substrate becomes coordinated to the metal ion. The water molecule is polarised by the combined action of the metal ion and the acidic amino acid, which leads to the addition of a hydroxyl group to the carbonyl group of the substrate peptide bond. Transfer of the proton from the acidic amino acid to the amino group of the substrate completes hydrolysis. Several members of this class possess activity in a pH range between pH 6 and 9.

C. Impact of Proteases on *Can. albicans* Pathogenesis

Genome annotation revealed that *Can. albicans* possesses 47 different proteases (Merops database, at http://merops.sanger.ac.uk/), among them ten secreted proteases which all belong to the class of aspartic proteases (Sap1–Sap10). No secreted serine-, cysteine-, or metalloproteases have been detected. Sap1–Sap6 and Sap8 are all believed to be secreted, whereas the location of Sap7 is unknown. Sap9 and Sap10 were shown to be glycophosphatidylinositol-anchored proteins (Albrecht et al. 2006). As discussed below, especially Sap1–Sap6 seem to contribute to the virulence of *Can. albicans*. Sap1–Sap3 (identity up to 67%) as well as Sap4–Sap6 (identity up to 89%) cluster in closely related groups (Monod et al. 1998; Naglik et al. 2003; Naglik et al. 2004). Sap7 does not seem to play a major role in virulence (Taylor et al. 2005), and Sap9 and Sap10 target proteins necessary for both cellular processes and host–pathogen interactions (Albrecht et al. 2006).

The proteolytic activity of *Can. albicans* was discovered early by growing the yeast in the presence of serum proteins as a nitrogen source (Staib 1965) and it was discovered that the overall extracellular protease activity of single strains seems to correlate directly with virulence. Furthermore, *Candida* strains from HIV-positive patients with oral or vaginal candidiasis showed enhanced proteolytic activity compared to isolates from HIV-negative humans, as reviewed by Naglik et al. (2003). Comparison of protease activity of different *Candida* species revealed the order: *Can. albicans* > *Can. tropicalis* > *Can. kefyr* > *Can. lusitaniae* > *Can. krusei* (Capobianco et al. 1992).

The secreted proteases, especially Sap2 and SAP4–SAP6, are important for growth when proteins are supplied as the sole nitrogen source (Hube et al. 1997; Sanglard et al. 1997), whereby SAP4–SAP6 seem to regulate the activity of SAP2 (Hube et al. 1994). The role of the secreted proteases in tissue penetration was investigated in a model of oral candidosis based on reconstituted human epithelium (RHE). Different single, double, and triple mutants of *Can. albicans* were tested for the formation of lesions within the epithelial tissue. All single mutants of Sap1–Sap3 showed a reduced lesion formation and this attenuation was even enhanced in a Sap1/Sap3 double mutant. By contrast, a *SAP4/5/6* triple mutant showed comparable lesions to that of the corresponding wild-type strain (Schaller et al. 1999). Therefore it was concluded that Sap1–Sap3 mainly contribute to the tissue damage, which is required to penetrate the human oral epithelium. However, the role of the different proteases seems also to depend on the tissue environment. When tissue penetration was investigated in a model of vaginal candidiasis based on reconstituted human vaginal epithelium (RHVE), it turned out that Sap1 and Sap2 seem to be the most important key players, because deletion of *SAP3*, *SAP4*, or *SAP6* did not reduce the virulence of the respective strains (Schaller et al. 2003).

Nevertheless, both groups of Saps (Sap1–Sap3, Sap4–Sap6) seem to be important for the virulence of *Can. albicans*. Deletion of the single *SAP1*–SAP3 genes led, in all cases, to increased survival of infected animals (Hube et al. 1997). In another study a triple mutant of *SAP4*–*SAP6* was investigated in animal infection models and revealed that the triple mutation led to a significantly longer survival of the animals and the cell counts from infected organs displayed a significant reduction of yeast cells compared to the parental wild-type strain (Sanglard et al. 1997).

Substrate specificity was best studied for Sap2, which is the most highly expressed protease under in vitro conditions when proteins are supplied as the sole source of nitrogen. Sap 2 degrades a broad spectrum of proteins, including human proteins at the mucosal site, like keratin, collagen, vimentin, fibronectin, laminin, and mucin. Additionally Sap2 can hydrolyse, cleave, or activate several proteins from the innate and adaptive immune response, like most immunoglobulins, complement, interleukin-1β, $α_2$-macroglubulin and others (Naglik et al. 2004). Due to the high similarity of Sap1–Sap3, it is assumed that they hydrolyse similar substrates. Hydrolysis of host proteins may not only enable the fungus to enter the tissues and to evade the immune system, but can also provide nutrients, at least, in the initial phase of infection. After the yeast to hyphae switch the activity of Sap4–Sap6 may mainly contribute to proteolytic activity of the fungus. In contrast to Sap1–Sap3, which show highest activity at acidic pH values, Sap4–Sap6 are almost exclusively expressed during hyphal formation at near neutral pH values (Hube et al. 1994; White and Agabian 1995) and the optimum pH is in a range between pH 5 and 7. Sap4–Sap6 seem to play a key role in resistance and evasion to macrophage attack. *SAP4*–SAP6 triple mutants were approximately 50% more effectively eliminated by macrophages than a wild-type strain (Borg-von Zepelin et al. 1998). This suggests that Sap4–Sap6 recognize target proteins in macrophages, which are not recognised by Sap1–Sap3. Furthermore, Sap4–Sap6 are associated with the hyphae and hyphal growth is essential to escape the macrophage. Some studies have also proven a strong impact of Sap4–Sap6 in systemic infections, whereby Sap6 strongly contributes to tissue damage of liver and other parenchymal organs (Felk et al. 2002).

In conclusion, proteolytic activity deriving from several secreted aspartic proteases seems to be indispensable for penetration of host tissues, immune evasion, and dissemination in systemic infections. Therefore, the inhibition of proteolytic activity at the sites of infection by specific Sap inhibitors may be helpful in controlling *Can. albicans* infections.

D. Proteases from *Cry. neoformans*

Cry. neoformans strains possess approximately 78 different proteases (http://merops.sanger.ac.uk/). However, in contrast to *Can. albicans*: (a) the number of aspartic proteases is significantly lower, (b) most of the proteases are still unassigned, (c) the impact of the proteases on virulence has hardly been studied.

For a long time it was believed that *Cry. neoformans* is non-proteolytic. The first report on protease secretion, which was confirmed by others, was published in 1972 (Müller and Sethi 1972). During incubation of *Cry. neoformans* in vitro on media containing bovine serum albumin (BSA) as a source of carbon and nitrogen some strains produced proteases. The addition of alternative nitrogen sources or the addition of glucose suppresses the proteolytic activity (Brueske 1986; Aoki et al. 1994). Later, different components of the immune response were tested for their suitability to act as a substrate for proteolytic enzymes from *Cry. neoformans*. Mouse immunoglobulin G1 (an antibody directed against the capsular polysaccharide), bovine immunoglobulin G and human complement factor 5 supported growth when given as the sole source of carbon and nitrogen. Analysis of proteases in gelatine zymogram gels revealed three major proteolytic bands, with the highest intensities at 200 kDa and 100 kDa, with a minor intensity at 50 kDa (Chen et al. 1996). In contrast to the observations by Brueske (1986), protease activity was also observed when glucose was used as the carbon source and glycine as the nitrogen source, implying that different *Cry. neoformans* strains regulate protease production differently. This assumption was confirmed by another investigation which used a tannic acid based procedure to visualise protein hydrolysis after growth on gelatine or BSA-containing media. The tannic acid procedure allowed a simple and rapid staining with a good evidence for the clearing zones (Ruma-Haynes et al. 2000). Some 32 strains of *Cry. neoformans* var. *gattii* and 31 isolates of *Cry. neoformans* var. *neoformans* were tested for growth and clearing zones; and it turned out that the proteolytic activity of independent strains differs, regardless of the variants, but all strains were able to use proteins, at least, as a nitrogen source. In the presence of glucose the zones of protein hydrolysis were not clearly visible, which implies that glucose reduces protease production and/or secretion (Ruma-Haynes et al. 2000). Nevertheless, *Cry. neoformans* var. *neoformans* produced annuli around the colonies which were significantly larger than those from var. *gattii*; and it was speculated that the reduced dissemination of *Cry. neoformans* var. *gattii* in comparison to var. *neoformans* might be caused by the lower protease excretion.

Other studies investigated the impact of protease excretion on the host immune response. A given wild-type strain of *Cry. neoformans* was UV-radiated independently in different laboratories. The resulting strains displayed several differences concerning proteolytic activity, doubling time, and capsule thickness. All strains showed differences in virulence when tested in mouse infection models. Although it was difficult to attribute the attenuation to a specific gene, it was observed that a strain with low proteolytic activity produced a strong inflammatory response. Lung yeast counts of this protease "low-producer" strain were decreased and the inflammatory response of the host immune system eliminated the cells from the infected lung tissue. By contrast, a strain with high proteolytic activity (which was growing even slower under in vitro conditions than the strain with low proteolytic activity) caused only a weak granulomatous response, displayed high CFUs, and disseminated throughout the alveoli of the lung (Chen and Casadevall 1999). This result was explained by the ability of the excreted protease to cleave several proteins involved in the host immune response (Chen et al. 1996). Furthermore, it was shown that a serine protease from *Cry. neoformans* is involved in the cleavage of human fibronectin, which is present as a polymeric fibrillar network in the extracellular matrix. Since opsonisation of pathogens with fibronectin enhances phagocytosis via both complement and Fc receptors, this serine protease may be involved in escaping phagocytosis (Rodrigues et al. 2003). In this respect a serine protease was purified to homogeneity from a selected *Cry. neoformans* isolate with a high proteolytic activity (Yoo Ji et al. 2004). The purified protein displayed a molecular mass of 43 kDa, which coincides with the proteolytic band observed from above (Chen et al. 1996). However, the purified enzyme was not tested for cleavage of immunoglobulins, complement, and fibronectin. Nevertheless, it turned out that proteolytic activity of *Cry. neoformans* is mainly attributed to serine proteases, since in all inhibitor studies the inhibitors of serine proteases were able to repress the proteolytic activity. Finally, from recent studies it was concluded that proteases are involved in weakening of the host immune response, may be involved in tissue penetration and also in dissemination (Vidotto et al. 2005).

E. Proteases and Their Role in *A. fumigatus* Pathogenesis

The MEROPS database (http://merops.sanger.ac.uk/) reveals that *A. fumigatus* contains 116 known or putative sequences coding for proteases. The number of proteases within different aspergilli seems to vary strongly, although not all genomes have yet been sequenced, published or added to the database. *A. nidulans*, which is known to be much less pathogenic compared to *A. fumigatus*, contains 205 proteases, *A. niger* 455, *A. oryzae* 75, and *A. terreus* 88 proteases. However, a function was attributed only to a small number of these proteases and it is, therefore, not obvious whether the assumed number of proteases is sufficient to predict for the virulence of a strain, especially since *A. nidulans* is known to be hardly virulent.

The literature on the impact of proteases on pathogenesis of *A. fumigatus* also gives no clear direction. Even more, it is a matter of debate whether tissue hydrolysis involves a single type of protease or a mixture of several protease family members. The fact that *A. fumigatus* is able to secrete a number of different proteases in response to the available carbon sources may be the reason why the investigation of the impact on virulence by these enzymes is problematic.

The main research on proteolytic activity of *A. fumigatus* strains started in the 1990s, when it was recognised that the lung tissue contains large amounts of elastin, which was thought to become hydrolysed by *A. fumigatus* during invasive tissue penetration. An alkaline serine protease (Alp) with an apparent molecular mass of approximately 32 kDa was purified from a clinical *A. fumigatus* isolate and revealed elastinolytic activity (Reichard et al. 1990). The purified enzyme was also able to detach Vero cells from a plastic surface, which suggests that the protease was able to act on proteins presented at the outer cell membrane. Independently, an Alp with an apparent molecular mass of 33 kDa was purified from *A. fumigatus* grown in the presence of collagen as sole carbon and nitrogen source. Due to the high collagenolytic activity and the ability of the enzyme to digest elastin, a role of the enzyme in tissue penetration was assumed (Monod et al. 1991). Three other groups also purified this 32–33 kDa alkaline serine protease by using different growth conditions and purification strategies (Frosco et al. 1992; Larcher et al. 1992; Kolattukudy et al. 1993) and a chemical mutagenesis was performed to identify strains which were deficient in elastase production. Virulence studies with an elastase-deficient strain revealed a strong attenuation in virulence. This attenuation was attributed to the specific disruption of elastase activity; but chemical mutagenesis may not necessarily target a single gene. However, immunogold localisation of the elastolytic protease of wild-type strains during lung infection showed that the enzyme is present during the germination of inhaled conidia. In contrast, when the gene coding for Alp was deleted, resulting in an elastase-deficient strain, virulence was not attenuated. However, it was noteworthy that elastinolysis within the infected lungs was observed regardless whether a mutant or wild-type strain was investigated (Tang et al. 1992; Tang et al. 1993). This experiment was confirmed by an independent study (Monod et al. 1993b). However, another protease was identified in culture supernatans of Alp-negative strains and the mutants were still able to grow on collagen medium when supplied as sole carbon and nitrogen source. This purified protease showed activity with collagen as a substrate but no elastinolytic activity. Inhibitor studies revealed that the purified 40-kDa protein behaved like a metalloprotease and was therefore named MEP. All serum samples of patients with an aspergilloma tested for antibodies directed against MEP turned out to be positive, whereas controls of patients with candidiasis remained negative. That implied that the secreted MEP is antigenic and produced during infection (Monod et al. 1993a). However, when the gene coding for MEP was disrupted from the genome of both a wild-type and an Alp-disruption strain, all strains remained virulent (Jaton-Ogay et al. 1994). As probably expected, other proteases were found to be produced by *A. fumigatus*. An aspartic protease was found to be secreted and N-terminal sequencing revealed a high identity to aspergillopepsin A from *A. niger* (Reichard et al. 1994, 1995). Immunofluorescence analysis showed that the enzyme is mainly found in developing conidiophores, in submerged mycelia and on the tips of growing aerial mycelia (Reichard et al. 1996) but no attenuation in virulence was detected when a mutant was tested in a mouse model of invasive aspergillosis (Reichard et al. 1997). However, indications for another protease were found. As can be seen from these investigations, *A. fumigatus* seems to display a large redundancy in protease function. In the meantime several other proteases, among them two dipeptidyl-peptidases, which also seem to be antigenic during infection (Beauvais et al. 1997a, b), were detected. Other proteases produced by *A. fumigatus* seem to be bound to the cell wall of hyphae and especially to the growing hyphal tip, like PEP2 (Reichard 1998; Reichard et al. 2000). They may also be involved in protein degradation during invasive growth. Further literature on the role and identification of proteases is presented in several reviews (Monod et al. 2002; Rementeria et al. 2005). Noteworthy, recently a class of sedolisins was discovered which represent serine proteases of the subtilisin subfamily and display maximum activity at lower pH. Secretion of the sedolosins is induced when *A. fumigatus* is grown in the presence of haemoglobin as the sole nitrogen source. Heterologous expression of four of the identified *sed* genes (*sedA*–*sedD*) in *Pichia pastoris* revealed that SedA acts as an endopeptidase whereas SedB, SedC, and SedD cleave tripeptides as known from tripeptidylpeptidases (TPP). Due to the preference of an acidic environment for full activity of the *sed* gene products it is proposed that the combined action of Pep1 and SedA (acting as endoproteases) and SedB, SedC, and SedD (acting as exoproteases) can efficiently provide nitrogen for assimilation in different acidic micro-environments. However, whether one or all of the *sed* genes have an impact on virulence has not been studied yet (Reichard et al. 2006).

It can be concluded that *A. fumigatus* possesses a large toolbox of proteases to degrade proteins. The ability to secrete proteases which are active either in an acidic, a neutral, or an alkaline condition seems to give an advantage for rapid adaptation to environmental changes and may be another reason for *A. fumigatus* being a perfect saprophyte. Although up to now no clear evidence has been provided that proteases are involved in invasive growth of *A. fumigatus*, it does not mean that proteases are not of special importance. To study their impact, it may be more suitable to study the effect of the accumulation of toxic compounds which derive from the degradation of single amino acids. One example for such an approach is the accumulation of toxic propionyl-CoA, which derives from the degradation of methionine, valine, and isoleucine (Brock and Buckel 2004; Zhang et al. 2004). Propionyl-CoA is generally removed by the methylcitrate cycle, which involves the condensation of propionyl-CoA with oxaloacetate to generate methylcitrate via a specific methylcitrate synthase. It has been shown that disruption of the methylcitrate synthase leads to reduced virulence in an insect (Maerker et al. 2005) as well as in a mouse infection model (Ibrahim-Granet et al. 2008). Histopathologic sections of infected mouse lungs revealed a reduced number of inflamed lesions; and mice that survived the first phase of acute infection were able to clear the mycelium from the tissue. Although this study provided new insights on the importance of proteins as a source of nutrients during infection, this investigation did not resolve the question of which proteins may act as main nutrients and which proteases may be involved in the provision of the single amino acids. Therefore, further studies on secreted proteases are urgently needed.

V. Role of Carbohydrate and Nitrogen Metabolism in the Virulence of Pathogenic Fungi

Lipids and proteins are assumed to provide, at least to some extent, nutrients for the growth and maintenance of pathogenic micro organisms.

However, especially during dissemination, glycolysis and carbohydrate metabolism may also play an essential role in nutrient supply since blood serum contains approximately 0.7 g/l glucose. By contrast, during growth within macrophages or tissues, gluconeogenesis may be of higher importance. Until now, most studies on the impact of carbohydrate metabolism have been performed on the yeasts *S. cerevisiae* and *Can. albicans*.

In order to study the impact of the ability to utilise different carbon and nitrogen sources during infection, mutants of *S. cerevisiae* were generated and injected into the tail vein of mice. The survival of the mutants was followed over a period of 14 days. It turned out that *S. cerevisiae*, at least when present in the blood stream and in the brain, is not dependent on a single nitrogen source. Deletion of genes involved in ammonium transport, urea transport and metabolism, nitrogen sensing, and the transport of some selected amino acids did not significantly affect survival of the mutants, which implies some metabolic redundancy. However, some amino acids seem to be taken up from the surrounding media, whereas others are synthesised de novo. When tyrosine and phenylalanine biosynthesis genes were deleted, a strong reduction in yeast counts was observed at 4 h post-infection. By contrast, when a gene involved in lysine biosynthesis (here the saccharopine dehydrogenase) was disrupted, no effect was observed. This indicates that lysine may be efficiently taken up from the host, whereas other amino acids are not (Kingsbury et al. 2006). However, for *A. fumigatus* another picture can be drawn. Deletion of the gene coding for homoaconitase, which is essential for the de novo synthesis of lysine, led to an avirulent phenotype, indicating that lysine is not sufficiently provided by the host to allow the germination and growth of the inhaled conidia (Liebmann et al. 2004). Nevertheless, whether lysine can be taken up sufficiently from protein degradation in later stages of *A. fumigatus* infections has not yet been studied.

Besides the importance of the de novo synthesis of some amino acids to support growth of fungal pathogens, other metabolites and cofactors also have to be synthesised by the fungi, because the levels of these compounds in host tissues and serum are either too low or cannot efficiently be taken up by the fungi. Examples are given by the need for the de novo synthesis of uridine and *para*-aminobenzoic acid by *Can. albicans* and *A. fumigatus*. Deletion of one of the key enzymes of uridine synthesis, the orotidine-5´-monophosphate decarboxylase (*ura3* for *Can. albicans*, *pyrG* for *A. fumigatus*), led to a strongly attenuated virulence (Kirsch and Whitney 1991; D'Enfert et al. 1996). However, in the case of *A. fumigatus*, supplementation of the drinking water with uridine restored the virulence of the *pyrG* mutant, which indicates the presence of a highly efficient uridine-uptake system.

Folate contains *para*-aminobenzoic acid (PABA) as a core molecule in its structure (Cossins and Chen 1997). Mutants of different *Aspergillus* spp., which carry a defective *pabaA*-gene which codes for the *para*-aminobenzoic acid synthetase, are auxotrophic for the supplementation of PABA. When different *Aspergillus pabaA* mutants were tested for their virulence in mouse infection models, they turned out to be avirulent. This implies that the freely available concentration of either folate or PABA within the host is not sufficient to support fungal growth. Nevertheless, PABA supplementation of the drinking water restored virulence of the mutants (Sandhu et al. 1976; Tang et al. 1994; Brown et al. 2000). Implications for the importance of folate synthesis on the virulence of *Can. albicans* have also been reported. The folate synthesis of *Can. albicans* can be inhibited by antifolates such as pyrimethamine or methotrexate. The addition of these drugs together with azole antifungal compounds strongly increased the efficiency of the inhibition of ergosterol biosynthesis (Navarro-Martínez et al. 2006). Therefore, inhibitors of folate biosynthesis may be, at least, suitable for a combination therapy against fungal infections.

In addition to the importance of nitrogen supply from amino acids and the need for the de novo synthesis of uridine and folate, the effect of the disruption of carbohydrate metabolism was studied in some fungal species. In *S. cerevisiae* the disruption of enzymes involved in glucose sensing (especially low-glucose sensing), high-affinity hexose transport, glycolysis (by targeting hexokinase 2), and glucose signalling (a site-specific mutation in the hexokinase 2) led to a strong attenuation of the survival rate, indicating that *S. cerevisiae* depends on glucose as a carbon source (Kingsbury et al. 2006). However, these results only considered cells which are present in the blood stream or within the brain, loci which may contain a higher glucose concentration than found in other tissues. Furthermore, the ability to utilise trehalose was found to be essential for the survival of *S. cerevisiae*, an observation which was shared by investigations on *Cry. neoformans* (Petzold et al. 2006). The disaccharide trehalose is an important stress protectant and is additionally used as a carbon storage compound in the conidia of *Aspergillus* and other fungal species (d'Enfert et al. 1999; Fillinger et al. 2001). Several studies on *S. cerevisiae* have shown that trehalose mediates heat resistance, prevents dehydration, desiccation, the aggregation of denatured proteins, and can also stabilise membranes. Nearly all mutants of *S. cerevisiae* with defects in trehalose utilisation showed a severely decreased cell count after re-isolation, which was mainly attributed to the important function in stress response and re-naturation of proteins during stress recovery (Kingsbury et al. 2006). In *Cry. neoformans* especially trehalose-6-phosphate synthase (TPS1) seemed to be highly important. Mice infected with a *TPS1* mutant survived for a period of 40 days, whereas all mice infected with the wild type died within 16 days post-infection, underlining the role of trehalose in stress resistance. Nevertheless, in a *Caenorhabditis elegans* model, which was not performed at high temperature, the *TPS1* mutant was again much less virulent, implicating a role of TPS1 beyond heat-stress response (Petzold et al. 2006). In the case of *Can. albicans*, disruption of the gene coding for the neutral trehalase did not alter the resistance to oxidative stress (Pendreno et al. 2006). Therefore, it is not clear whether trehalose breakdown also plays a role in virulence of *Can. albicans*. However, recent studies revealed an interesting niche-specific regulation of central metabolic pathways in *Can. albicans*, mainly concerning gluconeogenesis versus glycolysis. As mentioned before, the glyoxylate cycle in *Can. albicans* is active when cells are phagocytosed by macrophages, attacked by neutrophils, and also in some cells growing within tissues. This expression pattern coincided with that of the gluconeogenic marker enzyme phosphoenolpyruvate carboxykinase. In contrast, genes from the glycolysis were present but not induced after phagocytosis by macrophages and enzymes from glycolysis were always active regardless the status of infection. The importance of glycolysis was confirmed by the deletion of the genes coding for pyruvate kinase and phosphofructokinase. Interestingly, the mutants were not only strongly attenuated in virulence but additionally were less virulent than a gluconeogenic or glyoxylate cycle mutant. Furthermore, the fungal burdens recovered from infected tissues were significantly lower with the glycolysis mutants than those obtained with the other two mutants. Therefore, carbohydrate metabolism seems also to play an essential role in the virulence of *Can. albicans* (Barelle et al. 2006). Interestingly, glycolytic enzymes such as the glycerlaldehyde-3-phosphate dehydrogenase (Gil-Navarro et al. 1997) as well as the 3-phosphoglycerate kinase (Alloush et al. 1997) are presented on the outer surface of *Can. albicans* cells and may contribute to recognition of yeast cells by the host immune system. Furthermore, these proteins seem to be antigenic, because sera of patients with candidiasis showed an antibody response against these proteins. In general, it is difficult to understand why these enzymes are presented on the outer surface but it was assumed that such enzymes might also act on immune modulation of the host response since it has been shown that *Can. albicans* presents proteins on the outer surface which are able to bind complement regulators such as factor H and C4b (Meri et al. 2002; Meri et al. 2004). Whether glycolytic enzymes play a role in this binding is, at the moment, unclear.

Data confirming or refuting an impact of glycolysis on the virulence of *A. fumigatus* have not been collected or published yet. Therefore, the role of carbohydrate metabolism displays another interesting field to elucidate mechanisms contributing to the virulence of fungi.

VI. Conclusion

Free-living but pathogenic fungi possess a versatile metabolic capacity. This feature makes it likely that these micro-organisms are independent of specific nutrients within the host. Nearly every carbon and nitrogen source can be consumed and used for the synthesis of building blocks which are not provided by the host. Nevertheless, some specific adaptations may have occurred that enable some fungi to grow within a host, whereas others do not. The large number of secreted aspartic proteases of *Can. albicans* seems to result from gene duplications and allows the degradation of proteins in acidic as well as in neutral and alkaline environments. Due to the fact that, at least, SAP1–SAP6 play a role in virulence makes such an adaptation to the host environment likely. Furthermore, the ability of fungi to produce prostaglandins from host arachidonic acid may reflect an adaptation to the host environment but may also be important during specific environmental conditions. As soil micro-organisms, these fungi also represent nutriments for predators such as nematodes or amoeba. The ability to lyse host cells or to survive and evade the acidic environment of phagolysosomes may perfectly fit into the requirements needed to act as a human opportunistic pathogen. The assumption that the ability to act as a pathogen derives from an adaptation to the environmental niche is supported by the fact that infection models such as *Drosophila melanogaster*, *Caenorhabditis elegans*, *Acanthamoeba castellanii*, *Dictyostelium discoideum*, and *Galleria mellonella* have been successfully used for identification of genes and molecules involved in pathogenesis (Fuchs and Mylonakis 2006). This, of course, does not exclude a specific adaptation and the need for specialised virulence determinants, because every fungus having its natural habitat within the soil has to defend itself against predators, but only a few of these fungi act as pathogens. Therefore, the investigation of virulence determinants in mammalian infection models will remain an indispensable experimental system for the elucidation of virulence-mediating determinants and for the identification of new antifungal drug targets.

Acknowledgements. The work in the author's laboratory is supported by the Deutsche Forschungsgemeinschaft and the Leibniz Institute for Natural Product Research and Infection Biology (HKI).

References

Albrecht A, Felk A, Pichova I, Naglik JR, Schaller M, Groot P de, Maccalum D, Odds FC, Schäfer W, Klis F, Monod M, Hube B (2006) Glycosylphosphatidylinositol-anchored proteases of *Candida albicans* target proteins necessary for both cellular processes and host-pathogen interactions. J Biol Chem 281:688–694

Alloush HM, Lopez-Ribot JL, Masten BJ, Chaffin WL (1997) 3-Phosphoglycerate kinase: a glycolytic enzyme protein present in the cell wall of *Candida albicans*. Microbiology 143:321–330

Aoki S, Ito-Kuwa S, Nakamura K, Kato J, Ninomiya K, Vidotto V (1994) Extracellular proteolytic activity of *Cryptococcus neoformans*. Mycopathologia 128:143–150

Barelle CJ, Priest CL, MacCallum DM, Gow NA, Odds FC, Brown AJ (2006) Niche-specific regulation of central metabolic pathways in a fungal pathogen. Cell Microbiol 8:961–971

Barrett AJ, Rawlings ND (1995) Families and clans of serine peptidases. Arch Biochem Biophys 318:247–250

Beauvais A, Monod M, Debeaupuis JP, Diaquin M, Kobayashi H, Latgé JP (1997a) Biochemical and antigenic characterization of a new dipeptidyl-peptidase isolated from *Aspergillus fumigatus*. J Biol Chem 272:6238–6244

Beauvais A, Monod M, Debeaupuis JP, Grouzmann E, Brakch N, Svab J, Hovanessian AG, Latgé JP (1997b) Dipeptidyl-peptidase IV secreted by *Aspergillus fumigatus*, a fungus pathogenic to humans. Infect Immun 65:3042–3047

Bernheimer AW, Linder R, Weinstein SA, Kim KS (1987) Isolation and characterization of a phospholipase B from venom of Collett's snake, *Pseudechis colletti*. Toxicon 25:547–554

Birch M, Robson G, Law D, Denning DW (1996) Evidence of multiple extracellular phospholipase activities of *Aspergillus fumigatus*. Infect Immun 64:751–755

Birch M, Denning DW, Robson GD (2004) Comparison of extracellular phospholipase activities in clinical and environmental *Aspergillus fumigatus* isolates. Med Mycol 42:81–86

Borg-von Zepelin M, Beggah S, Boggian K, Sanglard D, Monod M (1998) The expression of the secreted aspartyl proteinases Sap4 to Sap6 from *Candida albicans* in murine macrophages. Mol Microbiol 28:543–554

Brock M, Buckel W (2004) On the mechanism of action of the antifungal agent propionate. Eur J Biochem 271:3227–3241

Brown JS, Aufauvre-Brown A, Brown J, Jennings JM, Arst H Jr, Holden DW (2000) Signature-tagged and directed mutagenesis identify PABA synthetase as essential for *Aspergillus fumigatus* pathogenicity. Mol Microbiol 36:1371–1380

Brueske CH (1986) Proteolytic activity of a clinical isolate of *Cryptococcus neoformans*. J Clin Microbiol 23:631–633

Capobianco JO, Lerner CG, Goldman RC (1992) Application of a fluorogenic substrate in the assay of proteolytic activity and in the discovery of a potent inhibitor of *Candida albicans* aspartic proteinase. Anal Biochem 204:96–102

Chen LC, Casadevall A (1999) Variants of a *Cryptococcus neoformans* strain elicit different inflammatory responses in mice. Clin Diagn Lab Immunol 6:266–268

Chen LC, Blank ES, Casadevall A (1996) Extracellular proteinase activity of *Cryptococcus neoformans*. Clin Diagn Lab Immunol 3:570–574

Chen SC, Müller M, Zhou JZ, Wright LC, Sorrell TC (1997) Phospholipase activity in *Cryptococcus neoformans*: a new virulence factor? J Infect Dis 175:414–420

Chen SC, Wright LC, Golding JC, Sorrell TC (2000) Purification and characterization of secretory phospholipase B, lysophospholipase and lysophospholipase/transacylase from a virulent strain of the pathogenic fungus *Cryptococcus neoformans*. Biochem J 347:431–439

Cossins EA, Chen L (1997) Folates and one-carbon metabolism in plants and fungi. Phytochemistry 45:437–452

Cox GM, McDade HC, Chen SC, Tucker SC, Gottfredsson M, Wright LC, Sorrell TC, Leidich SD, Casadevall A, Ghannoum MA, Perfect JR (2001) Extracellular phospholipase activity is a virulence factor for *Cryptococcus neoformans*. Mol Microbiol 39:166–175 d'Enfert C, Diaquin M, Delit A, Wuscher N, Debeaupuis JP, Huerre M, Latgé JP (1996) Attenuated virulence of uridine–uracil auxotrophs of *Aspergillus fumigatus*. Infect Immun 64:4401–4405

d'Enfert C, Bonini BM, Zapella PD, Fontaine T, Silva AM da, Terenzi HF (1999) Neutral trehalases catalyse intracellular trehalose breakdown in the filamentous fungi *Aspergillus nidulans* and *Neurospora crassa*. Mol Microbiol 32:471–483

Djordjevic JT, Del Poeta M, Sorrell TC, Turner KM, Wright LC (2005) Secretion of cryptococcal phospholipase B1 (PLB1) is regulated by a glycosylphosphatidylinositol (GPI) anchor. Biochem J 389:803–812

Dolan JW, Bell AC, Hube B, Schaller M, Warner TF, Balish E (2004) *Candida albicans* PLD I activity is required for full virulence. Med Mycol 42:439–447

Ebel F, Schwienbacher M, Beyer J, Heesemann J, Brakhage AA, Brock M (2006) Analysis of the regulation, expression, and localisation of the isocitrate lyase from *Aspergillus fumigatus*, a potential target for antifungal drug development. Fungal Genet Biol 43:476–489

Felk A, Kretschmar M, Albrecht A, Schaller M, Beinhauer S, Nichterlein T, Sanglard D, Korting HC, Schäfer W, Hube B (2002) *Candida albicans* hyphal formation and the expression of the Efg1-regulated proteinases Sap4 to Sap6 are required for the invasion of parenchymal organs. Infect Immun 70:3689–3700

Fillinger S, Chaveroche MK, Dijck P van, Vries R de, Ruijter G, Thevelein J, d'Enfert C (2001) Trehalose is required for the acquisition of tolerance to a variety of stresses in the filamentous fungus *Aspergillus nidulans*. Microbiology 147:1851–1862

Fradin C, Kretschmar M, Nichterlein T, Gaillardin C, d'Enfert C, Hube B (2003) Stage-specific gene expression of *Candida albicans* in human blood. Mol Microbiol 47:1523–1543

Fradin C, De Groot P, MacCallum D, Schaller M, Klis F, Odds FC, Hube B (2005) Granulocytes govern the transcriptional response, morphology and proliferation of *Candida albicans* in human blood. Mol Microbiol 56:397–415

Frosco M, Chase T Jr, Macmillan JD (1992) Purification and properties of the elastase from *Aspergillus fumigatus*. Infect Immun 60:728–734

Fuchs BB, Mylonakis E (2006) Using non-mammalian hosts to study fungal virulence and host defense. Curr Opin Microbiol 9:346–351

Ghannoum MA (2000) Potential role of phospholipases in virulence and fungal pathogenesis. Clin Microbiol Rev 13:122–143

Giddey K, Favre B, Quadroni M, Monod M (2007) Closely related dermatophyte species produce different patterns of secreted proteins. FEMS Microbiol Lett 267:95–101

Gil-Navarro I, Gil ML, Casanova M, O'Connor JE, Martinez JP, Gozalbo D (1997) The glycolytic enzyme glyceraldehyde-3-phosphate dehydrogenase of *Candida albicans* is a surface antigen. J Bacteriol 179:4992–4999

Greenspan P, Mayer EP, Fowler SD (1985) Nile red: a selective fluorescent stain for intracellular lipid droplets. J Cell Biol 100:965–973

Heung LJ, Kaiser AE, Luberto C, Del Poeta M (2005) The role and mechanism of diacylglycerol-protein kinase C1 signaling in melanogenesis by *Cryptococcus neoformans*. J Biol Chem 280:28547–28555

Hube B, Naglik J (2001) *Candida albicans* proteinases: resolving the mystery of a gene family. Microbiology 147:1997–2005

Hube B, Monod M, Schofield DA, Brown AJ, Gow NA (1994) Expression of seven members of the gene family encoding secretory aspartyl proteinases in *Candida albicans*. Mol Microbiol 14:87–99

Hube B, Sanglard D, Odds FC, Hess D, Monod M, Schäfer W, Brown AJ, Gow NA (1997) Disruption of each of the secreted aspartyl proteinase genes SAP1, SAP2, and SAP3 of *Candida albicans* attenuates virulence. Infect Immun 65:3529–3538

Hube B, Hess D, Baker CA, Schaller M, Schäfer W, Dolan JW (2001) The role and relevance of phospholipase D1 during growth and dimorphism of *Candida albicans*. Microbiology 147:879–889

Ibrahim AS, Mirbod F, Filler SG, Banno Y, Cole GT, Kitajima Y, Edwards JE Jr, Nozawa Y, Ghannoum MA (1995) Evidence implicating phospholipase as a virulence factor of *Candida albicans*. Infect Immun 63:1993–1998

Ibrahim-Granet O, Dubourdeau M, Latgé JP, Ave P, Huerre M, Brakhage AA, Brock M (2008) Methylcitrate synthase from *Aspergillus fumigatus* is essential for manifestation of invasive aspergillosis. Cell Microbiol 10:134–148

Jahn B, Boukhallouk F, Lotz J, Langfelder K, Wanner G, Brakhage AA (2000) Interaction of human phagocytes with pigmentless *Aspergillus* conidia. Infect Immun 68:3736–3739

Jahn B, Langfelder K, Schneider U, Schindel C, Brakhage AA (2002) PKSP-dependent reduction of phagolysosome fusion and intracellular kill of *Aspergillus fumigatus* conidia by human monocyte-derived macrophages. Cell Microbiol 4:793–803

Jaton-Ogay K, Paris S, Huerre M, Quadroni M, Falchetto R, Togni G, Latgé JP, Monod M (1994) Cloning and disruption of the gene encoding an extracellular metalloprotease of *Aspergillus fumigatus*. Mol Microbiol 14:917–928

Kingsbury JM, Goldstein AL, McCusker JH (2006) Role of nitrogen and carbon transport, regulation, and metabolism genes for *Saccharomyces cerevisiae* survival in vivo. Eukaryot Cell 5:816–824

Kirsch DR, Whitney RR (1991) Pathogenicity of *Candida albicans* auxotrophic mutants in experimental infections. Infect Immun 59:3297–3300

Knechtle P, Goyard S, Brachat S, Ibrahim-Granet O, d'Enfert C (2005) Phosphatidylinositol-dependent phospholipases C Plc2 and Plc3 of *Candida albicans* are dispensable for morphogenesis and host-pathogen interaction. Res Microbiol 156:822–829

Kolattukudy PE, Lee JD, Rogers LM, Zimmerman P, Celeski S, Fox B, Stein B, Copelan EA (1993) Evidence for possible involvement of an elastolytic serine protease in aspergillosis. Infect Immun 61:2357–2368

Kunze D, Melzer I, Bennett D, Sanglard D, MacCallum D, Nörskau J, Coleman DC, Odds FC, Schäfer W, Hube B (2005) Functional analysis of the phospholipase C gene CaPLC1 and two unusual phospholipase C genes, CaPLC2 and CaPLC3, of *Candida albicans*. Microbiology 151:3381–3394

Kvaal C, Lachke SA, Srikantha T, Daniels K, McCoy J, Soll DR (1999) Misexpression of the opaque-phase-specific gene PEP1 (SAP1) in the white phase of *Candida albicans* confers increased virulence in a mouse model of cutaneous infection. Infect Immun 67:6652–6662

Langfelder K, Jahn B, Gehringer H, Schmidt A, Wanner G, Brakhage AA (1998) Identification of a polyketide synthase gene (*pksP*) of *Aspergillus fumigatus* involved in conidial pigment biosynthesis and virulence. Med Microbiol Immunol 187:79–89

Larcher G, Bouchara JP, Annaix V, Symoens F, Chabasse D, Tronchin G (1992) Purification and characterization of a fibrinogenolytic serine proteinase from *Aspergillus fumigatus* culture filtrate. FEBS Lett 308:65–69

Leidich SD, Ibrahim AS, Fu Y, Koul A, Jessup C, Vitullo J, Fonzi W, Mirbod F, Nakashima S, Nozawa Y, Ghannoum MA (1998) Cloning and disruption of caPLB1, a phospholipase B gene involved in the pathogenicity of *Candida albicans*. J Biol Chem 273:26078–26086

Liebmann B, Mühleisen TW, Müller M, Hecht M, Weidner G, Braun A, Brock M, Brakhage AA (2004) Deletion of the *Aspergillus fumigatus* lysine biosynthesis gene *lysF* encoding homoaconitase leads to attenuated virulence in a low-dose mouse infection model of invasive aspergillosis. Arch Microbiol 181:378–383

Lorenz MC, Fink GR (2001) The glyoxylate cycle is required for fungal virulence. Nature 412:83–86

Lorenz MC, Fink GR (2002) Life and death in a macrophage: role of the glyoxylate cycle in virulence. Eukaryot Cell 1:657–662

Lorenz MC, Bender JA, Fink GR (2004) Transcriptional response of *Candida albicans* upon internalization by macrophages. Eukaryot Cell 3:1076–1087

Maerker C, Rohde M, Brakhage AA, Brock M (2005) Methylcitrate synthase from *Aspergillus fumigatus*. Propionyl-CoA affects polyketide synthesis, growth and morphology of conidia. FEBS J 272:3615–3630

Mare L, Iatta R, Montagna MT, Luberto C, Del Poeta M (2005) APP1 transcription is regulated by inositol-phosphorylceramide synthase 1-diacylglycerol pathway and is controlled by ATF2 transcription factor in *Cryptococcus neoformans*. J Biol Chem 280:36055–36064

Meri T, Hartmann A, Lenk D, Eck R, Würzner R, Hellwage J, Meri S, Zipfel PF (2002) The yeast *Candida albicans* binds complement regulators factor H and FHL-1. Infect Immun 70:5185–5192

Meri T, Blom AM, Hartmann A, Lenk D, Meri S, Zipfel PF (2004) The hyphal and yeast forms of *Candida albicans* bind the complement regulator C4b-binding protein. Infect Immun 72:6633–6641

Monod M, Togni G, Rahalison L, Frenk E (1991) Isolation and characterisation of an extracellular alkaline protease of *Aspergillus fumigatus*. J Med Microbiol 35:23–28

Monod M, Paris S, Sanglard D, Jaton-Ogay K, Bille J, Latgé JP (1993a) Isolation and characterization of a secreted metalloprotease of *Aspergillus fumigatus*. Infect Immun 61:4099–4104

Monod M, Paris S, Sarfati J, Jaton-Ogay K, Ave P, Latgé JP (1993b) Virulence of alkaline protease-deficient mutants of *Aspergillus fumigatus*. FEMS Microbiol Lett 106:39–46

Monod M, Hube B, Hess D, Sanglard D (1998) Differential regulation of SAP8 and SAP9, which encode two new members of the secreted aspartic proteinase family in *Candida albicans*. Microbiology 144:2731–2737

Monod M, Capoccia S, Lechenne B, Zaugg C, Holdom M, Jousson O (2002) Secreted proteases from pathogenic fungi. Int J Med Microbiol 292:405–419

Mukherjee PK, Seshan KR, Leidich SD, Chandra J, Cole GT, Ghannoum MA (2001) Reintroduction of the PLB1 gene into *Candida albicans* restores virulence in vivo. Microbiology 147:2585–2597

Müller HE, Sethi KK (1972) Proteolytic activity of *Cryptococcus neoformans* against human plasma proteins. Med Microbiol Immunol 158:129–134

Naglik J, Challacombe S, Hube B (2003) *Candida albicans* secreted aspartyl proteinases in virulence and pathogenesis. Microbiol Mol Biol Rev 67:400–428

Naglik J, Albrecht A, Bader O, Hube B (2004) *Candida albicans* proteinases and host/pathogen interactions. Cell Microbiol 6:915–926

Navarro-Martínez MD, Cabezas-Herrera J, Rodríguez-Lopez JN (2006) Antifolates as antimycotics? Connection between the folic acid cycle and the ergosterol biosynthesis pathway in *Candida albicans*. Int J Antimicrob Agents 28:560–567

Noverr MC, Cox GM, Perfect JR, Huffnagle GB (2003) Role of PLB1 in pulmonary inflammation and cryptococcal eicosanoid production. Infect Immun 71:1538–1547

Pendreno Y, Gonzalez-Parraga P, Conesa S, Martinez-Esperaza M, Aguinaga A, Hernandez JA, Arguelles JC (2006) The cellular resistance against oxidative stress (H_2O_2) is independent of neutral trehalase (Ntc1p) activity in *Candida albicans*. FEMS Yeast Res 6:57–62

Petzold EW, Himmelreich U, Mylonakis E, Rude T, Toffaletti D, Cox GM, Miller JL, Perfect JR (2006) Characterization and regulation of the trehalose synthesis pathway and its importance in the pathogenicity of *Cryptococcus neoformans*. Infect Immun 74:5877–5887

Price MF, Wilkinson ID, Gentry LO (1982) Plate method for detection of phospholipase activity in *Candida albicans*. Sabouraudia 20:7–14

Rawlings ND, Morton FR, Barrett AJ (2006) MEROPS: the peptidase database. Nucleic Acids Res 34:D270–D272

Reichard U (1998) The significance of secretory and structure-associated proteases of *Aspergillus fumigatus* for the pathogenesis of invasive aspergillosis. Mycoses 41[Suppl 1]:78–82

Reichard U, Büttner S, Eiffert H, Staib F, Rüchel R (1990) Purification and characterisation of an extracellular serine proteinase from *Aspergillus fumigatus* and its detection in tissue. J Med Microbiol 33:243–251

Reichard U, Eiffert H, Rüchel R (1994) Purification and characterization of an extracellular aspartic proteinase from *Aspergillus fumigatus*. J Med Vet Mycol 32:427–436

Reichard U, Monod M, Rüchel R (1995) Molecular cloning and sequencing of the gene encoding an extracellular aspartic proteinase from *Aspergillus fumigatus*. FEMS Microbiol Lett 130:69–74

Reichard U, Monod M, Rüchel R (1996) Expression pattern of aspartic proteinase antigens in aspergilli. Mycoses 39:99–101

Reichard U, Monod M, Odds F, Rüchel R (1997) Virulence of an aspergillopepsin-deficient mutant of *Aspergillus fumigatus* and evidence for another aspartic proteinase linked to the fungal cell wall. J Med Vet Mycol 35:189–196

Reichard U, Cole GT, Rüchel R, Monod M (2000) Molecular cloning and targeted deletion of PEP2 which encodes a novel aspartic proteinase from *Aspergillus fumigatus*. Int J Med Microbiol 290:85–96

Reichard U, Lechenne B, Asif AR, Streit F, Grouzmann E, Jousson O, Monod M (2006) Sedolisins, a new class of secreted proteases from *Aspergillus fumigatus* with endoprotease or tripeptidyl-peptidase activity at acidic pHs. Appl Environ Microbiol 72:1739–1748

Rementeria A, Lopez-Molina N, Ludwig A, Vivanco AB, Bikandi J, Ponton J, Graizar J (2005) Genes and molecules involved in *Aspergillus fumigatus* virulence. Rev Iberoam Micol 22:1–23

Rodrigues ML, dos Reis FC, Puccia R, Travassos LR, Alviano CS (2003) Cleavage of human fibronectin and other basement membrane-associated proteins by a *Cryptococcus neoformans* serine proteinase. Microb Pathol 34:65–71

Rude TH, Toffaletti DL, Cox GM, Perfect JR (2002) Relationship of the glyoxylate pathway to the pathogenesis of *Cryptococcus neoformans*. Infect Immun 70:5684–5694

Ruma-Haynes P, Brownlee AG, Sorrell TC (2000) A rapid method for detecting extracellular proteinase activity in *Cryptococcus neoformans* and a survey of 63 isolates. J Med Microbiol 49:733–737

Sandhu DK, Sandhu RS, Khan ZU, Damodaran VN (1976) Conditional virulence of a *p*-aminobenzoic acid-requiring mutant of *Aspergillus fumigatus*. Infect Immun 13:527–532

Sanglard D, Hube B, Monod M, Odds FC, Gow NA (1997) A triple deletion of the secreted aspartyl proteinase genes SAP4, SAP5, and SAP6 of *Candida albicans* causes attenuated virulence. Infect Immun 65:3539–3546

Santangelo R, Zoellner H, Sorrell T, Wilson C, Donald C, Djordjevic J, Shounan Y, Wright L (2004) Role of extracellular phospholipases and mononuclear phagocytes in dissemination of cryptococcosis in a murine model. Infect Immun 72:2229–2239

Schaller M, Korting HC, Schäfer W, Bastert J, Chen W, Hube B (1999) Secreted aspartic proteinase (Sap) activity contributes to tissue damage in a model of human oral candidosis. Mol Microbiol 34:169–180

Schaller M, Bein M, Korting HC, Baur S, Hamm G, Monod M, Beinhauer B, Hube B (2003) The secreted aspartyl proteinases Sap1 and Sap2 cause tissue damage in an in vitro model of vaginal candidiasis based on reconstituted human vaginal epithelium. Infect Immun 71:3227–3234

Schöbel F, Ibrahim-Granet O, Ave P, Latgé JP, Brakhage AA, Brock M (2007) *Aspergillus fumigatus* does not require fatty acid metabolism *via* isocitrate lyase for development of invasive aspergillosis. Infect Immun 75:1237–1244

Shea JM, Kechichian TB, Luberto C, Del Poeta M (2006) The cryptococcal enzyme inositol phosphosphingolipid-phospholipase C confers resistance to the antifungal effects of macrophages and promotes fungal dissemination to the central nervous system. Infect Immun 74:5977–5988

Shen DK, Noodeh AD, Kazemi A, Grillot R, Robson G, Brugere JF (2004) Characterisation and expression of phospholipases B from the opportunistic fungus *Aspergillus fumigatus*. FEMS Microbiol Lett 239:87–93

Shiloah J, Klibansky C, Vries A de, Berger A (1973) Phospholipase B activity of a purified phospholipase A from V*ipera palestinae* venom. J Lipid Res 14:267–278

Siafakas AR, Wright LC, Sorrell TC, Djordjevic JT (2006) Lipid rafts in *Cryptococcus neoformans* concentrate the virulence determinants phospholipase B1 and Cu/Zn superoxide dismutase. Eukaryot Cell 5:488–498

Siezen RJ, Leunissen JA (1997) Subtilases: the superfamily of subtilisin-like serine proteases. Protein Sci 6:501–523

Staib F (1965) Serum-proteins as nitrogen source for yeast-like fungi. Sabouraudia 4:187–193

Steenbergen JN, Casadevall A (2003) The origin and maintenance of virulence for the human pathogenic fungus *Cryptococcus neoformans*. Microbes Infect 5:667–675

Takasaki C, Tamiya N (1982) Isolation and properties of lysophospholipases from the venom of an Australian elapid snake, *Pseudechis australis*. Biochem J 203:269–276

Tang CM, Cohen J, Holden DW (1992) An *Aspergillus fumigatus* alkaline protease mutant constructed by gene disruption is deficient in extracellular elastase activity. Mol Microbiol 6:1663–1671

Tang CM, Cohen J, Krausz T, Van Noorden S, Holden DW (1993) The alkaline protease of *Aspergillus fumigatus* is not a virulence determinant in two murine models of invasive pulmonary aspergillosis. Infect Immun 61:1650–1656

Tang CM, Smith JM, Arst HN Jr, Holden DW (1994) Virulence studies of *Aspergillus nidulans* mutants requiring lysine or *p*-aminobenzoic acid in invasive pulmonary aspergillosis. Infect Immun 62:5255–5260

Taylor BN, Hannemann H, Sehnal M, Biesemeier A, Schweizer A, Rollinghoff M, Schröppel K (2005) Induction of SAP7 correlates with virulence in an intravenous infection model of candidiasis but not in a vaginal infection model in mice. Infect Immun 73:7061–7063

Tsitsigiannis DI, Bok JW, Andes D, Nielsen KF, Frisvad JC, Keller NP (2005) *Aspergillus* cyclooxygenase-like enzymes are associated with prostaglandin production and virulence. Infect Immun 73:4548–4559

Vidotto V, Sinicco A, Di Fraia D, Cardaropoli S, Aoki S, Ito-Kuwa S (1996) Phospholipase activity in *Cryptococcus neoformans*. Mycopathologia 136:119–123

Vidotto V, Melhem M, Pukinskas S, Aoki S, Carrara C, Pugliese A (2005) Extracellular enzymatic activity and serotype of *Cryptococcus neoformans* strains isolated from AIDS patients in Brazil. Rev Iberoam Micol 22:29–33

White TC, Agabian N (1995) *Candida albicans* secreted aspartyl proteinases: isoenzyme pattern is determined by cell type, and levels are determined by environmental factors. J Bacteriol 177:5215–5221

Wright LC, Santangelo RM, Ganendren R, Payne J, Djordjevic JT, Sorrell TC (2007) Cryptococcal lipid metabolism: phospholipase B1 is implicated in transcellular metabolism of macrophage-derived lipids. Eukaryot Cell 6:37–47

Yoo Ji J, Lee YS, Song CY, Kim BS (2004) Purification and characterization of a 43-kilodalton extracellular serine proteinase from *Cryptococcus neoformans*. J Clin Microbiol 42:722–726

Zhang YQ, Brock M, Keller NP (2004) Connection of propionyl-CoA metabolism to polyketide biosynthesis in *Aspergillus nidulans*. Genetics 168:785–794

5 CO_2 Sensing and Virulence of *Candida albicans*

Estelle Mogensen[1,2], Fritz A. Mühlschlegel[1]

CONTENTS

I. Introduction......................... 83
   A. *Candida albicans* and Candidiasis 83
   B. Virulence Determinants in *C. albicans* . 83
   C. Environmental Sensing in *C. albicans*.. 84
II. CO_2 Sensing and Signalling 84
   A. Role of CO_2 as a Signalling Molecule in *C. albicans* 85
   B. Metabolism of CO_2 in *C. albicans*...... 85
      1. Carbonic Anhydrase............... 85
      2. Pathological Growth of *C. albicans* Under Low Levels of CO_2 86
III. Chemosensing of CO_2/Bicarbonate 86
   A. Adenylyl Cyclase 86
   B. Role of Adenylyl Cyclase in CO_2-Dependant Filamentation 86
   C. Direct Activation of Adenylyl Cyclase by Bicarbonate 87
IV. Signalling Pathways Involving Adenylyl Cyclase 87
   A. Mitogen-Activated Protein Kinases Pathway.................... 87
   B. Cyclic AMP/Protein Kinase A Pathway.. 89
   C. Regulation of Morphogenesis Mediated by pH 90
V. Potential CO_2 Transporters or Receptors 90
VI. Integration of Sensing and Metabolism 91
VII. Conclusions 91
   References........................... 91

I. Introduction

The genus *Candida* includes more than 150 species, among which six are most frequently isolated from candidiasis-suffering patients. Although *C. glabrata*, *C. parapsilosis*, *C. tropicalis*, *C. krusei*, and *C. lusitaniae* are clinically prevalent species, *C. albicans* is the main causative agent of invasive candidiasis, affecting 50% of patients worldwide. In addition, it is the most virulent of all species (Ben-Abraham et al. 2004; Tortorano et al. 2004; Almirante et al. 2005; Avila-Aguero et al. 2005; Martin et al. 2005; Fridkin et al. 2006).

A. *Candida albicans* and Candidiasis

Candida albicans is the main fungal pathogen of humans. This commensal yeast belongs to the normal flora of skin as well as gastro-intestinal and genital tracts of healthy individuals. *Candida* infections are frequently superficial and initiated when the epithelial barrier functions are impaired. However, in immunocompromised patients (premature newborns, elderly individuals, chemotherapy-treated patients, HIV patients, transplant recipients), *C. albicans* can enter the bloodstream and infect almost all internal organs, causing life-threatening systemic infections (Odds 1988; Calderone 2002). Disseminated candidiasis can reach mortality rates up to 40% (Rangel-Frausto et al. 1999; MacPhail et al. 2002; Kibbler et al. 2003). The population of patients with immune dysfunction is currently increasing and recent reports have shown that *Candida* is simultaneously developing resistance to azoles and polyene antibiotics, the most common treatments administered against fungal infections (Sanglard and Odds 2002; Tortorano et al. 2004; Almirante et al. 2005; Richter et al. 2005). Therefore, there is a need for identifying new drug targets and improving diagnostic procedures to combat this devastating situation. To achieve these objectives, the scientific and the medical communities require an improved understanding of the biology of *C. albicans* and in particular how this pathogen responds and adapts to host environmental cues.

B. Virulence Determinants in *C. albicans*

Virulence of *C. albicans* is triggered by several factors, including the secretion of adhesins required for host recognition, mannoproteins and integrin-like

---
[1] Biomedical Science Group, Department of Biosciences, University of Kent, Canterbury, Kent, CT2 7NJ, United Kingdom; e-mail: F.A.Muhlschlegel@kent.ac.uk
[2] Present address: Institut Pasteur, Unité de Mycologie Moléculaire; CNRS, URA3012, F-75015, Paris, France

proteins necessary for adherence of the pathogen to epithelial cells, in addition to proteolytic and lipolytic enzymes which promote tissue invasiveness (Calderone and Fonzi 2001). One key characteristic of *C. albicans* is its ability to switch between yeast and filamentous (hyphae, pseudohyphae) forms (Odds 1988). This reversible morphogenetic transition also enables *C. albicans* to escape the immune system, and hence it is considered as a major virulence attribute (Lo et al. 1997; Leberer et al. 2001; Saville et al. 2003).

C. Environmental Sensing in *C. albicans*

As a commensal yeast, *C. albicans* is able to survive in various anatomic niches of its mammalian host. During its life cycle, it is exposed to a multitude of environmental cues that fluctuate dramatically depending on the host niches essentially temperature, pH, serum, amino acids, sugar availability and carbon dioxide. *C. albicans* is able to sense these cues and, upon exposure to physiological levels of such factors, can respond by switching from yeast to hyphal growth forms. This morphological change is caused by the expression of target genes regulated by various signal transduction pathways (Eckert et al. 2007).

This chapter focuses on the filamentation of *C. albicans* in response to the key signalling molecule carbon dioxide (CO_2). The CO_2 concentration in mammals (5%) is above 150-fold higher than in atmospheric air (0.036%). Levels of CO_2 on the mammalian host's skin are low as a result from equilibration with the atmosphere (Frame et al. 1972). Consequently, *C. albicans* must adapt to the dramatic variations in CO_2 concentrations encountered during skin infection and systemic infection. It is of crucial relevance for the pathogen to sense and respond to such variations. Therefore, this chapter aims to give the latest insights gained about the mechanism of CO_2 sensing in *C. albicans* at a molecular level and reveals the importance of this process for the virulence of the pathogen.

II. CO_2 Sensing and Signalling

Carbon dioxide plays a vital role in most ecosystems. Micro-organisms and mammals produce CO_2 as the final product of fermentation and/or cellular respiration. Moreover, photoautotrophic organisms fix atmospheric CO_2 to produce glucose by means of photosynthesis, which creates a cycle of CO_2.

CO_2 has also been identified as a key signalling molecule in various organisms. For instance, female mosquitos locate their host by detecting the level of released CO_2 (Dekker et al. 2005), and avoidance behaviour in *Drosophila* results from the activation of sensory neurons by CO_2 via a G protein-coupled receptor (Suh et al. 2004).

Importantly, CO_2 also plays a role in bacterial virulence. High concentrations of CO_2 induce the synthesis of a tripartite exotoxin (encoded by the genes *pagA*, *lef* and *cya*) and of an anti-phagocytic polysaccharide capsule (encoded by the polycistronic operon *capBCAD*) in *Bacillus anthracis*. Transcription of the toxin and capsule encoding genes is activated by the two *trans*acting regulatory proteins AtxA and AcpA, encoded by genes located on two virulence plasmids (Guignot et al. 1997; Uchida et al 1997). Drysdale et al. (2005) have shown that 5% CO_2 enhances the transcription of *atxA*. Deletion of this gene interferes with the virulence of *B. anthracis* in a mouse model of anthrax and reduces the immunological response to toxins in infected mice (Dai et al. 1995). Furthermore, it has been reported in group A *Streptococci* that elevated levels of CO_2 enhance the synthesis of the regulatory protein Mga which activates the transcription of two virulence genes, *emm* and *scpA*, encoding the antiphagocytic M protein and the C5 a endopeptidase Scp, respectively (Okada et al. 1993; McIver et al. 1995). However, no CO_2 sensor proteins have been so far identified in this model.

It has recently been shown that environmental CO_2 also strongly influences growth and morphogenesis of the pathogenic yeast *Cryptococcus neoformans* (Bahn et al. 2005; Klengel et al. 2005; Mogensen et al. 2006; Bahn and Mühlschlegel 2006). *C. neoformans* is a leading cause of central nervous system infections affecting immunosuppressed patients (Perfect and Casadevall 2002). The ubiquitous fungus is exposed to atmospheric CO_2 levels during growth in its natural habitat. Upon inhalation and subsequent lung and brain infection of its host, *C. neoformans* responds to elevated concentrations of CO_2 by producing a polysaccharide capsule surrounding its cell wall. This virulence factor interferes with phagocytosis and clearance by the immune system (Bose et al. 2003).

A. Role of CO_2 as a Signalling Molecule in *C. albicans*

CO_2 has been proven to be a powerful signal affecting dimorphism in *C. albicans* (Sims 1986). Sheth et al. (2005) recently extended this observation by investigating the response of different *Candida* species to CO_2. The authors showed that, out of 13 species tested, only *C. albicans* filaments (Fig. 5.1) when exposed to physiological concentrations of CO_2, demonstrating that filamentation in response to CO_2 is specific to *C. albicans*.

The next objectives were to determine at a molecular level how CO_2 is metabolised in the cell, what sensors detect fluctuations of CO_2 concentrations in the environment, and how the signal is transduced in *C. albicans*.

B. Metabolism of CO_2 in *C. albicans*

1. Carbonic Anhydrase

After diffusing from the surrounding environment into the cell cytoplasm, carbon dioxide is spontaneously hydrated to bicarbonate, but this reaction is accelerated by carbonic anhydrase (see Eq. 5.1). This zinc metalloenzyme is ubiquitous and has been classified in three main evolutionarily independent classes (Tripp et al. 2001). All mammalian carbonic anhydrases, as well as a few bacterial isozymes, belong to the α-class. The β-class is found in bacteria, archaea, algae and plants and has recently been identified in the basidiomycete *Cryptococcus neoformans* (Bahn et al. 2005; Mogensen et al. 2006). To date, the γ-class only consists of the archeon *Methanosarcina thermophila* enzyme (Smith and Ferry 2000).

Despite their unrelated origins, all three classes share a similar two-step enzymatic mechanism. The first step consists in the nucleophilic attack of CO_2 by a zinc-bound hydroxide ion (see Eq. 5.2). Zinc is ligated by one histidine and two cystein residues. During the second step, the active site is restored by ionisation of a zinc-bound water molecule and elimination of a proton from the active site (see Eq. 5.3; Lindskog 1997). Although this process is conserved among all characterised enzymes, the prokaryote *M. thermophila* has been shown to contain a γ-class carbonic anhydrase which is activated by iron instead of zinc (Tripp et al. 2004).

$$CO_2 + H_2O \Leftrightarrow H^+ + HCO_3^- \qquad (5.1)$$

$$E-Zn_2^+-OH^- + CO_2 + H_2O \Leftrightarrow E-Zn_2^+-HCO_3^- + H_2O \qquad (5.2)$$

$$E - Zn_2^+ - HCO_3^- + H_2O \Leftrightarrow E - Zn_2^+$$
$$-H_2O + HCO_3^- \Leftrightarrow E - Zn_2^+ - OH^- + H^+ \qquad (5.3)$$

Multiple copies of carbonic anhydrase-encoding genes are commonly found in many organisms, including fungal species such as *Neurospora crassa* and *Magnaporthe grisea*. Most fungal carbonic anhydrases belong to the β-class but within each organism the enzymes encountered may belong

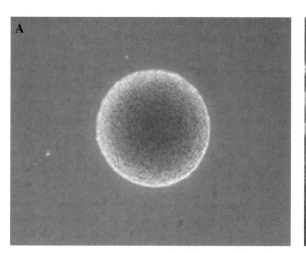
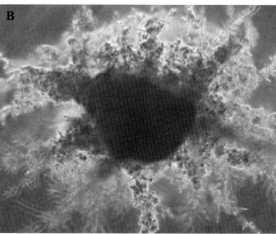

Fig. 5.1. CO_2 mediated filamentation in *Candida albicans* observed on DMEM medium at pH 7. When exposed to atmospheric air (0.033% CO_2) *C. albicans* is found as a yeast form (**A**), but physiological concentrations of CO_2 trigger filamentation (**B**)

to different families (Hewett-Emmett 2000). For instance, *Aspergillus fumigatus*, a major airborne fungal pathogen causing pulmonary infections, contains three β-CA homologues. However, it is still unknown which enzyme is required for virulence of this pathogen (Bahn and Mühlschlegel 2006).

2. Pathological Growth of *C. albicans* Under Low Levels of CO_2

The carbonic anhydrase encoding-gene *NCE103* has been cloned and characterised in *Candida albicans* (Klengel et al. 2005). It encodes a β-class enzyme, similar to those of the two yeasts *Saccharomyces cerevisiae* and *Cryptococcus neoformans* (Bahn et al. 2005; Mogensen et al. 2006). Carbonic anhydrase double deletion mutants are unable to grow in atmospheric concentrations of CO_2. Exposure to 5% CO_2 totally restores their growth defect. More importantly, carbonic anhydrase mutants of *Candida albicans* fail to invade human reconstituted epithelium in air concentrations of CO_2 but recover a phenotype similar to that of wild-type and revertant strains under physiological levels of CO_2. Moreover, *nce103* null mutants are as virulent as the revertants in a murine model of systemic infection. Interestingly, the *C. albicans* Nce103 is constitutively expressed (Klengel et al. 2005), whereas the expression of the *S. cerevisiae* ortholog is induced by low CO_2 levels (Amoroso et al. 2005). Taken together, these data demonstrate that carbonic anhydrase functions as a CO_2 scavenger and is essential for the survival of *C. albicans* in host niches where the CO_2 is limiting such as on the skin, but dispensable during systemic infection where physiological concentrations of CO_2 are encountered (Klengel et al. 2005). As CO_2 is present as concentration gradients within various niches of the host, Nce103 may be differentially expressed during colonisation and/or subsequent invasion. Consistent with these results, Allen and King (1978) have shown that *Candida* infections were enhanced when CO_2 concentrations increased on the skin surface of patients.

III. Chemosensing of CO_2/Bicarbonate

A. Adenylyl Cyclase

The second messenger cyclic adenosine 3′-5′-monophosphate (cAMP) is synthesised by adenylyl cyclase. Mammalian cells contain two classes of adenylyl cyclases which harbour a conserved catalytic domain but differ in their subcellular localisations and their regulation mechanism (Kamenetsky et al. 2006). Transmembrane cyclases are encoded by nine distinct genes and are activated by heterotrimeric G proteins (Hanoune and Defer 2001). The more recently discovered soluble adenylyl cyclase is present in various intracellular compartments, including mitochondria and the nucleus. Its activity is directly regulated by bicarbonate (HCO_3^-) and calcium (Chen et al. 2000; Jaiswal and Conti 2003; Litvin et al. 2003). The catalytic domains of the soluble cyclase have been shown to be more closely related to those of cyanobacteria than to those of other eukaryotes, revealing a link between prokaryotic and eukaryotic signal transduction processes (Litvin et al. 2003).

The adenylyl cyclase-encoding-gene *CYR1* (formerly named *CDC35*) was identified and cloned by Rocha and collaborators (2001). The authors showed that *cyr1* null mutants were unable to filament when incubated in various inducing media (serum, low ammonium, Spider or Lee's medium). Such mutants also presented a defect in hyphal formation when co-cultured with macrophages, preventing them from escaping from these cells. Moreover, the authors reported that *cyr1* deletion strains lost their capability to cause infection in a mouse model. These results demonstrate the essential role of adenylyl cyclase in hyphal development and virulence in *C. albicans*.

B. Role of Adenylyl Cyclase in CO_2-Dependant Filamentation

The molecular sensor and the signal transduction pathways activated by CO_2 have only recently been described in *C. albicans*. Klengel et al. (2005) demonstrated that *ras1* null mutants were able to filament and to invade agar medium upon exposure to 5% CO_2, whereas *cyr1* null mutants were not. These findings reveal that CO_2 signalling in *C. albicans* requires Cyr1 and bypasses Ras1. The adenylyl-cyclase encoding gene *CYR1*, expressed under the control of its own promoter or the constitutive strong *TEF2* promoter, was serially truncated and reintroduced into a *cyr1Δ/cyr1Δ* strain of *C. albicans*. The transformants were phenotypically screened for filamentation in response to physiological concentrations of CO_2. The minimal functional fragment enabling CO_2-induced filamentation was identified as containing 120 amino acids only, showing that the core catalytic domain of Cyr1 is sufficient for CO_2/HCO_3^- activation (Eckert et al. 2007).

C. Direct Activation of Adenylyl Cyclase by Bicarbonate

Bacterial and mammalian cells contain soluble adenylyl cyclases which are directly activated by physiological concentrations of bicarbonate (Chen et al. 2000; Wuttke et al. 2001; Zippin et al. 2001). The catalytic domain of the *C. albicans* adenylyl cyclase was purified and assayed in the presence of a range of concentrations of sodium bicarbonate. Cyclase activity was stimulated more than 20-fold, demonstrating that the cellular effect of CO_2 can be mediated by its hydrated form, bicarbonate, and that bicarbonate directly activates adenylyl cyclase (Klengel et al. 2005). At physiological levels of CO_2, the intracellular bicarbonate concentration is equilibrated at 25 mM in the absence of carbonic anhydrase activity. At this concentration, the *C. albicans* cyclase Cyr1 has reached its maximal activity to induce filamentation. Therefore, this result is in agreement with the finding that carbonic anhydrase is dispensable for hyphal growth of *C. albicans* under physiological concentrations of CO_2.

Interestingly, the activity of the purified adenylyl cyclase from the closely related pathogen *Cryptococcus neoformans* increased more than six-fold in the presence of bicarbonate, hence presenting a similar pattern of direct enzymatic activation (Klengel et al. 2005; Mogensen et al. 2006). Moreover, soluble-like adenylyl cyclases have been shown to present a similar mode of activation by bicarbonate in cyanobacteria and eubacteria (Litvin et al. 2003; Zippin et al. 2004), mycobacteria (Cann et al. 2003) and the malaria-causing parasite *Plasmodium falciparum* (Levin and Buck, personal communication). These findings reveal an evolutionary link between cAMP signalling and CO_2/HCO_3^- sensing which is conserved across kingdoms.

Recent structural studies proved that bicarbonate stimulates bacterial and mammalian soluble-like adenylyl cyclases by inducing a conformational change which facilitates catalysis. Binding of bicarbonate provokes a shift of the α1 helix and of the β7–β8 loop in the same direction, causing a closure of the active site. These shifts force the β- and γ-phosphates of the ATP analog out of its binding site, which facilitate the release of the reaction product pyrophosphate from the cAMP (Steegborn et al. 2005). Nevertheless, although a number of point mutations have been introduced in the *Candida albicans* Cyr1, it is still unknown what key residues are required for the activation mechanism of the adenylyl cyclase by CO_2/HCO_3^- (Cann et al. 2003; Steegborn et al. 2005). Moreover, Hammer et al. (2006) have identified two prokaryotic cyclases isolated from *Synechocystis* and *Anabaena* which are directly activated by molecular carbon dioxide. Therefore, it remains to be elucidated which inorganic carbon species (CO_2 or/and HCO_3^-) directly activates the *C. albicans* adenylyl cyclase.

IV. Signalling Pathways Involving Adenylyl Cyclase

The dimorphic transition in *C. albicans* is regulated by two signal transduction pathways. This dual control includes the mitogen-activated protein kinases (MAPK) pathway and the cyclic AMP/protein kinase A (cAMP/PKA) pathway that are interconnected by the GTP-binding protein Ras1 (Leberer et al. 2001; Fig. 5.2).

A. Mitogen-Activated Protein Kinases Pathway

The first signalling cascade that was characterised in *C. albicans* is the MAPK pathway. In this pathway, hyphae-specific genes are activated by the transcription factor Cph1, phosphorylation of which is sequentially regulated by the kinases Cst20, Ste11, Hst7 and Cek1 (for reviews, see Brown and Gow 1999; Lengeler et al. 2000; Monge et al. 2006). Null mutations in any of the genes encoding these regulatory proteins (except *ste11*) conferred filamentation defects on synthetic low ammonium dextrose medium, Spider and solid Lee's media; however, they did not affect hyphal development in medium containing serum (liquid and solid) as well as in liquid Lee's medium (Kohler and Fink 1996; Leberer et al. 1996; Csank et al. 1998). Interestingly, *cek1Δ/cek1Δ* and *cst20Δ/cst20Δ* deletion strains of *C. albicans* present a reduced virulence compared to the wild-type strain (Leberer et al. 1996; Csank et al. 1998; Guhad et al. 1998). However, *hst7* and *cph1* null mutants are able to cause systemic candidiasis in

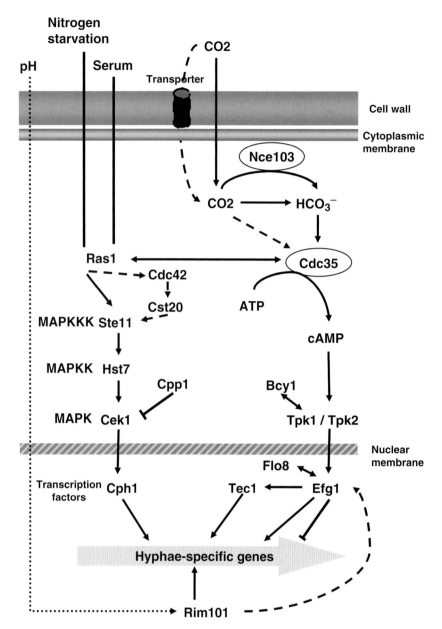

Fig. 5.2. Schematic representation of the CO_2 signal transduction pathway activating filamentation in *Candida albicans* and its connection to both the MAPK and Rim pathways. Physiological concentrations of CO_2 activate the cAMP/PKA pathway via adenylyl cyclase, leading to the expression of hyphae-specific genes. In addition, Cyr1 is activated by Ras1 which also activates the MAPK signalling cascade. The link between the Rim101 pathway and the cAMP/PKA pathway still remains hypothetical. *Dashed lines* indicate hypothesised pathways and effectors. *Dotted lines* show indirect activations. *Lines with double arrows* indicate protein–protein interactions

a mouse model (Leberer et al. 1996; Lo et al. 1997). An additional component, the MAPK phosphatase Cpp1, has been shown to inhibit Cek1. Disruption of both *cpp1* alleles reduces virulence (Csank et al. 1997; Guhad et al. 1998). In addition, the Rho-type GTP-binding protein Cdc42 and its exchange factor Cdc24 are required for hyphal development in *C. albicans* (Ushinsky et al. 2002; Bassilana et al. 2003; Vandenberg et al. 2004). Leberer et al. (2001) hypothesised that Cdc42 acts downstream Ras1 and activates the kinase Cst20.

In a study performed to determine the role of filamentation in *C. albicans* virulence, Lo et al. (1997) have shown that a single mutation in either the *cph1* or the *efg1* gene did not affect the virulence of *C. albicans*, whereas *cph1 efg1* double mutants were shown to be avirulent. This result suggests that *cph1* and *efg1* are involved in virulence related to filamentation via two different signalling pathways (Section IV B).

B. Cyclic AMP/Protein Kinase A Pathway

Another pathway controlling development in *C. albicans* includes cAMP and the protein kinase A. The cascade consists of Ras1, adenylyl cyclase, the two isoforms Tpk1 and Tpk2 of the protein kinase A (PKA), the transcription factors Efg1 (which belongs to an important group of helix–loop–helix transcription factors controlling development in fungi and vertebrates; Stoldt et al. 1997), and Tec1 (Fig. 5.2).

While the MAPK pathway is activated mainly by starvation conditions, the cAMP/PKA cascade is stimulated by serum, N-acetylglucosamine and CO_2/HCO_3^- in *C. albicans*.

The CO_2/HCO_3^- activation of Cyr1 has been shown to bypass Ras1 (see Section III B). However, Feng et al. (1999) showed that *ras1* double deletion mutants were unable to form hyphae in response to serum, demonstrating that Ras1 is required for serum-induced filamentation of *C. albicans*. The interaction between Ras1 and Cyr1 has been proved to be critical for increasing the levels of cAMP, hence for filamentation (Fang and Wang 2006). Using yeast two-hybrid and binding assays, the authors showed that this GTP-dependant interaction is direct and that Ras1 binds to Cyr1 at a unique Ras association domain. In the latter, at least two conserved residues (one lysine, one leucine) are critical for the interaction of Ras1 with Cyr1.

C. albicans strains deleted for both *tpk1* alleles presented a defect in hyphal formation on solid media but their morphology was only slightly affected in liquid media. In contrast, partial filamentation of *tpk2* deletion mutants was observed on solid media whereas their hyphal morphology was totally inhibited in liquid cultures. Only *tpk2* mutants lost their capability to invade solid media (starvation and serum containing-media; Sonneborn et al. 2000). Moreover, homozygous *tpk1Δ/tpk1Δ tpk2Δ/tpk2Δ* mutants grew as much as the wild-type, but a conditional *tpk2Δ/tpk2Δ* strain containing an allele of *tpk1* under the expression of a regulatable promoter grew very slowly and was impaired in hyphal differentiation (Bockmühl et al. 2001). These results suggest that both catalytic subunits of the protein kinase A share growth function but have distinctive roles in filamentation. Tpk1 seems to induce hyphal development on solid media but is not essential for agar invasion, whereas Tpk2 is required for both filamentation and invasiveness. Construction of hybrid genes was used to demonstrate that the catalytic domains of the isoforms are responsible for morphogenesis while invasion is mediated by the N-terminal domain of Tpk2 (Bockmühl et al. 2001).

Moreover, *tpk1* deletion mutants present a delay in germ-tube formation compared to *tpk2Δ* and wild-type strains in various inducing media, which suggests that *tpk1* plays a role in early filamentation response (Souto et al. 2006). More importantly, Sonneborn et al. (2000) proved that *tpk2* deletion mutants presented a reduction in virulence. Consistent with these results, Park et al. (2005) demonstrated that homozygous *tpk2* deletion mutants present a reduced ability to invade and damage reconstituted oral epithelium compared to *tpk1* double deletion strains and to the wild-type strain. These findings suggest that Tpk2 is essential for virulence of *C. albicans*. Souto et al. (2006) have recently shown that *C. albicans* produces two *TPK2* transcripts, a major one of 1.8 kb and a minor one of 1.4 kb. They observed that transcript levels of *TPK1* are lower than those of *TPK2* at all time during vegetative growth of the pathogen. The mRNA levels of either gene were similar in *tpk1* and *tpk2* mutants compared to the wild-type strain, suggesting that each PKA isoform does not compensate the loss of the other one. Furthermore, Cassola et al. (2004) cloned Bcy1, the regulatory subunit of PKA. They determined that Tpk1 was localised mostly to the nucleus in both the wild-type strain and the *tpk2* null mutant, whereas it was disseminated throughout the cell in a *bcy1 tpk2* double mutant. Moreover, they proved that Bcy1 interacts with Tpk1, suggesting that Bcy1 triggers the nucleic localisation of Tpk1.

The invasion defect observed with *tpk2* deletion mutants was reversed by overexpressing *EFG1* or *CEK1*, whereas the *efg1* phenotype was not suppressed by overexpressing *TPK2*, suggesting that Efg1 is a downstream target of Tpk2. Interestingly, Efg1 contains a potential site of phosphorylation by PKA (Sonneborn et al. 2000).

Lo et al. (1997) have reported that the response of *efg1* mutants to serum was greatly attenuated. A *C. albicans EFG1/efg1Δ* heterozygous strain develops hyphae but is less virulent than the wild-type strain, suggesting that the cAMP/PKA pathway regulates not only morphogenesis but also the expression of other virulence factors. In addition, the cAMP/PKA cascade is also involved in programmed cell death of *C. albicans* (Phillips et al. 2006). Moreover, *hst7* and *cph1* deletion mutants present mating and hyphal development defects, demonstrating that the MAPK pathway controls mating and sporulation in addition to invasiveness (Csank et al. 1998; Chen et al. 2002; Magee et al. 2002). The finding that the MAPK and the cAMP/PKA pathways have a pleiotropic effect elucidates why both pathways are essential for filamentation in *C. albicans*. Moreover, these two pathways are complementary as they respond to different environmental signals. For instance, serum and CO_2/HCO_3^- activates specifically the cAMP/PKA pathway.

Phan et al. (2000) showed that *cph1* and *efg1* deletion mutants present a defect in invading and damage endothelial cells, demonstrating that the two genes trigger filamentation and virulence. Moreover, *cph1Δ/cph1Δ efg1Δ/efg1Δ* mutants did not stimulate leukocyte production, whereas single mutants of either gene did. Therefore, both Cph1 and Efg1 are required to induce a proinflammatory response in endothelial cells. It is of importance to mention that the transcription factor Flo8 plays an crucial role in invasive and filamentous growth of *S. cerevisiae*. Interestingly, Cao et al. (2006) have recently cloned the *C. albicans* homolog and demonstrated that *flo8* deletion mutants failed to express hyphae-specific genes. These mutants were shown to be avirulent in infected mice. It was suggested that Flo8, via its interaction with Efg1, functions downstream the cAMP/PKA pathway and regulates filamentation and virulence in *C. albicans*.

C. Regulation of Morphogenesis Mediated by pH

The pathway controlling the pH response has been characterised at the molecular level. Environmental pH modulates the expression of *PHR1* and *PHR2* encoding proteins related to glycosidases which are required for cell wall assembly (Fonzi 1999). The expression of these two genes follow an opposite pattern. Indeed, *PHR1* is highly expressed at alkaline pH whereas *PHR2* is activated by acidic pH (Saporito-Irwin et al. 1995; Mühlschlegel and Fonzi 1997). Consistent with this result, in a mouse model, *C. albicans phr1* homozygous mutant is virulent in vaginal (where pH is acidic) but not systemic (where alkaline pH is found) infections.

Opposite results were observed with *phr2* mutants (De Bernardis et al. 1998). The expression of both *PHR1* and *PHR2* is regulated by the zing finger transcription factor Rim101, which is the last effector in the pH-induced pathway. El Barkani et al. (2000) demonstrated that hyphal development induced by Rim101 is Efg1-dependant. Although it is still unknown whether these two proteins belong to the same signalling pathway regulating filamentation, the authors have proposed that Rim101 acts as an upstream regulator of Efg1. As *C. albicans* is able to sense and respond to variations in pH and CO_2 levels, it was possible to speculate that these two environmental factors regulate common target genes which affect morphogenesis of the pathogen. It has been recently reported that (Sheth et al. 2008), physiological levels of pH and CO_2 coregulate *HSP12* in a Rim101 and cAMP-dependent manner, respectively (Sheth et al. 2008).

V. Potential CO_2 Transporters or Receptors

Adenylyl cyclase has been identified as a CO_2 chemosensor in *C. albicans* but the mechanism of CO_2 diffusion and/or transport into the cell still remains unknown. The apolar properties of CO_2 indicate that it may spontaneously diffuse through the cell membrane.

Aquaporins are transmembrane proteins. Although they function mainly as water channels which prevent permeation of ions, they have also been shown to be involved in the transport of CO_2 across the membrane of plants and human erythrocytes (Tyerman et al. 2002; Blank and Ehmke 2003; Uehlein et al. 2003; Endeward et al. 2006). Interestingly, Carbrey et al. (2001) have identified in *C. albicans* a single gene encoding aquaporin (*AQY1*), but *aqy1* deletion mutants retain their ability to filament in response to 5% CO_2, revealing that aquaporin is dispensable for CO_2 transport (Klengel et al. 2005).

Furthermore, rhesus proteins, which were originally discovered as ammonium transporters, have more recently been discovered in *Chlamydomonas reinhardtii* as being CO_2 channels. Their expression has been shown to be enhanced under elevated CO_2 conditions. Moreover, rhesus mutants present a growth defect at high CO_2 compared to the wild-type strain as a result of intracellular CO_2 limitation (Soupene et al. 2004; Kustu & Inwood 2006). These proteins are structurally related to the Mep/Amt family of ammonium transporters which have been identified in the fungal species *Saccharomyces cerevisiae* and *Hebeloma cylindrosporum* (Javelle et al. 2003; Marini et al. 2006). The role of these proteins as CO_2 channels in *C. albicans* remains to be investigated.

VI. Integration of Sensing and Metabolism

Besides being involved in sensing and signal transduction, bicarbonate is also a substrate for vital carboxylation reactions leading to the synthesis of phospholipids, arginine, purines and pyrimidines, and ATP via the citric acid cycle. Although the addition of carboxylation reaction products or metabolic cycle intermediates such as adenine, oleate or citrate to minimal growth medium does not restore the growth defect of carbonic anhydrase mutants, it is believed that there is a direct link between CO_2 signalling and metabolism (Klengel et al. 2005). Interestingly, carbonic anhydrase deletion strains of *C. neoformans* present a growth defect in its natural environment where CO_2 levels are low. Growth can be partially restored by addition of the fatty acid palmitate (Bahn et al. 2005). Therefore, due to the crucial role of the various biosynthetic pathways which require bicarbonate, it is hypothesised that carbonic anhydrase mutants of *C. albicans* and *Ccryptococcus neoformans* require a combination of metabolic intermediates or products to be able to survive in niches where CO_2 concentrations are limited.

VII. Conclusions

The success of *Candida albicans* in colonising and causing infection in its mammalian hosts is largely attributed to its ability to switch from hyphal to filamentous form. This trait allows the pathogen to adapt to environmental cues that dramatically change upon the sites of invasion, in particular carbon dioxide. Physiological concentrations of CO_2 directly activate the chemosensor adenylyl cyclase, which regulates the cAMP/PKA signalling cascade and the expression of hyphae-specific genes. These findings demonstrate the importance of the role of CO_2 sensing in host-pathogen interaction and virulence.

C. albicans is able to colonise diverse host niches where it senses and responds to a multitude of signals. Carbon dioxide has been proven to be a key signalling molecule affecting growth and morphogenesis of the pathogen. In addition, growth requires various metabolic processes and morphogenesis is also mediated by pH. Therefore, the integration of multiple signals seems crucial to the survival and the virulence of *C. albicans*. The connection between CO_2 signalling and metabolism as well as the potential co-regulation of CO_2 and pH sensing in *C. albicans* requires further investigation.

The increased resistance of *C. albicans* to antifungal agents is now a real concern and drives the medical community to finding new efficient treatments. The main effectors of the signalling cascades and metabolic pathways that are involved in and connected to CO_2 sensing may be promising targets for drug development.

Acknowledgements. We wish to thank all colleagues whose important work we read but could not cite due to space restrictions. This work in the FAM laboratory was funded by the Wellcome Trust, the Biotechnology and Biological Sciences Research Council (BBSRC) and the European Union under the Interreg IIIA programme

References

Allen AM, King RD (1978) Occlusion, carbon dioxide, and fungal skin infections. Lancet 1:360–362

Almirante B, Rodriguez D, Park BJ, Cuenca-Estrella M, Planes AM, Almela M, Mensa J, Sanchez F, Ayats J, Gimenez M, Saballs P, Fridkin SK, Morgan J, Rodriguez-Tudela JL, Warnock DW, Pahissa A (2005) Epidemiology and predictors of mortality in cases of *Candida* bloodstream infection: results from population-based surveillance, Barcelona, Spain, from 2002 to 2003. J Clin Microbiol 43:1829–1835

Amoroso G, Morell-Avrahov L, Muller D, Klug K, Sultemeyer D (2005) The gene *NCE103* (*YNL036w*) from *Saccharomyces cerevisiae* encodes a functional carbonic anhydrase and its transcription is regulated by the concentration of inorganic carbon in the medium. Mol Microbiol 56:549–558

Avila-Aguero ML, Canas-Coto A, Ulloa-Gutierrez R, Caro MA, Alfaro B, Paris MM (2005) Risk factors for *Candida* infections in a neonatal intensive care unit in Costa Rica. Int J Infect Dis 9:90–95

Bahn YS, Mühlschlegel FA (2006) CO2 sensing in fungi and beyond. Curr Opin Microbiol 9:572–578

Bahn YS, Cox GM, Perfect JR, Heitman J (2005) Carbonic anhydrase and CO2 sensing during *Cryptococcus neoformans* growth, differentiation, and virulence. Curr Biol 15:2013–2020

Bassilana M, Blyth J, Arkowitz RA (2003) Cdc24, the GDP-GTP exchange factor for Cdc42, is required for invasive hyphal growth of *Candida albicans*. Eukaryot Cell 2:9–18

Ben-Abraham R, Keller N, Teodorovitch N, Barzilai A, Harel R, Barzilay Z, Paret G (2004) Predictors of adverse outcome from candidal infection in a tertiary care hospital. J Infect 49:317–323

Blank ME, Ehmke H (2003) Aquaporin-1 and HCO3(-)-Cl- transporter-mediated transport of CO2 across the human erythrocyte membrane. J Physiol 550:419–429

Bockmühl DP, Krishnamurthy S, Gerads M, Sonneborn A, Ernst JF (2001) Distinct and redundant roles of the two protein kinase A isoforms Tpk1p and Tpk2p in morphogenesis and growth of *Candida albicans*. Mol Microbiol 42:1243–1257

Bose I, Reese AJ, Ory JJ, Janbon G, Doering TL (2003) A yeast under cover: the capsule of *Cryptococcus neoformans*. Eukaryot Cell 2:655–663

Brown AJ, Gow NA (1999) Regulatory networks controlling *Candida albicans* morphogenesis. Trends Microbiol 7:333–338

Calderone RA (2002) *Candida* and candidiasis. ASM Press, Washington, D.C.

Calderone RA, Fonzi WA (2001) Virulence factors of *Candida albicans*. Trends Microbiol 9:327–335

Cann MJ, Hammer A, Zhou J, Kanacher T (2003) A defined subset of adenylyl cyclases is regulated by bicarbonate ion. J Biol Chem 278:35033–35038

Cao F, Lane S, Raniga PP, Lu Y, Zhou Z, Ramon K, Chen J, Liu H (2006) The Flo8 transcription factor is essential for hyphal development and virulence in *Candida albicans*. Mol Biol Cell 17:295–307

Carbrey JM, Cormack BP, Agre P (2001) Aquaporin in *Candida*: characterization of a functional water channel protein. Yeast 18:1391–1396

Cassola A, Parrot M, Silberstein S, Magee BB, Passeron S, Giasson L, Cantore ML (2004) *Candida albicans* lacking the gene encoding the regulatory subunit of protein kinase A displays a defect in hyphal formation and an altered localization of the catalytic subunit. Eukaryot Cell 3:190–199

Chen J, Chen J, Lane S, Liu H (2002) A conserved mitogen-activated protein kinase pathway is required for mating in *Candida albicans*. Mol Microbiol 46:1335–1344

Chen Y, Cann MJ, Litvin TN, Iourgenko V, Sinclair ML, Levin LR, Buck J (2000) Soluble adenylyl cyclase as an evolutionarily conserved bicarbonate sensor. Science 289:625–628

Csank C, Schroppel K, Leberer E, Harcus D, Mohamed O, Meloche S, Thomas DY, Whiteway M (1998) Roles of the *Candida albicans* mitogen-activated protein kinase homolog, Cek1p, in hyphal development and systemic candidiasis. Infect Immun 66:2713–2721

Dai Z, Sirard JC, Mock M, Koehler TM (1995) The atxA gene product activates transcription of the anthrax toxin genes and is essential for virulence. Mol Microbiol 16:1171–1181

De Bernardis F, Mühlschlegel FA, Cassone A, Fonzi WA (1998) The pH of the host niche controls gene expression in and virulence of *Candida albicans*. Infect Immun 66:3317–3325

Dekker T, Geier M, Carde RT (2005) Carbon dioxide instantly sensitizes female yellow fever mosquitoes to human skin odours. J Exp Biol 208:2963–2972

Drysdale M, Bourgogne A, Koehler TM (2005) Transcriptional analysis of the *Bacillus anthracis* capsule regulators. J Bacteriol 187:5108–5114

Eckert SE, Sheth CC, Mühlschlegel (2007) Regulation of morphogenesis in *Candida* species. In: Hube B, d'Enfert C (eds) Candida: comparative and fungal genomics. Caister Academic Press, London, pp 263–293

El Barkani A, Kurzai O, Fonzi WA, Ramon A, Porta A, Frosch M, Mühlschlegel FA (2000) Dominant active alleles of *RIM101* (*PRR2*) bypass the pH restriction on filamentation of *Candida albicans*. Mol Cell Biol 20:4635–4647

Endeward V, Musa-Aziz R, Cooper GJ, Chen LM, Pelletier MF, Virkki LV, Supuran CT, King LS, Boron WF, Gros G (2006) Evidence that aquaporin 1 is a major pathway for CO2 transport across the human erythrocyte membrane. Faseb J 20:1974–1981

Fang HM, Wang Y (2006) RA domain-mediated interaction of Cyr1 with Ras1 is essential for increasing cellular cAMP level for *Candida albicans* hyphal development. Mol Microbiol 61:484–496

Feng Q, Summers E, Guo B, Fink G (1999) Ras signaling is required for serum-induced hyphal differentiation in *Candida albicans*. J Bacteriol 181:6339–6346

Fonzi WA (1999) *PHR1* and *PHR2* of *Candida albicans* encode putative glycosidases required for proper cross-linking of beta-1,3- and beta-1,6-glucans. J Bacteriol 181:7070–7079

Frame GW, Strauss WG, Maibach HI (1972) Carbon dioxide emission of the human arm and hand. J Invest Dermatol 59:155–159

Fridkin SK, Kaufman D, Edwards JR, Shetty S, Horan T (2006) Changing incidence of *Candida* bloodstream infections among NICU patients in the United States: 1995–2004. Pediatrics 117:1680–1687

Guhad FA, Jensen HE, Aalbaek B, Csank C, Mohamed O, Harcus D, Thomas DY, Whiteway M, Hau J (1998) Mitogen-activated protein kinase-defective *Candida albicans* is avirulent in a novel model of localized murine candidiasis. FEMS Microbiol Lett 166:135–139

Guignot J, Mock M, Fouet A (1997) AtxA activates the transcription of genes harbored by both *Bacillus anthracis* virulence plasmids. FEMS Microbiol Lett 147:203–207

Hammer A, Hodgson DR, Cann MJ (2006) Regulation of prokaryotic adenylyl cyclases by CO2. Biochem J 396:215–218

Hanoune J, Defer N (2001) Regulation and role of adenylyl cyclase isoforms. Annu Rev Pharmacol Toxicol 41:145–174

Hewett-Emmett D (2000) Evolution and distribution of the carbonic anhydrase gene families. In: Chegwidden WR, Carter ND, Edwards YH (eds) The carbonic anhydrases: new horizons. Birkhäuser, Basel, pp 29–76

Jaiswal BS, Conti M (2003) Calcium regulation of the soluble adenylyl cyclase expressed in mammalian spermatozoa. Proc Natl Acad Sci USA 100:10676–10681

Javelle A, Morel M, Rodriguez-Pastrana BR, Botton B, Andre B, Marini AM, Brun A, Chalot M (2003) Molecular characterization, function and regulation of ammonium transporters (Amt) and ammonium-metabolizing enzymes (GS, NADP-GDH) in the ectomycorrhizal fungus *Hebeloma cylindrosporum*. Mol Microbiol 47:411–430

Kamenetsky M, Middelhaufe S, Bank EM, Levin LR, Buck J, Steegborn C (2006) Molecular details of cAMP generation in Mammalian cells: a tale of two systems. J Mol Biol 362:623–639

Kibbler CC, Seaton S, Barnes RA, Gransden WR, Holliman RE, Johnson EM, Perry JD, Sullivan DJ, Wilson JA (2003) Management and outcome of bloodstream infections due to *Candida* species in England and Wales. J Hosp Infect 54:18–24

Klengel T, Liang WJ, Chaloupka J, Ruoff C, Schroppel K, Naglik JR, Eckert SE, Mogensen EG, Haynes K, Tuite MF, Levin LR, Buck J, Mühlschlegel FA (2005) Fungal adenylyl cyclase integrates CO_2 sensing with cAMP signaling and virulence. Curr Biol 15:2021–2026

Kohler JR, Fink GR (1996) *Candida albicans* strains heterozygous and homozygous for mutations in mitogen-activated protein kinase signaling components have defects in hyphal development. Proc Natl Acad Sci USA 93:13223–13228

Kustu S, Inwood W (2006) Biological gas channels for NH_3 and CO_2: evidence that Rh (Rhesus) proteins are CO_2 channels. Transfus Clin Biol 13:103–110

Leberer E, Harcus D, Dignard D, Johnson L, Ushinsky S, Thomas DY, Schroppel K (2001) Ras links cellular morphogenesis to virulence by regulation of the MAP kinase and cAMP signalling pathways in the pathogenic fungus *Candida albicans*. Mol Microbiol 42:673–687

Lengeler KB, Davidson RC, D'Souza C, Harashima T, Shen WC, Wang P, Pan X, Waugh M, Heitman J (2000) Signal transduction cascades regulating fungal development and virulence. Microbiol Mol Biol Rev 64:746–785

Lindskog S (1997) Structure and mechanism of carbonic anhydrase. Pharmacol Ther 74:1–20

Litvin TN, Kamenetsky M, Zarifyan A, Buck J, Levin LR (2003) Kinetic properties of "soluble" adenylyl cyclase. Synergism between calcium and bicarbonate. J Biol Chem 278:15922–15926

Lo HJ, Kohler JR, DiDomenico B, Loebenberg D, Cacciapuoti A, Fink GR (1997) Nonfilamentous *C. albicans* mutants are avirulent. Cell 90:939–949

McIver KS, Heath AS, Scott JR (1995) Regulation of virulence by environmental signals in group A *streptococci*: influence of osmolarity, temperature, gas exchange, and iron limitation on emm transcription. Infect Immun 63:4540–4542

MacPhail GL, Taylor GD, Buchanan-Chell M, Ross C, Wilson S, Kureishi A (2002) Epidemiology, treatment and outcome of candidemia: a five-year review at three Canadian hospitals. Mycoses 45:141–145

Magee BB, Legrand M, Alarco AM, Raymond M, Magee PT (2002) Many of the genes required for mating in *Saccharomyces cerevisiae* are also required for mating in *Candida albicans*. Mol Microbiol 46:1345–1351

Marini AM, Boeckstaens M, Andre B (2006) From yeast ammonium transporters to Rhesus proteins, isolation and functional characterization. Transfus Clin Biol 13:95–96

Martin D, Persat F, Piens MA, Picot S (2005) *Candida* species distribution in bloodstream cultures in Lyon, France, 1998-2001. Eur J Clin Microbiol Infect Dis 24:329–333

Mogensen EG, Janbon G, Chaloupka J, Steegborn C, Fu MS, Moyrand F, Klengel T, Pearson DS, Geeves MA, Buck J, Levin LR, Mühlschlegel FA (2006) *Cryptococcus neoformans* senses CO_2 through the carbonic anhydrase Can2 and the adenylyl cyclase Cac1. Eukaryot Cell 5:103–111

Monge RA, Roman E, Nombela C, Pla J (2006) The MAP kinase signal transduction network in *Candida albicans*. Microbiology 152:905–912

Mühlschlegel FA, Fonzi WA (1997) *PHR2* of *Candida albicans* encodes a functional homolog of the pH-regulated gene *PHR1* with an inverted pattern of pH-dependent expression. Mol Cell Biol 17:5960–5967

Okada N, Geist RT, Caparon MG (1993) Positive transcriptional control of *mry* regulates virulence in the group A *streptococcus*. Mol Microbiol 7:893–903

Odds FC (1988) *Candida* and Candidosis. Balliere Tindall, London

Park H, Myers CL, Sheppard DC, Phan QT, Sanchez AA, Edwards JE Jr, Filler SG (2005) Role of the fungal Ras-protein kinase A pathway in governing epithelial cell interactions during oropharyngeal candidiasis. Cell Microbiol 7:499–510

Perfect JR, Casadevall A (2002) Cryptococcosis. Infect Dis Clin North Am 16:837–874, v–vi

Phan QT, Belanger PH, Filler SG (2000) Role of hyphal formation in interactions of *Candida albicans* with endothelial cells. Infect Immun 68:3485–3490

Phillips AJ, Crowe JD, Ramsdale M (2006) Ras pathway signaling accelerates programmed cell death in the pathogenic fungus *Candida albicans*. Proc Natl Acad Sci USA 103:726–731

Rangel-Frausto MS, Wiblin T, Blumberg HM, Saiman L, Patterson J, Rinaldi M, Pfaller M, Edwards JE Jr, Jarvis W, Dawson J, Wenzel RP (1999) National epidemiology of mycoses survey (NEMIS): variations in rates of bloodstream infections due to *Candida* species in seven surgical intensive care units and six neonatal intensive care units. Clin Infect Dis 29:253–258

Richter SS, Galask RP, Messer SA, Hollis RJ, Diekema DJ, Pfaller MA (2005) Antifungal susceptibilities of *Candida* species causing vulvovaginitis and epidemiology of recurrent cases. J Clin Microbiol 43:2155–2162

Rocha CR, Schroppel K, Harcus D, Marcil A, Dignard D, Taylor BN, Thomas DY, Whiteway M, Leberer E (2001) Signaling through adenylyl cyclase is essential for hyphal growth and virulence in the pathogenic fungus *Candida albicans*. Mol Biol Cell 12:3631–3643

Sanglard D, Odds FC (2002) Resistance of *Candida* species to antifungal agents: molecular mechanisms and clinical consequences. Lancet Infect Dis 2:73–85

Saporito-Irwin SM, Birse CE, Sypherd PS, Fonzi WA (1995) *PHR1*, a pH-regulated gene of *Candida albicans*, is required for morphogenesis. Mol Cell Biol 15:601–613

Saville SP, Lazzell AL, Monteagudo C, Lopez-Ribot JL (2003) Engineered control of cell morphology in vivo reveals distinct roles for yeast and filamentous forms of *Candida albicans* during infection. Eukaryot Cell 2:1053–1060

Sheth CC, Johnson E, Baker ME, Haynes K, Mühlschlegel FA (2005) Phenotypic identification of *Candida albicans* by growth on chocolate agar. Med Mycol 43:735–738

Sheth CC, Mogensen EG, Fu MS, Blomfield IC, Muhlschlegel FA (2008) *Candida albicans* HSP12 is co-regulated by Physiological CO_2 and pH. Fungal Genet Biol In Press

Sims W (1986) Effect of carbon dioxide on the growth and form of *Candida albicans*. J Med Microbiol 22:203–208

Smith KS, Ferry JG (2000) Prokaryotic carbonic anhydrases. FEMS Microbiol Rev 24:335–366

Sonneborn A, Bockmühl DP, Gerads M, Kurpanek K, Sanglard D, Ernst, JF (2000) Protein kinase A encoded by *TPK2* regulates dimorphism of *Candida albicans*. Mol Microbiol 35:386–396

Soupene E, Inwood W, Kustu S (2004) Lack of the Rhesus protein Rh1 impairs growth of the green alga *Chlamydomonas reinhardtii* at high CO_2. Proc Natl Acad Sci USA 101:7787–7792

Souto G, Giacometti R, Silberstein S, Giasson L, Cantore ML, Passeron S (2006) Expression of *TPK1* and *TPK2* genes encoding PKA catalytic subunits during growth and morphogenesis in *Candida albicans*. Yeast 23:591–603

Steegborn C, Litvin TN, Levin LR, Buck J, Wu H (2005) Bicarbonate activation of adenylyl cyclase via promotion of catalytic active site closure and metal recruitment. Nat Struct Mol Biol 12:32–37

Stoldt VR, Sonneborn A, Leuker CE, Ernst JF (1997) Efg1p, an essential regulator of morphogenesis of the human pathogen *Candida albicans*, is a member of a conserved class of bHLH proteins regulating morphogenetic processes in fungi. Embo J 16:1982–1991

Suh GS, Wong AM, Hergarden AC, Wang JW, Simon AF, Benzer S, Axel R, Anderson DJ (2004) A single population of olfactory sensory neurons mediates an innate avoidance behaviour in *Drosophila*. Nature 431:854–859

Tortorano AM, Peman J, Bernhardt H, Klingspor L, Kibbler CC, Faure O, Biraghi E, Canton E, Zimmermann K, Seaton S, Grillot R (2004) Epidemiology of candidaemia in Europe: results of 28-month European Confederation of Medical Mycology (ECMM) hospital-based surveillance study. Eur J Clin Microbiol Infect Dis 23:317–322

Tripp BC, Smith K, Ferry JG (2001) Carbonic anhydrase: new insights for an ancient enzyme. J Biol Chem 276:48615–48618

Tripp BC, Bell CB 3rd, Cruz F, Krebs C, Ferry JG (2004) A role for iron in an ancient carbonic anhydrase. J Biol Chem 279:6683–6687

Tyerman SD, Niemietz CM, Bramley H (2002) Plant aquaporins: multifunctional water and solute channels with expanding roles. Plant Cell Environ 25:173–194

Uchida I, Makino S, Sekizaki T, Terakado N (1997) Crosstalk to the genes for *Bacillus anthracis* capsule synthesis by *atxA*, the gene encoding the transactivator of anthrax toxin synthesis. Mol Microbiol 23:1229–1240

Uehlein N, Lovisolo C, Siefritz F, Kaldenhoff R (2003) The tobacco aquaporin NtAQP1 is a membrane CO_2 pore with physiological functions. Nature 425:734–737

Ushinsky SC, Harcus D, Ash J, Dignard D, Marcil A, Morchhauser J, Thomas DY, Whiteway M, Leberer E (2002) *CDC42* is required for polarized growth in human pathogen *Candida albicans*. Eukaryot Cell 1:95–104

Vandenberg AL, Ibrahim AS, Edwards JE Jr, Toenjes KA, Johnson DI (2004) Cdc42p GTPase regulates the budded-to-hyphal-form transition and expression of hypha-specific transcripts in *Candida albicans*. Eukaryot Cell 3:724–734

Wuttke MS, Buck J, Levin LR (2001) Bicarbonate-regulated soluble adenylyl cyclase. J Pancreas 2:154–158

Zippin JH, Levin LR, Buck J (2001) CO(2)/HCO(3)(-)-responsive soluble adenylyl cyclase as a putative metabolic sensor. Trends Endocrinol Metab 12:366–370

6 Hyphal Growth and Virulence in *Candida albicans*

ANDREA WALTHER[1], JÜRGEN WENDLAND[1]

CONTENTS

I. Introduction.......................... 95
II. Comparison of Yeast
    and Hyphal Growth................... 95
    A. Yeast growth in *Saccharomyces
       cerevisiae* and *Candida albicans*...... 96
    B. Hyphal Growth in *Candida albicans*... 97
    C. Different Colony Morphologies
       and Biofilm Formation.............. 99
III. Signal Transduction Pathways
     Leading to Hyphal Growth............. 102
    A. Extracellular Signals that Promote
       Morphogenetic Events in *C. albicans*.. 102
    B. The Cyclic AMP Pathway............ 103
    C. The Pheromone Response
       MAP-Kinase Cascade............... 104
    D. The Role of CO_2 and the Rim101
       Pathway............................ 105
    E. Organization of the Actin
       Cytoskeleton....................... 106
    F. Repression of Filamentation
       in *C. albicans*.................... 106
IV. Role of Hyphal Growth
    as a Virulence Factor................. 107
    A. Differential Gene Expression
       During the Yeast-to-Hyphal Switch.... 107
    B. Role of Adhesion in Virulence
       of *C. albicans*.................... 108
    C. Penetration of Tissues and Immune
       Evasion............................ 109
V. Conclusions.......................... 109
   References........................... 110

I. Introduction

Fungi grow either as unicellular yeasts or form elongated tubes known as hyphae that, by branching, can form large mycelia. Yeast-like growth, as seen, e.g. in the unicellular brewer's and baker's yeast *Saccharomyces cerevisiae*, includes an active step of cell separation after mitosis. Such a cytokinesis requires a partial degradation of the cell wall and the chitin-rich septum at the mother daughter cell junction to allow separation of both cells. In contrast, hyphae consist of multiple concatenated cells which are compartmentalized by septation but which are not fragmented. Dimorphic fungi, such as the human pathogen *Candida albicans* can switch growth modes between yeast and hyphal stages. This versatility allows conquering different environmental or host niches and in *C. albicans* contributes to the successful colonization and infection of its host. The pathogenicity of *C. albicans* is brought about in concert with other virulence factors such as the production of secreted aspartic proteases and lipases or the phase-specific expression of genes, as well as the reduced ability of the host to fight off infections due to a compromised immune system. In this chapter we discuss differences on the molecular level between yeast and hyphal growth by comparison of *C. albicans* with *S. cerevisiae*. From there we review the signals that induce filamentation in *C. albicans*, the signal transduction cascades used to process these signals, and the output in terms of changes at the transcriptional level that induce phase-specific gene expression.

II. Comparison of Yeast and Hyphal Growth

Cell growth at some time point requires the generation of an axis of polarity at which to direct vesicle delivery to initiate polarized growth. This initiates two distinct growth phases, namely the establishment of cell polarity and the maintenance of polarized cell growth. While yeast cells show a cell cycle-dependent stop of polarized growth at the tip of the newly formed bud to prepare for cytokinesis, filamentous fungi keep their growth polarized at the hyphal tip. Several morphological and cytological differences can be found that promote this process. Most notably, in filamentous fungi an organelle has been observed

---

[1] Carlsberg Laboratory, Yeast Biology, 2500 Valby, Copenhagen, Denmark; e-mail: juergen.wendland@crc.dk

at the hyphal apex that can act as a vesicle supply center; and this has been termed Spitzenkörper (Girbardt 1957).

A. Yeast Growth in *Saccharomyces cerevisiae* and *Candida albicans*

In *S. cerevisiae* morphogenesis is controlled by cell cycle events. In their seminal paper Lew and Reed (1993) showed that regulation of Cdc28 by cyclins results in differential activation or inactivation of polarized cell growth. This can be monitored by following the distribution of actin cortical patches in these cells using, e.g. rhodamine-phalloidin staining. Activation of the Cdc28 (which is the homolog of the mammalian Cdc2) by the G1-cyclins triggers START and the polarization of the actin cytoskeleton to a selected site at the cell cortex. Conversely, activation of Cdc28 by the G2-cyclins results in the depolarization of the actin cytoskeleton. Thus in yeast cells the polarized growth phase at the bud tip is restricted to only a small period of the cell cycle (Fig. 6.1A). There is ample evidence that downstream of Cdc2 Rho-type GTPases control the organization of the actin cytoskeleton (Bishop and Hall 2000; Casamayor and Snyder 2002). Particularly the Cdc42-GTPase has been shown to responsible for cell polarity establishment, since inactivation of Cdc42 (or its guanine nucleotide exchange factor Cdc24) gives rise to cells that are unable to form buds in *S. cerevisiae* and *C. albicans* (Bender and Pringle 1991; Ushinsky et al. 2002; Bassilana et al. 2003). Similarly deletion of *CDC42/CDC24* in the filamentous ascomycete *Ashbya gossypii* resulted in round germ cells that were not able to generate germ tubes (Wendland and Philippsen 2001). In *S. cerevisiae* several mechanisms are required to ensure switching of growth modes from polar to non-polar (isotropic; Fig. 6.1B). Particularly the use of feedback loops is employed: a positive feedback loop in which activated Cdc24 leads to loading of Cdc42 with GTP, thereby activating Bem1 which in turn helps to stabilize Cdc24 at the site of polarized growth (Butty et al. 2002). In contrast, a negative feedback loop results in the down-regulation of Cdc42. This is achieved by Cdc42-GTP activating the p21-activated kinase Cla4, which then phosphorylates Cdc24, leading to the dissociation of Cdc24 from Bem1 (Gulli et al. 2000). Activation of the Cdc42-GTPase module itself may be by the

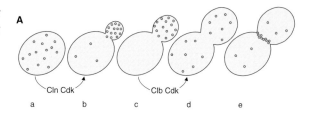

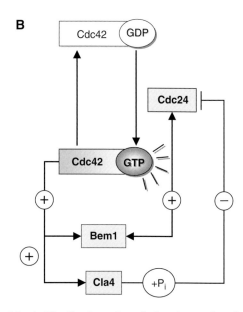

Fig. 6.1. **A** Distribution of cortical actin patches during the *Saccharomyces cerevisiae* cell cycle. In *S. cerevisiae* isotropic growth in the G1 phase of the cell cycle is characterized by a random distribution of cortical actin (*a*). Cyclin-dependent kinase (CDK) complexed with G1 cyclins (Cln) promote clustering of cortical actin patches at the incipient bud-site and in the bud at the time of bud emergence and during the initial phases of bud growth (*b*). Depolarization of the actin patches, indicating isotropic growth of the daughter cell is brought about by the G2–cyclin complexes (Clb-CDK) (*c, d*). Repolarization of the actin cytoskeleton at the end of mitosis to the bud neck region occurs prior to cytokinesis (*e*). **B** Regulatory circuits controlling Cdc42-activity. Cdc42 activity is controlled by a positive feedback loop in which Bem1 stabilizes Cdc24 at sites of polarized growth. Cdc24 acts as guanine nucleotide exchange factor on Cdc42 reinforcing this feedback loop. A negative feedback loop can be used to interrupt Cdc42 activation. Here, the Cla4-kinase phosphorylates Cdc24, which leads to a weakening of its interaction with Bem1

ras-related GTPase Bud1 or in its absence by a stochastic mechanism (Michelitch and Chant 1996; Park et al. 1997; Wedlich-Söldner et al. 2003). To actually gain control on the actin cytoskeleton Rho-GTPases activate another set of effector proteins. One class comprises the formin homologs Bni1 and Bnr1. Formins are actin nucleators involved in the formation of actin cables which are required for directed delivery of secretory vesicles to sites of growth (Dong et al. 2003). Bni1 functions as part of a complex that is termed the polarisome and includes Spa2 and Bud6 (Ozaki-Kuroda et al. 2001; Bidlingmaier and Snyder 2004).)

In *C. albicans* homologs of the *S. cerevisiae* BEM1, BNI1, BNR1, BUD1, BUD6, and SPA2 genes are present in the genome. Their molecular study has been carried out recently and phenotypes quite similar to the *S. cerevisiae* mutants could be detected. *CaBEM1* is an essential gene (Michel et al. 2002). Deletions of the polarisome components BNI1, SPA2, or BUD6 give rise to *C. albicans* mutant cells which are enlarged and develop a widened bud neck indicating defects in polarity (Zheng et al. 2003; Crampin et al. 2005; Li et al. 2005; Martin et al. 2005; Song and Kim 2006). Deletion of *CaBNR1* resultsd in a mild cellular defect with increased cell length and defects in cell separation (Martin et al. 2005).Deletion of *CaBUD1/RSR1* results in defects in yeast growth, particularly in random budding (Yaar et al. 1997; Hausauer et al. 2005). This demonstrates that similar networks are used to promote polarized cell growth in *S. cerevisiae* and *C. albicans*. Nevertheless, these protein networks have evolved differently and other players may be part of the species-specific networks. An interesting example is given by the cyclins. Cyclins are subunits that interact with the Cdc28 cyclin-dependent kinase and by elaborate mechanisms control the timing of events in the cell cycle. The *C. albicans* G2-cyclins negatively regulate polarized cell growth, as in *S. cerevisiae* (Bensen et al. 2005). There are, however, differences concerning the G1-cyclins between *C. albicans* and *S. cerevisiae*. Molecular analysis of *CaCLN3* showed that this gene is essential for budding. Depletion of Cln3 resulted in unbudded enlarged cells which eventually produced filaments (Bachewich and Whiteway 2005; Chapa y Lazo et al. 2005). Surprisingly, *C. albicans* possesses a hypha-specific G1-cyclin, encoded by *HGC1*, which is specifically expressed in hyphal stages and is essential for hyphal morphogenesis (Zheng et al. 2004). This indicates that cell cycle regulation in *C. albicans* requires specific cyclin/Cdc28 complexes during different growth modes. In contrast, overexpression of *HGC1* is not sufficient to generate hyphal growth in *C. albicans*, showing that Hgc1 does not control the yeast-to-hyphal switch (see below). In summary, work on the *C. albicans* cyclins indicates that a block in cell cycle progression leads to elongated cell growth, which is reminiscent of *S. cerevisiae* cells in which, e.g. overexpression of the G1-cyclins also results in hyperpolarized growth. Similarly, a delay in cell cycle progression, for example, by a deletion of the *C. albicans* dynein heavy chain encoding gene *DYN1*, also results in elongated cell phenotypes (Martin et al. 2004).

B. Hyphal Growth in *Candida albicans*

The drastic differences between yeast and hyphal growth become apparent in *C. albicans* yeast cells induced for hyphal formation (see Section IIIA). Upon induction *C. albicans* yeast cells form germ tubes that extend by polarized hyphal growth; and after several hours they form branched mycelia as seen in other non-dimorphic filamentous fungi (Fig. 6.2). Several other morphological and cytological features distinguish hyphae from yeast cells in *C. albicans*: (a) the organization of the actin cytoskeleton, (b) the positioning of the first septum after germ tube emergence, and (c) the presence of a Spitzenkörper in the hyphal tips.

As discussed above, during yeast growth the actin cytoskeleton is polarized to the bud tip and subsequently depolarized in a cell cycle-dependent manner. During hyphal growth, however, the actin cytoskeleton consisting of actin cables and cortical actin patches are continuously polarized to the hyphal tip, thus maintaining polarized delivery of secretory vesicles to the tip (Fig. 6.3). It is not known how this switch is brought about and maintained on the molecular level. But it seems likely that a continuous positive feedback establishes Cdc42 signaling at the hyphal tip. Constitutively activating Cdc42 in *C. albicans* by mutations that keep Cdc42 in the GTP-bound form ($CDC42^{G12V}$) proved to be lethal, indicating that a cycling of Cdc42 between GTP- and GDP-bound forms is required (Ushinsky et al. 2002).

One peculiarity in germ tube formation in *C. albicans* distinguishes not only hyphal from pseudohyphal growth in *C. albicans* but also germ tube emergence in *C. albicans* from that in other filamentous ascomycetes, such as *Ashbya gossypii* and *Aspergillus nidulans*. This is concerned with the placement of the first septum after germ tube emergence in *C. albicans*. This septum is placed at the germ cell–germ tube junction in *Ash. gossypii* and *Asp. nidulans*. In *C. albicans*, however this septal site is placed within the germ tube some 10–15 μm away from the germ cell (Sudbery 2001).

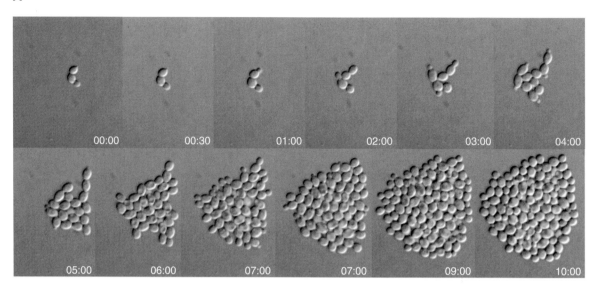

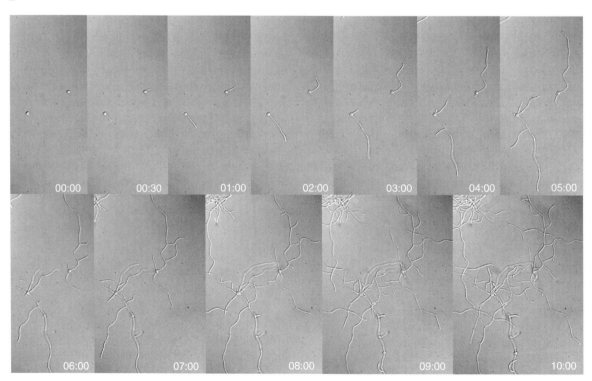

Fig. 6.2. A Time lapse recording of *Candida albicans* yeast growth. Images of cells grown at 30 °C on a microscope slide were acquired at the indicated time-points (*hours:minutes*). Cells proliferate by budding and separate from each other by cytokinesis **B** Time lapse analysis of *C. albicans* cells grown under hypha-inducing conditions. Images of cells grown at 37 °C on a microscope slide containing a medium supplemented with 10% serum were acquired at the indicated time-points. A *C. albicans* yeast cell is induced to form filaments that branch and form septa. After 10 h a mycelium was formed, in contrast to colonial growth of yeast cells

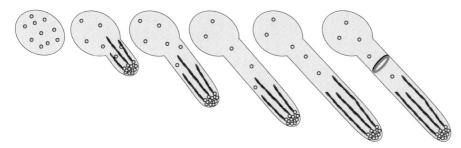

Fig. 6.3. Organization of the actin cytoskeleton during germ tube formation of *Candida albicans*. Similar events as during yeast like growth lead to the accumulation of cortical actin patches at the tip of the germ tube. At the hyphal tip actin cables are nucleated that serve as tracks for the delivery of secretory vesicles. In contrast to yeast cell growth, the organization of the actin cytoskeleton is constantly polarized in the tip during hyphal growth, thus uncoupling morphogenetic events from the cell cycle. The position of the actin ring marking the first septal site after germ tube formation is within the germ tube and not at the mother cell–germ tube neck in *C. albicans*

Furthermore, the first nuclear division after germ tube emergence takes place in the germ tube across the septal site, with subsequent transport of the mother nucleus back into the germ cell. This is different from nuclear division in *C. albicans* yeast cells, in which the mother nucleus does not leave the mother cell (Fig. 6.4). Nuclear migration required for this process is dependent on the dynein motor protein; and mycelium formation is blocked in *dyn1* mutants due to failure in the delivery of nuclei into the hyphal filaments and hyphal tips (Martin et al. 2004).

Filamentous ascomycetes such as *Asp. nidulans* and *Neurospora crassa* show a stainable structure at the hyphal apex that is termed the Spitzenkörper (Girbardt 1957; Harris et al. 2005; Harris 2006). *C. albicans* yeast cells do not generate a Spitzenkörper. However, recent evidence using *C. albicans* filaments stained by the lipophilic dye FM4-64 show Spitzenkörper at the hyphal tips (Crampin et al. 2005; Martin et al. 2005) (Fig. 6.5). A distinction between the polarisome and the Spitzenkörper can be made, suggesting that the polarisome components are positioned as a cap close to or within the hyphal tip membrane, whereas the Spitzenkörper is localized as a ball structure just subapical of the tip. This indicates the different roles for both complexes: the polarisome is a complex that establishes cell polarity, whereas the Spitzenkörper is a vesicle supply center collecting secretory vesicles and recycling vesicles to be transported into the hyphal tip.

C. Different Colony Morphologies and Biofilm Formation

The colony morphology in *C. albicans* is dependent on the growth mode of the cells and differs based on the media used. Yeast cells grow in round, shiny, and smooth colonies, while filamentous forms show wrinkled colony morphologies with protruding filaments forming a corona of hyphae at the colony edges (Fig. 6.6). Other distinct morphological stages are elongated "pseudohyphal" cells as well as "opaque" cells generated from the yeast or "white" forms which are mating-competent stages in *C. albicans* (Miller and Johnson 2002; Sudbery et al. 2004). A particular solid medium, referred to as Staib agar, is generally used to distinguish between the closely related species *C. albicans* and *C. dubliniensis*. This distinction is based on the growth of *C. albicans* as smooth colonies on Staib agar, while *C. dubliniensis* colonies are rough on this medium, indicative of filament formation (Staib and Morschhäuser 1999; Al Mosaid et al. 2001).

Yeast and filamentous cell forms are not mutually exclusive during colonial growth. One peculiar form of growth which also includes both cell forms is three-dimensional growth on solid surfaces, termed biofilm. The ability of *C. albicans* to form biofilms on medical devices such as catheters and prosthetic devices causes severe problems, particularly due to resistance of these biofilms and their cells to most of the antifungals in use (d'Enfert 2006). Therefore, any undetected biofilm forma-

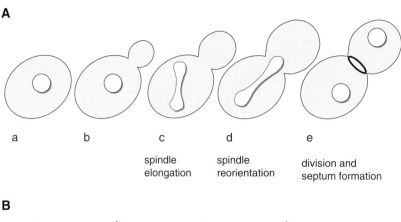

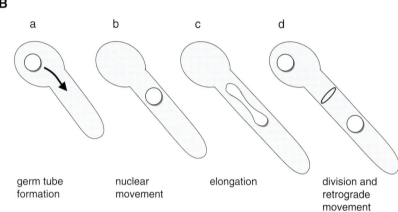

Fig. 6.4. Nuclear migration in *Candida albicans* yeast and hyphal stages. **A** Nuclear dynamics during the cell cycle of a yeast cell (*a–e*) include elongation of the spindle (*c*), realignment of the elongated spindle with the mother–bud axis (*d*), and migration of the daughter nucleus into the bud which is followed by septum formation and cell separation (*e*). **B** During germ tube formation (*a*) the mother nucleus moves into the germ tube (*b*) where the spindle elongates (*c*) and divides across a predefined septal site (*d*). This is followed by nuclear migration of the mother nucleus back into the germ cell

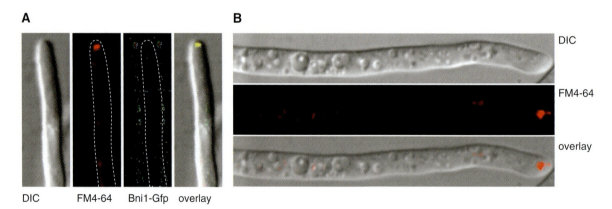

Fig. 6.5. Filamentous fungi possess a Spitzenkörper at the hyphal tip. **A** In *Candida albicans* hyphae a Spitzenkörper can be stained by FM4-64. The forming Bni1 co-localizes with this structure. **B** In the filamentous ascomycete *Ashbya gossypii* FM4-64 stains a similar structure at the hyphal apex. Since FM4-64 is used to visualize endocytic vesicles, the Spitzenkörper therefore contains apparently both endocytic and exocytotic vesicles

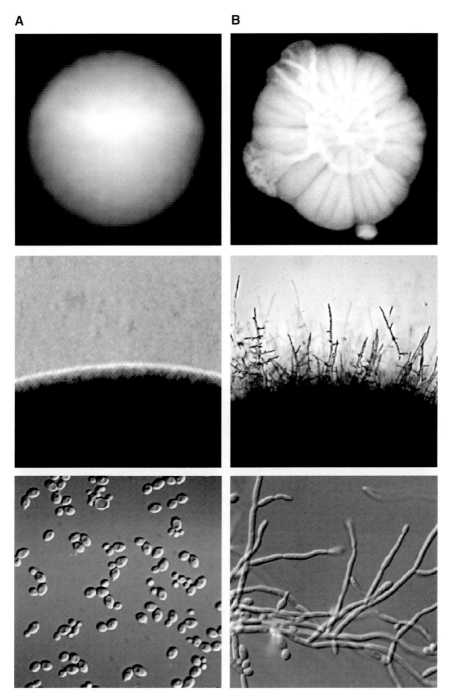

Fig. 6.6. Colony morphology in *Candida albicans*. **A** *C. albicans* cells grown at 30 °C form smooth, shiny colonies (*upper panel*). The colony edge shows a rim of yeast cells (*middle panel*). Microscopic observation of a colony sample shows only yeast cells (*lower panel*). **B** *C. albicans* cells grown at 37 °C in the presence of serum form wrinkled colonies (*upper panel*). The colony edge of such a colony shows a corona of hyphae protruding into the surrounding medium (*middle panel*). Microscopy of a sample of such a colony shows abundant hyphal filaments and also yeast cells (*lower panel*)

tion on such medical devices present a threat of severe systemic infection of patient. Biofilms are not only composed of different cell types, but these cell types are embedded in an extracellular polymeric matrix. The generation of a biofilm (which may vary in size between 25 μm and 500 μm) may undergo the successive steps of attachment of yeast cells to a surface, initial growth that generates a yeast colony which at in the next stage also includes hyphae (Ramage et al. 2005; Nobile and Mitchell 2006). And, finally after about 24–48 h, an established biofilm can be generated, embedded in its matrix and relatively shielded from external assaults such as drug treatments (Kumamoto and Vinces 2005). The genetics of biofilm formation include several of the components required for the yeast-to-hyphal switch which will be described below. In particular, mutants defective in hyphal morphogenesis generate only poor biofilms, consisting, e.g. only of few yeast cell layers (Ramage et al. 2002). The sensing of cell–cell contacts seems to be important for biofilm formation. The conserved cell wall integrity MAP kinase cascade serves as a signal transduction cascade for biofilm formation. Upon biofilm formation the respective MAP kinase cascade Mkc1 (a homolog of the *S. cerevisiae* Mpk1/Slt2) is activated and, conversely, deletion of *MKC1* leads to mutant cells that are defective in biofilm formation (Kumamoto 2005). Another feature that was revealed using expression profiling is the upregulated expression of genes involved in amino acid biosynthesis. Deletion of a key regulatory gene, *GCN4*, led to a reduction in biofilm biomass production (Garcia-Sanchez et al. 2004). Transcriptional regulation of biofilm formation relies on the transcription factor genes *TEC1*, which is involved in hyphal formation, and *BCR1* (biofilm and cell wall regulator 1), which regulates cell adhesion properties, e.g. via *ALS3* (Nobile and Mitchell 2005; Nobile et al. 2006).

III. Signal Transduction Pathways Leading to Hyphal Growth

In this chapter we discuss the molecular biology of events that ultimately lead to hyphal growth in *C. albicans*. Previously, we noted that intrinsic cues may lead to cell polarization in *S. cerevisiae* and *C. albicans*, specifically any delays in the cell cycle. However, hyphal growth in *C. albicans* relies on a variety of extracellular signals that trigger signal transduction pathways leading to filamentation and biofilm formation (Brown and Gow 1999). While a lot of progress has been made in recent years to understand the signaling network required for hyphal growth promotion, we are still only at the beginning of understanding how these signals are perceived and how *C. albicans* manages to generate a specific response adapted to the environmental stresses or topologies.

A. Extracellular Signals that Promote Morphogenetic Events in *C. albicans*

C. albicans has quite a variety of host niches that force it to adapt to these locations, most likely through the regulated expression of a set of niche-specific genes. For example, different entry pathways into the human body present different environments and also different other microorganisms with which *C. albicans* has to cope. In the laboratory environment different cues are used with which *C. albicans* can be induced to form filaments. While yeast growth is generally observed at low temperatures (~30 °C) hyphal growth is most often triggered at elevated temperatures (>34 °C, usually 37 °C). Temperature as such is not sufficient to induce hyphal morphogenesis but at least another condition needs to be favorable. Among the most potent inducers is serum. However, the relevant component in serum that acts as the inducing substance has not been identified. Recently, it was suggested that glucose is one of two active substances (Hudson et al. 2004). Glucose sensing via the Hgt4 protein (which is an ortholog of the *S. cerevisiae* Rgt2 and Snf3 sensors) is important for growth decisions. A *hgt4* mutant strain is slightly attenuated in virulence and less filamentous than the wild type, whereas a constitutive signaling mutant is hyper-filamenting (Brown et al. 2006). Solid media used for monitoring hyphal growth are Spider medium and Lee's medium (Lee et al. 1975; Liu et al. 1994). Spider medium contains mannitol as a carbon source while Lee's medium uses amino acids as inducing substances (e.g. proline or arginine). Interestingly, both media contain large amounts of potassium. Potassium (and also proline) can act as a compatible solute and may be accumulated upon osmotic shock (Kempf and Bremer 1998). Furthermore, a connection between hyphal morphogenesis and intracellular potassium concentration was suggested in a study which showed that valinomycin and miconazole promoted potassium leakage from the cells and inhibited hyphal growth (Watanabe et al. 2006). Another carbon source, *n*-acetyl glucosamine, was also shown to be a filament-inducing substance (Mattia et al. 1982).

Filamentation of *C. albicans* cells was shown to be inhibited at low pH values, e.g. pH4, whereas at alkaline pH hyphal formation can occur. Such behavior may be critically linked with the ability to cause infections (Davis et al. 2000a). Indeed it was shown that mutants in the pH-responsive gene PHR1, which is expressed at pH >5.5 are avirulent in a systemic infection model (the pH of blood is neutral to alkaline) but readily caused vaginal infections (acidic environment). Conversely, mutants in *PHR2* which is expressed at acidic pH were virulent in systemic infections but avirulent in vaginal infection assays (de Bernardis et al. 1998). This indicates that *C. albicans* is able to respond to pH (and also to changes in pH) with a morphogenetic response that may enhance its virulence (see below).

Two other hyphal inducing conditions have been described that trigger filamentation in *C. albicans* cells, namely growth under embedded conditions and growth in a CO_2-rich atmosphere. *C. albicans* cells embedded in an agar matrix rapidly induce filamentation at the even lower temperature of 25 °C (Brown et al. 1999). Hyphal growth induction under embedded conditions was, however, not dependent on the matrix itself nor on the media composition. These embedded conditions may, in part, resemble conditions *C. albicans* cells find themselves in during the generation of a biofilm. A zinc-finger transcription factor, encoded by *CZF1* (*Candida* zinc finger protein 1), was isolated in a forward genetic screen monitoring-enhanced filamentation under embedded conditions (Brown et al. 1999). Czf1 appears to function rather specifically as a transcription factor inducing morphogenesis under these special conditions but is not sufficient to trigger filamentation in standard solid- or liquid-phase assays.

Signaling via CO_2 came into the limelight of research efforts due to the inadvertent discovery that *C. albicans* (but none of the other *Candida* species tested) was able to filament under an atmosphere containing 6% CO_2 at 37 °C (Sheth et al. 2005). This may be very useful for the routine clinical determination of a pathogen in order to give early directions for a specific treatment.

These experiments reveal that *C. albicans* is able to integrate a variety of extracellular signals resulting into filamentation. This versatility may be key to survival in the different host niches and to the success of *C. albicans* as a pathogen. However, there are also mechanisms that inhibit filamentation in *C. albicans*. Phospholipase D, which hydrolyses membrane phospholipids, has been implicated in various processes such as endocytosis and motility in eukaryotic cells. Butan-1-ol has been shown to be a specific inhibitor of phospholoipase D in *Dictyostelium discoideum* resulting in strong defects in the organization of the actin cytoskeleton (Zouwail et al. 2005). In *C. albicans* propanolol inhibits germ tube formation without affecting yeast growth; and cells lacking phospholipase D, via deletion of *PLD1*, are more sensitive to propanolol (McLain and Dolan 1997; Baker et al. 2002).

Interestingly, *C. albicans* itself produces a substance that regulates morphogenesis by inhibiting germ tube emergence at high cell densities. This quorum sensing is brought about by the production and secretion of farnesol (Hornby et al. 2001). Farnesol also inhibits the formation of a biofilm in a dose-dependent manner (Ramage et al. 2002). Another quorum-sensing molecule in *C. albicans* is tyrosol. Tyrosol allows the shortening of the lag phase of cells when inoculated at lower densities into fresh medium. Furthermore, under conditions that induce filament formation, tyrosol accelerates the morphogenetic switch in low-density cultures without being itself an inducer of polarized morphogenesis (Chen et al. 2004).

B. The Cyclic AMP Pathway

The conserved cAMP pathway plays a major role in hyphal morphogenesis in *C. albicans*. The pathway integrates most of the extracellular inducing signals via the central Ras-GTPase Ras1; for example, nitrogen starvation triggers its activation via the ammonium transporter Mep2 (Biswas and Morschhäuser 2005). Activation of Ras1 by its guanine nucleotide exchange factors (Cdc25/Scd25) triggers activation of the adenylate cyclase Cdc35 (Lengeler et al. 2000). This is achieved by direct interaction of Ras1 with Cdc35 via the Ras association domain of Cdc35. Deletion of this domain abolishes this interaction and results in defective hyphal morphogenesis as does deletion of either the *RAS1* or *CDC35* genes (Feng et al. 1999; Rocha et al. 2001; Fang and Wang 2006). Adenylate cyclase catalyzes the production of cAMP. Cyclic AMP activates the protein kinase A. Protein kinase A consists of regulatory and catalytic subunits, encoded by *BCY1* and *TPK1/TPK2*, respectively (Sonneborn et al. 2000; Bockmühl et al. 2001). The catalytic subunits of PKA are activated by dissociation from Bcy1 which is triggered by cAMP-binding to Bcy1 (Lengeler et al. 2000). Deletion of *BCY1* in *Candida* could not be achieved, suggesting that constitutive signaling via this pathway is lethal (Cassola et al. 2004). Conversely, overexpression of the Tpk2 catalytic subunit leads to hyphal development under non-inducing conditions, demonstrating a positive role of PKA in filamentation (Sonneborn et al. 2000). In contrast, deletion of the phosphodiesterase *PDE2* results in abnormal hyphal development, elevated cAMP levels, and a reduced *EFG1* transcription (Jung and Stateva 2003). *EFG1* is the major transcriptional regulator required for hyphal morphogenesis in *C. albicans*. Overexpression of *EFG1* (enhancer of filamentous growth 1) results in enhanced pseudohyphal/hyphal growth (Stoldt et al. 1997). Expression of *EFG1* also has phenotypic consequences on white-opaque switching. The "white" cells are the standard yeast form, whereas "opaque" cells are elongated (Slutsky et al. 1987). Cells in the opaque phase are the mating-competent form of *C. albicans* (Miller and Johnson 2002). *EFG1* expression occurs in white cells but not in opaque cells and, if induced in opaque cells, promotes switching to the white phase (Sonneborn et al. 1999; Srikantha et al. 2000). Conversely, a recently identified regulator of white–opaque switching, Wor1, is highly expressed in opaque cells, but not expressed in white cells. Deletion of *WOR1* blocks opaque cell formation and ectopic expression of *WOR1* induces opaque cell formation (Huang et al. 2006; Zordan et al. 2006). Evidence that Efg1 is a downstream target of PKA was indicated by the identification of a putative PKA phosphorylation site at position T206. Mutations of T206, T206A, and T206E led to a block of hyphal formation or caused hyperfilamentation, respectively (Bockmühl and Ernst 2001). Interestingly, Czf1 (see above) is required to mediate relief of Efg1-mediated repression of genes to allow invasive hyphal growth at low temperatures under embedded conditions while expression of *CZF1* itself requires Efg1 (Giusani et al. 2002; Vinces et al. 2006). Efg1 is a transcriptional regulator that binds to the promoters of regulated genes (E-box = 5 - CAnnTG-3 ; Leng et al. 2001). Efg1 participates in a negative feedback loop in which Efg1 activated via PKA downregulates *EFG1*-transcription (Tebarth et al. 2003). In fact, Efg1 was shown to be a repressor while a related protein Efh1, belonging to the same class of bHLH-APSES-domain containing proteins, was shown to be an activator that modulates the function of Efg1 (Doedt et al. 2004). However, *EFH1*, e.g. via overexpression, cannot suppress the block in filamentation in *efg1* mutants. One of the Efg1-regulated genes is Hgc1 (the hyphal-specific cyclin; see above). Efg1 controls hyphal specific genes together with another transcriptional regulator, encoded by *FLO8*. Consistent with this Efg1 and Flo8 were shown to interact with each other (Cao et al. 2006).

C. The Pheromone Response MAP-Kinase Cascade

Detailed knowledge of the *S. cerevisiae* pheromone response MAP-kinase cascade helped in shaping the understanding of the homologous pathway in *C. albicans* and may have been one reason that attracted researchers to start their studies with *C. albicans*. In the classic screen for non-mating mutants done by Vivian MacKay and Thomas Manney (1974a, b) sterile mutants of yeast were isolated. A conserved MAP-kinase cascade is used to relay the pheromone response to a transcription factor encoded by *STE12* in *S. cerevisiae* (Lengeler et al. 2000). The cascade is conserved in *C. albicans*, although the components have not all been analyzed to date (e.g. the *STE11* homolog; Fig. 6.7). The discovery of mating-type loci in *C. albicans* and the demonstration of mating in suitable strains triggered the analysis of the involvement of the pheromone response MAP-kinase cascade in *Candida* mating (Hull and Johnson 1999; Hull et al. 2000; Magee and Magee 2000; Johnson 2003). Disruptions in the *C. albicans STE7* and *STE12* homologs, encoded by *HST7* and *CPH1*, respectively, blocked mating in both cell types while mutations in the *STE20* and *FUS3/KSS1* homologs, encoded by *CST20* and *CEK1*, respectively, reduced mating efficiency; and in case of the *CEK1/CEK2* double deletion abolished mating (Chen et al. 2002; Magee et al. 2002). Deletion of *CPH1* suppressed hyphal morphogenesis, at least under some conditions on solid medium, but allowed filamentation in liquid culture and upon serum induction (Liu et al. 1994). The *efg1/cph1* double mutant of *C. albicans* is defective in filamentous growth even when stimulated via serum, and, furthermore, this mutant strain is avirulent in a mouse model (Lo et al. 1997). Activation of Cek1 through the MAP-kinase cascade results in activation of downstream effectors, e.g. Cph1, via phosphorylation. To terminate such activation, dephosphorylation via a protein phosphatase may

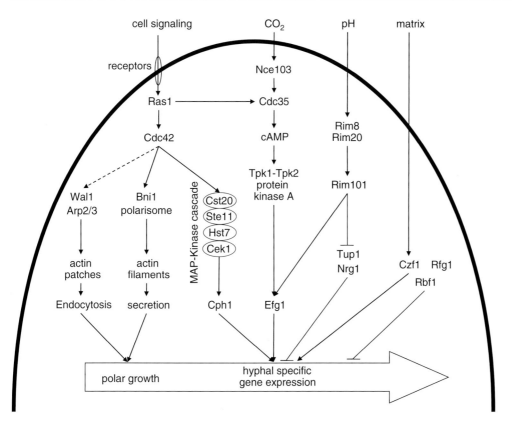

Fig. 6.7. Signal transduction pathways involved in regulating morphogenesis in *Candida albicans*. Extracellular signals serve as input to activate various pathways. Signalling via membrane receptors may activate the Ras1 GTPase which activates, e.g. the MAP kinase cascade and the cAMP pathway that signal to the transcription factors Cph1 and Efg1, respectively. Positive and negative regulation allows the control of hyphal specific gene expression. Reorganization of the actin cytoskeleton is required to promote polarized growth

be used. In *C. albicans* the Cpp1 phosphatase was found to act as a MAP-kinases phosphatase. Deletion of *CPP1* in *C. albicans* results in a hyper-filamentous phenotype at room temperature. This phenotype can be suppressed by concomitant deletion of *CEK1*. Thus Cpp1 acts as a negative regulator of Cek1-induced morphogenesis (Csank et al. 1997; Schröppel et al. 2000). In *S. cerevisiae* Ste5p acts as a scaffold of the MAP-kinases of the kinase cascade. *C. albicans*, however, seems to lack a *STE5* homolog which may allow a differential regulation of the cascade as compared with yeast.

D. The Role of CO_2 and the Rim101 Pathway

Physiological levels of CO_2 in the host environment are about 5%. This level is sufficient to trigger filamentation in *C. albicans* and capsule formation in *Cryptococcus neoformans* (Bahn et al. 2005; Klengel et al. 2005; Mitchell 2005; Mogensen et al. 2006; McFadden et al. 2006). Both of these processes are dependent on the activity of adenylate cyclase, and thus the cAMP pathway. In the cells, the conversion of CO_2 to bicarbonate (HCO_3^-) is catalyzed by carbonic anhydrases. The action of carbonic anhydrase is essential for *C. albicans* virulence in host niches with limited CO_2 levels (Klengel et al. 2005). The catalytic domain of adenylate cyclase is sufficient to mediate the response to physiological levels of CO_2 and result in filamentation (Mogensen et al. 2006). This is further supported by studies of Fang and Wang (2006) who report evidence that the Ras1-adenylate cyclase interaction domain (RA) of adanylate cyclase (and thus interaction of Ras1 with Cdc35) is not required for CO_2-induced filamentation.

Alkaline pH levels were found to induce hyphal growth by activating the Rim101 pathway (Davis et al. 2000a; El Barkani et al. 2000). Full-length Rim101 is processed under alkaline pH conditions.

C-Terminal cleavage and thus the activation of Rim101 is done by Rim8/Rim20 and other factors including Rim13. In its activated state Rim101 induced the expression of a set of downstream target genes (Davis et al. 2000a; Li et al. 2004). Dominant active alleles of *RIM101* that resulted in the C-terminal truncation of the open reading frame bypassed the pH requirement for filamentation and allowed hyphal morphogenesis at acidic pH. Additionally overexpression of *RIM101* was shown to alleviate the temperature requirement for filamentation and allowed filamentation at 29 °C. The Rim101 response is dependent on the Efg1 transcriptional regulator (El Barkani et al. 2000). Deletion of *RIM101* leads to attenuated virulence in the mouse model of hematogenously disseminated systemic candidiasis and results in strains that generate decreased damage to endothelial tissues (Davis et al. 2000b). Regulation of gene expression in response to ambient pH is a general feature of microorganisms and has been studied in other systems, most notably *Asp. nidulans* (Davis 2003). Recent evidence suggested that although the Rim101 pathway is conserved, regulation of target gene expression has diverged between *Asp. nidulans* and *C. albicans*. This has been shown by the analysis of target-binding sequences of the Rim101 transcription factor, which corresponds to 5′-GCCARG-3′ in *Asp. nidulans* but to 5′-CCAA-GAAA-3′ in *C. albicans* (Ramon and Fonzi 2003). A link between the Rim101 pathway and endocytosis was established by showing that mutations in ESCRT (endosomal sorting complexes required for transport) components are needed for signaling through Rim101 (Cornet et al. 2005).

E. Organization of the Actin Cytoskeleton

Polarized cell growth is a basic cellular feature and is required for unicellular yeast growth as well as for highly elongated filamentous growth. As discussed above, the Cdc42 cell polarity-establishment protein plays a central role downstream of Ras1 to promote polarized growth. Recent evidence has shown that, next to polarized delivery of secretory vesicles to the Spitzenkörper and hyphal tip, endocytosis is also required for hyphal morphogenesis in *C. albicans* (Walther and Wendland 2004; Wendland and Walther 2005). Deletion of the *C. albicans* homolog, *WAL1*, of the human Wiskott–Aldrich syndrome protein (WASP) resulted in delayed endocytosis, vacuolar fragmentation, and abolished mycelium formation. Endocytosis may be followed microscopically via the presence of cortical actin patches which are clustered into the growing hyphal tip surrounding the Spitzenkörper. Wal1 is a multi-domain protein containing several proline-rich regions that allow for complex formation, e.g. with SH3 domain-containing proteins. Several proteins are involved in actin patch assembly and function. One of them, CaMyo5, was also found to be required for polarized morphogenesis. Interestingly, a constitutive active form of Myo5, Myo5^{S366D}, allows for polarized hyphal growth even in the absence of polarized positioning of cortical actin patches (Oberholzer et al. 2002, 2004). This phosphorylation site may be recognized by Ste20/Cla4-protein kinases, as was shown for *S. cerevisiae* (Wu et al. 1997). Similarity in mutant phenotypes between myo5, wal1, and cla4 suggest a conserved link between exocytosis and endocytosis regulatated via Cdc42 and its interaction with Cla4 in ascomycetous fungi (Leberer et al. 1997; Walther and Wendland 2004).

Other actin patch components and the endocytotic pathway need to be analyzed in *C. albicans* to clarify to route of membrane transport and membrane sorting and their link to sustained polarized hyphal growth and virulence.

F. Repression of Filamentation in *C. albicans*

As with other things 'what goes up must come down' is also valid for filamentation in *C. albicans*. Therefore, several mechanisms are active in *C. albicans* to repress filamentation. Intriguingly *C. albicans* produces quorum-sensing molecules that allow for the control of morphogenesis on a colony level. One of them, farnesol, is excreted by yeast cells and prevents the yeast-to-mycelium transition at micromolar concentrations produced at high cell densities (Hornby et al. 2001). Because of this action, farnesol also inhibits the formation of elaborate biofilms. Farnesol reduces the expression of *HWP1* which encodes a hyphae-specific cell wall protein but induces, for example, the G1-cyclin *CCN1*, and the chitinases *CHT2* and *CHT3* (Ramage et al. 2002; Cao et al. 2005). Furthermore, farnesol was suggested to suppress the MAP-kinase cascade by decreasing the expression of *HST7* and *CPH1* (Sato et al. 2004). Other evidence indicated that a two-component pathway involving Chk1 is required for the transduction of the farnesol signal (Kruppa et al. 2004).

C. albicans also produces another quorum-sensing molecule, tyrosol, which reduces the lag phase after dilution into fresh medium but also promotes hyphal morphogenesis under inducing conditions (Chen et al. 2004).

As recent results indicated that Efg1, an enhance of filamentous growth acts rather like a suppressor it is of importance to more clearly define the regulatory circuits which are involved in fine-tuning morphogenetic decisions in *C. albicans*. Other regulators of hyphae-specific gene expression were identified in the *TUP1, NRG1, RBF1, and RFG1* genes. Deletion of *TUP1* results in constitutive hyphal growth (Braun and Johnson 1997). Using subtractive hybridization and DNA-array technology the set of genes repressed by Tup1 were identified. Among them are GPI-anchored proteins, e.g. *HWP1, RBT1, RBT5*, and *WAP1*, as well as a cell-surface iron reductase, encoded by *FRE10/RBT2* (Braun et al. 2000; Kadosh and Johnson 2005). Kadosh and Johnson (2005) also showed via DNA-arrays that a large set of genes are simultaneously controlled by Tup1, Nrg1, and Rfg1, indicating that relief of this repression plays a central role in filamentation. The function of the zinc-finger transcription factor Nrg1 as a repressor, in contrast, may require the presence of Tup1 and Ssn6, two proteins that in *S. cerevisiae* form a co-repressor dimer (Garcia-Sanchez et al. 2005). As with *tup1* cells, *nrg1* cells are filamentous under non-inducing conditions. Furthermore induction of hyphal growth results in the downregulation of *NRG1* expression (Braun et al. 2001; Murad et al. 2001). Rfg1 (repressor of filamentous growth 1) in *C. albicans* is the homolog of the *S. cerevisiae* Rox1. Rox1 in yeast recruits, again, the Tup1/Ssn6 repressor complex, to achieve repression of a specific gene set. In *C. albicans* deletion of *RFG1* resulted in filamentous growth, similar to the deletion of *NRG1* or *TUP1* (Khalaf and Zitomer 2001).

Another transcription factor, Rbf1 which shows telomere-binding activity in *C. albicans* is also somehow involved in the regulation of hyphal-specific gene expression (Ishii et al. 1997).

IV. Role of Hyphal Growth as a Virulence Factor

Virulence is defined as the ability of a microorganism to cause disease. Virulence describes the degree of pathogenicity indicating that there are strains that are more or less virulent than others. Pathogenicity itself is the mere ability to cause disease. Virulence is often determined by multiple factors. Specific host factors involved in *C. albicans* virulence are, for example, a compromised immune system. Several virulence factors of *C. albicans* have been noted, e.g. the expression of secreted aspartyl proteases, lipases, agglutinin-like sequence genes of the *ALS*-gene family, and, of course, the yeast-to-hyphal switch accompanied by the expression of a large number of hyphae-specific cell surface proteins (Berman and Sudbery 2002). Hyphal growth per se has been described as a virulence factor, due to the demonstration that non-filamentous *C. albicans* mutants were avirulent (Lo et al. 1997). This does not indicate, however, that constitutively filamentous mutants were hypervirulent. Rather, for full virulence both growth forms (and maybe even other morphologies such as white–opaque switching) are required. This can be understood in light of the different host niches that can be occupied by *C. albicans*: adhesion to tissue can be promoted via hyphal growth whereas dissemination in the blood stream can be accomplished by yeast cells.

In this last section we discuss changes when the expression level during the yeast-to-hyphal switch and the role of hyphal growth during the different stages of colonization of a host niche involving adhesion to a substrate and the penetration of epithelia. At other stages engulfment by macrophages can trigger hyphal growth resulting in the evasion of the immune response.

A. Differential Gene Expression During the Yeast-to-Hyphal Switch

Using DNA-microarrays differentially expressed genes during the induction of hyphal growth were analyzed (Lane et al. 2001; Nantel et al. 2002; Sohn et al. 2003). This indicated that multiple pathways regulated the same gene set allowing for crosstalk between different signaling pathways and concerted activation of target genes. Nrg1 was found to repress hyphae-specific genes (Murad et al. 2001). Efg1 was found to be the major regulator of cell wall genes acting either as an inducer or repressor or gene expression (Sohn et al. 2003). Although Efg1 is a target of the Ras1 signaling pathway, transcript profiles showed that there is a ras-independent gene set controlled by adenylate cyclase, which is consistent, for example, with the ras-independent induction of adenylate cyclase via CO_2 (Harcus et al. 2004). Analyses on other transcription factors revealed specific functions for Cph2, Tec1, and Flo8 for hyphae-specific gene expression (Schweizer et al. 2000; Lane et al. 2001; Cao et al. 2006). Deletion of *FLO8* blocked hyphal morphogenesis and the mutant strain was shown to be avirulent. Flo8 interacts with Efg1 and controls a subset of Efg1-regulated genes (Cao et al. 2006). Cph2 is necessary for transcriptional induction of *TEC1*, which is expressed in the hyphal form. *TEC1* overexpression can partially suppress the *efg1* defects in filamentation, indicating that Tec1 can be placed downstream of Efg1 (Schweizer et al. 2000; Lane et al. 2001).

B. Role of Adhesion in Virulence of *C. albicans*

C. albicans has a sense of touch which makes hyphae grow along grooves on a substrate (Gow 1997). Contact formation in itself could trigger hyphal morphogenesis which has been described as filamentation under embedded conditions (Brown et al. 1999). Interestingly, even the *efg1/cph1* mutant, which is non-filamentous under most of the hyphal inducing conditions, can form filaments under these conditions. The *INT1* gene has been described providing a linkage between adhesion, filamentation, and virulence (Gale et al. 1998). Int1 has some similarity to vertebrate integrins and the yeast Bud4. The hyphae-specific hyphal wall protein, encoded by *HWP1*, plays a role in promoting adhesion to epithelial cells. Deletion of *HWP1* leads to the inability to form stable attachments to epithelial cells and also to reduced virulence (Staab et al. 1999). Filamentation was also found to be a pre-requisite of adhesion to porcine intestinal epithelium (Wendland et al. 2006; Fig. 6.8). A large family of cell surface glycoproteins, encoded by the *ALS* genes, plays an important role in the adhesion of *C. albicans* to epithelia (Hoyer et al. 2001). Als proteins resemble *S. cerevisiae* flocculins in that they can generate genetic variability based on internal repetitive sequence elements. *ALS* genes are differentially regulated and the variety of Als proteins may provide *C. albicans* with the ability to adhere to different substrates. The cell wall plays of course a critical role in this process. *C. albicans* encodes more than 100 cell surface GPI-anchored proteins. A functional GPI-anchoring machine was shown to be required in *C. albicans* for full virulence and hyphal formation (Richard et al. 2002). One of these GPI-anchored proteins, Eap1, is involved in epithelial adhesion. *EAP1* expression is regulated by Efg1, providing a link between hyphal morphogenesis and an altered cell wall structure (Li and Palecek 2003).

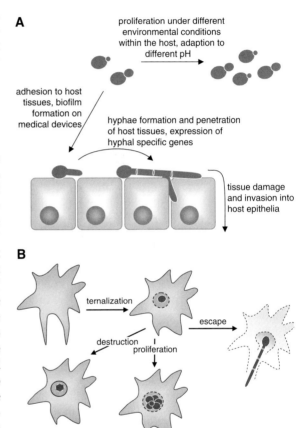

Fig. 6.8. Hyphal growth as a virulence factor in *Candida albicans*. **A** *C. albicans* yeast can adapt to different host environments, e.g. different pH values. Upon contact with host epithelia hyphal growth is initiated, resulting in adhesion to and colonization of the epithelial surface. On medical devices *C. albicans* may form biofilms. Colonization of epithelia and the expression of specific virulence factors fosters the penetration and destruction of host epithelial cell layers which leads to the invasion of host tissues and dissemination of *C. albicans* cells via the blood stream. **B** Upon contact of *C. albicans* yeast cells with the cellular immune system, the yeast cells may be internalized by macrophages. This may result in either destruction of the pathogen, proliferation of *C. albicans* within the macrophage, or germ tube formation of the yeast cells which finally results in the destruction of the macrophage and the escape of *C. albicans* cells from macrophage attack

C. Penetration of Tissues and Immune Evasion

To assay the virulence of *C. albicans* mutant strains, a mouse model of hematogenously disseminated candidiasis is generally used. This allows distinguishing between fully virulent strains, strains attenuated in virulence, and avirulent strains. However, injecting *C. albicans* blastospores directly into the bloodstream eliminates natural defence barriers of the host, since the early stages of colonization by *C. albicans* require adhesion to epithelia, colonization, and penetration of tissues.

Recently, reconstituted human tissues have been described in their use to monitor adhesion and invasion of *C. albicans* (Korting et al. 1998; Dieterich et al. 2002; Schaller et al. 2003). These were used for example to demonstrate that the secreted aspartyl proteases encoded by the *SAP1* and *SAP2* genes are involved in causing tissue damage in a reconstituted human vaginal epithelium (Schaller et al. 2003). The Efg1-regulated proteinases Sap4–Sap6 are required for invasive growth in tissues but may be dispensable for filament formation (Felk et al. 2002). The five isoforms of protein mannosyltransferases (Pmt1, Pmt2, Pmt4–Pmt6) were analyzed using reconstituted epithelia or in a muse model. No single deletion of a PMT gene conveyed a drastic phenotype, suggesting that the encoding proteins fulfil overlapping functions (Rouabhia et al. 2005). In a model of reconstructed intestinal epithelium a non-filamentous strain bearing deletions in *CPH1* and *EFG1* was shown to be unable to adhere or penetrate (Dieterich et al. 2002). This highlights the importance of a concerted activation of hyphal morphogenesis, cell wall restructuring, the co-expression of other virulence factors such as adhesins and lipases and proteases to overcome the host defense barriers and establish a successful infection site (Kumamoto and Vinces 2005).

An early study showed that neutrophils can mediate the protection of endothelial cells from damage by *C. albicans* hyphae by selectively killing the hyphae (Edwards et al. 1987). However, hyphal growth can not only generate the force to penetrate tissues but can also force the cells out after ingestion by a macrophage.

Normally when a phagocyte recognizes a *C. albicans* cell, it engulfs it into a phagosome which can be fused to lysosomes to charge the pathogen with the host defenses. *C. albicans* may overcome this hostile attack by switching to hyphal morphogenesis, destroying the macrophage from the inside (Fig. 6.8; Lorenz and Fink 2002).

The general ability to form hyphae and the readiness to trigger morphological changes accompanied by altering the transcriptional profile can turn *C. albicans* into such a powerful pathogen (Lorenz et al. 2004). The idea that *C. albicans* normally acts as a commensal, e.g. in our gastrointestinal tract (Rozell et al. 2006), is hard to reconcile with this behavior. Rather it seems that the counteractions taken by our constantly vigilant immune system fight off all *C. albicans* attacks, leaving little room for *C. albicans* to survive in the human host. Yet this little amount of room and the balance in the fight may be sufficient to remain in place long enough to wait for an opportunity to strike back. One way of hiding from the immune system is by masking the beta-glucan of the cell wall (which is a strong proinflammatory substance) with mannoproteins (Wheeler and Fink 2006). This may explain the synthesis of hyphae-specific wall proteins.

V. Conclusions

Hyphal morphogenesis in *C. albicans* is a very attractive biological system. Studies on the genetics, molecular biology, and the underlying signal transduction pathways have opened a view on how *C. albicans* manages to recognize and process a variety of diverse extracellular signals into an output that allows survival in different host environments. This is a particularly troublesome task, considering the host defenses that counteract such an invading pathogen. Thus the notion of *C. albicans* as an harmless commensal may only be based upon a strong host defense. *C. albicans* has a rewired set of conserved pathways, namely the MAP-kinase cascade homologous to the *S. cerevisiae* pheromone response/pseudohyphal growth pathway, the cAMP pathway and others, e.g. the pH regulatory pathway, to result in the concerted activation of virulence genes and the morphogenetic program. Due to the completed genome sequence and due to global transcript profiling many new genes, often genes without homologs in *S. cerevisiae* await further analysis. Thus the new challenges are to proceed with the functional analysis of the *C. albicans* genome and to generate a protein network that allows us to define the core machinery that promotes its virulence. This will generate many new concepts of antifungal therapy that may not only prove useful in the defense against *C. albicans* but may generate novel broad-spectrum antimycotica.

Acknowledgements. J.W. is supported by the Deutsche Forschungsgemeinschaft (We2634/5-2) and the Carlsberg Laboratory. We thank Pete Sudbery for sharing unpublished data.

References

Al Mosaid A, Sullivan D, Salkin IF, Shanley D, Coleman DC (2001) Differentiation of *Candida dubliniensis* from *Candida albicans* on staib agar and caffeic acid-ferric citrate agar. J Clin Microbiol 39:323–327

Bachewich C, Whiteway M (2005) Cyclin Cln3p links G1 progression to hyphal and pseudohyphal development in *Candida albicans*. Eukaryot Cell 4:95–102

Bahn YS, Cox GM, Perfect JR, Heitman J (2005) Carbonic anhydrase and CO_2-sensing during *Cryptococcus neoformans* growth, differentiation, and virulence. Curr Biol 15:2013–2020

Baker CA, Desrosiers K, Dolan JW (2002) Propranolol inhibits hyphal development in *Candida albicans*. Antimicrob Agents Chemother 46:3617–3620

Bassilana M, Blyth J, Arkowitz RA (2003) Cdc24, the GDP-GTP exchange factor for Cdc42, is required for invasive hyphal growth of *Candida albicans*. Eukaryot Cell 2:9–18

Bender A, Pringle JR (1991) Use of a screen for synthetic lethal and multicopy suppressee mutants to identify two new genes involved in morphogenesis in *Saccharomyces cerevisiae*. Mol Cell Biol 11:1295–1305

Bensen ES, Clemente-Blanco A, Finley KR, Correa-Bordes J, Berman J (2005) The mitotic cyclins Clb2p and Clb4p affect morphogenesis in *Candida albicans*. Mol Biol Cell 16:3387–3400

Berman J, Sudbery PE (2002) *Candida albicans*: a molecular revolution built on lessons from budding yeast. Nat Rev Genet 3:918–930

Bidlingmaier S, Snyder M (2004) Regulation of polarized growth initiation and termination cycles by the polarisome and Cdc42 regulators. The J Cell Biol 164:207–218

Bishop AL, Hall A (2000) Rho GTPases and their effector proteins. Biochem J 348:241–255

Biswas K, Morschhauser J (2005) The Mep2p ammonium permease controls nitrogen starvation-induced filamentous growth in *Candida albicans*. Mol Microbiol 56:649–669

Bockmuhl DP, Ernst JF (2001) A potential phosphorylation site for an A-type kinase in the Efg1 regulator protein contributes to hyphal morphogenesis of *Candida albicans*. Genetics 157:1523–1530

Bockmuhl DP, Krishnamurthy S, Gerads M, Sonneborn A, Ernst JF (2001) Distinct and redundant roles of the two protein kinase A isoforms Tpk1p and Tpk2p in morphogenesis and growth of *Candida albicans*. Mol Microbiol 42:1243–1257

Braun BR, Johnson AD (1997) Control of filament formation in *Candida albicans* by the transcriptional repressor *TUP1*. Science 277:105–109

Braun BR, Head WS, Wang MX, Johnson AD (2000) Identification and characterization of *TUP1*-regulated genes in *Candida albicans*. Genetics 156:31–44

Braun BR, Kadosh D, Johnson AD (2001) *NRG1*, a repressor of filamentous growth in *C.albicans*, is down-regulated during filament induction. EMBO J 20:4753–4761

Brown AJ, Gow NA (1999) Regulatory networks controlling *Candida albicans* morphogenesis. Trends Microbiol 7:333–338

Brown DH Jr, Giusani AD, Chen X, Kumamoto CA (1999) Filamentous growth of *Candida albicans* in response to physical environmental cues and its regulation by the unique *CZF1* gene. Mol Microbiol 34:651–662

Brown V, Sexton JA, Johnston M (2006) A Glucose Sensor in *Candida albicans*. Eukaryot Cell 5:1726–1737

Butty AC, Perrinjaquet N, Petit A, Jaquenoud M, Segall JE, Hofmann K, Zwahlen C, Peter M (2002) A positive feedback loop stabilizes the guanine-nucleotide exchange factor Cdc24 at sites of polarization. EMBO J 21:1565–1576

Cao F, Lane S, Raniga PP, Lu Y, Zhou Z, Ramon K, Chen J, Liu H (2006) The Flo8 transcription factor is essential for hyphal development and virulence in *Candida albicans*. Mol Biol Cell 17:295–307

Cao YY, Cao YB, Xu Z, Ying K, Li Y, Xie Y, Zhu ZY, Chen WS, Jiang YY (2005) cDNA microarray analysis of differential gene expression in *Candida albicans* biofilm exposed to farnesol. Antimicrob Agents Chemother 49:584–589

Casamayor A, Snyder M (2002) Bud-site selection and cell polarity in budding yeast. Curr Opin Microbiol 5:179–186

Cassola A, Parrot M, Silberstein S, Magee BB, Passeron S, Giasson L, Cantore ML (2004) *Candida albicans* lacking the gene encoding the regulatory subunit of protein kinase A displays a defect in hyphal formation and an altered localization of the catalytic subunit. Eukaryot Cell 3:190–199

Chapa y Lazo B, Bates S, Sudbery P (2005) The G1 cyclin Cln3 regulates morphogenesis in *Candida albicans*. Eukaryot Cell 4:90–94

Chen H, Fujita M, Feng Q, Clardy J, Fink GR (2004) Tyrosol is a quorum-sensing molecule in *Candida albicans*. Proc Natl Acad Sci USA 101:5048–5052

Chen J, Chen J, Lane S, Liu H (2002) A conserved mitogen-activated protein kinase pathway is required for mating in *Candida albicans*. Mol Microbiol 46:1335–1344

Cornet M, Bidard F, Schwarz P, Da Costa G, Blanchin-Roland S, Dromer F, Gaillardin C (2005) Deletions of endocytic components *VPS28* and *VPS32* affect growth at alkaline pH and virulence through both *RIM101*-dependent and *RIM101*-independent pathways in *Candida albicans*. Infect Immun 73:7977–7987

Crampin H, Finley K, Gerami-Nejad M, Court H, Gale C, Berman J, Sudbery P (2005) *Candida albicans* hyphae have a Spitzenkörper that is distinct from the polarisome found in yeast and pseudohyphae. J Cell Sci 118:2935–2947

Csank C, Makris C, Meloche S, Schroppel K, Rollinghoff M, Dignard D, Thomas DY, Whiteway M (1997) Derepressed hyphal growth and reduced virulence in a

VH1 family-related protein phosphatase mutant of the human pathogen *Candida albicans*. Mol Biol Cell 8:2539–2551

Davis D (2003) Adaptation to environmental pH in *Candida albicans* and its relation to pathogenesis. Curr Genet 44:1–7

Davis D, Edwards JE Jr, Mitchell AP, Ibrahim AS (2000a) *Candida albicans RIM101* pH response pathway is required for host–pathogen interactions. Infect Immun 68:5953–5959

Davis D, Wilson RB, Mitchell AP (2000b) *RIM101*-dependent and -independent pathways govern pH responses in *Candida albicans*. Mol Cell Biol 20:971–978

De Bernardis F, Muhlschlegel FA, Cassone A, Fonzi WA (1998) The pH of the host niche controls gene expression in and virulence of *Candida albicans*. Infect Immun 66:3317–3325

d'Enfert C (2006) Biofilms and their role in the resistance of pathogenic *Candida* to antifungal agents. Curr Drug Target 7:465–470

Dieterich C, Schandar M, Noll M, Johannes FJ, Brunner H, Graeve T, Rupp S (2002) *In vitro* reconstructed human epithelia reveal contributions of *Candida albicans EFG1* and *CPH1* to adhesion and invasion. Microbiology 148:497–506

Doedt T, Krishnamurthy S, Bockmuhl DP, Tebarth B, Stempel C, Russell CL, Brown AJ, Ernst JF (2004) APSES proteins regulate morphogenesis and metabolism in *Candida albicans*. Mol Biol Cell 15:3167–3180

Dong Y, Pruyne D, Bretscher A (2003) Formin-dependent actin assembly is regulated by distinct modes of Rho signaling in yeast. J Cell Biol 161:1081–1092

Edwards JE Jr, Rotrosen D, Fontaine JW, Haudenschild CC, Diamond RD (1987) Neutrophil-mediated protection of cultured human vascular endothelial cells from damage by growing *Candida albicans* hyphae. Blood 69:1450–1457

El Barkani A, Kurzai O, Fonzi WA, Ramon A, Porta A, Frosch M, Muhlschlegel FA (2000) Dominant active alleles of *RIM101* (*PRR2*) bypass the pH restriction on filamentation of *Candida albicans*. Mol Cell Biol 20:4635–4647

Fang HM, Wang Y (2006) RA domain-mediated interaction of Cdc35 with Ras1 is essential for increasing cellular cAMP level for *Candida albicans* hyphal development. Mol Microbiol 61:484–496

Felk A, Kretschmar M, Albrecht A, Schaller M, Beinhauer S, Nichterlein T, Sanglard D, Korting HC, Schafer W, Hube B (2002) *Candida albicans* hyphal formation and the expression of the Efg1-regulated proteinases Sap4 to Sap6 are required for the invasion of parenchymal organs. Infect Immun 70:3689–3700

Feng Q, Summers E, Guo B, Fink G (1999) Ras signaling is required for serum-induced hyphal differentiation in *Candida albicans*. J Bacteriol 182:6339–6346

Gale CA, Bendel CM, McClellan M, Hauser M, Becker JM, Berman J, Hostetter MK (1998) Linkage of adhesion, filamentous growth, and virulence in *Candida albicans* to a single gene, *INT1*. Science 279:1355–1358

Garcia-Sanchez S, Aubert S, Iraqui I, Janbon G, Ghigo JM, d'Enfert C (2004) *Candida albicans* biofilms: a developmental state associated with specific and stable gene expression patterns. Eukaryot Cell 3:536–545

Garcia-Sanchez S, Mavor AL, Russell CL, Argimon S, Dennison P, Enjalbert B, Brown AJ (2005) Global roles of Ssn6 in Tup1- and Nrg1-dependent gene regulation in the fungal pathogen, *Candida albicans*. Mol Biol Cell 16:2913–2925

Giusani AD, Vinces M, Kumamoto CA (2002) Invasive filamentous growth of *Candida albicans* is promoted by Czf1p-dependent relief of Efg1p-mediated repression. Genetics 160:1749–1753

Gow NA (1997) Germ tube growth of *Candida albicans*. Curr Top Med Mycol 8:43–55

Gulli MP, Jaquenoud M, Shimada Y, Niederhauser G, Wiget P, Peter M (2000) Phosphorylation of the Cdc42 exchange factor Cdc24 by the PAK-like kinase Cla4 may regulate polarized growth in yeast. Mol Cell 6:1155–1167

Harcus D, Nantel A, Marcil A, Rigby T, Whiteway M (2004) Transcription profiling of cyclic AMP signaling in *Candida albicans*. Mol Biol Cell 15:4490–4499

Harris SD (2006) Cell polarity in filamentous fungi: shaping the mold. Intl Rev Cytol 251:41–77

Harris SD, Read ND, Roberson RW, Shaw B, Seiler S, Plamann M, Momany M (2005) Polarisome meets spitzenkorper: microscopy, genetics, and genomics converge. Eukaryot Cell 4:225–229

Hausauer DL, Gerami-Nejad M, Kistler-Anderson C, Gale CA (2005) Hyphal guidance and invasive growth in *Candida albicans* require the Ras-like GTPase Rsr1p and its GTPase-activating protein Bud2p. Eukaryot Cell 4:1273–1286

Hornby JM, Jensen EC, Lisec AD, Tasto JJ, Jahnke B, Shoemaker R, Dussault P, Nickerson KW (2001) Quorum sensing in the dimorphic fungus *Candida albicans* is mediated by farnesol. Appl Environ Microbiol 67:2982–2992

Hoyer LL, Fundyga R, Hecht JE, Kapteyn JC, Klis FM, Arnold J (2001) Characterization of agglutinin-like sequence genes from non-albicans *Candida* and phylogenetic analysis of the ALS family. Genetics 157:1555–1567

Huang G, Wang H, Chou S, Nie X, Chen J, Liu H (2006) Bistable expression of *WOR1*, a master regulator of white-opaque switching in *Candida albicans*. Proc Natl Acad Sci USA 103:12813–12818

Hudson DA, Sciascia QL, Sanders RJ, Norris GE, Edwards PJ, Sullivan PA, Farley PC (2004) Identification of the dialysable serum inducer of germ-tube formation in *Candida albicans*. Microbiology 150:3041–3049

Hull CM, Johnson AD (1999) Identification of a mating type-like locus in the asexual pathogenic yeast *Candida albicans*. Science 285:1271–1275

Hull CM, Raisner RM, Johnson AD (2000) Evidence for mating of the "asexual" yeast *Candida albicans* in a mammalian host. Science 289:307–310

Ishii N, Yamamoto M, Yoshihara F, Arisawa M, Aoki Y (1997) Biochemical and genetic characterization of Rbf1p, a putative transcription factor of *Candida albicans*. Microbiology 143:429–435

Johnson A (2003) The biology of mating in *Candida albicans*. Nat Rev Microbiol 1:106–116

Jung WH, Stateva LI (2003) The cAMP phosphodiesterase encoded by *CaPDE2* is required for hyphal development in *Candida albicans*. Microbiology 149:2961–2976

Kadosh D, Johnson AD (2005) Induction of the *Candida albicans* filamentous growth program by relief of transcriptional repression: a genome-wide analysis. Mol Biol Cell 16:2903–2912

Kempf B, Bremer E (1998) Uptake and synthesis of compatible solutes as microbial stress responses to high-osmolality environments. Arch Microbiol 170:319–330

Khalaf RA, Zitomer RS (2001) The DNA binding protein Rfg1 is a repressor of filamentation in *Candida albicans*. Genetics 157:1503–1512

Klengel T, Liang WJ, Chaloupka J, Ruoff C, Schroppel K, Naglik JR, Eckert SE, Mogensen EG, Haynes K, Tuite MF, Levin LR, Buck J, Muhlschlegel FA (2005) Fungal adenylyl cyclase integrates CO_2 sensing with cAMP signaling and virulence. Curr Biol 15:2021–2026

Korting HC, Patzak U, Schaller M, Maibach HI (1998) A model of human cutaneous candidosis based on reconstructed human epidermis for the light and electron microscopic study of pathogenesis and treatment. J Infect 36:259–267

Kruppa M, Krom BP, Chauhan N, Bambach AV, Cihlar RL, Calderone RA (2004) The two-component signal transduction protein Chk1p regulates quorum sensing in *Candida albicans*. Eukaryot Cell 3:1062–1065

Kumamoto CA (2005) A contact-activated kinase signals *Candida albicans* invasive growth and biofilm development. Proc Natl Acad Sci USA 102:5576–5581

Kumamoto CA, Vinces MD (2005) Alternative *Candida albicans* lifestyles: growth on surfaces. Annu Rev Microbiol 59:113–133

Lane S, Birse C, Zhou S, Matson R, Liu H (2001) DNA array studies demonstrate convergent regulation of virulence factors by Cph1, Cph2, and Efg1 in *Candida albicans*. J Biol Chem 276:48988–48996

Leberer E, Ziegelbauer K, Schmidt A, Harcus D, Dignard D, Ash J, Johnson L, Thomas DY (1997) Virulence and hyphal formation of *Candida albicans* require the Ste20p-like protein kinase CaCla4p. Curr Biol 7:539–546

Lee KL, Buckley HR, Campbell CC (1975) An amino acid liquid synthetic medium for the development of mycelial and yeast forms of *Candida albicans*. Sabouraudia 13:148–153

Leng P, Lee PR, Wu H, Brown AJ (2001) Efg1, a morphogenetic regulator in *Candida albicans*, is a sequence-specific DNA binding protein. J Bacteriol 183:4090–4093

Lengeler KB, Davidson RC, D'Souza C, Harashima T, Shen WC, Wang P, Pan X, Waugh M, Heitman J (2000) Signal transduction cascades regulating fungal development and virulence. Microbiol Mol Biol Rev 64:746–785

Lew DJ, Reed SI (1993) Morphogenesis in the yeast cell cycle: regulation by Cdc28 and cyclins. J Cell Biology 120:1305–1320

Li CR, Wang YM, De Zheng X, Liang HY, Tang JC, Wang Y (2005) The formin family protein CaBni1p has a role in cell polarity control during both yeast and hyphal growth in *Candida albicans*. J Cell Sci 118:2637–2648

Li F, Palecek SP (2003) EAP1, a *Candida albicans* gene involved in binding human epithelial cells. Eukaryot Cell 2:1266–1273

Li M, Martin SJ, Bruno VM, Mitchell AP, Davis DA (2004) *Candida albicans* Rim13p, a protease required for Rim101p processing at acidic and alkaline pHs. Eukaryot Cell 3:741–751

Liu H, Kohler J, Fink GR (1994) Suppression of hyphal formation in *Candida albicans* by mutation of a *STE12* homolog. Science 266:1723–1726

Lo HJ, Kohler JR, DiDomenico B, Loebenberg D, Cacciapuoti A, Fink GR (1997) Nonfilamentous *C. albicans* mutants are avirulent. Cell 90:939–949

Lorenz MC, Bender JA, Fink GR (2004) Transcriptional response of *Candida albicans* upon internalization by macrophages. Eukaryot cell 3:1076–1087

Lorenz MC, Fink GR (2002) Life and death in a macrophage: role of the glyoxylate cycle in virulence. Eukaryot Cell 1:657–662

Mackay V, Manney TR (1974a) Mutations affecting sexual conjugation and related processes in *Saccharomyces cerevisiae*. I. Isolation and phenotypic characterization of nonmating mutants. Genetics 76:255–271

Mackay V, Manney TR (1974b) Mutations affecting sexual conjugation and related processes in *Saccharomyces cerevisiae*. II. Genetic analysis of nonmating mutants. Genetics 76:273–288

Magee BB, Magee PT (2000) Induction of mating in *Candida albicans* by construction of MTLa and MTLalpha strains. Science 289:310–313

Magee BB, Legrand M, Alarco AM, Raymond M, Magee PT (2002) Many of the genes required for mating in Saccharomyces cerevisiae are also required for mating in *Candida albicans*. Mol Microbiol 46:1345–1351

Martin R, Walther A, Wendland J (2004) Deletion of the dynein heavy-chain gene *DYN1* leads to aberrant nuclear positioning and defective hyphal development in *Candida albicans*. Eukaryot Cell 3:1574–1588

Martin R, Walther A, Wendland J (2005) Ras1-induced hyphal development in *Candida albicans* requires the formin Bni1. Eukaryot Cell 4:1712–1724

Mattia E, Carruba G, Angiolella L, Cassone A (1982) Induction of germ tube formation by N-acetyl-D-glucosamine in *Candida albicans*: uptake of inducer and germinative response. J Bacteriology 152:555–562

McFadden D, Zaragoza O, Casadevall A (2006) The capsular dynamics of *Cryptococcus neoformans*. Trends Microbiol 14:497–505

McLain N, Dolan JW (1997) Phospholipase D activity is required for dimorphic transition in *Candida albicans*. Microbiology 143:3521–3526

Michel S, Ushinsky S, Klebl B, Leberer E, Thomas D, Whiteway M, Morschhauser J (2002) Generation of conditional lethal *Candida albicans* mutants by inducible deletion of essential genes. Mol Microbiol 46:269–280

Michelitch M, Chant J (1996) A mechanism of Bud1p GTPase action suggested by mutational analysis and immunolocalization. Curr Biol 6:446–454

Miller MG, Johnson AD (2002) White-opaque switching in *Candida albicans* is controlled by mating-type locus homeodomain proteins and allows efficient mating. Cell 110:293–302

Mitchell AP (2005) Fungal CO_2 sensing: a breath of fresh air. Curr Biol 15:R934–R936

Mogensen EG, Janbon G, Chaloupka J, Steegborn C, Fu MS, Moyrand F, Klengel T, Pearson DS, Geeves MA, Buck J, Levin LR, Muhlschlegel FA (2006) *Cryptococcus* neoformans senses CO_2 through the carbonic anhydrase Can2 and the adenylyl cyclase Cac1. Eukaryot Cell 5:103–111

Murad AM, d'Enfert C, Gaillardin C, Tournu H, Tekaia F, Talibi D, Marechal D, Marchais V, Cottin J, Brown AJ (2001) Transcript profiling in Candida albicans reveals new cellular functions for the transcriptional repressors CaTup1, CaMig1 and CaNrg1. Mol Microbiol 42:981–993

Nantel A, Dignard D, Bachewich C, Harcus D, Marcil A, Bouin AP, Sensen CW, Hogues H, van het Hoog M, Gordon P, Rigby T, Benoit F, Tessier DC, Thomas DY, Whiteway M (2002) Transcription profiling of Candida albicans cells undergoing the yeast-to-hyphal transition. Mol Biol Cell 13:3452–3465

Nobile CJ, Mitchell AP (2005) Regulation of cell-surface genes and biofilm formation by the C. albicans transcription factor Bcr1p. Curr Biol 15:1150–1155

Nobile CJ, Mitchell AP (2006) Genetics and genomics of Candida albicans biofilm formation. Cell Microbiol 8:1382–1391

Nobile CJ, Andes DR, Nett JE, Smith FJ, Yue F, Phan QT, Edwards JE, Filler SG, Mitchell AP (2006) Critical role of Bcr1-dependent adhesins in C. albicans biofilm formation in vitro and in vivo. PLoS Pathog 2:e63

Oberholzer U, Marcil A, Leberer E, Thomas DY, Whiteway M (2002) Myosin I is required for hypha formation in Candida albicans. Eukaryot Cell 1:213–228

Oberholzer U, Iouk TL, Thomas DY, Whiteway M (2004) Functional characterization of myosin I tail regions in Candida albicans. Eukaryot Cell 3:1272–1286

Ozaki-Kuroda K, Yamamoto Y, Nohara H, Kinoshita M, Fujiwara T, Irie K, Takai Y (2001) Dynamic localization and function of Bni1p at the sites of directed growth in Saccharomyces cerevisiae. Mol Cell Biol 21:827–839

Park HO, Bi E, Pringle JR, Herskowitz I (1997) Two active states of the Ras-related Bud1/Rsr1 protein bind to different effectors to determine yeast cell polarity. Proc Natl Acad Sci USA 94:4463–4468

Ramage G, Saville SP, Wickes BL, Lopez-Ribot JL (2002) Inhibition of Candida albicans biofilm formation by farnesol, a quorum-sensing molecule. Appl Environ Microbiol 68:5459–5463

Ramage G, Saville SP, Thomas DP, Lopez-Ribot JL (2005) Candida biofilms: an update. Eukaryot Cell 4:633–638

Ramon AM, Fonzi WA (2003) Diverged binding specificity of Rim101p, the Candida albicans ortholog of PacC. Eukaryot Cell 2:718–728

Richard M, Ibata-Ombetta S, Dromer F, Bordon-Pallier F, Jouault T, Gaillardin C (2002) Complete glycosylphosphatidylinositol anchors are required in Candida albicans for full morphogenesis, virulence and resistance to macrophages. Mol Microbiol 44:841–853

Rocha CR, Schroppel K, Harcus D, Marcil A, Dignard D, Taylor BN, Thomas DY, Whiteway M, Leberer E (2001) Signaling through adenylyl cyclase is essential for hyphal growth and virulence in the pathogenic fungus Candida albicans. Mol Biol Cell 12:3631–3643

Rouabhia M, Schaller M, Corbucci C, Vecchiarelli A, Prill SK, Giasson L, Ernst JF (2005) Virulence of the fungal pathogen Candida albicans requires the five isoforms of protein mannosyltransferases. Infect Immun 73:4571–4580

Rozell B, Ljungdahl PO, Martinez P (2006) Host–pathogen interactions and the pathological consequences of acute systemic Candida albicans infections in mice. Curr Drug Targets 7:483–494

Sato T, Watanabe T, Mikami T, Matsumoto T (2004) Farnesol, a morphogenetic autoregulatory substance in the dimorphic fungus Candida albicans, inhibits hyphae growth through suppression of a mitogen-activated protein kinase cascade. Biol Pharm Bull 27:751–752

Schaller M, Bein M, Korting HC, Baur S, Hamm G, Monod M, Beinhauer S, Hube B (2003) The secreted aspartyl proteinases Sap1 and Sap2 cause tissue damage in an in vitro model of vaginal candidiasis based on reconstituted human vaginal epithelium. Infect Immun 71:3227–3234

Schroppel K, Sprosser K, Whiteway M, Thomas DY, Rollinghoff M, Csank C (2000) Repression of hyphal proteinase expression by the mitogen-activated protein (MAP) kinase phosphatase Cpp1p of Candida albicans is independent of the MAP kinase Cek1p. Infect Immun 68:7159–7161

Schweizer A, Rupp S, Taylor BN, Rollinghoff M, Schroppel K (2000) The TEA/ATTS transcription factor CaTec1p regulates hyphal development and virulence in Candida albicans. Mol Microbiol 38:435–445

Sheth CC, Johnson E, Baker ME, Haynes K, Muhlschlegel FA (2005) Phenotypic identification of Candida albicans by growth on chocolate agar. Med Mycol 43:735–738

Slutsky B, Staebell M, Anderson J, Risen L, Pfaller M, Soll DR (1987) "White-opaque transition": a second high-frequency switching system in Candida albicans. J Bacteriol 169:189–197

Sohn K, Urban C, Brunner H, Rupp S (2003) EFG1 is a major regulator of cell wall dynamics in Candida albicans as revealed by DNA microarrays. Mol Microbiol 47:89–102

Song Y, Kim JY (2006) Role of CaBud6p in the polarized growth of Candida albicans. J Microbiol 44:311–319

Sonneborn A, Tebarth B, Ernst JF (1999) Control of white-opaque phenotypic switching in Candida albicans by the Efg1p morphogenetic regulator. Infect Immun 67:4655–4660

Sonneborn A, Bockmuhl DP, Gerads M, Kurpanek K, Sanglard D, Ernst JF (2000) Protein kinase A encoded by TPK2 regulates dimorphism of Candida albicans. Mol Microbiology 35:386–396

Srikantha T, Tsai LK, Daniels K, Soll DR (2000) EFG1 null mutants of Candida albicans switch but cannot express the complete phenotype of white-phase budding cells. J Bacteriol 182:1580–1591

Staab JF, Bradway SD, Fidel PL, Sundstrom P (1999) Adhesive and mammalian transglutaminase substrate properties of Candida albicans Hwp1. Science 283:1535–1538

Staib P, Morschhauser J (1999) Chlamydospore formation on Staib agar as a species-specific characteristic of Candida dubliniensis. Mycoses 42:521–524

Stoldt VR, Sonneborn A, Leuker CE, Ernst JF (1997) Efg1p, an essential regulator of morphogenesis of the human pathogen Candida albicans, is a member of a conserved class of bHLH proteins regulating

Sudbery PE (2001) The germ tubes of *Candida albicans* hyphae and pseudohyphae show different patterns of septin ring localization. Mol Microbiol 41:19–31

Sudbery P, Gow N, Berman J (2004) The distinct morphogenic states of *Candida albicans*. Trends Microbiol 12:317–324

Tebarth B, Doedt T, Krishnamurthy S, Weide M, Monterola F, Dominguez A, Ernst JF (2003) Adaptation of the Efg1p morphogenetic pathway in *Candida albicans* by negative autoregulation and PKA-dependent repression of the EFG1 gene. J Mol Biol 329:949–962

Ushinsky SC, Harcus D, Ash J, Dignard D, Marcil A, Morchhauser J, Thomas DY, Whiteway M, Leberer E (2002) CDC42 is required for polarized growth in human pathogen *Candida albicans*. Eukaryot Cell 1:95–104

Vinces MD, Haas C, Kumamoto CA (2006) Expression of the *Candida albicans* morphogenesis regulator gene *CZF1* and its regulation by Efg1p and Czf1p. Eukaryot Cell 5:825–835

Walther A, Wendland J (2004) Polarized hyphal growth in *Candida albicans* requires the Wiskott-Aldrich Syndrome protein homolog Wal1p. Eukaryot Cell 3:471–482

Watanabe H, Azuma M, Igarashi K, Ooshima H (2006) Relationship between cell morphology and intracellular potassium concentration in *Candida albicans*. J Antibiot 59:281–287

Wedlich-Soldner R, Altschuler S, Wu L, Li R (2003) Spontaneous cell polarization through actomyosin-based delivery of the Cdc42 GTPase. Science 299:1231–1235

Wendland J, Philippsen P (2001) Cell polarity and hyphal morphogenesis are controlled by multiple rho-protein modules in the filamentous ascomycete *Ashbya gossypii*. Genetics 157:601–610

Wendland J, Walther A (2005) *Ashbya gossypii*: a model for fungal developmental biology. Nat Rev Microbiol 3:421–429

Wheeler RT, Fink GR (2006) A drug-sensitive genetic network masks fungi from the immune system. PLoS Pathog 2:e35

Wu C, Lytvyn V, Thomas DY, Leberer E (1997) The phosphorylation site for Ste20p-like protein kinases is essential for the function of myosin-I in yeast. J Biol Chem 272:30623–30626

Yaar L, Mevarech M, Koltin Y (1997) A *Candida albicans* RAS-related gene (*CaRSR1*) is involved in budding, cell morphogenesis and hypha development. Microbiology 143:3033–3044

Zheng X, Wang Y, Wang Y (2003) *CaSPA2* is important for polarity establishment and maintenance in *Candida albicans*. Mol Microbiol 49:1391–1405

Zheng X, Wang Y, Wang Y (2004) Hgc1, a novel hypha-specific G1 cyclin-related protein regulates *Candida albicans* hyphal morphogenesis. EMBO J 23:1845–1856

Zordan RE, Galgoczy DJ, Johnson AD (2006) Epigenetic properties of white-opaque switching in *Candida albicans* are based on a self-sustaining transcriptional feedback loop. Proc Natl Acad Sci USA 103:12807–12812

Zouwail S, Pettitt TR, Dove SK, Chibalina MV, Powner DJ, Haynes L, Wakelam MJ, Insall RH (2005) Phospholipase D activity is essential for actin localization and actin-based motility in *Dictyostelium*. Biochem J 389:207–214

7 Pathogenicity of *Malassezia* Yeasts

Peter A. Mayser[1], Sarah K. Lang[1], Wiebke Hort[1]

CONTENTS

I. Introduction – Historical and Current Taxonomy	115
II. Phylogeny and Identification	116
A. Morphology of *Malassezia* Yeasts	116
B. Culture and Differentiation of *Malassezia* Yeasts	116
C. Molecular Differentiation of *Malassezia* Yeasts	117
III. Epidemiology in Man and in Animals	118
A. Epidemiology of *Malassezia* Yeasts in Animals	118
IV. Physiology and Biochemistry	119
A. Nutritional Requirements	119
1. Lipid Metabolism	119
B. Cellular Envelope	119
C. Production of Filaments	119
D. Enzymatic Activities	120
1. Lipase	120
2. Lipoxygenase	120
3. Azelaic Acid	121
4. Gamma Lactone	121
E. Production of Pigments	121
1. Melanin	121
2. Tryptophan-Derived Indole Pigments	122
3. New Findings on the Biosynthesis of *Malassezia* Compounds	125
V. Host–Pathogen Interactions	127
A. Antigens in *Malassezia* spp. and Their Characterization	127
1. Characterization of the Antigens	127
B. *Malassezia* spp. and Immunity	130
1. Innate Immune Response	130
a) Complement System	130
b) Cellular Response and Cytokine Production	130
c) Interaction of *Malassezia* with Cutaneous Cells	131
2. Adaptive Immune Response	133
a) Humoral Response	133
b) Cellular Response	133
C. Models for *Malassezia*-Associated Diseases	134
VI. *Malassezia* Yeasts in Human and Animal Disease	134
A. Pityriasis Versicolor	134
B. Seborrhoic Eczema	135
C. *Malassezia* Folliculitis	136
D. *Malassezia* Sepsis	136
E. Atopic Eczema/Dermatitis Syndrome	136
F. Psoriasis	137
G. *Malassezia* Yeasts as Pathogens in Animals	139
H. Human–Animal Crosslinking	140
VII. Conclusion	140
References	141

I. Introduction – Historical and Current Taxonomy

Lipohilic yeasts of the genus *Malassezia* are found as members of the normal microflora of the skin in humans and in a variety of warm-blooded animal species. Howewer, these yeasts are associated with a number of diseases in humans and animals. Eichstedt (1846) was the first to recognize the association of *Malassezia* yeasts with pityriasis versicolor (PV). Difficulties in the isolation and culture of *Malassezia* yeasts – mainly because of ignorance of the obligatory lipid-dependence of the organism – led to a long-lasting nomenclatural and clinical controversy, resulting in a multitude of names and differentiations, which are listed in Table 7.1.

Historically, confusion was mainly caused by the creation of two different genera for the same microorganism resulting from its hyphal and yeast form. Because of the lack of possibilities to culture the organism it was not recognized that both features were produced by a single microorganism. So Baillon (1889) created the genus *Malassezia*, describing the species *M. furfur* as the agent of PV in which he observed yeasts and hyphae. The genus *Pityrosporum* was created to accommodate similar yeasts observed, without any filaments, in the scales of the human scalp (*P. ovale*, Castellani and Chalmers 1913), rhinoceros skin (*P. pachydermatis*, Weidman 1925) and dog ear (*P. canis*, Gustafson 1955).

---

[1] Zentrum für Dermatologie und Andrologie, Justus Liebig Universität, Gaffkystrasse 14, 35385 Giessen, Germany; e-mail: petermayser@derma.med.uni-giessen.de

Table 7.1. Historical names for yeasts of the genus *Malassezia* (modified according to Ingham and Cunningham 1993)

Name	Author	Year	Name	Author	Year
Microsporon furfur	Robin	1853	*Pityrosporum cantlieni*	Castellani	1908
Cryptococcus psoriasis	Rivolta	1873	*Microsporon macfadyeni*	Castellani	1908
Saccharomyces ovalis	Bizzozero	1884	*Dermatophyton malassezi*	Dold	1910
Saccharomyces sphericus	Bizzozero	1884	*Pityrosporum ovale*	Castellani	1913
Sacharomyces capillitii	Oudemans	1885	*Malassezia tropica*	Panja	1927
Malassezia furfur	Baillon	1889	*Malassezia ovalis*	Panja	1927
"*Flaschenbazillus*"	Unna	1894	*Cryptococcus malassezi*	Benedek	1930
Pityrosporum malassezii	Sabouraud	1904	*Monilia furfur*	Vuillemin	1931
Microsporon tropica	Castellani	1908	*Pityrosporum orbiculare*	Gordon	1951

Table 7.2. *Malassezia* spp. in 2006

Species	Described by
M. furfur (Robin)	Baillon (1889)
M. pachydermatis (Weidman)	Dodge (1935)
M. sympodialis	Simmons and Guého (1990)
M. globosa	Guého et al. (1996)
M. obtusa	Guého et al. (1996)
M. restricta	Guého et al. (1996)
M. sloofiae	Guého et al. (1996)
M. dermatis[a]	Sugita et al. (2002)
M. equi[a]	Nell et al. (2002)
M. japonica[a]	Sugita et al. (2003b)
M. yamatoensis[a]	Sugita et al. (2004)
M. nana[a]	Hirai et al. (2004)

[a] Based mainly on molecular data, also see text

Not until 1933 was the obligatory need for lipids of *Malassezia* yeasts recognized by Ota and Huang, rendering culture of the yeasts possible. Conventional systems of differentiation could not be applied because of lipid dependence. This resulted in the fact that in 1984 the genus *Malassezia* was still restricted to two species, namely *M. furfur*, a species that was thought to be pathogenic for humans, and *M. pachydermatis*, which was easier to cultivate, growing without addition of lipids, and thought to be mainly a pathogen in animals (Yarrow and Ahearn 1984). A third species, *M. sympodialis*, was differentiated in 1990 because of genomic differences.

Based on physiological characteristics (assimilation of different lipid combinations) and genomic data, a reclassification was done in 1996, which resulted in the discrimination of seven species (Gueho et al. 1996). Since then, further species have been differentiated, mainly according to molecular data (Table 7.2).

II. Phylogeny and Identification

The sexual form (teleomorph) of *Malassezia* yeasts is still unknown, but they are thought to belong to the basidiomycetes showing characteristics such as multilayered cell wall, formation of urease and red staining with diazonium blue. The classification of *Malassezia* yeasts into the basidiomycetes within the family of *Filobasidiaceae* has been genetically confirmed in the meantime.

A. Morphology of *Malassezia* Yeasts

Malassezia yeasts show monopolar or sympodial budding with the formation of a collarette (Slooff 1970; Yarrow and Ahearn 1984; Simmons and Ahearn 1987; Simmons and Guého 1990). This "scar" resulting from phialidic conidiogenesis can occasionally be seen by light microscopy. The yeast cells are round (2.5–5.0 μm in diameter), oval or cylindric (1.5–3.0 × 2.5–8.0 μm) in shape. The formation of hyphae in some species during culture was described by Guého (1996), but is only occasionally observed.

B. Culture and Differentiation of *Malassezia* Yeasts

Culture conditions are not identical for all species, especially *M. globosa*, *M. obtusa* and *M. restricta* are known to be difficult to culture. With the exception of *M. pachydermatis*, *Malassezia* species need long-chain fatty acids. Therefore media, e.g. Sabouraud–glucose agar 4%, must be overlaid with olive oil, or special media such as Dixon agar have to be used. Optimal growth conditions are between 32 °C and

37 °C, at pH 5.5–6.5. Cycloheximide at a concentration of 400–1000 ppm should be added in order to avoid contamination of the culture. Conventional differentiation of *Malassezia* yeasts is performed on the basis of physiological and morphological criteria, including assimilation of Tween 20, 40, 60 and 80, catalase reaction, cleavage of esculin and glucose (Guillot et al. 1996; Mayser et al. 1997; Guého et al. 1998; Batra et al. 2005). If conventional differentiation fails to give clear results, molecular characterization should be performed.

C. Molecular Differentiation of *Malassezia* Yeasts

For years, taxonomy of *Malassezia* yeasts was based in particular on morphologic criteria. 1996 the technique of characterization of the large subunit of ribosomal DNA (D1/D2 region) was applied to *Malassezia*, which led to reclassification of the genus. At that time seven species were proposed, which are now widely accepted: *M. furfur*, *M. pachydermatis*, *M. sympodialis*, *M. obtusa*, *M. globosa*, *M. restricta* and *M. slooffiae* (Guillot and Guého 1995; Guého et al. 1996). Since then, further molecular studies have led to the proposal of additional species such as *M. equi* (Nell et al. 2002), *M. dermatis* (Sugita et al. 2002), *M. japonica* (Sugita et al. 2003b), *M. nana* (Hirai et al. 2004) and *M. yamatoensis* (Sugita et al. 2004). Different molecular techniques have been applied to *Malassezia* yeasts (Table 7.3). Molecular techniques should be chosen according to the aim of the respective study (Table 7.4).

Reclassification of the genus *Malassezia* with the seven above-mentioned species relying on sequence analysis of the LSU region was confirmed by pulsed field gel electrophoresis (PGFE), which showed no intraspecific variation in karyotypes with the exception of *M. furfur* (Boekhout et al. 1998) As PGFE is a relatively time-consuming procedure, it is not appropriate for routine identification of species. Analysis of the ITS1 region of ribosomal DNA was described by Makimura et al. (2002). For this technique, differentiation of the seven above-mentioned species as well as strain typing could be shown. New species were proposed because of genetic variations in the ITS region, for example *M. dermatis* (Sugita et al. 2002). Restriction fragment length polymorphism was described by different authors. Recently, Mirhendi et al. (2005) described a PCR-RFLP method for identification and differentiation of 11 *Malassezia* species, including *M. dermatis*, *M. nana*, *M. japonica* and *M. yamatoensis*, by amplification of the 26SrDNA sequence and use of only two restriction enzymes (*Cfo*I and *Bst*F51; Mirhendi et al. 2005). By use of AFLP the seven *Malassezia* species could be differentiated; moreover, intraspecific variations can be shown in the species of *M. furfur*, of which one variant is proposed to be of special clinical importance in that it causes systemic diseases (Theelen et al. 2001). By applying the technique of randomly amplified polymorphic DNA (RAPD), genetic variations in each of the seven species could be found (Boekhout et al. 1998). Therefore, this method is applicable to strain typing for epidemiologic

Table 7.3. Molecular techniques used for differentiation of *Malassezia* yeasts

Pulsed field gel electrophoresis (PGFE)	Boekhout et al. (1998)
Randomly amplified polymorphic DNA (RAPD)	Boekhout et al. (1998)
Amplified fragment length polymorphism (AFLP)	Theelen et al. (2001)
Denaturing gradient gel elecrophoresis (DGGE)	Theelen et al. (2001)
Sequencing analysis:	References:
a) Ribosomal DNA analysis (D1/D2 region of the large subunit of ribosomal DNA	a) Guého et al. (1995) b) Makimura et al. (2002) c) Sugita et al. (2003)
b) Analysis of internal transcribed spacer regions (ITS1/2) of the rDNA	Gaitanis et al. (2002) Gemmer et al. (2002) Mirhendi et al. (2005)
c) Analysis of intergenic spacer region (IGS) of the rDNA	
Restriction analysis of PCR amplicons:	
Restriction fragment length polymorphism (RFLP)	
Terminal fragment length polymorphism (tFLP)	
Chitin synthase gene sequence analysis	Aizawa et al. (1999, 2001)

Table 7.4. Appropiate molecular methods for differerent aims of a study

Aim of study	Appropriate molecular methods
Species identification	Sequencing of LSU (D1/D2); PGFE, sequencing of rDNA (IGS, ITS1/ITS2), PCR-restriction endonuclease analysis (REA), RFLP
Strain typing for epidemiologic survey	RFLP, AFLP, RAPD
In-vivo diagnosis	Nested PCR for amplification of the ITS region and RFLP

purposes, but is unsuitable for species identification. For in vivo diagnosis restriction fragment length polymorphism (RFLP) and terminal fragment length polymorphism (tFLP) can be used. Both techniques do not require any (pre)culture, thus eliminating the problem of "culture bias" by varying cultivation requirements of the different species.

III. Epidemiology in Man and in Animals

Since the reclassification of *Malassezia* yeasts in 1996, several studies have been performed in order to figure out the distribution of different species on normal and diseased skin. Varying results in these studies may be partly due to examination of different body sites, different means of sample taking, as well as diverse methods of culture and differentiation. Furthermore, there seem to be variations in the distribution of the species in individual ethnic groups or different geographic areas. It appears that *M. sympodialis, M. globosa* and *M. restricta* are most frequently isolated from healthy human skin (Table 7.5). *M. pachydermatis* is found mainly in animals – in human skin it can only be observed as part of the transient flora (Midgley 1989; Chen and Hill 2005).

Relationships of different species with diseases have been investigated as well. A correlation between *M. globosa* and PV has been supposed by some authors because it was isolated most frequently from lesions of PV (Crespo-Erchiga et al. 2000; Nakabayashi et al. 2000; Aspiroz et al. 2002). Nevertheless, as shown in Table 7.5, it can also be found in normal skin – alone or in combination with *M. sympodialis, M. restricta* and *M. furfur* (Sugita et al. 2001). In the remaining dermatological disorders associated with *Malassezia* yeasts, their role is controversial. Most studies have focused on immunological aspects in the context of *Malassezia* yeasts in the head and neck form of atopic dermatitis and seborrhoic eczema, which are mentioned in the corresponding parts of this chapter.

A. Epidemiology of *Malassezia* Yeasts in Animals

As part of the resident flora of the skin and the meatus acusticus externus, *M. pachydermatis* is the predominant *Malassezia* species in mammals and birds (Gustafson 1959, 1960; Baxter 1976; Kennis et al. 1996). *M. globosa, M. sympodialis* and *M. furfur* and even *M. sloofiae* and *M. restricta*, alone or in combination, have also been isolated from animals (Bond et al. 1996a, b, 1997; Raabe et al. 1998; Crespo et al. 2002a, b; Duarte et al. 2003; Nardoni et al. 2004, 2005; Cafarchia et al. 2005; Garau et al. 2005; Coutinho et al. 2006; White et al. 2006). *M. nana*, a new *Malassezia* species, has been isolated from animal skin (Hirai et al. 2004). Nell et al. (2002) isolated a *M. sympodialis*-related species from equine skin, which they named "*M. equi*". Because of the level of sequence divergence of the 26S ribosomal DNA D1/D2 sequence, they supposed it to be a new species (Nell et al. 2002). Further investigation is needed for validation of the isolate as a new species (Cabanes et al. 2005). *M. pachydermatis* was first isolated from an Indian rhinoceros with exfoliative dermatitis (Weidman 1925). Guillot et al. (1997) discriminated several variants (named Ia–Ig) of *M. pachydermatis* by partial LSU rRNA sequencing,

Table 7.5. *Malassezia* species isolated on healthy human skin (%). *n.e.* not examined

Study	No. of individuals examined	Individuals with *M.* spp	*M. globosa*	*M. restricta*	*M. sympodialis*	*M. furfur*	*M. sloofiae*	*M. obtusa*	*M. pachydermatis*
Aspiroz et al. (1999)	38	100	52.0	22.0	22.0	n.e.	n.e.	n.e.	0
Nakabayashi et al. (2000)	35	n.e.	22.0	n.e.	10.0	3.0	n.e.	n.e.	n.e.
Salah et al. (2005)	30	31.2	21.1	3.3	10.0	8.9	1.1	0	0
Rincón (2005)	37	n.e.	16.2	5.4	29.7	21.6	0	2.7	5.4
Sandström et al. (2005)	31	84.0	12.0	0	69.0	0	4.0	15.0	0
Sugita et al. (2001)	18	77.8	44.4	61.1	50.0	11.1	0	0	0
Tarazooie et al. (2004)	100	60.0	41.7	3.3	25.0	23.3	6.7	0	0

of which some apparently showed host specificity (Ic – rhinoceros, Id – dogs, Ig – ferrets; Guillot et al. 1997). Aizawa et al. (1999, 2001) classified *M. pachydermatis* by RAPD and chitin synthase 2 (CHS2) gene sequence analyses and suggested four sequence types (A–D); according to the CHS2 sequence analysis, *M. pachydermatis* type B is genetically closely related to *M. furfur*. Some *M. pachydermatis* strains produce pigment on media containing tryptophan as sole nitrogen source, although to a lesser extent and significantly delayed compared to *M. furfur* (Mayser et al. 2004b). Despite the close genetic relationship between *M. pachydermatis* strain B classified by Aizawa et al. (2001) and *M. furfur* none of the type B strains produced pigment (Hossain et al. 2006). In the same study, 28 genotypes of *M. pachydermatis* were differentiated by RAPD. Twelve of these clusters contained only pigment-producing strains and two contained both pigment producers and non-pigment producers. High genetic variability among the *M. pachydermatis* strains was supported by the observation by Chryssanthou et al. (2001), who detected six different RAPD band patterns in six *M. pachydermatis* blood isolates, and confirmed the observation by Kurtzmann and Robnett (1994) that a high rate of nucleotide substitution exists among *Malassezia* spp., which far exceeds the levels of other yeast genera. These findings support the hypothesis of a rapid "molecular clock", resulting in enormous polymorphism in the genus.

IV. Physiology and Biochemistry

A. Nutritional Requirements

With the exception of *M. pachydermatis*, all *Malassezia* species require an external lipid source for growth, i.e. they are obligatorily lipid-dependent. However, these lipids are thought to be important virulence factors of the fungus, as the lipid layer seems to protect *Malassezia* from phagocytosis (2002) and down-regulates the inflammatory immune response (Kesavan et al. 2000). In addition, adhesion to host cells may be mediated by the hydrophobicity of the lipid-rich cell wall.

1. Lipid Metabolism

Lipid dependence is known to be based on a defect in the synthesis of myristic acid, which serves as the precursor of long-chained fatty acids (Shifrine and Marr 1963; Wilde and Stewart 1968; Porro et al. 1976). Longer-chain fatty acids can be synthesized from these mid-chain fatty acids. Transformation from saturated fatty acids into unsaturated ones is possible (Porro et al. 1976). Nontheless, the lipid composition of these yeasts is not constant but reflects to a great extent the nutritive lipid offer (Shiffrine 1963; Nazzarro-Porro et al. 1976; Huang et al. 1993). In most cases, triacylglycerides, free fatty acids or polyoxysorbitan fatty acid esters (Tween) are used as substrates. Apart from hydrolysis of neutral fat, which has been known for a long time, assimilation of phospholipids has also been described (Riciputo et al. 1996). Assimilation of steroids has been demonstrated for cholesterol and cholesterol esters (Nazzaro-Porro et al. 1977), leading to the formation of hyphae.

B. Cellular Envelope

The cell wall of *Malassezia* spp. shows a unique structure (Simmons and Ahearn 1987; Mittag 1994, 1995). It is 0.12 µm thick, multilayered and exhibits a spiral structure (Barfatani et al. 1964; Swift and Dunbar 1965; Keddie 1966; Breathnach et al. 1976; Mittag 1995), consisting of the outer lamella, the multilayered, multilamellar wall and the plasma membrane. A characteristic feature is the striated pattern in the inner wall sections, corresponding to a spiral structure. This striation is visible by light microscopy (Slooff 1970). Functions of single compounds of the cell wall have not yet been defined.

With a lipid percentage of 15% the cell wall of *Malassezia* spp. shows a significantly higher portion than that of *Saccharomyces* spp. (1–2%; Thompson and Colvin 1970).

The construction of the whole cell sheath with a high lipid percentage is presumably responsible for the great resistance to extraneous influences, e.g. for high mechanical stability and osmoresistance (Brotherton 1967).

C. Production of Filaments

The hyphal stage observed by Eichstedt, i.e. dimorphism, is difficult to induce by culture. Dorn and Roehnert (1977) were successful, by adding glycine to the medium, as were Nazzaro-Porro et al. (1977), using a medium that contained cholesterol and cholesterol esters. Faergemann and Bernander (1981) induced hyphal growth under increased carbon dioxide tension and observed filamentous growth of *P. ovale* on human stratum corneum in vitro. Most hyphae were produced by the *P. ovale* strains ATCC 44341 and ATCC 44031, both of which are now classified as *M. sympodialis*. This is in line with a more recent finding by Saadatzadeh et al. (2001). These authors tested several *Malassezia* strains for their capability of producing mycelia. Only strains of serovar A/*M. sympodialis* produced hyphae. Although *M. globosa* is supposed to be the causative agent for PV (Crespo-Erchiga et al. 2000), mycelial growth has not yet been induced in vitro.

Initial hyphae formation is occasionally observed in primary cultures (Gordon 1951a, b; Burke 1961; Roberts 1969; Slooff 1970; Caprilli et al. 1973; Faergemann and Bernander 1981; Gueho et al. 1996). However little is known about the hyphal stage. The hyphae are now considered to be pseudohyphae, as they have no dolipores which interrupt the true filaments in basidiomycetes (Guillot and Guého 1995). However, the hyphal stage is thought to play a role in the pathogenesis of pityriasis versicolor, as hyphae are abundantly demonstrable in scales taken from lesions (Gordon 1951a, b). The frequency of hyphae in patients with pityriasis versicolor is 100% in lesions, 42% in non-lesional skin of the trunk and 50% on the head. In 6–7% hyphae are also observed on healthy skin (McGinley et al. 1975).

D. Enzymatic Activities

1. Lipase

In *Malassezia* yeasts, hydrolysis of triacylglycerides during release of glycerol and free fatty acids was demonstrated (Weary 1970; Caprilli et al. 1973). A membrane-bound extracellullar lipase was shown in in vitro studies by Catterall et al. (1978). It is extracellularly located on the membrane surface after generation within the cell and extrusion through the membrane. However, identical techniques of histochemistry and electron microscopy failed to demonstrate this enzyme in vivo – in scales of pityriasis versicolor.

Using different fatty acid (mono-)esters Mayser et al. (1995) found that yeast-dependent hydrolysis of these synthetic fatty acid (mono-)esters was critically dependent on alcohol moiety, while growth promotion was best stimulated by unsaturated fatty acids. They observed greater extracellular lipase activity and more pronounced production of fatty acid ethyl esters in *M. sympodialis* compared to *M. furfur* (Mayser et al. 1998c). In 1997 they described differences in the lipid metabolism of the individual *Malassezia* species, in particular specificity of *M. furfur* to assimilate ricinoleic acid (12-hydroxy oleic acid) and PEG-35 castor oil (Cremophor EL), which is now used for species differentiation. Ran et al. (1993) did not detect lipases in the supernatant but localized the main lipolytic activity in the insoluble fraction of cell extracts. In contrast, Plotkin et al. (1996) demonstrated lipolytic activity in the supernatant and in intracellular soluble and insoluble extracts of *M. furfur* and further characterized three different lipolytic activities in the soluble fraction.

Cloning and characterization of the first gene encoding a secreted lipase of *M. furfur* was reported by Brunke and Hube (2006). The gene, *MfLIP1*, shows high sequence similarities to other known extracellular lipases, but is not a member of a lipase gene family in *M. furfur*. *MfLIP1* consists of 1464 base pairs coding for a protein with a molecular mass of 54.3 kDa, a conserved lipase motif and an N-terminal signal peptide of 26 amino acids. The cDNA of MfLIP1 was expressed in *Pichia pastoris* and the biochemical properties of the recombinant lipase were analyzed. MfLip1 is most active at 40 °C and the pH optimum was found to be pH 5.8. The lipase hydrolyzed lipids such as Tweens, which are frequently used as the sole source of carbon in media for *Malassezia*, and had minor phospholipase and esterase activities.

2. Lipoxygenase

Increased levels of lipid peroxides and their derivatives were demonstrated in lesional, but not in non-lesional skin of patients with pityriasis versicolor (Nazzaro-Porro et al. 1986). Therefore, they were thought to be significant for the disease. Nonetheless the enzyme has not yet been synthesized. De Luca et al. (1996) observed increased formation of lipid peroxides when the medium had been enriched with polyunsaturated fatty acids. Depigmentation accompanying pityriasis versicolor was mostly explained by the formation of lipid peroxides, which might have toxic effects on melanocytes.

3. Azelaic Acid

In a medium containing olive oil, *Malassezia* yeasts produced C_9–C_{11} dicarboxylic acids (Nazzarro-Porro and Passi 1978), especially azelaic acid, a C_9 dicarboxylic acid [HOOC–$(CH_2)_7$–COOH], competitively inhibits tyrosinase, an enzyme that is involved in the production of melanin. However, subsequent studies showed that these fatty acids did not influence melanocytes in vivo and in vitro (Breathnach et al. 1984; Robins et al. 1987). Furthermore, their therapeutic use in the treatment of hyperpigmentation and melanoma, based on tyrosinase inhibition, yielded disappointing results (Breathnach 1984). Because of its antimicrobial activity, azelaic acid is now predominantly used in the therapy of acne and inflammatory rosacea (Elewski and Thiboutot 2006).

Little is known about the growth conditions and metabolic activity outside of lipid metabolism. As standard assimilation tests cannot be performed because of the lipid dependence, only few studies are available. Carbohydrate assimilation has been determined solely for *M. pachydermatitis*. *M. pachydermatis* can assimilate mannite, glycerol and sorbitol (Slooff 1970; Hossain et al. 2006). The capability of fermenting sugar has not been demonstrated in any of the *Malassezia* species (Slooff 1970). *Malassezia/Pityrosporum* yeast cells were able to grow in vitamin-free basal media that contained fat as the sole carbon source (Porro et al. 1976). Enhanced growth was observed after addition of asparagine, thiamine and pyridoxine, but these are not essential growth agents (Benham 1945; Porro et al. 1976).

Inorganic sulfate and sulfite were not assimilated (Brotherton 1967). *Malassezia* yeasts utilize both organic and inorganic nitrogen sources (Slooff 1970) but are unable to assimilate potassium nitrate.

4. Gamma Lactone

The characteristic odor of *Malassezia* cultures was first described by van Abbe (1964). Later these volatile substances were identified as γ-decalactone (Labows et al. 1979). They are thought to be responsible for the fruity odor of the cultures.

E. Production of Pigments

The feature of pigment formation in higher fungi, particularly in macromycetes, has long been used for taxonomic purposes. However, elucidation of some of the complex and instabile compounds, their synthesis pathways and thus their industrial usage was not achieved until the past 30 years. The investigation of pigment-free mutants, especially those of *Aspergillus fumigatus*, showed that pigment formation can be correlated with pathogenicity (for an excellent review, see Langfelder et al. 2003). Two metabolic pathways leading to pigment synthesis have so far been demonstrated in *Malassezia* yeasts.

1. Melanin

Melanin production is a well studied pathogenetic mechanism in fungi (Langfelder et al. 2003). Many properties related to the evasion of the host immune system and antifungal drug resistance have been attributed to melanin or melanin-like pigments. Gaitanis et al. (2005a) tested all *Malassezia* strains semiquantitatively for their abilities to produce pigment when grown in lipid-supplemented and lipid-depleted L-DOPA and tyrosine agars. No oxidation of tyrosine was detected when *Malassezia* yeasts had been grown on tyrosine agar, indicating that melanogenesis may occur via a tyrosinase-independent pathway. By contrast, *Malassezia* strains tested on L-DOPA agar produced a pigment with various melanization intensities. *M. dermatis* strains demonstrated maximum and *M. furfur* demonstrated minimum pigment production. However, the L-DOPA substrate was oxidized only after *Malassezia* membrane disruption, suggesting that phenoloxidase, the enzyme mediating melanin production, is not secreted, but is either attached to the cell wall or bound to the membrane, as in *Cryptococcus neoformans*. The fact that skin scales originating from hyperpigmented PV lesions, even those from patients displaying both types of lesions, showed *Malassezia* cells and hyphae staining positive for Masson–Fontana led the authors to conclude that melanization takes place in vivo and may be responsible for hyperpigmentation observed within the lesions. Thus, *M. sympodialis* was isolated from the hyperpigmented lesions and *M. furfur* from the hypopigmented lesions of the same patient. No melanin-like pigment was detected in yeast cells and hyphae in skin scales from the hypopigmented PV lesions.

2. Tryptophan-Derived Indole Pigments

During investigations on the metabolic requirements of *Malassezia* species (Mayser et al. 1998a) a subpopulation of *Malassezia* yeasts was distinguished that produces a brownish pigment diffusing into agar when cultured on a selective agar medium consisting of the amino acid tryptophan (Trp) and a lipid source. Under UV light the produced pigment showed green-yellowish fluorescence (Mayser et al. 1998b). Pigment production is specifically dependent on the amino acid Trp, but is independent of its optical activity (D or L). The lipid source is exchangeable. Thus, there are similarities to the pigment synthesis of another basidiomycetic yeast, *Filobasidiella neoformans* (better known as the imperfect form *Cryptococcus neoformans*) in which, depending on the nitrogen source (diphenyls, particularly dopa), a secondary metabolism can be induced which results in pigment synthesis (melanin; Polacheck et al. 1988). By induction of this pathway it is possible to demonstrate increased pathogenicity and, in particular, inhibition of phagocytosis (Polak 1990).

The ability to produce pigment is typical of the species *M. furfur* and applies to reference strains as well as wild strains isolated from foci. Some *M. pachydermatis* strains were also found to produce pigment, though after markedly longer incubation periods, with a lower yield and a limited "color spectrum". (Mayser et al. 2004b; Hossain et al. 2006). Chromatography shows a complex pigment composition with a variety of differently colored single bands and numerous fluorochromes.

The chemistry and pharmacology of the compounds were found to be especially interesting in that they might explain phenomena of pityriasis versicolor. Because of the extremely complex composition of the pigment, priority was given to the isolation of biologically active metabolites detected by means of bioassays that were based on the clinical phenomena of pityriasis versicolor.

Pityriasis versicolor is one of the commonest human skin diseases, which is characterized by lesions of varying color, a phenomenon that is still unclear (Gupta et al. 2003). Given the characteristic occurrence of PV only under special conditions such as high air humidity and excessive sweating and the fact that *Malassezia* yeasts belong to the normal cutaneous flora, the regularly produced dicarboxylic acids cannot explain why pityriasis versicolor (alba) does not occur as a rule when the organism is found on the skin (Thoma et al. 2005).

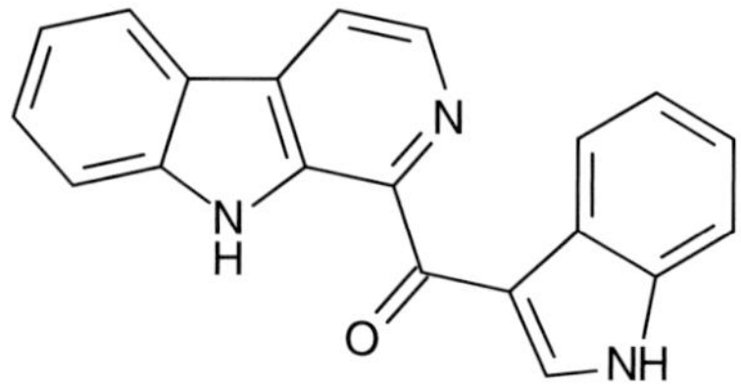

$C_{20}H_{13}N_3O$
MW 311.34

Fig. 7.1. Pityriacitrin

The following substances have been characterized, which may help clarify the pathogenesis of PV and allow first insights into pathogenetic factors especially of *M. furfur*.

Pityriacitrin (Fig. 7.1), a yellow compound eluting from the column with 64% acetonitrile, was found to be a potent UV filter due to its broad absorption (λ_{max} 389, 315, 289, 212 nm). It is an indole derivative {(9H-pyrido[3,4-b]indol-1-yl)(1H-indol-3-yl)methanon; $C_{20}H_{13}N_3O$; pityriacitrin}, which has also been demonstrated in bacteria as a potent UV filter. Its UV-protective properties were confirmed by means of a yeast model and in humans (Mayser and Pape 1998; Mayser et al. 2002; Machowinski et al. 2006). Pityriacitrin produced by *M. furfur* might explain UV protection in depigmented foci of PV alba (Thoma et al. 2005; de Almeida et al. 2006).

Pityrialactone (Fig. 7.2) was found to be a fluorescent, hitherto unknown bisindole compound which, due to its mesomeric structure, might serve as a radical scavenger, but also as a light protector. Remarkably, the substance shows blue fluorescence in a lipophilic environment and yellow fluorescence in watery milieu. This might explain the multicolor fluorescence of PV lesions described in the

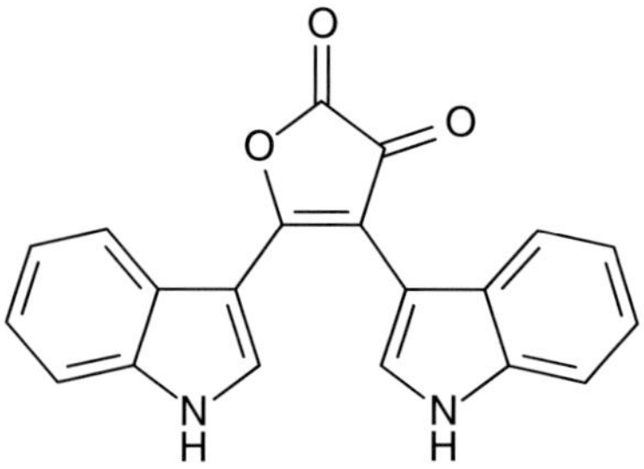

$C_{20}H_{12}N_2O_3$
MW 328.32

Fig. 7.2. Pityrialactone

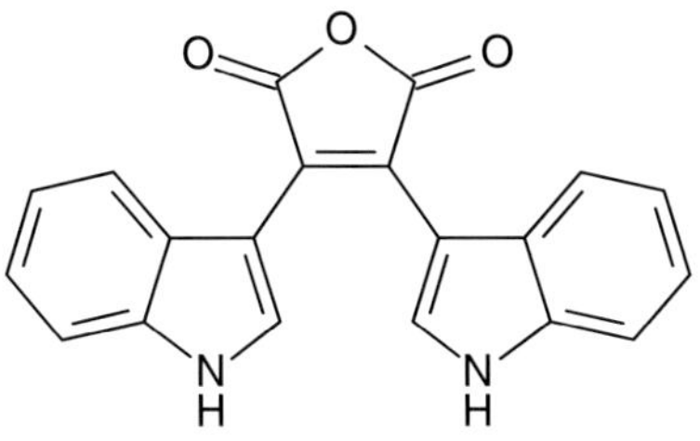

Fig. 7.3. Pityriaanhydride

literature (from yellow-orange to green-yellow) depending on whether the substance is dissolved in sweat or epidermal lipids. With the isomeric structure, the brick-red pityriaanhydride (Fig. 7.3), another bisindole compound, was demonstrated for the first time in nature (Mayser et al. 2003a, b).

The likewise brick-red pityriarubins (Figs. 7.4–7.6) are the first compounds of the new substance class of bisindolyl cyclo-pentendiones (Irlinger et al. 2004). In their basic structure they are related to pityriaanhydride, but also to bisindolylmaleinimides and indolocarbazoles, which have gained much interest as protein kinase inhibitors (Davis et al. 1992).

The isolated pityriarubins A, B and C (Figs. 7.4–7.6) can suppress the release of reactive oxygen species (ROS, "burst") from activated granulocytes (Krämer et al. 2005a). Inhibition was observed on activation with A23187, FMLP and interleukin 3. No inhibition occurred on activation with the classic protein kinase C (PKC) activators phorbol ester and dioctanoyl-glycerol, and also on unspecific activation of G proteins with sodium fluoride via the alternative complement way using zymosan. In all these cases, apart from zymosan, the unspecific arcyriarubin A (as a representative of the class of bisindolylmaleinimides) showed inhibition. Pityriarubins had dose-effectiveness curves similar to those of the highly potent arcyriarubin A, so that a comparably potent, but very specific effect on the same target(s) can be assumed (IC_{50} between $2\,\mu M$ and $5\,\mu M$). These effects of the described substances might explain the missing granulocyte infiltrate in lesions of pityriasis versicolor despite high fungal load (Wroblewski et al. 2005), as they may help the yeast to escape the immune response. Based on their specific mechanism of action they might therefore present a new anti-inflammatory principle.

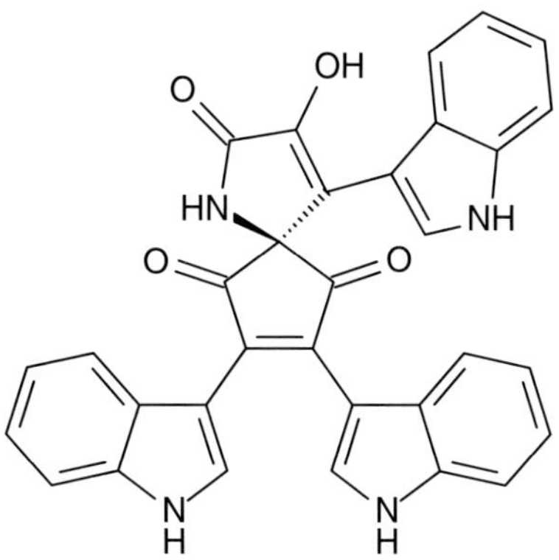

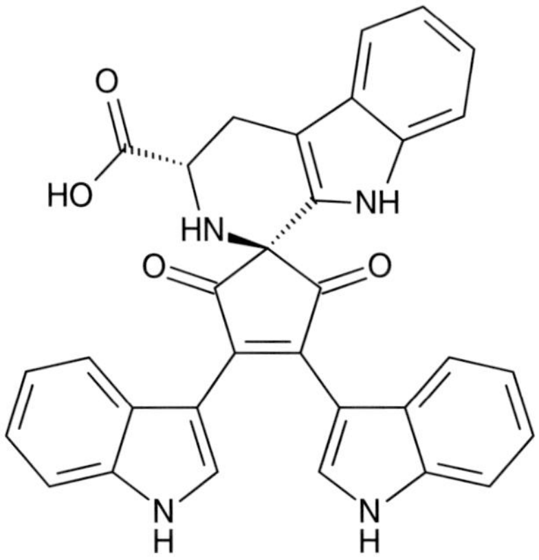

$C_{32}H_{20}N_4O_4$
MW 524.53

Fig. 7.5. Pityriarubin B

$C_{32}H_{22}N_4O_4$
MW 526.55

Fig. 7.4. Pityriarubin A

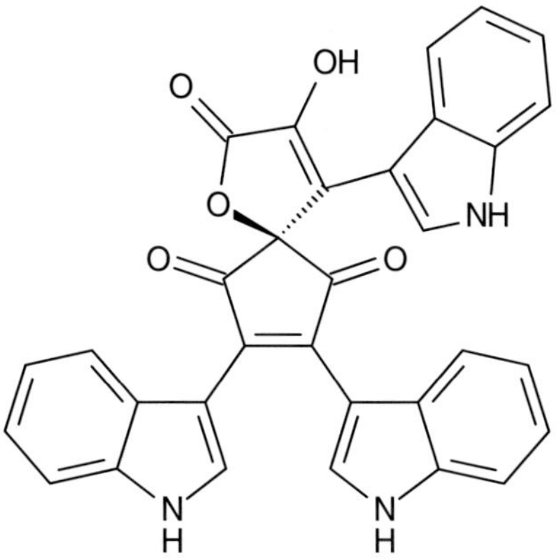

$C_{32}H_{19}N_3O_5$
MW 525.51

Fig. 7.6. Pityriarubin C

Malassezin (Fig. 7.7) is a colorless substance which shows the feature of an aryl hydrocarbon receptor (AHR) agonist ($EC_{50Erod} = 1.6 \times 10^{-6}$) and induces cytochrome P450 ($EC_{50} = 1.57\,\mu M$) in cultures of rat hepatocytes (Wille et al. 2001). Because of its acyclic structure malassezin does not meet the criteria of a potent (highly affinic) AHR agonist such as indolo[3,2-*b*]carbazole. Remarkably, with the yellow to orange-red malassezia carbazoles A–D [Fig. 7.8; malassezia carbazol A (**18**), B (**20**), C (**21**), D (**19**)] compounds fulfilling these criteria have meanwhile been isolated from the raw extract of *M. furfur* (Irlinger et al. 2005). Of these, compounds **18**, **19** and **21** show high similarity to synthetic indolo[3,2-*b*]carbazol (**22**), which is thought to be the best ligand for the aryl hydrocarbon receptor (AHR; Wille et al. 2001). They have not yet been investigated for their ability to act as AHR agonists because of insufficient amounts of substance. Interestingly, malassezin causes dose-dependent induction of apoptosis in cultivated human melanocytes, which may explain the damage to melanocytes in lesions of PV alba (Krämer 2005b). Cytoskeletal changes have also been observed. By activity determination of caspases 8 and 9, the apoptosis pathway was characterized as being "intrinsic". It is assumed that changes in the cytoskeletal structure lead to impaired melanosome transport. Both apoptosis induction and cytoskeletal changes may be connected with the depigmentation that is observed in pityriasis versicolor alba. In addition, by interaction with the AHR receptor, malassezin might induce changes in the differentiation and growth behavior of keratinocytes, which would be responsible for the characteristic desquamation observed with PV.

Furthermore, with its unusual azepine structure malassezia indole A (Fig. 7.9) shows a dose-dependent inhibition of tyrosinase, which is

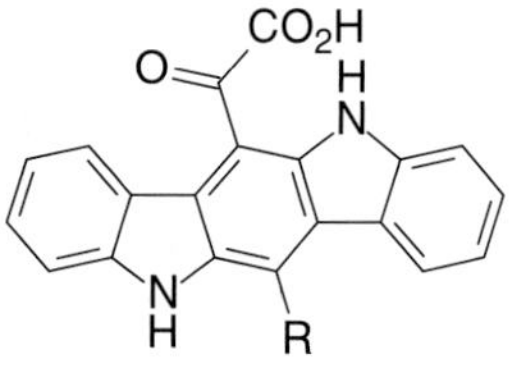

18, R = H
19, R = CHO

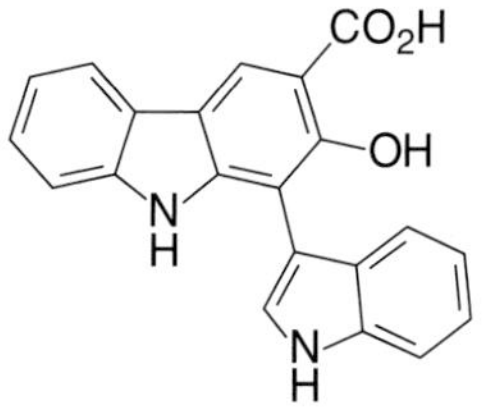

20

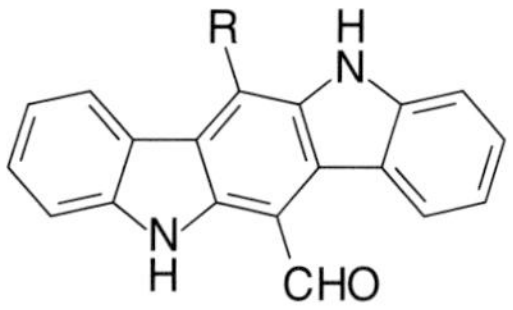

21, R = CO_2H
22, R = CHO

Fig. 7.8. Malassezia carbazoles A–D (**18–21**) and indolo[3,2-*b*]carbazol (**22**). Malassezia carbazol A (**18**), B (**20**), C (**21**) and D (**19**)

a key enzyme of melanin synthesis, so that both effects may complement each other in their depigmenting mechanism (Dahms et al. 2002).

A remarkable feature of the two substances O52, a so far unknown 1,3-Bis(indol-3-yl)acetone, and keto-malassezin (Fig. 7.10), which were isolated

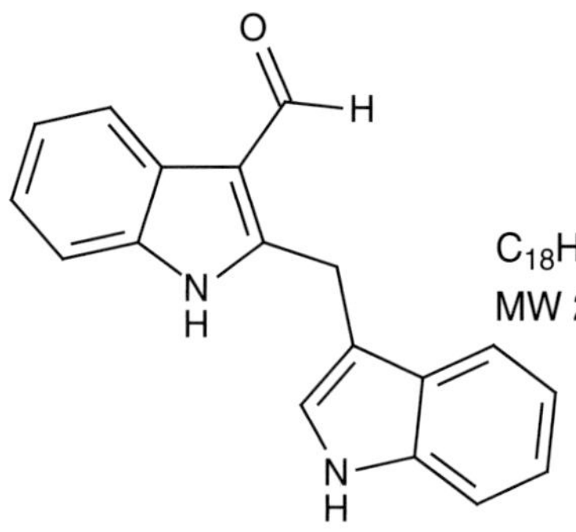

$C_{18}H_{14}N_2O$
MW 274.32

Fig. 7.7. Malassezin

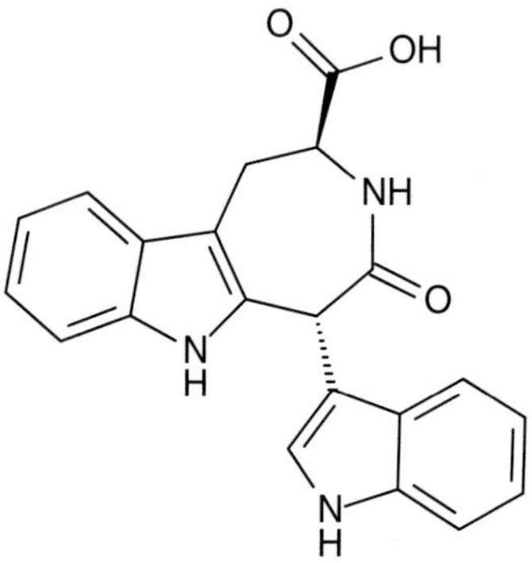

$C_{21}H_{17}N_3O_3$
MW 359.38

Fig. 7.9. Malassezia indole A

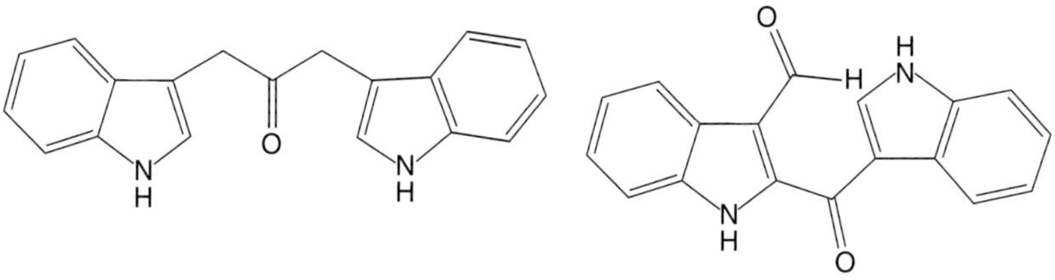

Fig. 7.10. O52 [a so far unknown 1,3-bis(indol-3-yl)acetone] and keto-malassezin

and described by us for the first time, is their capability of inhibiting the dopa reaction on human epidermal melanocytes in situ. They are probably also tyrosinase inhibitors (Dahms et al. 2002).

Keto-malassezin especially represents a very interesting substance: as a malassezin derivative it combines the features of an AHR agonist with those of a tyrosinase inhibitor and thus appears most interesting with regard to its effects on melanocytes (apoptosis induction, additionally inhibition of melanin synthesis).

Further metabolites, which have not yet been assigned to specific effects, have been summarized by Irlinger et al. (2005). The hitherto most complicated and structurally most interesting indole alkaloid from cultures of *M. furfur* is malasseziacitrin (Fig. 7.11). It is probably biosynthetically produced by three tryptophan molecules.

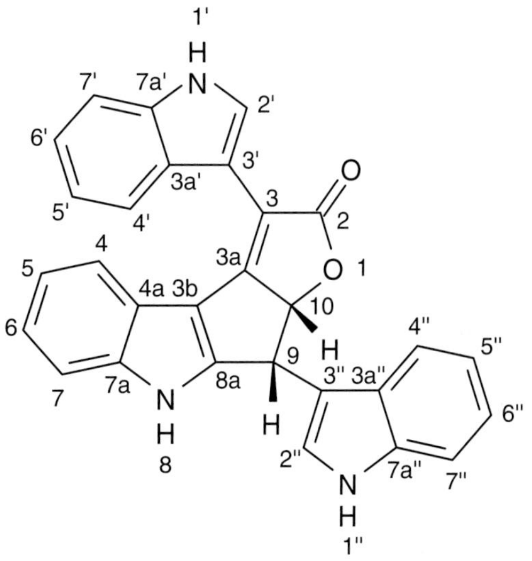

Fig. 7.11. Malasseziacitrin

3. New Findings on the Biosynthesis of *Malassezia* Compounds

By feeding with [1-^{13}C]-DL-tryptophan, it could be clarified whether the spiro carbon atom of the pityriarubins originates from the carboxyl group of tryptophan (Fig. 7.12; Irlinger et al. 2004). The labeled precursor [1-^{13}C]-DL-tryptophan was demonstrated according to Erlenmeyer's azlactone synthesis. In both pityrialacton and pityriarubins incorporation of [1-^{13}C]-DL-tryptophan was confirmed by means of ^{13}C-NMR spectra (Fig. 7.13). All three pityriarubins show ^{13}C enrichment of the spiro carbon atom and also labeling of the functions of carboxyl, lactam and lactone. Pityriacitrin and pityriaanhydride fail to show ^{13}C enrichment, which is in accordance with the proposed synthesis pathway. Thus it has been proved that the spiro carbon atom originates from the carboxyl group. Lack of optical activity in pityriarubin A and malasseziaindol A indicates that the configuration of tryptophan does not play a role in biosynthesis.

Altogether, the Trp-derived pathway found in *M. furfur* and in some strains of *M. pachydermatis* comprises an extremely complex and hitherto unknown metabolic pathway which might explain some phenomena of pityriasis versicolor and help understand the pathogenetic factors of *Malassezia* yeasts. *M. furfur* produces a variety of unknown substances and most of the secondary metabolites that have been found are hitherto unknown structures which may act as factors of pathogenicity. However, *M. furfur* is rarely isolated from lesions of pityriasis versicolor, so the impact of the indolic pathway for the disease has not been clearly assessed so far. As the pigment itself has not yet been demonstrated in the lesions of PV because of poor and contaminated material, an interesting approach is to identify genes associated with pigment production and to demonstrate the expression of these genes in an in vitro skin model and later in the lesions of PV. By means of

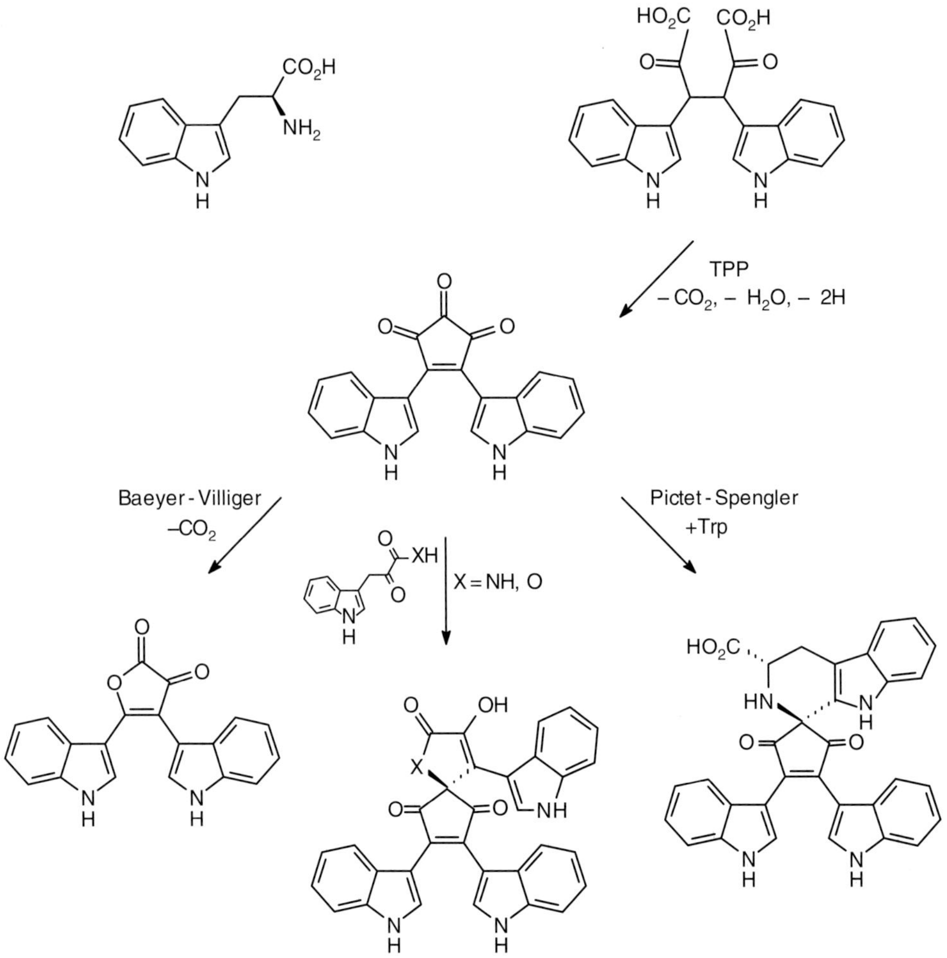

Fig. 7.12. Proposal on the biosynthesis of pityriarubins and pityrialactons

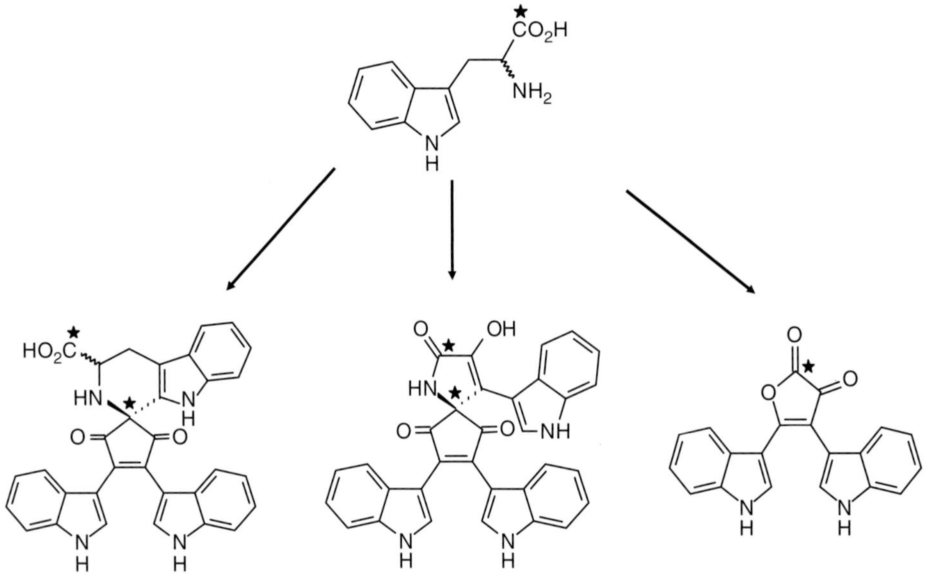

Fig. 7.13. Confirmation of the proposed biosynthesis by feeding with racemic [1–13C] tryptophan

the cDNA subtraction technology according to Diatchenko et al. (1996), several genes identified in the Trp culture have been sequenced and compared with already annotated genes in gene data banks (Hort et al. 2005). First results indicate that tryptophan apparently induces an almost toxin-like stress metabolism in *M. furfur*, since several genes involved in detoxifying singulet oxygen and nitrosative stress were identified. It is possible that the new, hitherto scarcely investigated pigment pathway represents an energy-consuming attempt of *M. furfur* to transform tryptophan into "less toxic" metabolites. This would be in line with the observation that production of pigments is associated with slow growth (Barchmann et al. 2005). The relevance of these genes in the disease is to be investigated by demonstration of the corresponding transcripts in skin samples of patients with PV and in an in vitro infection model. Moreover, a surprisingly high number of fungi are able to synthesize the mentioned compounds, indicating a hitherto unknown common origin. Studies on *Ustilago maydis*, the closest related fungus sequenced so far, have revealed several genes that are putatively involved in pigment synthesis. Among pigment-positive ascomycetes the haploid species *Candida glabrata* (Mayser et al. 2007) is suitable for the identification of pigment-associated genes. Identification of pigment-associated genes in these fungi will help provide an insight into the evolution of a widespread pathway and, by knockout mutants, to determine whether these are factors of pathogenicity in human, animal or plant disease.

V. Host–Pathogen Interactions

A. Antigens in *Malassezia* spp. and Their Characterization

Before single species could be distinguished by means of molecular biology, *Malassezia* yeasts had mainly been classified by cellular and cultural morphologies and physiological properties. To find a relation between *P. ovale*, *P. orbiculare*, *M. furfur* and their hyphal forms, the distribution of specific antigens was investigated (Sternberg and Keddie 1961; Tanaka and Imamura 1979; Faergeman et al. 1982; for elegant reviews, see Ashbee and Evans 2002: Ashbee 2006). The finding of common antigens within the genus *Malassezia* importantly influenced the classification of the several species, leading to the unification of *P. ovale*, *P. orbiculare* and *M. furfur* within the species *M. furfur* (Cannon 1986). However, Midgley (1989) redefined the *Malassezia* yeasts according to her findings of specific antigens in the morphological variants of *Malassezia*. Based on the detection of specific antigens in each group, Cunningham et al. (1990) reclassified the genus *Malassezia* into the three serovars of *M. furfur*.

Topical studies revealed the distribution of common and variable antigens within the genus *Malassezia* (Koyama et al. 2001; Andersson et al. 2003; Gaitanis et al. 2003, 2005b).

Because of the existence of easily applicable biochemical and molecular biological techniques, antigens have lost their relevance in the classification and differentiation of *Malassezia* spp.

1. Characterization of the Antigens

Several serological studies revealed a large variety of Ig-binding components of major and minor relevance, with molecular masses of the antigenic proteins ranging from 9kDa to 150kDa (Johansson and Karlström 1991; Jensen-Jarolim et al. 1992; Squiquera et al. 1994; Zargari et al. 1994; Huang et al. 1995; Lintu et al. 1997; Silva et al. 1997; Nissen et al. 1998; Mayser and Gross 2000; Gandra et al. 2001; Arzumanyan et al. 2003). Several workers identified high-molecular carbohydrates and glycoproteins such as mannans and mannoproteins as major antigenic agents, which may be missed by standard serological methods (Savolainen and Broberg 1992; Doekes et al. 1993a, b; Lintu et al. 1997; Kosonen et al. 2005). Major allergens are defined as antigens reacting with >50% of patients' sera with *Malassezia*-associated diseases. Antigens with molecular masses of 86, 76, 67, 28, 17 and 13kDa have been described by Johansson and Karlström (1991), the 67-kDa protein being the most important allergen. By means of monoclonal antibodies Zargari et al. (1994) identified three major antigens, including a 67-kDa antigen. The 67-kDa component and the 37-kDa component were proteins, while a 14-kDa component was most probably of carbohydrate origin. Lintu et al. (1997, 1999) found a protein of 9kDa, in agreement with a previous finding by Jensen-Jarolim et al. (1992), a protein of 96 kDa and mannan, all of which are major antigens. Other workers identified antigens of similar sizes, which may be identical antigens, but several found antigens of completely different molecular masses. These variations may result from different methods applied in the individual studies and deviations in determining the molecular masses. According to a study by Zargari et al. (1995), the incubation time of *Malassezia* yeasts essentially influences the prevailing antigens. After incubation of more than 4 days, most of the protein antigens were lost. A 37-kDa protein lasted longest, while the carbohydrates remained at a relatively constant level. In contrast, Gandra et al. (2001) suggested the use of 28-day-old strains for an optimal yield of IgE-binding antigens. They compared the prevalence of protein and carbohydrate antigens in crude extracts of *M. furfur* after 2, 6, 10 and 28 days of incubation. Most proteins and carbohydrates were isolated after 10 days, while most antigenic compounds were detected after 28 days (Gandra et al. 2001).

Another influence factor, especially affecting skin prick results, is the storage of the antigens. Most antigens degrade easily when stored at room temperature even if glycerol is added as stabilizer. Storage of antigen extracts at 4 °C improves their stability. After 6 months, most of the important allergens could still be detected, although signs of degradation were observed. After 1 year, only low-molecular-mass proteins remained. Accordingly, the reliability of skin prick tests decreases with the age of the antigenic extracts (Lintu et al. 1998).

Also, the strain used for antigen preparation affects the results profoundly. A study by Mayser and Gross (2000) was among the first to examine IgE-mediated sensitization against *Malassezia* species according to the new classification. Antigen preparations were obtained from the two reference strains CBS 1878 *M. furfur* and CBS 7222 *M. sympodialis*. For comparison, the commercially available radioallergosorbent test (RAST) against *P. orbiculare*, ImmunoCAP "m70" from Pharmacia (Uppsala, Sweden), was used. According to the manufacturer's specification, this test is based on the strain *M. furfur/P. orbiculare* ATCC 42132. The molecular masses of the most important bands of *M. sympodialis* and *M. furfur* were, respectively, 15, 22, 30, 37, 40, 58, 79, 92, 99, 124 kDa, and 15, 25, 27, 43, 58, 92, 99, 107 kDa. The control sera were negative. The 15-kDa band disappeared after pretreatment of *M. furfur*-positive sera with *M. sympodialis*, indicating cross-reactivity of the antibodies. Results obtained with the ATCC 42132 strain were extremely similar to the results obtained with *M sympodialis*. ATCC 42132 could be reclassified as *M. sympodialis* by several methods for differentiation (Guillot et al. 1996; Mayser et al. 1997, 1998c), later confirmed by Faergeman (2002). As the strain ATCC 42132 was used in several of the studies mentioned above, many of the described antigens were mistakenly attributed to *M. furfur*. Other researchers used reference strains of *M. furfur*, but some used strains from clinical isolates typed according to the former nomenclatures. The different and similar antigens detected in all these studies indicate the existence of common and species-specific antigens. Some of the major antigens were further analyzed. To date, 13 antigens have been annotated (Table 7.6).

Schmidt et al. (1977) were the first to elucidate the protein and cDNA sequence of a *Malassezia* allergen. The allergens Mala s 5–11 were detected by phage surface display (Lindborg et al. 1999). Mala s 12 and 13 have not yet been described in the literature, but their sequences were directly submitted to the NCBI (http://www.ncbi.nlm.nih.gov/) and are available under accession numbers CAI43283 and CAI78451, respectively. Since the detection of the first allergen the nomenclature of the *Malassezia* antigens has been changed from Mal to Mala. In 2000 the allergens Mala f 1 and Mala f 5–9 were redefined as Mala s 1 and Mala s 5–9 after Mayser and Gross had identified the strain ATCC 42132 used for identification as *M. sympodialis*. The allergens Mala f 2–4 were isolated from strain 2782 *M. furfur* (TIMM, Teikyo University Institute of Medical Mycology).

Recombinant allergens rMala s 1 and rMala s 5–9 possess attributes equal to those of natural proteins (Zargari et al. 2001). This will be advantageous for standardization of serological tests with *Malassezia* antigens. Only little is known about antigens of further *Malassezia* strains. Koyama et al. (2000) detected three major allergens in *M. globosa* and named them Mala g 46a, Mala g 46b and Mala g 67 in accordance with their molecular masses. Both Mala g 46a and Mala g 46b bind most strongly to IgE. In the lectin blot, both Mala g 46a and Mala g 46b reacted with concavalin A, indicating that they contain mannan. Mala g 67 seemed to be a protein. In *M. pachydermatis* major allergens

Table 7.6. Annotated allergens of *Malassezia* spp.

Allergen	Molecular weight	Homology	Author
Mala s 1	37 kDa	No homology with known proteins, maltose-binding protein	Schmidt et al. (1997)
Mala f 2	21 kDa (reduced); 42 kDa (non-reduced)	Peroxisomal membrane protein; putative thioredoxin reductase (*C. boidinii*); *Aspergillus fumigatus* allergen, Asp f 3	Yasueda et al. (1998)
Mala f 3	20 kDa (reduced); 40 kDa (non-reduced)	Peroxisomal membrane protein; putative thioredoxin reductase (*C. boidinii*); *Aspergillus fumigatus* allergen, Asp f 3	Yasueda et al. (1998)
Mala f 4	35 kDa	Mitochondrial malate dehydrogenase	Onishi et al. (1999)
Mala s 5	18 kDa	Mala f 2, Mala f 3	Lindborg et al. (1999)
Mala s 6	17 kDa	Cyclophilin (*S. pombe*)	Lindborg et al. (1999)
Mala s 7	16 kDa	No homology with known proteins	Rasool et al. (2000)
Mala s 8	19 kDa	No homology with known proteins	Rasool et al. (2000)
Mala s 9	14 kDa	No homology with known proteins	Rasool et al. (2000)
Mala s 10	86 kDa	Heat-shock protein	Andersson et al. (2004)
Mala s 11	22 kDa	Manganese superoxide dismutase	Andersson et al. (2004)
Mala s 12	67 kDa	GMC oxidoreductase	www.ncbi.nlm.nih.gov/entrez/viewer; fcgi?db=protein&val=78038796
Mala s 13		Thioredoxin	www.ncbi.nlm.nih.gov/entrez/viewer; fcgi?db=protein&val=91680611

with molecular masses of 45, 52, 56 and 63 kDa were found, which may play a role in canine atopic dermatitis (Chen et al. 2002a). In a recent study it was found that protein bands of 62 kDa and 49 kDa were recognized by IgG in all tested sera, proteins of bands of 98 kDa and 68 kDa by five out of six sera, and proteins of 188, 66, 58, 57, 38, 28 and 17 kDa only by high-titer sera (Habibah et al. 2005). Bond and Lloyd (2002) found most of the dogs within their study to be immunoreactive towards the proteins of 132, 66 and 50–54 kDa and usually also reactive towards the proteins of 219, 110, 71 and 42 kDa. Another study by Koyama et al. (2001) supports the assumption that both common and species-specific antigens exist within the genus *Malassezia*. In this study, 83% of sera of patients with atopic dermatitis reacted with one or more *Malassezia* species. IgE antibodies were most frequently present to *M. globosa*. Several lectin blots revealed species-specific antigens at protein level. Pretreatment of patients' sera with antigen extract of *M. globosa* partially inhibited reaction with the other *Malassezia* species, except for *M. furfur*. Zargari et al. (2003) evaluated the presence of immunoglobulin E (IgE) antibodies to seven different *Malassezia* spp. in atopic eczema/dermatitis syndrome (AEDS). Antibodies to at least one of the other *Malassezia* species were detected in 20% of patients' sera that were nonreactive to the ImmunoCAP "m70". Inhibition tests with *M. sympodialis* extract (ATCC 42132) showed significant inhibition of IgE binding to *M. furfur*, *M. obtusa* and *M. pachydermatis* and partial inhibition of IgE binding to *M. globosa* and *M. restricta*, indicating that *Malassezia* species "share antigenic determinants to a great extent" (Koyama et al. 2001). The distribution of the allergens within different *Malassezia* species was studied by Andersson et al. (2003). *M. furfur* expressed Mala f 2–4, *M. globosa* and *M. obtusa* only Mala s 6. *M. pachydermatis* expressed Mala f 4, Mala s 6 and Mala s 8, while *M. restricta* and *M. slooffiae* expressed Mala f 4 and Mala s 6. *M. sympodialis* expressed all the allergens, except Mala f 2 and Mala f 3. PCR amplification of genomic DNA and sequencing indicated that all the species apart from *M restricta* and *M. slooffiae* possessed genes for more allergens than they actually expressed. Only *M. sympodialis* seems to possess Mala s 1 (Gaitanis et al. 2003, 2005b). *Malassezia* apparently shares antigen determinants with other fungi (Savolainen and Broberg 1992; Doekes et al. 1993a; Huang et al. 1995; Lintu et al. 1999; Leino et al. 2006). Most of these antigens are glycoproteins or of carbohydrate origin (Savolainen and Broberg 1992; Doekes et al. 1993a; Lintu et al. 1999; Leino et al. 2006). These cross-reactions may lead to false-positive results in allergy diagnosis and must be taken into account. Important in the development of atopic diseases is the cross-reactivity of *Malassezia* antigens with human antigens (Flückiger et al. 2002; Schmid-Grendelmeier et al. 2005). Sensitization to such compounds may lead to persistent inflammatory response even beyond elimination of the pathogen. Molecular mimicry of allergens may cause formation of auto-antibodies. This is suggested to be one of the factors by which *Malassezia* could trigger AEDS.

B. *Malassezia* spp. and Immunity

1. Innate Immune Response

a) Complement System
The complement system includes more than 30 serum proteins involved in both innate and adaptive immunity. Activation occurs via three possible pathways: the classical pathway, mediated by immune complexes, the alternative pathway, mediated by pathogen antigens, or the recently discovered mannan-binding lectin complement pathway. Activation of the complement cascade leads to inflammation, lysis of bacterial cells and chemoattraction of phagocytic cells. A main function of the complement system is the opsonization of pathogens, in particular bacteria and fungi, to access them for phagocytosis. *Malassezia* acitvates the complement cascade via the classical pathway (Suzuki et al. 1998) and via the alternative pathway (Belew et al. 1980; Sohnle and Collins-Lech 1983; Suzuki et al. 1998). Complement factors, especially C3, were present in lesional skin in seborrheic dermatitis (SD) but absent in normal skin, indicating that the complement cascade may be involved in the inflammatory process of SD (Pierard-Franchimont et al. 1995).

b) Cellular Response and Cytokine Production
Phagocytosis plays an important role in the immediate defense against fungi, as leukopenia increases the risk of severe fungal infections (Kobayashi et al. 2005; Enoch et al. 2006). Incorporation of *Malassezia* cells by neutrophils occurs complement-dependently and persists after 40 min. Intercellular killing of *Malassezia* cells is poor but can be increased by pretreatment with ketoconazole (Richardson and Shankland 1991). The mechanism by which *Malassezia* yeasts evade killing by neutrophils is not fully understood. Akamatsu et al. (1991) reported that azelaic acid decreased peroxide formation in neutrophile granulocytes. Azelaic acid has been detected in *Malassezia* cultures (Nazzaro-Porro and Passi 1978). Recently the tryptophan-derived bisindolylspiran alkaloids pityriarubin A, B and C have been described in *M. furfur*. They reveal structural similarity to bisindolylmaleimides like arcyriarubin A, which strongly inhibits the release of reactive oxygen species (ROS) in neutrophil granulocytes. The pityriarubins, especially pityriarubin C, share the ability to inhibit the oxidative burst but are more specific to the stimulating agent (Krämer et al. 2005a). The pityriarubins have not yet been detected in situ in *Malassezia*-associated diseases but may represent a strategy to escape neutrophil attack. Suzuki et al. (1998) reported that *Malassezia* cells are recognized by the phagocytic human mononuclear cell line THP-1 via mannose receptor, β-glucan receptor and complement receptor type 3. Heat-killed cells were more easily incorporated than living cells. Incubation of the human phagocytic and granulocytic cell lines THP-1 and HP-60 induced the production of IL-8 in a dose- and time-dependent manner (Suzuki et al. 2000). Opsonized cells were the most stimulating ones, and living cells more than heat-killed cells. HP-60 cells produced more IL-8 and additionally IL-1α. IL-1α and IL-8 have proinflammatory and chemotactic action and activate lymphocytes and leukocytes. This suggests that *Malassezia* yeasts stimulate the immune system and maintain inflammation. A similar observation was made in two early studies concerning the immunomodulation by *Malassezia* (Takahashi et al. 1984, 1986). Mice were treated intraperitoneally with killed *Malassezia* cells in advance of inoculation with *S. typhimurium* or Ehrlich ascite carcinoma cells. Mice treated with *Malassezia* survived significantly longer than untreated mice. Apparently *Malassezia* stimulated the reticuloendothelial system and enhanced the release of superoxide of the macrophages.

In contrast to the findings described above, some studies found a decrease in IL-1β in peripheral blood mononuclear cells (PBMCs) after incubation with *Malassezia* cells, which may explain the missing or relatively mild inflammation in *Malassezia*-associated dermatoses (Walters et al. 1995; Kesavan et al. 1998). In addition, the latter authors found a depression of IL-6 and TNF-α in the supernatant of PBMCs treated with viable *Malassezia* cells and formalin-preserved yeast cells at a ratio of 20 yeast cells to one PBMC.

Malassezia cells formalin-preserved at exponential phase increased cytokine production at a yeast:PBMC ratio of 1:1. Both studies suppose the unusual lipid content of the cell wall to be responsible for the depression of cytokine production. Removal of the surface lipids led to an increase in the production of IL-1β, IL-6 and TNF-α (Kesavan et al. 2000). The authors suggested that an altered lipid composition of the *Malassezia* cells might be responsible for the inflammation observed in SD. Accordingly, the lipids may protect *Malassezia* cells from phagocytosis and killing by neutrophils via oxydative burst (Ashbee and Evans 2002). Similar findings exist for *Cryptococcus neoformans*. The capsule of *Cryptococcus* is an important virulence factor, suppressing cytokine production and playing a role in evading phagocytosis (Kozel et al. 1988; Cross and Bancroft 1995; Vecchiarelli 2000).

During recent years special emphasis has been placed on the participation of dendritic cells in the response to *Malassezia* yeasts. Dendritic cells are a crosslink between the unspecific and the specific immune system and play a crucial role in the immune response of the skin. Langerhans cells (LC) are the best characterized cutaneous dendritic cells. Two other subtypes of dendritic cells have been detected in dermatitis, inflammatory dendritic epidermal cells (IDEC) and plasmacytoid dendritic cells (pDC; Wollenberg et al. 1996, 2002). In inflamed skin, the dendritic cells express the high affinity IgE-receptor FcεRI. In contrast to most other immature dendritic cells, Langerhans cells, monocytes and mature dendritic cells do not express the macrophage mannose receptor. Immature dendritic cells can ingest pathogenic material via receptor-mediated endocytosis and pinocytosis, but are unable to stimulate T-cells. The uptake of antigenic material induces maturation. Inflammatory mediators promote maturation and the migration of mature dendritic cells to the lymph nodes, where they present antigen to naïve T-cells via MHC complex (for a review, see Banchereau and Steinman 1998). Buentke et al. (2000, 2001, 2002, 2004) performed several studies concerning the interaction between dendritic cells and *Malassezia*. The studies were performed with human immature CD1a+ dendritic cells derived from blood monocytes. These apparently ingest *Malassezia* extract and *Malassezia* mannan via the mannan receptor, rMal s 5 was taken up via pinocytosis (Buentke et al. 2000). The ingestion of *Malassezia* cells led to the expression of CD83, indicating maturation of dendritic cells and upregulation of CD80 and CD86. Production of TNF-α, IL-1β and IL-18 was increased. IL-12p70 and IL-10 levels were unaltered. Thus, the induction of a Th2-type immune response might be supported (Buentke et al. 2001), which is the favored immune reaction type in the acute exacerbation of AEDS. *Malassezia* directly influences the interaction between natural killer cells and dendritic cells. Dendritic cells cocultured with *Malassezia* became less susceptible to natural killer cell-induced cell death (Buentke et al. 2002). Furthermore, *Malassezia* enhances maturation of immature dendritic cells induced by natural killer cells. The expression of CD83 and CD86 increased in dendritic cells cocultured with *Malassezia* and stimulated with natural killer cells (Buentke et al. 2004).

c) Interaction of *Malassezia* with Cutaneous Cells
Of special interest is the interaction of *Malassezia* yeasts with cutaneous cells. Keratinocytes, the major cell type in the epidermis, have a barrier function towards intruding microorganisms and influence the skin immune system via cytokine production. A study by Walters et al. (1995) was the first to analyze the production of IL-1α of human keratinocytes cocultured with *Malassezia* yeasts. The authors found a constitutive level of IL-1α release by the human keratinocytes, which was not consistently elevated by *Malassezia*. Watanabe et al. (2001) investigated the influence of several reference strains of *M. furfur*, *M. sympodialis*, *M. sloofiae* and *M. pachydermatis* on cytokine production by human keratinocytes. Treatment of the keratinocytes with whole cells resulted in an increase of IL-1β, -6, -8 and TNFα in cytokine ELISA for all strains but *M. furfur*, which did not induce cytokine production at all. Monocyte chemotactic protein 1 was not elevated by any of the strains used. The major reaction was induced by *M. pachydermatis*, corresponding to the much higher severity of diseases induced in animals by this agent. Supernatants of the fungal cultures did not induce any cytokine production in human keratinocytes, indicating that cell contact is required for the stimulation of an inflammatory reaction (Watanabe et al. 2001). At the same time, Baroni et al. (2001a) presented a study concerning almost the same subject. They cocultured the human keratinocyte cell Line HaCat with a *P. ovale* strain (ATCC 12078) referred to as *M. furfur*. At a yeast cell/keratinocyte ratio of 50/1 the keratinocytes died from necrosis, while the application of lower

ratios avoided this effect. After 24 h the fungal cells adhered to the keratinocytes; and 30% invaded the cells after 48 h and endocytosed within the vacuole, but they were not killed intracellularly because phagolysosome fusion seemed to be suppressed. The study demonstrated a change in the actin filaments of the keratinocytes due to the invasion of *Malassezia*. TGase I expression decreased. The TGase I inter alia is responsible for keratin filament crosslinking and supports the integrity of the epidermis. An inhibition of this enzyme may cause a breach in the skin's barrier function. Reverse transcriptase PCR was used to determine the cytokine gene expression in HaCat cells. Production of the immunosuppressive cytokines TGFβ-1 and IL-10 was upregulated by coculture with *Malassezia*, while production of the proinflammatory cytokines IL-1α, IL-6 and TNFα was consecutively inhibited (Baroni et al. 2001a). Suppression of IL-1α, IL-6 and TNFα was also observed in PBMCs as described above, although different strains were used (Kesavan et al. 2000). This constellation leads to a depressed immune response and could enhance intracellular survival of *Malassezia*. Further studies of the same working group confirmed the increase of TGFβ-1 and IL-10 on mRNA and protein levels and an increased production of IL-8, which is important in chemotaxis (Donnarumma et al. 2004; Baroni et al. 2006). Additionally, the group investigated the expression of the human β-defensins HBD-1 and HBD-2. HBD-1 is expressed in several tissues such as the skin, the respiratory tract and the urinary tract. HBD-2 is expressed in the skin. HBD-2 is an important member of the innate immune response against bacteria, fungi and viruses. HBD-1 was not influenced by *Malassezia*, while it induced an elevated PKC-dependent expression in HBD-2. As described above, 30% of the cocultured *Malassezia* cells are taken up by keratinocytes, but intracellular killing did not occur. Instead, *Malassezia* prevails inside the cell inducing immunosuppression. The expression of HBD-2 probably hinders the invasion of further yeast cells and thereby promotes immunity against *Malassezia* yeasts (Donnarumma et al. 2004). The expression of HBD-2 and IL-8 appears to be mediated by the toll-like receptor type 2 (TLR2; Baroni et al. 2006). TLR2 belongs to a recently described protein family. They initiate the innate immune response and subsequently influence adaptive immune responses (Netea et al. 2005). TLR2 is broadly specific and is involved in the recognition of yeast and other pathogens (Kawai and Akira 2006).

Investigating the ability of *M. pachydermatis* to stimulate keratinocyte proliferation in canine skin in vitro it was found that neither extracts (Chen et al. 2002b) nor coculture with living yeast cells (Chen et al. 2004) were able to induce proliferation of canine keratinocytes. Thus, epidermal hyperplasia which is frequently observed in *Malassezia*-associated diseases, is probably not directly induced by *Malassezia* or its components. Instead, increasing numbers of yeast cells in the cocultures led to increased apoptosis of the canine keratinocytes (Chen et al. 2004).

Malassezia is apparently also internalized by human fibrocytes. After 4 h of coculture the yeast cells adhered to the fibroblasts and were taken up after 24 h, followed by phagosome and endosome fusion. Ingestion of the yeast cells by the fibroblast was found to be an active process, depending on F-actin, a cytoskeletal component, and resulted in membrane damage of the fibroblasts (Baroni et al. 2001b). As *Malassezia* yeasts in general live in the upper layers of the stratum corneum, they usually do not get in contact with fibroblasts, the predominant cell type of the dermis.

Unfortunately, there is only limited information about the interaction of *Malassezia* yeasts with melanocytes, although melanocytes, melanosomes and melanin are important members of the innate immune system, especially against fungal infections (Mackintosh 2001). Melanocytes are capable of phagocytosis, MHC-II mediated antigen presentation and production of the several immune mediators including IL-1 and IL-6. The melanosomes are related to the lysosomes of phagocytes. Besides the well known protection against UV irradiation, melanin and its precursors have antimicrobial properties (for a review, see Mackintosh 2001). Especially pityriasis versicolor is characterized by hyper- and depigmentation of the lesional areas, but also AEDS and other inflammatory diseases of the skin are frequently accompanied by hyperpigmentation. Azelaic acid was suggested to be the relevant metabolite of *Malassezia* responsible for the tyrosinase inhibition found in PV lesions (Nazzaro-Porro et al. 1986), although concentrations of this enzyme found in vivo are too low to block melanogenesis sufficiently (Breathnach et al. 1984; Robins et al. 1987; De Luca et al. 1996). The effects of lipoperoxides and by-products were suggested to cause damage to melanocytes (De Luca et al. 1996). Formation of lipoperoxides is regularly found in cultures of *Malassezia* yeasts. As these are part of the resident skin flora, depigmentation caused by lipoperoxides could occur in any human at any time. Another explanation is offered by the finding of several tryptophan-derived metabolites of *M. furfur* influencing melanogenesis in vitro (Wille et al. 2001; Dahms et al. 2002; Krämer et al. 2005b). Some of the described compounds inhibit the tyrosinase reaction, a key reaction of melanogenesis, and suppress the DOPA reaction in human epidermal melanocytes in situ (Thoma 2000a, b, 2001; Dahms et al. 2002). One of the compounds, malassezin, is a potent activator of the aryl hydrocarbon receptor and induces apoptosis in cultured human melanocytes (Podobinska et al. 2003; Krämer et al. 2005b). The induction of apoptosis in melanocytes and depression of melanogenesis may be a further strategy of *Malassezia*, especially *M. furfur*, to evade the cutaneous immune system. The induction of apoptosis in keratinocytes by malassezin has not yet been studied, but would be of interest as high numbers of *M. furfur* and *M. pachydermatis* apparently induce apoptosis in keratinocytes (Baroni et al. 2001a; Chen et al. 2004).

All these studies concerning the innate immune response indicate a pronounced ability of *Malassezia* spp. to modulate the cutaneous immune response. Simplistically, *M. furfur* rather depresses an immune response, while *M. pachydermatis* strongly promotes an immune reaction, and the other species lie in between. Conflicting findings may result from the use of different species or even strains within the individual studies. It is likely that different species (or strains) are predominant in single diseases associated with *Malassezia* yeasts. These diseases vary widely in the degree of inflammation, from none to minimal inflammation in PV to vivid inflammation in SD and AD.

2. Adaptive Immune Response

a) Humoral Response

Although B-cells are absent in human skin, immunglobulins specific to components of the commensal flora are present in healthy individuals (Tlaskalova-Hogenova et al. 2004). Antibodies of the classes IgM, IgG and IgA specific to *Malassezia* are present in healthy adult individuals (Sohnle et al. 1983; Bergbrant and Faergeman 1989; Bergbrant et al. 1991, Cunningham et al. 1992; Ashbee et al. 1994a; Arzumanyan et al. 2003). No difference was found between the yeast and the mycelial phase of different strains (Saadatzadeh et al. 2001). The titers of the several immunoglobulins correspond to the colonization of the skin by *Malassezia* yeasts in different periods of life. In general, lower titers of antibodies were found in children. Antibody titers rise until adulthood, with maximum titers until the fifth decade, and slowly decrease in the elderly (Faergemann 1983; Bergbrant and Faergemann 1988; Faggi et al. 1998). This reflects the colonization habit of *Malassezia* yeasts, as they are present in age groups with the highest sebaceous gland activity. The activity of the sebaceous glands is low in childhood and senium (Bergbrant and Faergemann 1988). The lack of contact of the immune system with *Malassezia* in these age groups may explain the reduced presence of antibodies against these yeasts. Accordingly, IgG levels are low in children, higher in adults and decrease with increasing age (Johannson and Faergemann 1990; Bergbrant 1991; Chua et al. 2003). IgM was detected in both children and adults but was lower in the elderly (Sohnle et al. 1983; Cunningham et al. 1992). IgA levels are generally low (Cunningham et al. 1992), indicating that sensitization does not essentially occur via mucosal membranes. Similar results were obtained in healthy dogs, which also presented positive IgG titers towards *M. pachydermatis* (Bond et al. 1998), indicating the development of specific antigens against commensal organisms.

b) Cellular Response

The adaptive cellular immune response is crucial to defending fungal infections. The incidence and severity of fungal diseases is increased in patients with AIDS, immunosuppressive therapies, malignoma and other immune deficiencies. Some *Malassezia*-associated diseases seem to occur more frequently in patients with reduced T-cell mediated immunity – *Malassezia* folliculitis appears in patients with maligancies and therapeutically or constitutionally suppressed immunity (Archer-Dubon et al. 1999; Alves et al. 2000; Rhie et al. 2000), the incidence PV is increased in patients with glucocorticoid therapy (Tatnall and Rycroft 1985) and organ transplantations (Freire-Ruano et al. 2000). SD is the most common skin manifestation in AIDS (Rigopoulos et al. 2004).

Apart from a study by Sohnle and Collins-Lech (1980), cellular immunity against *Malassezia* has not been investigated in detail in healthy individuals. Studies concerning cellular immunity in patients with *Malassezia*-associated diseases revealed significant specific immunity in the healthy control groups (Sohnle and Collins-Lech 1978, 1982; Ashbee et al. 1994b; Bergbant et al. 1999). Also, the mycelial phase of *Malassezia* spp. induces a positive leukocyte migration inhibition assay (Saadatzadeh et al. 2001) although mycelia are rare on the skin of healthy individuals.

As *Malassezia* is a commensal on human skin, healthy individuals can develop specific immunity against it. For this reason it is difficult to differentiate specific immune reactions of patients with *Malassezia*-associated diseases from those of healthy individuals with no history of skin disease.

C. Models for *Malassezia*-Associated Diseases

Several attempts have been made to develop animal and in vitro models of human diseases associated with *Malassezia* yeasts. Some studies showed that skin lesions resembling lesions in *Malassezia*-associated diseases could be evoked in healthy human skin, as well as in rabbit, mouse, guinea pig and dog skin. The fact that skin changes could only be evoked by incubation of *P. ovale* and/or *P. orbiculare* under occlusion and that the lesions healed spontaneously after the end of treatment indicated that more factors are involved in the pathogenesis of the disease than only fungal colonization (Faergemann 1979; Drouhet et al. 1980; Faergemann and Fredriksson 1981; Polonelli et al. 1986; Goodfield et al. 1987; Van Cutsem et al. 1990; Bond et al. 2004). Lober et al. (1982) induced psoriasiform lesions in patients with psoriasis and two of ten healthy control persons by application of heat-killed sonicated suspensions of *Malassezia*. Clinically and histologically the provoked lesions resembled psoriatic lesions. Bond et al. (2004) found the development of symptoms in saline- and yeast-treated skin sites in four out of ten dogs, the reaction to the yeasts being more severe than the reaction to the saline control treatment. Another working group induced psoriasiform lesions in rabbits by heavy scrubbing of the skin with *Pityrosporum* extracts (Rosenberg et al. 1980). A handicap of these animal models is the need for continuous manipulation to maintain the lesions. Faergemann (1989) developed a first in vitro model for *Malassezia* infections. Inoculation of stratum corneum with several *Malassezia* strains led to the formation of hyphae in some of the strains, especially ATCC 44341 and ATCC 44031, both of which are *M. sympodialis* strains according to the new nomenclature. Only short and few hyphae occurred in cultures grown in medium without skin contact (Faergemann 1989). No growth of hyphae was observed in a living skin equivalent, but *Malassezia* yeasts destroyed the upper layers of the stratum corneum (Bhattacharyya et al. 1998).

Based on the RHE (reconstituted human epidermis, SkinEthic), an in vitro model of pityriasis versicolor has recently been developed. *Malassezia furfur* strain CBS7019 was incubated in minimal medium containing olive oil and tryptophan as sole nitrogen source on top of the skin culture at 37 °C and 5% CO_2. Besides pigment production, a switch to mycelial growth and the invasion of hyphae in the stratum corneum were observed (Mayser, unpublished data), resembling the aspect of florid lesions of PV. The control culture containing arginin as sole nitrogen source lacked pigment production as well as production of hyphae and invasive growth. Interestingly, *M. furfur* cultures grown in the same tryptophan medium but without epidermis produced pigment but almost no hyphae. Accordingly, the switch of the nonpathogenic yeast form to the invasive mycelial form depends on the cooccurrence of an appropriate supply – or rather deprivation – of nutritives and factors deriving from the epidermis, which remain to be elucidated.

VI. *Malassezia* Yeasts in Human and Animal Disease

Although *Malassezia* yeasts belong to the resident flora of humans and different animals, they have also been associated with certain diseases (for an excellent review, see Gupta et al. 2004).

A. Pityriasis Versicolor

Pityriasis versicolor (PV), one of the most commonest skin diseases, is characterized by lesions of varying color, a phenomenon that has not yet been clarified. Especially in seborrhoic areas there are circumscribed or confluent, mostly brown/ochre, but also reddish erythematous, yellowish up to black macules showing fine-bran desquamation, which appear green-yellowish under the Wood lamp and may result in long-term depigmentation (pityriasis versicolor alba; PVa; Hay et al. 1998). Several predisposing factors to the disease have been described, of which hyperhydrosis is the most important one (Hay et al. 1998; Gupta et al. 2002, 2003). However, the pathogenesis of PV remains to be elucidated. Furthermore, it has to be clarified whether an individual species of the nine known *Malassezia* species is responsible for the disease, and if so, which in particular (Crespo-Erchiga et al. 2000; Nakabayashi et al. 2000; Gupta et al. 2002). Importance has been attached to *M. globosa*, which might explain the globose shape of cells in lesions of PV (Crespo-Erichiga et al. 2000; Nakabayashi et al. 2000; Crespo-Erchiga and Florencio 2006).

However, the pathogenesis of hyperpigmentation remains unclear. Hyperpigmentation was initially thought to be caused by inflammation and resulting postinflammatory hyperpigmentation (Dotz et al. 1985; Hay et al. 1998; Galadari et al. 1992). Histologically inflammatory processes could not be shown: in contrast lesions were characterized by lack of inflammation despite high fungal load, especially absence of neutrophilic granulocytes, which are usually observed in mycotic infections (Wroblewski et al. 2005). Another explanation of hyperpigmentation was the observation of abnormal big melanosomes in melanocytes and keratinocytes in lesions of PV (Allen et al. 1972; Charles et al. 1973). Other studies are contradictory to these findings and report equal size and number of melanosomes and melanocytes in hyperpigmented and normally pigmented skin (Konrad and Wolff 1973; Galadari et al. 1992). Reports about hyperpigmented PV in lesions of vitiligo (Dotz et al. 1985) are also inconsistent with the involvement of melanocytic structures as no melanocytes exist in lesions of vitiligo because of autoimmunologic processes. Other hypotheses, e.g. hyperpigmentation as a result of thickened stratum corneum or caused by the high number of fungal elements (Galadari et al. 1992) have been discussed but could not be proved. All these hypotheses fail to explain the yellow-green fluorescence of the lesions.

The pathogenesis of depigmention has not been established. A screening effect by skin scales and/or the fungal layer as well as toxic effects on pigment synthesis by fungal metabolites have been discussed. However, as depigmentation also occurs on areas that are not exposed to UV light and on black skin, a purely physical filter effect is an insufficient explanation. Furthermore, ultrastructural studies have demonstrated selective toxic damage to melanocytes in lesions of PV alba (Breathnach et al. 1984), also in association with impaired pigment transport (Oguchi 1982). Acelaic acid and other dicarboxylic acids have been discussed as tyrosinase inhibitors causing depigmentation (Nazzaro-Porro and Passi 1978; Breathnach et al. 1984). Jung and Bohnert (1976) demonstrated the inhibition of tyrosinase by an extract of scales of PV. The inhibitory effect was later ascribed to dicarboxylic acids which can be isolated from cultures of *Malassezia* yeasts, although these had actually not been demonstrated in the scale extract (Nazzaro-Porro et al. 1977; Bojanowsky et al. 1979). In particular, azelaic acid [$HOOC-(CH_2)_7-COOH$] inhibits tyrosinase in vitro and thus interferes with the first step of melanin formation. However, subsequent studies showed that concentrations of azelaic acid in vivo were too low for inhibition of tyrosinase and/or damage to melanocytes (Table 7.3; Breathnach et al. 1984; Robins et al. 1985; De Luca et al. 1996). The effects of lipoperoxides and byproducts were suggested to cause damage to melanocytes (De Luca et al. 1996). The formation of lipoperoxides is commonly found in cultures of *Malassezia* yeasts (DeLuca et al. 1996) As these are part of the resident skin flora, depigmentation caused by lipoperoxides would occur in any human at any time.

Biological properties of the metabolites of the above described tryptophan-dependent secondary metabolism of *M. furfur* offer an explanation for the pathogenesis of fluorescence, pigmentation and depigmentation in PV. Chromatographic separation of the tryptophan-derived pigment reveals a variety of differently colored components as well as fluorochromes resembling the different colours of clinical lesions of PV. Furthermore, quite a number of the isolated components were found to have pharmacological properties explaining clinical characteristics of PV/Pva, which are listed in Table 7.7.

B. Seborrhoic Eczema

Seborrhoic eczema, a common skin disease in immunocompetent individuals (Gupta and Bluhm

Table 7.7. Trp-derived indole compounds and their potential relationship to clinical phenomena in PV

Isolated compound	Pharmacological property	Phenomenon in PV
Pityriacitrin	UV protection	Lack of sunburn in depigmented areas of PVa
Pityrialactone	UV protection	Lack of sunburn in depigmented areas of PVa
Pityriarubins	Inhibition of oxidative burst in granulocytes	Lack of inflammation in lesions of PV
Malassezin, keto-malassezin	Induction of apoptosis in human melanocytes;	Depigmentation in PVa
Malassezia Indole A, keto-malassezin	Tyrosinase inhibition	Depigmentation in PVa
Fluorescent metabolites (e.g. pityrialactone)	Fluorescence	Fluorescence of lesions in PV
Diversity of pigmented compounds	Pigmentation	Pigmentation of lesions in PV

2004) with increased prevalence in immunocompromised patients (Smith et al. 1994), is characterized by yellowish-red lesions in seborrhoic areas covered with greasy scales and associated with mild itching. The pathogenesis of seborrhoic eczema is controversial. The relationship of SE with *Malassezia* yeasts is mainly based on the observation of amelioration by antimycotics (Ford et al. 1985; Gupta et al. 2004). It remains unclear whether colonization rates of the lesions are elevated (McGinley et al. 1975; Bergbrant and Faergemann 1989; Heng et al. 1999; Gupta et al. 2001) or if there is an altered response of the immune system as described by different authors (Bergbrant and Faergemann 1989; Bergbrant et al. 1991, 1999). Another hypothesis suggests an inflammatory effect of *Malassezia* yeasts by toxin production or cleavage triglycerides with release of free fatty acids (Parry and Sharpe 1998).

C. *Malassezia* Folliculitis

Malassezia folliculitis was first described in 1973 (Potter et al. 1973). Clinically, *Malassezia* folliculitis is characterized by itching papules and pustules of breast and back, scratching induces local urticaria (Faergemann and Meinhof 1988). Its pathogenesis remains unclear; presumably *Malassezia* yeasts invade deeper follicular parts, causing inflammation and retention of sebum (Bäck et al. 1985). The differentiation from other related diseases by histological criteria is difficult. The role of *Malassezia* folliculitis as an individual disease is controversial. The main argument for the classification of *Malassezia* folliculitis as an independent disease was the responsiveness to antimycotic treatment (Potter et al. 1973; Bojanovsky and Lischka 1977; Dompmartin and Drouhet 1977; Heid et al. 1978).

D. *Malassezia* Sepsis

Malassezia sepsis is a catheter-associated sepsis; the main risk factor for this disease is parenteral application of lipids in premature infants or patients with severe gastrointestinal diseases (Devlin 2006). However, the lipid emulsion factor is not an absolute condition, as the species *M. pachydermatis*, which is not obligatorily lipid-dependent, is also frequently demonstrated as the causative organism (Papavassilis et al. 1999).

E. Atopic Eczema/Dermatitis Syndrome

AEDS is a chronic recurrent pruritic inflammation of the skin. It has been suggested that microbial colonization of atopic skin may also be involved in the pathogenesis of the disease (Ring et al. 1992; Roll et al. 2004; Baker 2006). Beside *Staphylococcus aureus* (Neuber et al. 1993; Leung et al. 2004), *Candida* spp. (Savolainen et al. 1992, 1993, 1999; Tanaka et al. 1994; Faergemann 2002) and other fungi (Nissen et al. 1998; Fischer et al. 1999; Faergemann 2002; Sugita et al. 2003a), *Malassezia* spp. (Broberg et al. 1992; Faergemann 2002; Scheynius et al. 2002; Aspres and Anderson 2004; Schmid-Grendelmeier et al. 2006) are supposed to play a significant role in exacerbation and aggravation of the disease. It remains unclear which of the several *Malassezia* strains predominates in AEDS. Recent studies have postulated

M. globosa and/or *M. restricta* (Tajima 2005) and/or *M. furfur* (Rincón et al. 2005) to be predominant in AEDS (Sugita et al. 2006), while others have most frequently isolated *M. sympodialis* (Gupta et al. 2001; Sandstrom Falk et al. 2005). Two new *Malassezia* species, *M. dermatis* and *M. japonica*, have been isolated from the skin of patients with AEDS (Sugita et al. 2002, 2003b).

Malassezia yeasts seem to be associated mainly to the head-, neck- and face-type of AEDS (HND), corresponding to the numerous sebaceous glands found in this region (Waersted and Hjorth 1985; Svejgaard et al. 1989; Jensen-Jarolim et al. 1992; Kim et al. 1999; Devos and van der Valk 2000; Bayrou et al. 2005). Multiple studies demonstrated amelioration of HND by treatment with systemic or topical antimycotics (Table 7.8; Clemmensen and Hjorth 1983; Bäck et al. 1995; Broberg and Faergemann 1995; Bäck and Bartosik 2001; Lintu et al. 2001; Mayser et al. 2006). *Malassezia* spp. are thought to act as allergens via binding to the highly specific FcεRI-receptor, especially on Langerhans cells, inducing a cytokine profile favoring type 2 T-cell reaction and IgE production in vulnerable individuals (Scheynius et al. 2002; Johansson et al. 2003; Allam and Novak 2006). In fact, 15–80% of the patients show a positive skin prick test, positive atopy patch test and/or specific IgE antibodies against *Malassezia* (Young et al. 1989; Kieffer et al. 1990; Nordvall and Johansson 1990; Rokugo et al. 1990; Wessels et al. 1991; Broberg 1995; Tengvall-Linder et al. 2000; Arzumanyan et al. 2003; Johanson et al. 2003; for a review of additional literature, see Faergemann 2002). IgE production against *Malassezia* occurs almost exclusively in AEDS patients, but not in healthy individuals and very seldom in patients with other *Malassezia*-associated or atopic diseases (Kieffer et al. 1990; Wessels et al. 1991; Nordvall et al. 1992; Kawano and Nakagawa 1995; Koyama et al. 2000; Fischer Casagrande et al. 2006). Patients with HND showed markedly higher IgE titres than patients with predominant involvement of the extremities (Schrenker et al. 1989; Devos and van der Valk 2000; Mayser and Gross 2000).

Of special interest in the context with AEDS is the crossreactivity of some of the antigenic compounds with other fungal allergens (Savolainen and Broberg 1992; Doekes and van Ieperen-van Dijk 1993a; Huang et al. 1995; Lintu et al. 1999; Leino et al. 2006) and human antigens (Lindborg et al. 1999; Flückiger et al. 2002; Andersson et al. 2004; Schmid-Grendelmeyer et al. 2005: Glaser et al. 2006). Mala s 6, Mala s 10 and Mala s 11 show homologies to fungal and human cyclophilin (CyP) (Lindborg et al. 1999; Flückiger et al. 2002), heat-shock protein (HSP) or manganese superoxide dismutase (MnSOD; Andersson et al. 2004; Frealle et al. 2005). Autoimmunity resulting from sensitization to these allergens, especially against MnSOD and cyclophilin which are structurally related to human antigens, has been discussed as an important factor in exacerbation and maintenance of AEDS (Mittermann et al. 2004).

F. Psoriasis

Psoriasis is a chronic inflammatory disease of the skin specific to humans. It may also affect other organs (e.g. psoriatics arthritis). Some 1–3% of the population worldwide are afflicted. Heredity is an important factor in the pathogenesis. Psoriasis shows characteristics of autoimmunity as it is associated with certain HLA-alleles (Lee and Cooper 2006). The type 1 T-cell response is predominant in psoriatic lesions, characterized by the production of IFN-γ, IL-2 and TNF-α (Clark and Kupper 2006). Stimulation by inflammatory cytokines and activated T-cells leads to hyperkeratosis which is frequently observed in psoriasis. Microorganisms are supposed to be associated with psoriasis (Rosenberg et al. 1989; Noah 1990). As *Malassezia* yeasts were isolated from psoriatic lesions – mainly in the face and scalp region (Faergemann and Maibach 1984; Rosenberg et al. 1989) – and scalp lesions improved after antifungal treatment (Rosenberg and Belew 1982; Faergemann 1985; Farr et al. 1985; Alford et al. 1986) it was assumed that *Malassezia* yeasts are involved in the pathogenesis of psoriasis. *Malassezia* yeasts are supposed to play a role in the koebnerization of psoriatic patients by an increased chemoattraction of neutrophils specific of *Malassezia* yeasts (Bunse and Mahrle 1996). The exact mechanism has not yet been fully determined. Experimental research showed induction of psoriatic lesions by application of killed *malassezia* yeast cells in a rabbit model (Rosenberg et al. 1980) and by heat-killed sonicated suspensions in healthy skin of patients with psoriasis and healthy controls (Lober et al. 1982). Changes in the immunologic response to *Malassezia* yeasts by psoriatic patients are listed in Table 7.9.

Table 7.8. Cytokine profiles induced by *Malassezia* yeasts in AD patients. *PBMC* Peripheral blood mononuclear cells

Reference	*Malassezia* strain	Included subjects	Method	Result
Kröger et al. (1995)	*P. ovale*	8 patients with atopic eczema; 5 healthy non-atopic controls	Stimulation of PBMCs with antigen preparation	Elevated synthesis of TH2 (type 2 T-cell) related cytokines IL-4 and IL-10 in patients with RAST(+) against *Malassezia*
Tengvall-Linder et al. (1996)	*P. orbiculare*	10 AD patients with serum IgE antibodies against *P. orbiculare*; 6 healthy controls	Stimulation of TCCs from blood and skin of one patient with *P. orbiculare* extract	Expression of interleukin-5 (IL-5) mRNA and IL-13 mRNA in 5 out of 6 *P. orbiculare*-reactive clones favouring a TH2 profile
Tengvall-Linder et al. (1998)	*P. orbiculare*	12 patients with AD; 6 non-atopic healthy controls	Stimulation of *P.orbiculare*-reactive T-cell lines (TCL) and freshly isolated PBMCs with *P. orbiculare* extract	Increased IL-5 production in PBMCs incubated for 11 days with *P. orbiculare* extract. In TCLs higher prduction of IL-4 and IL-5 after anti-CD3 stimulation than in control group. High IFN-γ levels in AD and control groups. Higher IL-4/ IFN-γ ratio in AD patients. *P. orbiculare* stimulation favours type 2 T-cell response
Savolainen et al. (2001)	*P. ovale*; CBS 7854 (*M. furfur*)	15 AD patients and 7 healthy controls	Stimulation of PBMCs with antigenic preparations of *P. ovale*, *C. albicans* and with phytohemagglutinin (PHA)	Findings in patients with AD: – Increased IL-2 and IL-4 response by PHA-stimulation. – Increased IL-5 and IFN-γ response by *C. albicans* stimulation. – Increased IL-4 response and higher IL-4/ IFN-γ ratio by *P. ovale* stimulation. Stimulation with *P. ovale* favours TH2-response, stimulation with *C. albicans* TH1 response
Johansson et al. (2002)	*M. sympodialis* ATCC 42132	40 patients with mild to severe AD; 16 healthy subjects	Stimulation of PBMCs with *M. sympodialis* crude extract and rMal s 1, rMal s 5 and rMal s 6	Increased production of the T helper 2-related cytokines interleukins 4, 5, and 13; highest cytokine titres in SPT+/APT+ AD patients; no differences in IFN-γ production of stimulated and unstimulated cells and AD patients and healthy controls
Kanda et al. (2002)	*M. furfur*; extract purchased from Allergon AB (Angelholm, Sweden)	20 patients with AD; 15 normal healthy subjects	Stimulation of PBMCs with antigenic preparations of *M. furfur*, *C. albicans* and *T. rubrum*	*M. furfur* induced IL-4 and macrophage-derived cytokine (MDC) in only AD patients. *C. albicans* induced IL-4, MDC, IFN-γ and IP-10 in all groups, IL-4 and MDC higher in AD group. *T. rubrum* induced moderate IL-4 and MDC production only in AD group. All the fungi tested induced TH2-response in AD group, probably supported by PGE2-production induced by all three fungi

Table 7.9. Host reactions in psoriasis. *PBMC* Peripheral blood mononuclear cells

Reference	*Malassezia* strain	Subjects included	Method	Result
Kanda et al. (2002)	*M. furfur*: extract purchased from Allergon AB (Angelholm, Sweden)	15 patients with Psoriasis vulgaris; 15 normal healthy subjects	Stimulation iof PBMCs with antigenic preparations of *M. furfur*, *C. albicans* and *T. rubrum*	*Malassezia* induced IFN-γ, interferon inducible protein of 10 kDa (IP-10) only in psoriatic patients. Very few patients with psoriasis and healthy controls showed IFN-γ-response to *T. rubrum*. IFN-γ-response to C. albicans in all groups. *M. furfur* induced TH1-response in psoriatic patients
Baroni et al. (2004)	*M. furfur/P. ovale* ATCC 12078	10 skin biopsies from patients with psoriasis; skin biopsies from 5 healthy volunteers	Coculture of skin samples with *M. furfur* cells at a ratio of 30:1 yeast cells:keratinocytes	Coculture with *M. furfur* up-regulated TGF-β1, integrin chains and HSP-70 in human keratinocytes. Biopsies of *M. furfur*-positive psoriatic patients also showed an increase increase in TGF-b1, integrin chains, and HSP70 expression. *M. furfur* may induce the overproduction of molecules involved in cell migration and hyperproliferation, thereby favoring the exacerbation of psoriasis.

G. *Malassezia* Yeasts as Pathogens in Animals

Malassezia yeasts are associated with several diseases in animals, e.g. atopic diseases in dogs (Chen and Hill 2005), otitis externa in cats and dogs, but also ferrets (Dinsdale et al. 1995), pigs and dromedaries (Guillot and Bond 1999), and *Malassezia* dermatitis in dogs and, to a lower extent, in cats. Apart from its first description in an Indian rhinoceros, *M. pachydermatis* has been found in lesions of exfoliative dermatitis in a southern white rhinoceros (Bauwens et al. 1996). In contrast to the predominance of *M. pachydermatis* in other species, the prevalent yeasts in otitis in cattle are *M. globosa* and *M. sympodialis* (Duarte et al. 1999, 2003). Ceruminous otitis externa is often characterized by a waxy, moist, brown or yellow exudate with variable erythema and pruritus. It is the commonest *Malassezia*-associated disease in cats (Guillot and Bond 1999). Certain dog breeds such as West Highland white terriers, dachshunds, basset hounds and others are more susceptible to *Malassezia* dermatitis than others (for a review, see Carlotti 2001). *Malassezia* dermatitis may occur localized or generalized. Skin lesions are characterized by erythema, alopecia, greasy exudation and varying degrees of scaling. Chronic cases can have marked hyperpigmentation and lichenification. Pruritus varies from mild to extremely severe. Dogs with generalized lesions often have an offensive, rancid or yeasty odor (Chen and Hill 2005). For elegant reviews of the predisposing factors for *Malassezia*-associated skin diseases, see Chen and Hill (2005) and Carlotti (2001). Alterations of the cutaneous microclimate or host defense mechanisms allow *Malassezia* yeasts to multiply and to become pathogenic. Growth of *Malassezia* spp. is enhanced by excessive production of sebum and/or cerumen or a change in its composition and often accompanies idiopathic seborrhoea (Plant et al. 1992). *Malassezia* yeasts prefer cutaneous folds and an excess of moisture. Rupture of the epidermal barrier, which is often found in atopic dermatitis, facilitates invasion of *Malassezia*. Other infections of the skin, keratinization defects and endocrine disorders or malignancies support overgrowth with *Malassezia* yeasts (Plant et al. 1992; Bond et al. 1996a, b). Also, an association of

Malassezia dermatitis with antibiotic treatment (Plant et al. 1992) and/or glucocorticoid therapy has been suggested.

As described above for human diseases, *Malassezia*-associated diseases in animals are often correlated with a dysfunction in cellular immunity. Dogs with *Malassezia* dermatitis were found to have elevated IgG titres compared to healthy dogs (Bond et al. 1998; Bond and Lloyd 2002). Apparently the antibodies did not act protectively. Cellular response was reduced in affected dogs. In contrast, Morris et al. 2002 detected an increased lymphocyte blastogenic response to crude *M pachydermatis* extract in atopic dogs with *Malassezia* dermatitis, compared with clinically normal dogs and those with *Malassezia* otitis. Atopic control dogs did not differ significantly in their responses from atopic dogs with *Malassezia* dermatitis or otitis. Further studies revealed elevated reactivity in intradermal skin tests of atopic dogs with *Malassezia* dermatitis compared with atopic dogs without *Malassezia* dermatitis or healthy dogs (Morris et al. 1998; Farver et al. 2005). Bond et al. (2002) found an increased reactivity in some atopic dogs with or without *Malassezia* infection.

Atopic dogs showed significantly elevated IgE (Chen et al. 2002a) and IgG titers compared with healthy dogs and even dogs with *Malassezia* dermatitis (Nuttall and Halliwell 2001). There was no significant difference between atopic dogs with or without *Malassezia* dermatitis (Farver et al. 2005; Nuttall and Halliwell 2001). Morris and DeBoer (2003) demonstrated passive transfer of cutaneous anaphylaxis via atopic canine serum to normal dogs by the Prausnitz–Kustner test. This confirmed the suitability of canine anti-*Malassezia* IgE.

Patch test reactivity to *M. pachydermatis* may be positive in healthy dogs, but in contrast to delayed intradermal test reactivity is more frequent in dogs with *Malassezia* dermatitis (Bond et al. 2006). In cats, the association between *Malassezia* and allergic disorders has not been established. Generalized *Malassezia* dermatitis in cats is extremely rare.

Malassezia-associated skin diseases in cats may be correlated with immunodeficiency and malignancies such as thymoma (Mauldin et al. 2002). A correlation between *Malassezia pachydermatis* and feline acne has been reported (Jazic et al. 2006).

Especially animals with *Malassezia* dermatitis may profit from antifungal therapy, although underlying factors should not be disregarded.

H. Human–Animal Crosslinking

In contrast to preliminary findings, lipid-dependent *Malassezia* spp. other than *M. pachydermatis* can be found in animals (Bond et al. 1996a, b, 1997; Raabe et al. 1998; Crespo et al. 2002a, b; Nell et al. 2002; Duarte et al. 2003; Hirai et al. 2004; Nardoni et al. 2004, 2005; Cafarchia et al. 2005; Garau et al. 2005; Coutinho et al. 2006; White et al. 2006). Raabe et al. (1998) even detected *Malassezia* yeasts in animal feces and found that the yeasts tolerated a highly acidic milieu. Therefore animals may be a reservoir for *Malassezia* yeasts and probably support their distribution. This has also been assumed for *M. pachydermatis* infections in neonatal intensive care units (NICU) and patients with immunosuppression. *M. pachydermatis* is most probable transmitted from person to person (Chryassanthou et al. 2001; Welbel et al. 1994) and can possibly resist on incubator surfaces (van Belkum et al. 1994). Chang et al. (1998) found an epidemic outbreak of severe *M. pachydermatis* infection in premature neonates in a NICU. All babies were seized by the same *M. pachydermatis* strain, which could also be isolated from one nurse and three of the pet dogs owned by health care workers. Although the nurse was said not to have been in contact with one of the dogs, it might be assumed that *M. pachydermatis* has been brought to the NICU via a dog owner and distributed from patient to patient via hand contact. Adequate hand disinfection eliminated the occurrence of *M. pachydermatis*, as found out in a follow-up examination. *M. pachydermatis* infections may therefore be rather a sanitary problem than due to its virulence.

VII. Conclusion

Lipophilic yeasts of the genus *Malassezia* are found as members of the normal microflora of the skin in humans and in a variety of warm-blooded animal species. Further, they are associated with a number of diseases in humans and animals, but their pathogenesis has not been fully elucidated. Pathogenicity factors comprise the production of filaments, enzymatic activities such as lipase and lipoperoxidase and secondary metabolisms leading to the synthesis of melanin and tryptophan-derived indole pigments. Especially the latter may explain clinical phenomena of pityriasis versicolor, the commonest *Malassezia*-associated disease.

However, causality has to be proven, in particular by molecular approaches such as identification of pigment synthesis-associated genes. Furthermore, there are inconsistent data regarding the interaction of *Malassezia* spp. with the cellular and humoral immune system of man and animals. Both immunosuppressive and immunostimulating effects have been described, which might explain the broad clinical spectrum of *Malassezia*-associated diseases, but it still remain a wide field for further studies.

References

Aizawa T, Kano R, Nakamura Y, Watanabe S, Hasgawa A (1999) Molecular heterogeneity in clinical isolates of *Malassezia pachydermatis* from dogs. Vet Microbiol 70:67–75

Aizawa T, Kano R, Nakamura Y, Watanabe S, Hasgawa A (2001) The genetic diversity of clinical isolates of *Malassezia pachydermatis* from dogs and cats. Med Mycol 39:329–334

Akamatsu H, Komura J, Asada Y, Miyachi Y, Niwa Y (1991) Inhibitory effect of azelaic acid on neutrophil functions: a possible cause for its efficacy in treating pathogenetically unrelated diseases. Arch Dermatol Res 283:162–166

Alford RH, Vire CG, Cartwright BB, King LE Jr (1986) Ketoconazole's inhibition of fungal antigen-induced thymidine uptake by lymphocytes from patients with psoriasis. Am J Med Sci 291:75–80

Allam JP, Novak N (2006) The pathophysiology of atopic eczema. Clin Exp 31:89–93

Alves EV, Martins JE, Ribeiro EB, Sotto MN (2000) *Pityrosporum folliculitis*: renal transplantation case report. J Dermatol 27:49–51

Andersson A, Scheynius A, Rasool O (2003) Detection of Mala f and Mala s allergen sequences within the genus *Malassezia*. Med Mycol 41:479–485

Andersson A, Rasool O, Schmidt M, Kodzius R, Fluckiger S, Zargari A, Crameri R, Scheynius A (2004) Cloning, expression and characterization of two new IgE-binding proteins from the yeast *Malassezia sympodialis* with sequence similarities to heat shock proteins and manganese superoxide dismutase. Eur J Biochem 271:1885–1894

Archer-Dubon C, Icaza-Chivez ME, Orozco-Topete R, Reye E, Baez-Martinez R, Ponce de Leon S (1999) An epidemic outbreak of *Malassezia* folliculitis in three adult patients in an intensive care unit: a previously unrecognized nosocomial infection. Int J Dermatol 38:453–456

Arzumanyan VG, Serdyuk OA, Kozlova NN, Basnak'yan IA, Fedoseeva VN (2003) IgE and IgG antibodies to *Malassezia* spp. yeast extract in patients with atopic dermatitis. Bull Exp Biol Med 135:460–463

Ashbee HR (2006) Recent developments in the immunology and biology of *Malassezia* species. FEMS Immunol Med Microbiol 47:14–23

Ashbee HR, Evans EGV (2002) Immunology of diseases associated with *Malassezia* species. Clin Microbiol Rev 15:21–57

Ashbee HR, Frui A, Holland KT, Cunliffe WJ, Ingham E (1994a) Humoral immunity to *Malassezia furfur* serovars A, B and C in patients with pityriasis versicolor, seborrheic dermatitis and controls. Exp Dermatol 3:227–233

Ashbee HR, Ingham E, Holland KT, Cunliffe WJ (1994b) Cell-mediated immune responses to *Malassezia furfur* serovars A, B and C in patients with pityriasis versicolor, seborrheic dermatitis and controls. Exp Dermatol 3:106–112

Aspiroz C, Moreno LA, Rezusta A, Rubio C (1999) Differentiation of three biotypes of *Malassezia* species on human normal skin. correspondence with *M. globosa*, *M. sympodialis* and *M. restricta*. Mycopathologia 145: 69–74

Aspiroz C, Ara M, Varea M, Rezusta A, Rubio C (2002) Isolation of *Malassezia globosa* and M. sympodialis from patients with pityriasis versicolor in Spain. Mycopathologia 154:111–117

Aspres N, Anderson C (2004) *Malassezia* yeasts in the pathogenesis of atopic dermatitis. Australas J Dermatol 45:199–205

Bäck O, Bartosik J (2001) Systemic ketoconazole for yeast allergic patients with atopic dermatitis. J Eur Acad Dermatol Venereol 15: 34–38

Bäck O, Faergemann J, Hornqvist R (1985) *Pityrosporum* folliculitis: a common disease of the young and middle-aged. J Am Acad Dermatol 12:56–61

Bäck O, Scheynius A, Johansson SGO (1995) Ketoconazole in atopic dermatitis: therapeutic response is correlated with decrease in serum IgE. Arch Dermatol Res 287:448–451

Baillon EH (1889) Traité de botanique médicale cryptogamique suivi du tableau. Faculté de Médecine de Paris, p. 234

Baker BS (2006) The role of microorganisms in atopic dermatitis. Clin Exp Immunol 144:1–9

Banchereau J, Steinman RM (1998) Dendritic cells and the control of immunity. Nature 392:245–252

Barchmann T, Hort W, Mayser P (2005) Untersuchungen zur Regulation des Tryptophan-abhängigen Sekundärmetabolismus von *Malassezia furfur*. Mycoses 48:307

Barfatani M, Munn RJ, Schjeide OA (1964) An ultrastructure study of *Pityrosporum orbiculare*. J Invest Dermatol 43:231–233

Baroni A, Perfetto B, Paoletti I, Ruocco E, Canozo N, Orlando M, Buommino E (2001a) *Malassezia furfur* invasiveness in a keratinocyte cell line (HaCat): effects on cytoskeleton and on adhesion molecule and cytokine expression. Arch Dermatol Res 293:414–419

Baroni A, Perfetto B, Paoletti I, De Martino L, Buommino E, Ruocco E, Ruocco V (2001b) Uptake of *Malassezia furfur* by human dermal fibroblasts: effect of ketoconazole and cytoskeleton inhibitors. Arch Dermatol Res 293:407–413

Baroni A, Paoletti I, Ruocco E, Agozzino M, Tufano MA, Donnarumma G (2004) Possible role of *Malassezia furfur* in psoriasis: modulation of TGF-beta1, integrin, and HSP70 expression in human keratinocytes and in the skin of psoriasis-affected patients. J Cutan Pathol 31:35–42

Baroni A, Orlando M, Donnarumma G, Farro P, Iovene MR, Tufano MA, Buommino E (2006) Toll-like receptor 2 (TLR2) mediates intracellular signalling in human keratinocytes in response to *Malassezia furfur*. Arch Dermatol Res 297:280–288

Batra R, Boekhout T, Guého E, Cabanes FJ, Dawson TL Jr, Gupta AK (2005) *Malassezia Baillon*, emerging clinical yeasts. FEMS Yeast Res 5:1101–1113

Bauwens L, DeVroey C, DeMeurichy W (1996) A case of exfoliative dermatitis in a captive southern white rhinoceros (*Ceratotherium simum simum*). J Zoo Wildl Med 27:271–274

Baxter M (1976) The association of *Pityrosporum pachydermatis* with the normal external ear canal of dogs and cats. J Small Anim Pract 17:231–234

Bayrou O, Pecquet C, Flahault A, Artigou C, Abuaf N, Leynadier F (2005) Head and neck atopic dermatitis and *Malassezia-furfur*-specific IgE antibodies. Dermatology 211:107–113

Belew PW, Rosenberg EW, Jennings BR (1980) Activation of the alternative pathway of complement by *Malassezia ovalis (Pityrosporum ovale)*. Mycopathologia 70:187–191

Benedek T (1930) *Cryptococcus Malassez, Pityrosporum Malassez*: Sabouraud. Zentralbl Bakteriol Parasitenk Infektionskr 116: 317–332

Benham RW (1945) *Pityrosporum ovale*. A lipophilic fungus. Thiamin and oxalo-acetic acid as growth factors. Proc Soc Exp Biol Med 58: 99–201 (Br J Dermatol 137:208–213)

Bergbrant IM (1991) Seborrhoeic dermatitis and Pityrosporum ovale: cultural, immunological and clinical studies. Acta Derm Venereol Suppl (Stockh) 167:1–36

Bergbrant IM, Faergemann J (1988) Variations of *Pityrosporum orbiculare* in middle-aged and elderly individuals. Acta Derm Venereol 68:537–540

Bergbrant IM, Faergemann J (1989) Seborrhoeic dermatitis and *Pityrosporum ovale*: a cultural and immunological study. Acta Derm Venereol 69:332–335

Bergbrant IM, Johansson S, Robbins D, Bengtsson K, Faregemann J, Scheynius A, Soderstrom T (1991) The evaluation of various methods and antigens for the detection of antibodies against *Pityrosporum ovale* in patients with seborrhoeic dermatitis. Clin Exp Dermatol 16:339–343

Bergbrant IM, Andersson B, Faergemann J (1999) Cell-mediated immunity to *Malassezia furfur* in patients with seborrhoeic dermatitis and pityriasis versicolor. Clin Exp Dermatol 24:402–406

Bhattacharya T, Edward M, Cordery C, Richardson MD (1998) Colonization of living skin equivalents by *Malassezia furfur*. Med Mycol 36:15–19

Bizzozero J (1884) Über die Mikrophyten der normalen Oberhaut des Menschen. Virchow Arch Pathol Anat 98:441–459

Boekhout T, Kamp M, Guého E (1998) Molecular typing of *Malassezia* species with PFGE and RAPD. Med Mycol 36:365–372

Bojanovsky A, Lischka G (1977) *Pityrosporum orbiculare* bei akneiformen Eruptionen. Hautarzt 28:409–411

Bojanowsky A, Bohnert E, Jung EG (1979) Pityrosporum orbiculare: Erreger verschiedener klinischer Bilder und Modell einer Depigmentierungsart. Acta Dermatol 5:19–25

Bond R, Lloyd DH (2002) Immunoglobulin G responses to *Malassezia pachydermatis* in healthy dogs and dogs with *Malassezia* dermatitis. Vet Rec 150:509–512

Bond R, Anthony RM, Dodd M, Lloyd DH (1996a) Isolation of *Malassezia sympodialis* from feline skin. J Med Vet Mycol 34:145–147

Bond R, Ferguson EA, Curtis CF, Craig JM, Lloyd DH (1996b) Factors associated with elevated cutaneous *Malassezia pachydermatis* populations in dogs with pruritic skin disease. J Small Anim Pract 37:103

Bond R, Howell SA, Haywood PJ, Lloyd DH (1997) Isolation of *Malassezia sympodialis* and *Malassezia globosa* from healthy pet cats. Vet Rec 141:200–201

Bond R, Elwood CM, Littler RM, Pinter L, Lloys DH (1998) Humoral and cell-mediated responses to *Malassezia pachydermatis* in healthy dogs and dogs with *Malassezia* dermatitis. Vet Rec 143:381–384

Bond R, Patterson-Kane JC, Lloyd DH (2004) Clinical, histopathological and immunological effects of exposure of canine skin to *Malassezia pachydermatis*. Med Mycol 42:165–175

Bond R, Patterson-Kane JC, Perrins N, Lloyd DH (2006) Patch test responses to *Malassezia pachydermatis* in healthy basset hounds and in basset hounds with *Malassezia* dermatitis. Med Mycol 44:419–427

Breathnach AS, Gross M, Martin B (1976) Freeze-fracture replications of cultured *Pityrosporum orbiculare*. Sabouraudia 14:105–113

Breathnach AS, Nazzaro-Porro M, Passi S (1984) Azelaic acid. Br J Dermatol 111:115–120

Broberg A (1995) *Pityrosporum ovale* in healthy children, infantile seborrhoeic dermatitis and atopic dermatitis. Acta Derm Venereol Suppl (Stockh) 191:1–4

Broberg A, Faergemann J (1995) Topical antimycotic treatment of atopic dermatitis in the head/neck area. A double-blind randomized study. Acta Derm Venereol 75: 46–49

Broberg A, Faergemann J, Johansson SGO, Strannegard IL, Svejgaard E (1992) *Pityrosporum ovale* and atopic dermatitis in children and young adults. Acta Derm Venereol 72:187–192

Brotherton J (1967) Lack of swelling and shrinking of *Pityrosporum ovale* in media of different osmotic pressures and its relationship with survival in the relatively dry condition of the scalp. J Gen Microbiol 48:305–308

Brunke S, Hube B (2006) MfLIP1, a gene encoding an extracellular lipase of the lipid-dependent fungus *Malassezia furfur*. Microbiology 152:547–454

Buentke E, Zargari A, Heffler LC, Avila-Carino J, Savolainen J, Scheynius A (2000) Uptake of the yeast *Malassezia furfur* and its allergenic components by human immature CD1a+ dendritic cells. Clin Exp Allergy 30:1759–1770

Buentke E, Heffler LC, Wallin RP, Lofman C, Ljunggren HG, Scheynius A (2001) The allergenic yeast *Malassezia furfur* induces maturation of human dendritic cells. Clin Exp Allergy 31:1583–1593

Buentke E, Heffler LC, Wilson JL, Wallin RP, Lofman C, Chambers BJ, Ljunggren HG, Scheynius A (2002)

Natural killer and dendritic cell contact in lesional atopic dermatitis skin-*Malassezia*-influenced cell interaction. J Invest Dermatol 119:850–857

Buentke E, D'Amato M, Scheynius A (2004) *Malassezia* enhances natural killer cell-induced dendritic cell maturation. Scand J Immunol 59:511–516

Bunse T, Mahrle G (1996) Soluble Pityrosporum-derived chemoattractant for polymorphonuclear leukocytes of psoriatic patients. Acta Derm Venereol 76:10–12

Burke RC (1961) Tinea versicolor: susceptibility factors and experimental infections in human beings. J Invest Dermatol 36: 389–401

Cabanes FJ, Hernandez JJ, Castella G (2005) Molecular analysis of *Malassezia sympodialis*-related strains from domestic animals. J Clin Microbiol 43:277–283

Cafarchia C, Gallo S, Capelli G, Otranto D (2005) Occurrence and population size of *Malassezia* spp. in the external ear canal of dogs and cats both healthy and with otitis. Mycopathologia 160:143–149

Cannon PF (1986) International Commission on the taxonomy of fungi (ICTF): name changes in fungi of microbiological, industrial and medical importance. Microbiol Sci 3:285–287

Caprilli F, Mercantini R, Nazzaro-Porro M, Passi S, Tonolo A (1973) Studies of the genus *Pityrosporum* in submerged culture. Mycopathol Mycol Appl 51:171–189

Carlotti DN (2001) *Malassezia* dermatitis in the dog. World small animal veterinary association world congress 2001. http://www.vin.com/VINDBPub/SearchPB/Proceedings/PR05000/PR00097.htm

Castellani A (1908) Tropical forms of pityriasis versicolor. J Cut Dis 26:393–399

Castellani A, Chalmers A (1913) Manual of tropical medicine, 2nd edn. Baillière Tindall & Cox, London

Catterall MD, Ward MW, Jacobs P (1978) A reappraisal of the role of *Pityrosporum orbiculare* in pityriasis versicolor and the significance of extracellular lipase. J Invest Dermatol 71:398–401

Chang HJ, Miller HL, Watkins N, Arduino MJ, Ashford DA, Midgley G, Aguero SM, Pinto-Powell R, von Reyn CF, Edwards W, McNeil MM, Jarvis WR (1998) An epidemic of *Malassezia pachydermatis* in an intensive care nursery associated with colonization of health care workers' pet dogs. N Engl J Med 338:706–711

Charles CR, Sire DJ, Johnson BL, Beidler JG (1973) Hypopigmentation in tinea versicolor: a histochemical and electronmicroscopic study. Int J Dermatol 12:48–58

Chen TA, Hill PB (2005) The biology of *Malassezia* organisms and their ability to induce immune responses and skin disease. Vet Dermatol 16:4–26

Chen TA, Halliwell RE, Pemberton AD, Hill PB (2002a) Identification of major allergens of *Malassezia pachydermatis* in dogs with atopic dermatitis and *Malassezia* overgrowth. Vet Dermatol 13:141–150

Chen TA, Halliwell RE, Hill PB (2002b) Failure of extracts from *Malassezia pachydermatis* to stimulate canine keratinocyte proliferation in vitro. Vet Dermatol 13:323–329

Chen TA, Halliwell RE, Shaw DJ, Hill PB (2004) Assessment of the ability of *Malassezia pachydermatis* to stimulate proliferation of canine keratinocytes in vitro. Am J Vet Res 65:787–796

Chryssanthou E, Broberger U, Petrini B (2001) *Malassezia pachydermatis* fungaemia in a neonatal intensive care unit. Acta Paediatr 90:323–327

Chua KB, Devi S, Hooi PS, Chong KH, Phua KL, Mak JW (2003) Seroprevalence of *Malassezia furfur* in an urban population in Malaysia. Malays J Pathol 25:49–56

Clark RA, Kupper TS (2006) Misbehaving macrophages in the pathogenesis of psoriasis. J Clin Invest 116:2084–2087

Clemmensen O, Hjorth N (1983) Treatment of dermatitis of the head and neck with ketoconazole in patients with type I sensitivity to *Pityrosporum orbiculare*. Semin Dermatol 2: 26–29

Coutinho SD, Fedullo JD, Correa SH (2006) Isolation of *Malassezia* spp. from cerumen of wild felids. Med Mycol 44:383–387

Crespo MJ, Abarca ML, Cabanes FJ (2002a) Occurrence of *Malassezia* spp. in horses and domestic ruminants. Mycoses 45:333–337

Crespo MJ, Abarca ML, Cabanes FJ (2002b) Occurrence of *Malassezia* spp. in the external ear canals of dogs and cats with and without otitis externa. Med Mycol 40:115–121

Crespo-Erchiga V, Florencio VD (2006) *Malassezia* yeasts and pityriasis versicolor. Curr Opin Infect Dis 19:139–147

Crespo-Erchiga V, Ojeda Mertos A, Vra Casano A, Crespo-Erchiga A, Sanchez Fajardo F (2000) *Malassezia globosa* as the causative agent of pityriasis versicolor. Br J Dermatol 143:799–803

Cross CE, Bancroft GJ (1995) Ingestion of acapsular *Cryptococcus neoformans* occurs via mannose and beta-glucan receptors, resulting in cytokine production and increased phagocytosis of the encapsulated form. Infect Immun 63:2604–2611

Cunningham AC, Leeming JP, Ingham E, Gowland G (1990) Differentiation of three serovars of *Malassezia furfur*. J Appl Bacteriol 68:439–446

Cunningham AC, Ingham E, Gowland G (1992) Humoral responses to *Malassezia furfur* serovars A, B and C in normal individuals of various ages. Br J Dermatol 127:476–481

Dahms K, Krämer HJ, Thoma W, Mayser P (2002) Tyrosinaseinhibition durch Tryptophan-metabolite von *Malassezia furfur* in humaner Epidermis. Mycoses 45:230

Davis PD, Hill CH, Lawton G, Nixon JS, Wilkinson SE, Hurst StA, Keech E, Turner SE (1992) Inhibitors of protein kinase C. 1. 2,3-Bisindolyarylmaleimides. J Med Chem 35:177–184

De Almeida H Jr, Mayser P (2006) Absence of sunburn in lesions of pityriasis versicolor alba. Mycoses (in press)

De Luca C, Picardo M, Breathnach A, Passi S (1996) Lipoperoxidase activity of *Pityrosporum*: characterisation of by-products and possible role in pityriasis versicolor. Exp Dermatol 5:49–56

Devlin RK (2006) Invasive fungal infections caused by *Candida* and *Malassezia* species in the neonatal intensive care unit. Adv Neonatal Care 6:68–77

Devos SA, Valk PG van der (2000) The relevance of skin prick tests for *Pityrosporum ovale* in patients with head and neck dermatitis. Allergy 55:1056–1058

Diatchenko L, Lau YF, Campbell AP, Chenchik A, Moqadam F, Huang B, Lukyanov S, Lukyanov K, Gurskaya N, Sverdlov ED, Siebert PD (1996) Suppression subtractive hybridization: a method for generating differentially regulated or tissue-specific cDNA probes and libraries. Proc Natl Acad Sci USA 93:6025–6030

Dinsdale JR, Rest JR (1995) Yeast infection in ferrets. Vet Rec 137:647–648

Dompmartin D, Drouhet E (1977) Folliculitis à *Pityrosporum ovale*. Action de l'éconazole. Bull Soc Franc Mycol Med 6:15–20

Dodge CW (1935) Medical mycology. Mosby, St. Louis

Doekes G, Van Ieperen-Van Dijk AG (1993a) Allergens of *Pityrosporum ovale* and *Candida albicans*. I. Cross-reactivity of IgE-binding components. Allergy 48:394–400

Doekes G, Kaal MJH, Van Ieperen-Van Dijk AG (1993b) Allergens of *Pityrosporum ovale* and *Candida albicans*. II. Physiochemical characterization. Allergy 48:401–408

Dold H (1910) On the so-called Bottle Bacillus (*Dermatophyton Malassez*). Parasitology 3: 279–287

Donnarumma G, Paoletti I, Buommino E, Orlando M, Tufano MA, Baroni A (2004) *Malassezia furfur* induces the expression of beta-defensin-2 in human keratinocytes in a protein kinase C-dependent manner. Arch Dermatol Res 295:474–481

Dorn M, Roehnert K (1977) Dimorphism of *Pityrosporum orbiculare* in a defined culture medium. J Invest Dermatol 69: 224–248

Dotz WI, Henrikson DM, Yu GS, Galey CI (1985) Tinea versicolor: a light and electron microscopic study of hyperpigmented skin. J Am Acad Dermatol 12:580–581

Drouhet E, Dompmartin D, Papachristou-Moriati A, Ravisse P (1980) Experimental dermatitis caused by Pityrosporum ovale and (or) Pityrosporum orbiculare in the guinea pig and the mouse Sabouraudia 18:149–156

Duarte ER, Melo MM, Hahn RC, Hamdan JS (1999) Prevalence of *Malassezia* spp. in the ears of asymptomatic cattle and cattle with otitis in Brazil. Med Mycol 37:159–162

Duarte ER, Batista RD, Hahn RC, Hamdan JS (2003) Factors associated with the prevalence of *Malassezia* species in the external ears of cattle from the state of Minas Gerais, Brazil. Med Mycol 41:137–142

Eichstedt CF (1846) Pilzbildung in der Pityriasis versicolor. Froriep's Neue Notizen aus dem Gebiete der Natur- und Heilkunde 853:270–271

El-Hefnawi H, el-Gothamy T, Refai M (1972) Studies on pityriasis versicolor in Egypt. 3. Laboratory diagnosis and experimental infection. Mykosen 15:165–170

Elewski B, Thiboutot D (2006) A clinical overview of azelaic acid. Cutis 77[Suppl 2]:12–16

Enoch DA, Ludlam HA, Brown NM (2006) Invasive fungal infections: a review of epidemiology and management options. J Med Microbiol 55:809–818

Faergemann J (1979) Experimental tinea versicolor in rabbits and humans with *Pityrosporum orbiculare*. J Invest Dermatol 72:326–329

Faergemann J (1983) Antibodies to *Pityrosporum orbiculare* in patients with tinea versicolor and controls of various ages. J Invest Dermatol 80:133–135

Faergemann J (1985) Treatment of sebopsoriasis with itraconazole. Mykosen 28:612–618

Faergemann J (1989) A new model for growth and filament production of Pityrosporum ovale (orbiculare) on human stratum corneum in vitro. J Invest Dermatol 92:117–119

Faergemann J (2002) Atopic dermatitis and fungi. Clin Microbiol Rev 15:545–563

Faergemann J, Bernander S (1981) Micro-aerophilic and anaerobic growth of *Pityrosporum orbiculare*. Sabouraudia 17:171–179

Faergemann J, Fredriksson T (1981) Experimental infections in rabbits and humans with *Pityrosporum orbiculare* and P. ovale. J Invest Dermatol 1981 77:314–318

Faergemann J, Maibach HI (1984) The Pityrosporon yeasts. Their role as pathogens. Int J Dermatol. 23:463–465

Faergemann J, Meinhof W (1988) *Pityrosporum*-Folliculitis. Akt Dermatol 14:400–403

Faergemann J, Tjernlund U, Scheynius A, Bernander S (1982) Antigenic similarities and differences in genus *Pityrosporum*. J Investig Dermatol 78:28–31

Faggi E, Pini G, Campisis E, Gargani G (1998) Anti-*Malassezia furfur* antibodies in the population. Mycoses 41:273–275

Farr PM, Krause LB, Marks JM, Shuster S (1985) Response of scalp psoriasis to oral ketoconazole. Lancet 2:921–922

Farver K, Morris DO, Shofer F, Esch B (2005) Humoral measurement of type-1 hypersensitivity reactions to a commercial *Malassezia* allergen. Vet Dermatol 16:261–268

Fischer Casagrande B, Flückiger S, Linder MT, Johansson C, Scheynius A, Crameri R, Schmid-Grendelmeier P (2006) Sensitization to the yeast *Malassezia sympodialis* is specific for extrinsic and intrinsic atopic eczema. J Invest Dermatol (in press)

Flückiger S, Fijten H, Whitley P, Blaser K, Crameri R (2002) Cyclophilins, a new family of cross-reactive allergens. Eur J Immunol 32:10–17

Ford GP, Farr PM, Ive FA, Shuster S (1985) The response of seborrhoeic dermatitis to ketoconazole. Br J Derm 111:603–607

Frealle E, Noel C, Viscogliosi E, Camus D, Dei-Cas E, Delhaes L (2005) Manganese superoxide dismutase in pathogenic fungi: an issue with pathophysiological and phylogenetic involvements. FEMS Immunol Med Microbiol 45:411–422

Freire-Ruano A, Crespo-Leiro MG, Muniz J, Paniagua MJ, Almagro M, Castro-Beiras A (2000) Dermatologic complications after heart transplantation: incidence and prognosis. Med Clin (Barc) 115:208–210

Gaitanis G, Velegraki A, Velegraki A, Frangoulis E, Mitroussia A, Tsigonia A, Tzimogianni A, Katsambas A, Legatis NJ (2002) Identification of *Malassezia* species from skin scales by PCR-RFLP. Clin Microbiol Infect 8:162–173

Gaitanis G, Menounos P, Katsambas A, Velegraki A (2003) Detection and mutation screening of *Malassezia sympodialis* sequences coding for the Mal s 1 allergen implicated in atopic dermatitis. J Invest Dermatol 121:1559–1560

Gaitanis G, Chasapi V, Velegraki A (2005a) Novel application of the masson-fontana stain for demonstrating

Malassezia species melanin-like pigment production in vitro and in clinical specimens. J Clin Microbiol 43:4147–4151

Gaitanis G, Velegraki A, Rasool O, Scheynius A (2005b) Clarifications on conflicting results published on the amplification of the Mala s 1 allergen gene sequences. J Invest Dermatol 124:479

Galadari I, El Komy M, Mousa A, Hashimoto K, Mehregan AH (1992) Tinea versicolor: histologic and ultrastructural investigation of pigmentary changes. Int J Dermatol 31:253–256

Gandra RF, Melo TA, Matsumoto FE, Pires MFC, Croce J, Gambale W, Paula CR (2001) Allergenic evaluation of *Malassezia furfur* crude extracts. Mycopathologia 155:183–189

Garau M, Palacio A del, Garcia J (2005) Prevalence of *Malassezia* spp. in healthy pigs. Mycoses. 48:17–20

Gemmer CM, DeAngelis YM, Theelen B, Boekhout T, Dawson TL Jr (2002) Fast, noninvasisve method for molecular detection and differentiation of *Malassezia* species on human skin and application of the method to dandruff microbiology. J Clin Microbiol 40:3350–3357

Glaser AG, Limacher A, Flückiger S, Scheynius A, Scappozza L, Crameri R (2006) Analysis of the cross-reactivity and of the 1.5 A crystal structure of the *Malassezia sympodialis* Mala s 6 allergen, a member of the cyclophilin pan-allergen family. Biochem J 396:41–49

Goodfield MJD, Saihan EM, Crowley J (1987) Experimental folliculitis with *Pityrosporum orbiculare*: The influence of the host response. Acta Derm Venereol 67:445–447

Gordon MA (1951a) The lipophilic mycoflora of the skin. I. In vitro culture of *Pityrosporum orbiculare* n.sp. Mycologia 43:524–535

Gordon MA (1951b) Lipophilic yeast associated with tinea versicolor. J Invest Dermatol 17:267–272

Guého E, Midgley G, Guillot J (1996) The genus of *Malassezia* with description of four new species. Antonie von Leeuwenhock 69:337–355

Guého E, Boekhout T, Ashbee H R, Guillot J, Belkum A van, Faergemann J (1998) The role of *Malassezia* species in the ecology of human skin and as pathogens. Med Mycol 36:220–229

Guillot J, Bond R (1999) *Malassezia pachydermatis*: a review. Med Mycol 37:295–306

Guillot J, Guého E (1995) The diversitiy of *Malassezia* yeasts confirmed by rRNA sequence and nuclear DNA comparisons. Antonie van Leeuwenhock 67:297–314

Guillot J, Guého E, Lesourd M, Midgley G, Chevrier G, Dupont B (1996) Identification of *Malassezia* species, a practical approach. J Mycol Med 6:103–110

Guillot J, Guého E, Chevrier G, Chermette R (1997) Epidemiological analysis of *Malassezia pachydermatis* isolates by partial sequencing of the large subunit ribosomal RNA. Res Vet Sci 62:22–25

Gupta AK, Bluhm R (2004) Seborrheic dermatitis. J Eur Acad Dermatol Venereol 18:13–26

Gupta AK, Kohli Y, Summerbell RC, Faergemann J (2001) Quantitative culture of *Malassezia* species from different body sites of individuals with or without dermatoses. Med Mycol 39:243–251

Gupta AK, Bluhm R, Summerbell R (2002) Pityriasis versicolor. J Eur Acad Dermatol Venereol 16:19–33

Gupta AK, Batra R, Bluhm R, Faergemann J (2003) Pityriasis versicolor. Dermatol Clin 21:413–420

Gupta AK, Batra R, Bluhm R, Boekhout T, Dawson TL Jr (2004) Skin diseases associated with Malassezia species. J Am Acad Dermatol 51:785–798

Gustafson BA (1955) Otitis externa in the dog. A bacteriological and experimental study. Royal Veterinary College of Sweden, Stockholm

Gustafson BA (1959) Lipophilic yeasts belonging to genus *Pityrosporum*, found in swine. Acta Pathol Microbiol Scand 45:275–280

Gustafson BA (1960) The occurrence of yeasts belonging to genus *Pityrosporum* in different kinds of animals. Acta Pathol Microbiol Scand 48:51–55

Habibah A, Catchpole B, Bond R (2005) Canine serum immunoreactivity to *M. pachydermatis* in vitro is influenced by the phase of yeast growth. Vet Dermatol 16:147–152

Hay RJ, Roberts SOB, Mackenzie DWR (1998) Pityriasis versicolor. In: Champion RH, Burton JL, Ebling FJG (eds) Textbook of dermatology. vol 2, 6th edn. Blackwell, Oxford, pp 1286–1290

Heid E, Grosshans E, Provenchar D, Basset M (1978) Folliculites pityrosporiques. Ann Dermatol Venereol 105:133–138

Hirai A, Kano R, Makimura K, Duarte ER, Hamdan JS, Lachance MA, Yamaguchi H, Hasegawa A (2004) *Malassezia nana* sp. nov., a novel lipid-dependent yeast species isolated from animals. Int J Syst Evol Microbiol 54:623–627

Hort W, Fradin C, Hube B, Mayser P (2005) Identifikation Pigment-assoziierter Gene im Tryptophanstoffwechsel von *M. furfur*. JDDG 3[Suppl 1]:136

Hossain HM, Landgraf V, Weiß R, Mann M, Hayatpour J, Chakraborty T, Mayser P (2006) The genetic and biochemical characterization of the species *Malassezia pachydermatis* with particular attention on pigment-producing subgroups. Med Mycol (in press)

Huang HP, Little CJ, Fixter LM (1993) Effects of fatty acids on the growth and composition of *Malassezia pachydermatis* and their relevance to canine otitis externa. Res Vet Sci 55:119–123

Huang X, Johansson SG, Zargari A, Nordvall SL (1995) Allergen cross-reactivity between *Pityrosporum orbiculare* and *Candida albicans*. Allergy 50:648–656

Ingham E, Cunningham AC (1993) *Malassezia furfur*. J Med Vet Mycol 31:265–288

Irlinger B, Krämer HJ, Mayser P, Steglich W (2004) Pityriarubins, biologically active Bis(indolyl)spirans from cultures of the lipophilic yeast *Malassezia furfur*. Angew Chem Int Ed Engl 43:1098–1100

Irlinger B, Bartsch A, Krämer HJ, Mayser P, Steglich W (2005) New tryptophan metabolites from cultures of the lipophilic yeast *Malassezia furfur*. Helv Chim Acta 88:1472–1485

Jazic E, Coyner KS, Loeffler DG, Lewis TP (2006) An evaluation of the clinical, cytological, infectious and histopathological features of feline acne. Vet Dermatol 17:134–140

Jensen-Jarolim E, Poulsen LK, With H, Kieffer M, Ottevanger V, Stahl Skov P (1992) Atopic dermatitis of the face, scalp, and neck: type I reaction to the yeast *Pityrosporum ovale*? J Allergy Clin Immunol 89:44–51

Johansson C, Jeddi-Tehrani M, Grunewald J, Tengvall-Linder M, Bengtsson A, Hallden G, Scheynius A (1999) Peripheral blood T-cell receptor beta-chain V-repertoire in atopic dermatitis patients after in vitro exposure to *Pityrosporum orbiculare* extract. Scand J Immunol 49:293–301

Johansson C, Eshaghi H, Linder MT, Jakobson E, Scheynius A (2002) Positive atopy patch test reaction to *Malassezia furfur* in atopic dermatitis correlates with a T helper 2-like peripheral blood mononuclear cells response. J Invest Dermatol 118:1044–1051

Johansson C, Sandström MH, Bartosik J, Särnhult T, Christiansen J, Zargari A, Back O, Wahlgren CF, Faergemann J, Scheynius A, Tengvall-Linder M (2003) Atopy patch test reactions to *Malassezia* allergens differentiate subgroups of atopic dermatitis patients. Br J Dermatol 148:479–488

Johansson S, Faergemann J (1990) Enzyme-linked immunosorbent assay (ELISA) for detection of antibodies against *Pityrosporum orbiculare*. J Med Vet Mycol 28:257–260

Johansson S, Karlstrom K (1991) IgE binding components in *Pityrosporum orbiculare* identified by an immunoblotting technique. Acta Derm Venereol 71:11–16

Jung EG, Bohnert E (1976) Mechanism of depigmentation on pityriasis versicolor alba. Arch Dermatol Res 256:333–334

Kanda N, Tani K, Enomoto U, Nakai K, Watanabe S (2002) The skin fungus-induced Th1- and Th2-related cytokine, chemokine and prostaglandin E2 production in peripheral blood mononuclear cells from patients with atopic dermatitis and psoriasis vulgaris. Clin Exp Allergy 32:1243–1250

Kawai T, Akira S (2006) TLR signaling. Cell Death Differ 13:816–825

Kawano S, Nakagawa H (1995) The correlation between the levels of anti-*Malassezia furfur* IgE antibodies and severities of face and neck dermatitis of patients with atopic dermatitis. Arerugi 44:128–133

Keddie FM (1966) Electron microscopy of *Malassezia furfur* in tinea versicolor. Sabouraudia 5:134–137

Kennis RA, Rosser EJ Jr, Olivier NB, Walker RW (1996) Quantity and distribution of *Malassezia* organisms on the skin of clinically normal dogs. J Am Vet Med Assoc 208:1048–1051

Kesavan S, Walters CE, Holland KT, Ingham E (1998) The effects of *Malassezia* on pro-inflammatory cytokine production by human peripheral blood mononuclear cells in vitro. Med Mycol 36:97–106

Kesavan S, Holland KT, Ingham E (2000) The effects of lipid extraction on the immunomodulatory activity of *Malassezia* species in vitro. Med Mycol 38:239–247

Kieffer M, Bergbrant IM, Faergemann J, Jemec GB, Ottevanger V, Stahl Skov P, Svejgaard E (1990) Immune reactions to *Pityrosporum ovale* in adult patients with atopic and seborrheic dermatitis. J Am Acad Dermatol 22:739–742

Kim TY, Jang IG, Park YM, Kim HO, Kim CW (1999) Head and neck dermatitis: the role of *Malassezia furfur*, topical steroid use and environmental factors in its causation. Clin Exp Dermatol 24:226–231

Kobayashi SD, Voyich JM, Burlak C, DeLeo FR (2005) Neutrophils in the innate immune response. Arch Immunol Ther Exp (Warsz) 53:505–517

Konrad K, Wolff K (1973) Hyperpigmentation, melanosome size, and distribution patterns of melanosomes. Arch Dermatol 107:853–860

Kosonen J, Lintu P, Kortekangas-Savolainen o, Kalimo K, Terho EO (2005) Immediate hypersensitivity to *Malassezia furfur* and *Candida albicans* mannans in vivo and in vitro. Allergy 60:238–242

Koyama T, Kanbe T, Ishiguro A, Kikuchi A, Tomita Y (2000) Isolation and characterization of a major antigenic component of *Malassezia globosa* to IgE antibodies in sera of patients with atopic dermatitis. Microbiol Immunol 44:373–379

Koyama T, Kanbe T, Ishiguro A, Kikuchi A, Tomita Y (2001) Antigenic components of *Malassezia* species for immunoglobulin E antibodies in sera of patients with atopic dermatitis. J Dermatol Sci 26:201–208

Kozel TR, Pfrommer GS, Guerlain AS, Highison BA, Highison GJ (1988) Role of the capsule in phagocytosis of *Cryptococcus neoformans*. Rev Infect Dis 10[Suppl 2]:436–439

Krämer HJ, Kessler D, Hipler UC, Irlinger B, Hort W, Bodeker RH, Steglich W, Mayser P (2005a) Pityriarubins, novel highly selective inhibitors of respiratory burst from cultures of the yeast *Malassezia furfur*: comparison with the bisindolylmaleimide arcyriarubin A. ChemBioChem 6:2290–2297

Krämer HJ, Podobinska M, Bartsch A, Battmann A, Thoma W, Bernd A, Kummer W, Irlinger B, Steglich W, Mayser P (2005b) Malassezin, a novel agonist of the arylhydrocarbon receptor from the yeast *Malassezia furfur*, induces apoptosis in primary human melanocytes. ChemBioChem 6:860–865

Kröger S, Neuber K, Gruseck E, Ring J, Abeck D (1995) *Pityrosporum ovale* extracts increase interleukin-4, interleukin-10 and IgE synthesis in patients with atopic eczema. Acta Derm Venereol 75:357–360

Kurtzmann CP, Robnett CJ (1994) Orders and families of ascosporogenous yeasts and yeast-like taxa compared from ribosomal RNA sequence similarities. In: Hawksworth DL (ed) Ascomycete systematics: problems and perspectives in the nineties. Plenum, New York, pp 249–258

Labows JN, McGinley KJ, Leyden JJ, Webster GF (1979) Characteristic gamma-lactone odor production of the genus *Pityrosporum*. Appl Environ Microbiol 38:412–415

Langfelder K, Streibel M, Jahn B, Haase G, Brakhage AA (2003) Biosynthesis of fungal melanins and their importance for human pathogenic fungi. Fungal Genet Biol 38:143–158

Lee MR, Cooper AJ (2006) Immunopathogenesis of psoriasis. Australas J Dermatol 47:151–159

Leino M, Reijula K, Makinen-Kilijunen S, Haahtela T, Makela MJ, Alenius H (2006) *Cladosporium herbarum* and *Pityrosporum ovale* allergen extracts share cross-reacting glycoproteins. Int Arch Allergy Immunol 140:30–35

Leung DY, Boguniewicz H, Howell MD, Nomura I, Hamid QA (2004) New insights into atopic dermatitis. J Clin Invest 113:651–657

Lindborg M, Magnusson CGM, Zargari A, Schmidt M, Scheynius A, Crameri R, Whitley P (1999) Selective cloning of allergens from the skin colonizing yeast *Malassezia furfur* by phage surface display. J Invest Dermatol 113:156–161

Lintu P, Savolainen J, Kalimo K (1997) IgE antibodies to protein and mannan antigens of *Pityrosporum ovale* in atopic dermatitis. Clin Exp Allergy 27:87–95

Lintu P, Savolainen J, Kalimo K, Terho EO (1998) Stability of *Pityrosporum ovale* allergens during storage. Clin Exp Allergy 28:486–490

Lintu P, Savolainen J, Kalimo K, Kortekangas-Savolainen O, Nermes M, Terho EO (1999) Cross reacting IgE and IgG antibodies to *Pityrosporum ovale* mannan and other yeasts in atopic dermatitis. Allergy 54:1067–1073

Lintu P, Savolainen J, Kortekangas-Savolainen O, Kalimo K (2001) Systemic ketoconazole is an effective treatment of atopic dermatitis with IgE-mediated hypersensitivity to yeast. Allergy 56:512–517

Lober CW, Belew PW, Rosenberg EW, Bale G (1982) Patch tests with killed sonicated microflora in patients with psoriasis. Arch Dermatol 118:322–325

Machowinski A, Krämer HJ, Hort W, Mayser P (2006) Pityriacitrin – a potent UV filter produced by *Malassezia furfur* and its effect on human skin microflora. Mycoses 49:388–392

Mackintosh JA (2001) The antimicrobial properties of melanocytes, melanosomes and melanin and the evolution of black skin. J Theor Biol 211:101–113

Makimura K, Tamura Y, Kudo M, Uchida K, Saito H, Yamaguchi H (2002) Species identification and strain typing of *Malassezia* species stock strains and clinical isolates based on the DNA sequences of nuclear ribosomal internal transcribed spacer 1 regions. J Med Microbiol 49:29–35

Mauldin EA, Morris DO, Goldschmidt MH (2002) Retrospective study: the presence of *Malassezia* in feline skin biopsies. A clinicopathological study. Vet Dermatol 13:7–13

Mayser P, Gross A (2000) IgE antibodies to *Malassezia furfur*, *M. sympodialis* and *Pityrosporum orbiculare* in patients with atopic dermatitis, seborrheic eczema or pityriasis versicolor, and identification of respective allergens. Acta Derm Venereol 80:357–361

Mayser P, Pape B (1998) Decreased susceptibility of *Malassezia furfur* to UV light by synthesis of tryptophan derivatives. Antonie van Leeuwenhoek 73:315–319

Mayser P, Führer D, Schmidt R, Gründer K (1995) Hydrolysis of fatty acid esters by *Malassezia furfur*: different utilization depending on alcohol moiety. Acta Derm Venereol 75:105–109

Mayser P, Haze P, Papavassilis C, Pickel M, Gründer K, Guého E (1997) Differentiation of *Malassezia* spp. selectivity of cremophor EL, castor oil and ricinoleic acid for *Malassezia furfur*. Br J Dermatol 137:208–213

Mayser P, Imkampe A, Winkeler M, Papavassilis C (1998a) Growth requirements and nitrogen metabolism of *Malassezia furfur*. Arch Dermatol Res 290:277–282

Mayser P, Wille G, Imkampe A, Thoma W, Arnold N, Monsees T (1998b) Synthesis of fluorochromes and pigments in *Malassezia furfur* by use of tryptophan as single nitrogen source. Mycoses 41:265–271

Mayser P, Pickel M, Haze P, Erdmann F, Papavassilis C, Schmidt R (1998c) Different utilization of neutral lipids by *Malassezia furfur* and *Malassezia sympodialis*. Med Mycol 36:7–14

Mayser P, Schäfer U, Krämer HJ, Irlinger B, Steglich W (2002) Pityriacitrin – an ultraviolet-absorbing indole alkaloid from the yeast *Malassezia furfur*. Arch Dermatol Res 294:131–134

Mayser P, Stapelkamp H, Krämer HJ, Irlinger B, Steglich W (2003a) Pityrialacton – a new fluorescing indole alkaloid from the yeast *Malassezia furfur*. Antonie van Leeuwenhoek 84:185–191

Mayser P, Stapelkamp H, Thoma W, Krämer HJ, Irlinger B, Steglich W (2003b) Tryptophan-dependent synthesis of isomeric bisindolyl derivatives by *M. furfur*. Arch Dermatol Res 294:486

Mayser P, Begerow D, Kahmann R (2004a) Tryptophanabhängige Pigmentsynthese bei *Ustilago maydis*. Mycoses 47:379

Mayser P, Töws A, Krämer HJ, Weiß R (2004b) Further characterization of pigment-producing *Malassezia* strains. Mycoses 47:34–39

Mayser P, Kupfer J, Nemetz D, Schäfer U, Nilles M, Hort W, Gieler U (2006) Treatment of head and neck dermatitis with ciclopiroxolamine cream–results of a double-blind, placebo-controlled study. Skin Pharmacol Physiol 19:153–158

Mayser P, Wenzel M, Krämer HJ, Kindler BLJ, Spiteller P, Haase G (2007) Production of indole pigments by *Candida glabrata*. Med Mycol 45:519–524

McGinley KJ, Leyden LJ, Marples RR, Kligman AM (1975) Quantitative microbiology of the scalp in non-dandruff, dandruff and seborrheic dermatitis. J Invest Dermatol 64:401–405

Midgley G (1989) The diversity of *Pityrosporum* (*Malassezia*) yeasts in vivo and in vitro. Mycopathologia 106:143–155

Mirhendi H, Makumura K, Zomorodian K, Yamada T, Sugito T, Yamaguchi H (2005) A simple PCR-RFLP method for identification and differentiation of 11 *Malassezia* species. J Microbiol Methods 61:281–284

Mittag H (1994) Fine structural investigation of *Malassezia furfur*. I. Size and shape of the yeast cells and a consideration of their ploidy. Mycoses 37:393–399

Mittag H (1995) Fine structural investigation of *Malassezia furfur*. II. The envelope of the yeast cells. Mycoses 38:13–21

Mittermann I, Aichberger KJ, Bunder R, Mothes N, Renz H, Valenta R (2004) Autoimmunity and atopic dermatitis. Curr Opin Allergy Clin Immunol 4:367–371

Morris DO, Olivier NB, Rosser EJ (1998) Type-1 hypersensitivity reactions to *Malassezia pachydermatis* extracts in atopic dogs. Am J Vet Res 59:836–841

Nakabayashi A, Sei Y, Guillot J (2000) Identification of *Malassezia* species isolated from patients with seborrheic dermatitis, atopic dermatitis, pityriasis versicolor and normal subjects. Med Mycol 38: 337–341

Nardoni S, Mancianti F, Corazza M, Rum A (2004) Occurrence of *Malassezia* species in healthy and dermatologically diseased dogs. Mycopathologia 157:383–388

Nardoni S, Mancianti F, Rum A, Corazza M (2005) Isolation of *Malassezia* species from healthy cats and cats with otitis. J Feline Med Surg 7:141–145

Nazzaro-Porro M, Passi S (1978) Identification of tyrosinase inhibitors in cultures of *Pityrosporum*. J Invest Dermatol 71:205–208

Nazzaro-Porro M, Passi S, Caprilli F, Mercantini R (1977) Induction of hyphae in cultures of *Pityrosporum* by cholesterol and cholesterol esters. J Invest Dermatol 69:531–534

Nazzaro-Porro M, Passi S, Picardo M, Merrcantini R, Breathnach AS (1986) Lipoxygenase activity of *Pityrosporum* in vitro and vivo. J Invest Dermatol 87:108–112

Nell A, James SA, Bond CJ, Hunt B, Herrtage ME (2002) Identification and distribution of a novel *Malassezia* species yeast on normal equine skin. Vet Rec 150:395–398

Netea MG, Van der Meer JW, Sutmuller RP, Adema GJ, Kullberg BJ (2005) From the Th1/Th2 paradigm towards a Toll-like receptor/T-helper bias. Antimicrob Agents Chemother 49:3991–3996

Neuber K, König W, Ring J (1993) *Staphylococcus aureus* and atopic eczema. Hautarzt 44:135–142

Nissen D, Petersen LJ, Esch R, Svejgaard E, Skov PS, Poulsen LK, Nolte H (1998) IgE sensitization to cellular and culture filtrates of fungal extracts in patients with atopic dermatitis. Ann Allergy Asthma Immunol 81:247–255

Noah PW (1990) The role of microorganisms in psoriasis. Semin Dermatol 9:269–276

Nordvall SL, Johansson S (1990) IgE antibodies to *Pityrosporum orbiculare* in children with atopic disease. Acta Paediatr Scand 79:343–348

Nuttall TJ, Halliwell REW (2001) Serum antibodies to *Malassezia* yeasts in canine atopic dermatitis. Vet Dermatol 12:327–332

Oguchi M (1982) Electron microscopic studies of melanocytes in the affected skin of tinea versicolor. Acta Dermatol (Kyoto) 77:187–92

Onishi Y, Kuroda M, Yasueda H, Saito A, Sono-Koyama E, Tunasawa S, Hashida-Okado T, Yagihara T, Uchida K, Yamaguchi H, Akiyama K, Kato I, Takesako K (1999) Two-dimensional electrophoresis of *Malassezia* allergens for atopic dermatitis and isolation of Mal f 4 homologs with mitochondrial malate dehydrogenase. Eur J Biochem 261:148–154

Ota M, Huang PT (1933) Sur les champignons du genre *Pityrosporum* Sabouraud. Ann Parasit Hum Com 11:49–58

Oudemans CAJA, Pekelharing CA (1885) *Saccharomyces capillitii* Oudemans en Peklharing; ein spruitzwam van de behaarde hoofhuid. Ned Tijdschr Geneeskd 21:997–1005

Panja G (1927) The *Malassezia* of the skin, their cultivation, morphology and species. Trop Med 2:442–456

Papavassilis C, Mach KK, Mayser PA (1999) Medium-chain triglycerides inhibit growth of Malassezia: implications for prevention of systemic infection. Crit Care Med 27:1781–1786

Parry ME, Sharpe GR (1998) Seborrhoeic dermatitis is not caused by an altered immune response to *Malassezia* yeast. Br J Dermatol 139:254–263

Pierard-Franchimont C, Arrese JE, Pierard GE (1995) Immunohistochemical aspects of the link between *Malassezia ovalis* and seborrheic dermatitis. J Eur Acad Dermatol Venereol 4:14–19

Plant JD, Rosenkrantz WS, Griffin CE (1992) Factors associated with and prevalence of high *Malassezia pachydermatis* numbers on dog skin. J Am Vet Med Assoc 201:879–882

Plotkin LI, Squiquera L, Mathov I, Galimberti R, Leoni J (1996) Characterization of the lipase activity of *Malassezia furfur*. J Med Vet Mycol 34:43–48

Podobinska M, Krämer HJ, Bartsch A, Steglich W, Mayser P (2003) A cytotoxic metabolite from *Malassezia furfur*. Isolation and chemical identification as malassezin. Arch Dermatol Res 294:512

Polacheck I, Kwon-Chung KJ (1988) Melanogenesis in *Cryptococcus neoformans*. J Gen Microbiol 134:1037–1041

Polak A (1990) Melanin as a virulence factor in pathogenic fungi. Mycoses 33:215–224

Polonelli L, Lorenzini R, De Bernardis F, Morace G (1986) Potential therapeutic effect of yeast killer toxin. Mycopathologia 96:103–107

Porro MN, Passi S, Caprill F, Nazzaro P, Morpurgo G (1976) Growth requirement and lipid metabolism of *Pityrosporum orbiculare*. J Invest Dermatol 66:178–182

Potter BS, Burgoon CF, Johnson WC (1973) *Pityrosporum* folliculitis. Report of seven cases and review of the *Pityrosporum* organism relative to cutaneous disease. Arch Dermatol 107:388–391

Raabe P, Mayser P, Weiss R (1998) Demonstration of *Malassezia furfur* and *M. sympodialis* together with *M. pachydermatis* in veterinary specimens. Mycoses 41:493–500

Ran Y, Yoshike T, Ogawa H (1993) Lipase of *Malassezia furfur*: Some properties and their relationship to cell growth. J Med Vet Mycol 31:77–85

Rasool O, Zargari A, Almqvist J, Eshaghi H, Whitley P, Scheynius A (2000) Cloning, characterization and expression of complete coding sequences of three IgE binding *Malassezia furfur* allergens, Mal f 7, Mal f 8 and Mal f 9. Eur J Biochem 267:4355–4361

Rhie S, Turcios R, Buckley H, Suh B (2000) Clinical features and treatment of *Malassezia* folliculitis with fluconazole in orthotopic heart transplant recipients. J Heart Lung Transplant 19:215–219

Richardson MD, Shankland GS (1991) Enhanced phagocytosis and intracellular killing of *Pityrosporum ovale* by human neutrophils after exposure to ketoconazole is correlated to changes of the yeast cell surface. Mycoses 34:29–33

Riciputo RM, Oliveri S, Micali G, Sapuppo A (1996) Phospholipase activity in *Malassezia furfur* pathogeneic strains. Mycoses 39:233–235

Rigopoulos D, Paparizos V, Katsambas A (2004) Cutaneous markers of HIV infection. Clin Dermatol 22:487–498

Rincon S, Celis A, Sopo L, Motta A, Cepero de Garcia MC (2005) *Malassezia* yeast species isolated from patients with dermatologic lesions. Biomedica 25:189–195

Ring J, Abeck D, Neuber K (1992) Atopic eczema: role of microorganisms on the skin surface. Allergy 47: 265–269

Rivolta S (1873) In: DiGiulio Speiranai F (ed) Dei parasiti vegetali, 1st edn. Figli, Torino, pp 469–471

Roberts SOB (1969) Pityriasis versicolor: a clinical and mycological investigation. Br J Dermatol 81:315–326

Robin C (1853) Histoire naturelle des vegetaux parasites. Baillière, Paris

Robins EJ, Breathnach AS, Bennet D, Ward BJ, Bhasin Y, Ethridge L, Nazzaro-Porro M, Passi S, Picardo M (1985) Ultrastructural observations on the effect of azelaic acid on normal human melanocytes and human melanoma cell line in tissue culture. Br J Dermatol 113:687–697

Rokugo M, Tagami H, Usuba Y, Tomita Y (1990) Contact sensitivity to *Pityrosporum ovale* in patients with atopic dermatitis. Arch Dermatol 126:627–632

Roll A, Cozzio A, Fischer B, Schmid-Grendelmeier P (2004) Microbial colonization and atopic dermatitis. Curr Opin Allergy Clin Immunol 4:373–378

Rosenberg EW, Belew PW (1982) Improvement of psoriasis of the scalp with ketoconazole. Arch Dermatol 118:370–371

Rosenberg EW, Belew PW, Bale G (1980) Effect of topical applications of heavy suspensions of killed *Malassezia ovalis* on rabbit skin. Mycopathologia 72:147–154

Rosenberg EW, Noah PW, Skinner RB Jr, Vander Zwaag R, West SK, Browder JF (1989) Microbial associations of 167 patients with psoriasis. Acta Derm Venereol Suppl (Stockh) 146:72–74

Ruete AE (1933) Zur Frage der depigmentierenden Pityriasis versicolor. Dermatol Wochenschr 96:333–336

Saadatzadeh MR, Ashbee HR, Cunliffe WJ, Ingham E (2001) Cell-mediated immunity to the mycelial phase of *Malassezia* spp. in patients with pityriasis versicolor and controls. Br J Dermatol 144:77–84

Sabouraud R (1904) Maladies du cuir chevelu: II les maladies desquamatives. Masson, Paris

Salah SB, Makni F, Marrakchi S, Sellami H, Cheikhrouhou F, Bouassida S, Zahaf A, Ayadi A (2005) Identification of Malassezia species from Tunisian patients with pityriasis versicolor and normal subjects. Mycoses 48:242–245

Sandström Falk MH, Tengvall-Linder M, Johansson C, Bartosik J, Back O, Särnhult T, Wahlgren CF, Scheynius A, Faergemann J (2005) The prevalence of *Malassezia* yeasts in patients with atopic dermatitis, seborrhoeic dermatitis and healthy controls. Acta Derm Venereol 85:17–23

Savolainen J, Broberg A (1992) Crossreacting IgE antibodies to *Pityrosporum ovale* and *Candida albicans* in atopic children. Clin Exp Allergy 22:469–474

Savolainen J, Lammintausta K, Kalimo K, Viander M (1993) *Candida albicans* and atopic dermatitis. Clin Exp Allergy 23:332–339

Savolainen J, Kosonen J, Lintu P, Viander M, Pene J, Kalimo K, Terho EO (1999) *Candida albicans* mannan- and protein-induced humoral, cellular and cytokine responses in atopic dermatitis patients. Clin Exp Allergy 29:824–831

Savolainen J, Lintu P, Kosonen J, Kortekangas-Savolainen O, Viander M, Pene J, Kalimo K, Terho EO, Bousquet J (2001) *Pityrosporum* and *Candida* specific and non-specific humoral, cellular and cytokine responses in atopic dermatitis patients. Clin Exp Allergy 31:125–134

Scheynius A, Johansson C, Buentke E, Zargari A, Tengvall Linder M (2002) Atopic eczema/dermatitis syndrome and *Malassezia*. Int Arch Allergy Immunol 127:161–169

Schmid-Grendelmeier P, Flückiger S, Disch R, Trautmann A, Wuthrich B, Blaser K, Scheynius A, Crameri R (2005) IgE-mediated and T cell-mediated autoimmunity against manganese superoxide dismutase in atopic dermatitis. J Allergy Clin Immunol 115:1068–1075

Schmid-Grendelmeier P, Scheynius A, Crameri R (2006) The role of sensitization to *Malassezia sympodialis* in atopic eczema. Chem Immunol Allergy 91:98–109

Schmidt M, Zargari A, Holt P, Lindbom L, Hellman U, Whitley P, Ploeg I van der, Harfast B, Scheynius A (1997) The complete cDNA sequence and expression of the first major allergenic protein of *Malassezia furfur*, Mal f 1. Eur J Biochem 246:181–185

Schrenker T, Kalveram KJ, Hornstein OP, Hauck HD, Baurle G (1989) Specific IgE antibodies to *Pityrosporum orbiculare* in patients with atopic dermatitis. Z Hautkr 64:478–479

Shifrine M, Marr AG (1963) The requirement of fatty acids by *Pityrosporum ovale*. J Gen Microbiol 32:263–270

Silva V, Fischman O, de Camargo ZP (1997) Humoral immune response to *Malassezia furfur* in patients with pityriasis versicolor and seborrhoeic dermatitis. Mycopathologia 139:79–85

Simmons RB, Ahearn DG (1987) Cell wall ultrastructure and diazonium blue B reaction of *Sporopachydermia quercum*, *Bullera tsugae*, and *Malassezia* spp. Mycologia 79:38–43

Simmons RB, Guého E (1990) A new species of *Malassezia*. Mycol Res 94:1146–1149

Slooff WC (1970) Genus *Pityrosporum* Sabouraud. In: Lodder J (ed) The yeasts – a taxonomic study, 2nd edn. North-Holland, Amsterdam, pp 1167–1186

Smith KJ, Skelton HG, Yeager J, Ledsky R, McCarthy W, Baxter D, Turiansky GW, Wagner KF, Turianski G (1994) Cutaneous findings in HIV-1-positive patients: A 42-month prospective study. Military Medical Consortium for the Advancement of Retroviral Research (MMCARR). J Am Acad Dermatol 31:746–754

Sohnle PG, Collins-Lech C (1978) Cell-mediated immunity to *Pityrosporum orbiculare* in tinea versicolor. J Clin Invest 62:45–53

Sohnle PG, Collins-Lech C (1980) Relative antigenicity of *P. orbiculare* and *C. albicans*. J Invest Dermatol 75:279–283

Sohnle PG, Collins-Lech C (1982) Analysis of the lymphocyte transformation response to *Pityrosporum orbiculare* in patients with tinea versicolor. Clin Exp Immunol 49:559–564

Sohnle PG, Collins-Lech C (1983) Activation of complement by *Pityrosporum orbiculare*. J Invest Dermatol 80:93–97

Sohnle PG, Collins-Lech C, Huhta KE (1983) Class-specific antibodies in young and aged humans against organisms producing superficial fungal infections. Br J Dermatol 108:69–76

Squiquera L, Galimberti R, Morelli L, Plotkin L, Milicich R, Kowalckzuk A, Leoni J (1994) Antibodies to proteins from *Pityrosporum ovale* in the sera from patients with psoriasis. Clin Exp Dermatol 19:289–293

Sternberg TH, Keddie FM (1961) Immunofluorescence studies in tinea versicolor. Arch Dermatol 84:161–165

Sugita T, Suto H, Unno T, Tsuboi R, Ogawa H, Shinoda T, Nishikawa A (2001) Molecular analysis of Malassezia microflora on the skin of atopic dermatitis patients and healthy subjects. J Clin Microbiol 39:3486–3490

Sugita T, Takashima M, Shinoda T, Suto H, Unno T, Tsuboi R, Ogawa H, Nishikawa A (2002) New yeast species, *Malassezia dermatis*, isolated from patients with atopic dermatitis. J Clin Microbiol 40:1363–1367

Sugita T, Saito M, Ito T, Kato Y, Tsuboi R, Takeuchi S, Nishikawa A (2003a) The basidiomycetous yeasts *Cryptococcus diffluens* and *C. liquefaciens* colonize the skin of patients with atopic dermatitis. Microbiol Immunol 47:945–950

Sugita T, Takashima M, Kodama M, Tsuboi R, Nishikawa A (2003b) Description of a new yeast species, *Malassezia japonica*, and its detection in patients with atopic dermatitis and healthy subjects. J Clin Microbiol 41:4695–4699

Sugita T, Kodama M, Saito M, Ito T, Kato Y, Tsuboi R, Nishikawa A (2003c) Sequence diversity of the intergenic spacer region of the rRNA of *Malassezia globosa* colonizing the skin of patients with atopic dermatitis and healthy individuals. J Clin Microbiol 43: 3022–3027

Sugita T, Tajima M, Takashima M, Amaya M, Saito M, Tsuboi R, Nishikawa A (2004) A new yeast, *M. yamatoensis*, isolated from a patient with seborrhoic dermatitis, and its distribution in patients and healthy subjects. Microbiol Immunol 48: 579–583

Sugita T, Tajima M, Tsubuku H, Tsuboi R, Nishikawa A (2006) Quantitative analysis of cutaneous *malassezia* in atopic dermatitis patients using real-time PCR. Microbiol Immunol. 50:549–552

Suzuki T, Ohno N, Ohshima Y, Yadomae T (1998) Soluble mannan and beta-glucan inhibit the uptake of *Malassezia furfur* by human monocytic cell line, THP-1. FEMS Immunol Med Microbiol 21:223–230

Suzuki T, Tsuzuki A, Ohno N, Ohshima Y, Yadomae T (2000) Enhancement of IL-8 production from human monocytic and granulocytic cell lines, THP-1 and HL-60, stimulated with *Malassezia furfur*. FEMS Immunol Med Microbiol 28:157–162

Svejgaard E, Faergemann J, Jemec G, Kieffer M, Ottevanger V (1989) Recent investigations on the relationship between fungal skin diseases and atopic dermatitis. Acta Derm Venereol Suppl (Stockh) 144:140–142

Swift JA, Dunbar SF (1965) The ultrastructure of *Pityrosporum ovale* and *Pityrosporum canis*. Nature 206:1174–1175

Tajima M (2005) *Malassezia* species in patients with seborrheic dermatitis and atopic dermatitis. Nippon Ishinkin Gakkai Zasshi 46:163–167

Takahashi M, Ushijima T, Ozaki Y (1984) Biological activity of *Pityrosporum*. I. Enhancement of resistance in mice stimulated by *Pityrosporum* against *Salmonella typhimurium*. Immunology 51:697–702

Takahashi M, Ushijima T, Ozaki Y (1986) Biological activity of *Pityrosporum*. II. Antitumor and immune stimulating effect of *Pityrosporum* in mice. J Natl Cancer Inst 77:1093–1097

Tanaka M, Imamura S (1979) Immunological studies on *Pityrosporum* genus and *Malassezia furfur*. J Invest Dermatol 73:321–324

Tanaka M, Aiba S, Matsumura N, Aoyama H, Tabata N, Sekita Y, Takami H (1994) IgE-mediated hypersensitivity and contact sensitivity to multiple environmental allergens in atopic dermatitis. Arch Dermatol 130:1393–1401

Tarazooie B, Kordbacheh P, Zaini F, Zomorodian K, Saadat F, Zeraati H, Hallaji Z, Rezaie S (2004) Study of the distribution of *Malassezia* species in patients with pityriasis versicolor and healthy individuals in Tehran, Iran. BMC Dermatol 1:4–5

Tatnall FM, Rycroft RJ (1985) *Pityriasis versicolor* with cutaneous atrophy induced by topical steroid application. Clin Exp Dermatol 10:258–261

Tengvall Linder M, Johansson C, Zargari A, Bengtsson A, van der Ploeg I, Jones I, Harfast B, Scheynius A (1996) Detection of *Pityrosporum orbiculare* reactive T cells from skin and blood in atopic dermatitis and characterization of their cytokine profiles. Clin Exp Allergy 26:1286–1297

Tengvall Linder M, Johansson C, Bengtsson A, Holm L, Harfast B, Scheynius A (1998) *Pityrosporum orbiculare*-reactive T-cell lines in atopic dermatitis patients and healthy individuals. Scand J Immunol 47:152–158

Tengvall Linder M, Johansson C, Scheynius A, Wahlgren CF (2000) Positive atopy patch test reactions to *Pityrosporum orbiculare* in atopic dermatitis patients. Clin Exp Allergy 30:122–131

Theelen B, Silvestri M, Guého E, van Belkum A, Boekhout T (2001) Identification and typing of *Malassezia* yeasts using amplified fragment length polymorphism (AFLP^{Tm}), random amplified polymorphic DNA (RAPD) and denaturing gradient gel electrophoresis (DGGE). FEMS Yeast Res 1:79–86

Thoma W, Trinkaus M, Mayser P (2000a) Identifikation eines Tyrosinase-Inhibitiors im Stoffwechsel von *Malassezia furfur*. Mycoses 43:264

Thoma W, Trinkaus M, Mayser P (2000b) Tyrosinaseinhibition durch Stoffwechselprodukte von *M. furfur*. Haut Geschlechtskrankh 75:473

Thoma W, Dahms K, Krämer HJ, Steglich W, Irlinger B, Mayser P (2001) Tyrosinase-Inhibition durch KO27- einem Stoffwechselmetaboliten von *Malassezia furfur*. Mycoses 44:238

Thoma W, Krämer HJ, Mayser P (2005) Pityriasis versicolor alba. JEADV 19:147–152

Thompson E, Colvin JR (1970) Composition of the cell wall of *Pityrosporum ovale* (Bizzozero) Castellani and Chalmers. Can J Microbiol 16:263–265

Tlaskalova-Hogenova H, Stepankova R, Hudcovic T, Tuckova L, Cukrowska B, Lodinova-Zadnikova R, Kozakova H, Rossmann P, Bartova J, Sokol D, Funda DP, Borovska D, Rehakova Z, Sinkora J, Hofman J, Drastich P, Kokesova A (2004) Commensal bacteria (normal microflora), mucosal immunity and chronic inflammatory and autoimmune diseases. Immunol Lett 93:97–108

Unna PG (1894) Natürliche Reinkulturen der Oberhautpilze. Mh Prakt Derm 18:257–261

Van Abbe NJ (1964) The investigation of dandruff. J Soc Cosmet Chem 15:609–630

Van Belkum A, Boekhout T, Bosboom R (1994) Monitoring spread of *Malassezia* infections in a neonatal intensive care unit by PCR-mediated genetic typing. J Clin Microbiol 32:2528–2532

Van Cutsem J, Van Gerven F, Fransen J, Schrooten P, Janssen PA (1990) The in vitro antifungal activity of ketoconazole, zinc pyrithione, and selenium sulfide against *Pityrosporum* and their efficacy as a shampoo in the treatment of experimental pityrosporosis in guinea pigs. J Am Acad Dermatol 22:993–998

Vecchiarelli A (2000) Immunoregulation by capsular components of *Cryptococcus neoformans*. Med Mycol 38:407–417

Vuillemin P (1931) Les champignon parasites et les mycoses de l'homme. Masson, Paris

Waersted A, Hjorth N (1985) *Pityrosporum orbiculare*: a pathogenetic factor in atopic dermatitis of the face, scalp and neck? Acta Derm Venereol Suppl (Stockh) 114:146–148

Walters CE, Ingham E, Eady EA, Cove JH, Kearney JN, Cunliffe WJ (1995) In vitro modulation of keratinocyte-derived interleukin-1 alpha (IL-1 alpha) and peripheral blood mononuclear cell-derived IL-1 beta release in response to cutaneous commensal microorganisms. Infect Immun 63:1223–1228

Watanabe S, Kano R, Sato H, Nakamura Y, Hasegawa A (2001) The effects of *Malassezia* yeasts on cytokine production by human keratinocytes. J Invest Dermatol 116:769–773

Weary PE (1970) Comedogenic potential of the lipid extract of *Pityrosporum ovale*. Arch Dermatol 102:84–91

Weidman FD (1925) Exfoliative dermatitis in the Indian rhinoceros (*Rhinoceros unicornis*), with description of a new species: *Pityrosporum pachydermatis*. Rep Lab Mus Comp Zoo Soc Philadelphia 36–43

Welbel SF, McNeil MM, Pramanik A, Silberman R, Oberle AD, Midgley G, Crow S, Jarvis WR (1994) Nosocomial *Malassezia pachydermatis* bloodstream infections in a neonatal intensive care unit. Pediatr Infect Dis J 13:104–108

Wessels MV, Doekes G, Van Ieperen-Van Dijk AG, Koers WJ, Young E (1991) IgE antibodies to *Pityrosporum ovale* in atopic dermatitis. Br J Dermatol 125:227–232

White SD, Vandenabeele SI, Drazenovich NL, Foley JE (2006) *Malassezia* species isolated from the intermammary and preputial fossa areas of horses. J Vet Intern Med 20:395–398

Wilde PF, Stewart PS (1968) A study of the fatty acid metabolism of the yeast *Pityrosporum ovale*. Biochem J 108:225–231

Wille G, Mayser P, Thoma W, Monsees T, Baumgart A, Schmitz HJ, Schrenk D, Polborn K, Steglich W (2001) Malassezin – a novel agonist of the arylhydrocarbon receptor from the yeast *Malassezia furfur*. Bioorg Med Chem 9:955–960

Wollenberg A, Kraft S, Hanau D, Bieber T (1996) Immunomorphological and ultrastructural characterization of Langerhans cells and a novel, inflammatory dendritic epidermal cell (IDEC) population in lesional skin of atopic eczema. J Invest Dermatol 106:446–453

Wollenberg A, Wagner M, Gunther S, Towarowski A, Tuma E, Moderer M, Rothenfusser S, Wetzel S, Endres S, Hartmann G (2002) Plasmacytoid dendritic cells: a new cutaneous dendritic cell subset with distinct role in inflammatory skin diseases. J Invest Dermatol 119:1096–1102

Wroblewski N, Bär S, Mayser P (2005) Missing granulocytic infiltrate in pityriasis versicolor – indication of specific anti-inflammatory activity of the pathogen? Mycoses 48[Suppl 1]: 66–71

Yarrow D, Ahearn DG (1984) *Malassezia* Baillon. In: Kreger van Rij NJW (ed) The yeasts: a taxonomic study, 3rd edn. North Holland, Amsterdam, pp 882–885

Yasueda H, Hashida-Okado T, Saito A, Uchida K, Kuroda M, Onishi Y, Takahashi K, Yamaguchi H, Takesako K, Akiyama K (1998) Identification and cloning of two novel allergens from the lipophilic yeast, *Malassezia furfur*. Biochem Biophys Res Commun 248:240–244

Young E, Koers WJ, Berrens L (1989) Intracutaneous tests with *Pityrosporum* extract in atopic dermatitis. Acta Derm Venereol Suppl (Stockh) 144:122–124

Zargari A, Harfast B, Johansson SGO, Scheynius A (1994) Identification of allergen components of the opportunistic yeast *Pityrosporum orbiculare* by monoclonal antibodies. Allergy 49:50–56

Zargari A, Doekes G, Van Ieperen-Van Dijk AG, Landberg E, Harfast B, Scheynius A (1995) Influence of culture period on the allergenic composition of *Pityrosporum* extracts. Clin Exp Allergy 25:1235–1245

Zargari A, Eshaghi H, Back O, Johansson S, Scheynius A (2001) Serum IgE reactivity to *Malassezia furfur* extract and recombinant *M. furfur* allergens in patients with atopic dermatitis. Acta Derm Venereol 81:418–422

Zargari A, Midgley G, Back O, Johansson S, Scheynius A (2003) IgE-reactivity to seven *Malassezia* species. Allergy 58:306–311

Techniques

8 Proteomics and its Application to the Human-Pathogenic Fungi *Aspergillus fumigatus* and *Candida albicans*

OLAF KNIEMEYER[1], AXEL A. BRAKHAGE[1]

CONTENTS

I. Introduction..........................	155
II. Methods of Proteome Analysis..........	157
III. 2D Gel Electrophoresis.................	157
A. Evolution of the Technology.........	157
B. Sample Preparation.................	158
C. First Dimension: Isoelectric Focusing.........................	159
D. Second Dimension: SDS-PAGE.......	160
E. Detection of Proteins...............	160
F. Scanning and Image Analysis........	161
G. Tryptic Digestion and Mass Spectrometry Identification..........	161
H. Protein Identification by Mass Spectrometry......................	161
J. Peptide Mass Fingerprint and Tandem-Mass Spectrometry.........	163
IV. Gel-Independent Techniques............	163
A. Liquid Chromatography-Based MS Techniques.....................	163
B. Quantitative Proteomics Using LC-MS.......................	164
V. Sample Fractionation to Identify Low-Abundant Proteins................	165
A. Subcellular Fractionation............	165
B. Membrane Proteomics..............	165
C. Phosphoproteomics.................	166
1. Detection of Phosphoproteins with 2D-PAGE...................	166
2. Enrichment of Phosphoproteins....	167
3. Chromatographic Methods for Phosphoproteomics...........	167
VI. Standard Development.................	167
VII. Proteomics of Human-Pathogenic Fungi...............................	168
A. Sample Preparation.................	168
B. Proteomics Studies of Fungal Physiology........................	168
C. Drug-Induced Changes..............	171
D. Biofilm Formation.................	171
E. Proteomics of the Cell Wall..........	172
F. Immunoproteomics.................	173
G. Secretome.......................	175
H. Virulence Factors..................	175
J. Host–Pathogen Interaction..........	176
VIII. Conclusion: Towards Integrative Analysis–Systems Biology..............	177
References...........................	177

I. Introduction

The number of opportunistic fungal infections has increased significantly during the past decades, at least in part as the result of a rising number of immunocompromised patients. Individuals at risk for the development of a serious fungal infection include patients undergoing solid-organ, blood and bone marrow transplantation, cancer patients, patients of the acquired immunodeficiency syndrome (AIDS) and other patients receiving immunosuppressive treatment (Pfaller and Diekema 2004; Brakhage 2005). Today, invasive fungal infections are among the most challenging problems in haematology, oncology and intensive care medicine (Vandewoude et al. 2006). Among the approximately 140 000 known fungal species only a few cause human infections (Richardson 2003). The most predominant pathogens are the yeast *Candida albicans* and the filamentous fungus *Aspergillus fumigatus*, but also other fungal pathogens frequently cause systemic infections, such as the yeast species *C. glabrata, C. krusei, C. tropicalis, Cryptococcus* and *Trichosporon*, filamentous fungi such as *Aspergillus, Fusarium, Rhizopus* and *Mucor*, and dematiaceous hyphomycetes (Richardson 2005).

The dimorphic yeast *Candida albicans* is a commensal organism living on the mucosal surfaces of the gastrointestinal and urogenital tract, usually without any symptoms. Under certain circumstances, in particular due to high dose immunosuppressive therapy or intravascular catheters, *C. albicans* can turn into an opportunistic pathogen that causes severe bloodstream infections. Also less severe mucocutaneous infections such as oral thrush or vaginitis can occur, which are often

---

[1]Department of Molecular and Applied Microbiology, Leibniz Institute for Natural Product Research and Infection Biology–Hans Knöll Institute, Beutenbergstrasse 11a, 07745 Jena; and Friedrich Schiller University, Jena, Germany; e-mail: Olaf.Kniemeyer@hki-jena.de

caused by broad-spectrum antibiotic therapy. *C. albicans* blood infections have steadily increased and account for 8–15% of all hospital-acquired blood infections today (Richardson 2005). More and more patients with severe illness, but not immunosuppressed in the classic sense (e.g. no neutropenia, corticosteroid therapy, HIV infection, etc.), contract invasive *C. albicans* infections (for a review, see Perlroth et al. 2007). In comparison to *C. albicans*, the filamentous fungus *A. fumigatus* colonises a completely different ecological niche. It is a common saprophyte decaying organic and plant material (Tekaia and Latgé 2005). The ubiquitously distributed conidia of this fungus can be easily inhaled and they are small enough to reach the alveoli in the lung. If the conidia can overcome the immune defence of the host, they germinate and produce a mycelium that invades the lung tissues and disseminates hematogenously. The incidence of invasive aspergillosis has increased tremendously. Data from the Centers for Disease Control and Prevention indicate that the mortality associated with invasive aspergillosis has increased 357% since 1980 (Marr et al. 2002). Besides invasive aspergillosis, *A. fumigatus* can also cause diseases in immunocompetent individuals, such as allergies or a locally restricted colonisation of the lungs, a disease called aspergilloma (Brakhage 2005).

Despite the recently intensified research, fungal mycosis is still associated with a high mortality due to the lack of a sensitive, specific, quick and unambiguous diagnosis (Yeo and Wong 2002), a limited number of effective antifungal agents (Odds et al. 2003) and an increasing appearance of resistance to antifungal drugs (Kontoyiannis and Lewis 2002). For this reason there is a strong need to develop new diagnostic tools, to discover new therapeutic agents and to better understand the biology and the pathogenicity of *C. albicans* and *A. fumigatus*.

Both fungi lack a sexual cycle, but molecular biological tools have been established to characterise genes that are involved and are essential for the infection process (Brakhage and Langfelder 2002; Bader et al. 2006; da Silva Ferreira et al. 2006a; de Backer 2000b; Krappmann et. al 2006). For *C. albicans* several virulence determinants have been postulated, e.g. the ability to switch from a yeast form to a hyphal form, the ability to adhere to host tissues and the secretion of hydrolytic enzymes (for reviews, see Navarro-García et al. 2001; Sundstrom 2002; Hube 2006; Odds et al. 2006; Sundstrom 2006).

Until now, the iron acquisition system (Schrettl et al. 2004) and the dihydroxynaphtalene (DHN) melanin biosynthesis pathway (Langfelder et al. 1998) represent clear virulence determinants, which have been found in *A. fumigatus* (for reviews, see Latgé 2001; Brakhage 2005; Rementeria et al. 2005; Rhodes and Brakhage 2006). Nevertheless, the virulence of *C. albicans* and *A. fumigatus* does not seem to depend on a single gene. Instead, fungal virulence is a multi-factorial, complex process that requires the expression of many genes at different stages of the infection process and at different sites of infection (Odds et al. 2001).

The release of the genome sequences of *C. albicans* and *A. fumigatus* has been of great benefit for a more detailed insight into the evolution and pathogenesis of these fungi. The genome of *C. albicans* was one of the first of a human-pathogenic fungus to be sequenced. The diploid, heterozygous set of chromosomes made a sequence assembly difficult, since two alleles of many genes differ quite significantly, but the diploid genome sequence of strain SC5314 was published in 2004 (Jones et al. 2004) and, recently, an improved annotation was made publicly available (Braun et al. 2005). The *C. albicans* genome consists of eight chromosomes with 6354 genes, including 819 conserved hypothetical and 1218 hypothetical proteins with a total size of 14.88 Mb.

In contrast to *C. albicans*, *A. fumigatus* is haploid. The genome of the clinical isolate *A. fumigatus* Af293 consists of eight chromosomes with a total size of 29.2 Mb. This nuclear genome contains 9922 identified protein-encoding sequences of which about one-third could not be assigned to a gene ontology function. In comparison to the genomes of *A. nidulans* and *A. oryzae*, the pathogenic *A. fumigatus* possesses about 500 unique genes (Galagan et al. 2005).

With the completion and publication of the genomes of *C. albicans* and *A. fumigatus* (Jones et al. 2004; Nierman et al. 2005) it is feasible to gain a more detailed insight into the biology of both pathogenic fungi. Complete-genome comparative analyses (Hsiang and Baillie 2006) and global studies at the transcript and protein level have been made possible.

A variety of microarray studies have been carried out for *C. albicans* (for a review, see Magee and Magee 2005). In contrast to *C. albicans*, microarray experiments to study the gene expression of *A. fumigatus* are in their infancy. Nierman et al

(2005) studied the heat-shock response of the highly thermotolerant *A. fumigatus* and could show that the high-temperature response differ from the general stress response in yeast. Da Silva Ferreira et al. (2006b) analysed the response of *A. fumigatus* to voriconazole at the transcriptome level.

Nevertheless, to gain a global insight into the biology of development, stress response and pathogenicity of *C. albicans* and *A. fumigatus*, studies at the proteomic level are needed. Proteins are the active agents within the cell and proteomic techniques offer a global and integrated view of the entire proteins encoded by a genome. Only studies at the protein level allow the detection of protein isoforms and post-translational modifications. In addition, there is often a low correlation between the amount of transcript of a gene and the amount of its protein product (Griffin et al. 2002). Therefore, integrative studies at both levels are most desirable.

This chapter outlines a general overview of the two methods used for proteomic studies: 2D gel electrophoresis and liquid chromatography coupled to mass spectrometry, putting the main emphasis on gel-based proteomics. Since the number of publications about proteomics is tremendous and not all aspects can be discussed, the reader is referred in some cases to excellent reviews. Subsequently, recent advances in the proteomic analysis of *C. albicans* and *A. fumigatus* are discussed. Until now, there are only a few studies for *A. fumigatus*, but a lot of knowledge has already been gained for *C. albicans* and proteomic investigations were reviewed by Niimi et al. (1999), Pitarch et al. (2003, 2006b, c, 2007), Rupp (2004) and Thomas et al. (2006b).

II. Methods of Proteome Analysis

The word proteomics was invented in the mid-1990s and is defined as the study of the "protein complement of the genome" (Wilkins et al. 1995). Proteomic studies were initially just represented by quantitative 2D–polyacrylamide gel electrophoresis (2D-PAGE) with a subsequent identification of the proteins by Edman degradation or mass spectrometry analysis. Since then, a lot of "gel-free" methods have been established based on coupling of liquid chromatography with mass spectrometry. The term proteomics now covers all kind of studies on global changes of the protein pattern within a cell or tissue (expression proteomics) and surveys of protein complexes and its structures and functions (functional proteomics). Today, approaches like protein arrays, genome wide yeast two-hybrid approaches, co-immunoprecipitation assays, GFP-tagged gene libraries, NMR spectroscopy as well as X-ray crystallography are included, but they are not covered here (for reviews, see Phizicky et al. 2003; Zhu et al. 2003; de Hoog and Mann 2004).

All scientists working in the field of proteomics are confronted with one difficulty, which is the high diversity of proteins due to post-translational modifications, splice variations and the high dynamic range of proteins in biological systems. The dynamic range for eukaryotic cells is estimated to be in the range of 10^5–10^6 for eukaryotic cells and 10^{10} for serum (Patterson and Aebersold 2003). Many proteins are of low abundance, e.g. 50% of the protein content of yeast is the product of 100 genes and the rest of the proteins are just present at a low abundance ranging over 50–10 000 copies (Futcher et al. 1999). Unlike for DNA-based methods, the amount of starting material cannot be amplified for proteomic studies. Although a lot of progress has been made in the field of proteomics, no method is currently available that enables the detection of all protein components of a biological system (Patterson 2004).

III. 2D Gel Electrophoresis

A. Evolution of the Technology

Independently of each other O'Farrell (1975) and Klose (1975) invented 2D-PAGE in the mid-1970s. This technique is based on the separation of proteins according to their isoelectric point in the first dimension and their molecular mass in the second dimension. During the first separation step, i.e. isoelectric focusing (IEF), proteins migrate through a pH-gradient embedded in an IEF strip or tube gel by the application of an electric field. The proteins migrate in the electric field until they reach the pH at which their net charge is zero. Then they stop and focus. This procedure results in a banding pattern; and for a total cell lysate about 100 bands can be distinguished. The IEF strip or the tube gel is then transferred to the top of a large SDS-polyacrylamide slab gel and proteins are separated in the second dimension based on molecular mass differences according to the method of Laemmli (1970). By applying the combination of the two methods IEF and SDS-polyacrylamide gel electrophoresis 1000–2500 different proteins spots can be separated, corresponding to circa 300–1000 different proteins (Fig. 8.1). This method also allows the estimation of the pI and the M_r of each separated polypeptide.

B. Sample Preparation

Sample preparation for 2D-PAGE is the key step for obtaining good and reproducible results. The methods should be kept as simple as possible to ease reproducibility and to avoid protein loss. The proteins have to be extracted from the cells, denaturated and solubilised; and contaminants have to be removed.

Organisms with cell walls, like yeast or fungal cells, are often disrupted mechanically. Therefore, they are ground with mortar and pestle, but also French pressure cells and mechanical homogenisers are applied for yeast cells (Westermeyer and Naven 2002). Detergent lysis with SDS has been described as well (Harder et al. 1999; Blomberg 2002).

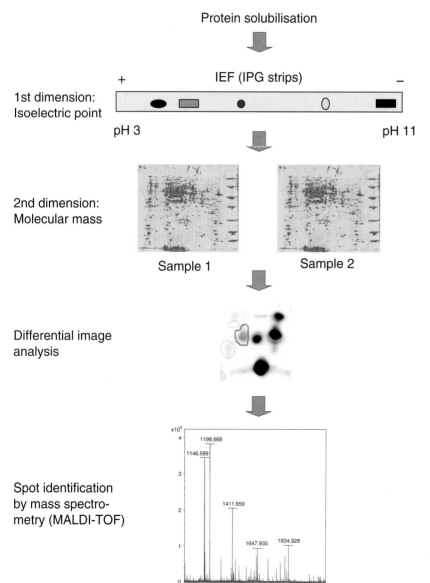

Fig. 8.1. Workflow for two-dimensional gel electrophoresis

A protein crude extract can be contaminated with salt ions, nucleic acid, lipids, polysaccharides and pigments, leading to disturbance in the spot pattern of the 2D gels. Nucleic acids can be removed with DNAse and RNAse treatment, but very often protein precipitation methods are applied to remove interfering substances. These methods often result in a better separation of protein spots. However, there is a risk of losing some proteins(Wessel and Flügge 1984; Damerval et al. 1986; Wang et al. 2003; Isaacson et al. 2006). There are also several commercial kits on the market for the removal of contaminating compounds, e.g. from GE Healthcare, Bio-Rad, Sigma-Aldrich, etc. (Stasyk et al. 2001).

After cell disruption or removal of contaminants the proteins are extracted with a "lysis buffer", which is often composed of high concentrations of a chaotrope, mostly urea (8–9 M), a detergent without net charge like CHAPS, a reducing agent like DTT, carrier ampholytes and traces of bromophenol blue. The lysis buffer denaturates, disaggregates, reduces and solubilises the proteins. Carrier ampholytes used in a concentration of 0.5–2.0% (v/v) are generating the pH gradient during the IEF. In addition, they improve the solubility of proteins and they can scavenge cyanate ions, which are produced during the decay of urea (Görg et al. 2004).

Before the IEF can be carried out, the amount of protein has to be quantified. Standard protein assays are often used. The Bradford assay (Bradford 1976) is less sensitive to additives such as urea, detergents and ampholytes than the BCA or Lowry assays. Nevertheless, low-dilution protein samples can be critical; and protein quantification kits are commercially available which are based on the precipitation of the protein, resolubilisation and subsequent detection (e.g. GE Healthcare, SERVA). A "one for all" sample preparation procedure does not exist yet, so cell breakage, lysis buffers, etc. have to be empirically optimised.

More comprehensive descriptions of sample preparations for 2D-PAGE can be found in Rabilloud (1996), Shaw and Riederer (2003), Görg et al. (2004), or are available from the web pages of laboratory suppliers like GE Healthcare (www.gehealthcare.com) or Bio-Rad (www.bio-rad.com).

C. First Dimension: Isoelectric Focusing

IEF is performed in a pH gradient. The techniques described by O' Farrel and Klose (1975) were based on carrier ampholytes, which are a heterogeneous mixture of chemically synthesised isomers of aliphatic oligoamino-oligocarboxylic acids. The carrier ampholytes possess a high buffering capacity and solubility at their isoelectric point. When an electric field is applied, the carrier ampholytes align themselves in between the anode and cathode according to their pI. In this way, the ampholytes form a pH gradient (Westemeyer and Naven 2002). The IEF gels were initially prepared by casting a mixture of acrylamide/bis-acrylamide solution and carrier ampholyte into glass tubes. This technique was developed further by Anderson and Anderson (1978b; Iso-Dalt system). Also Klose and co-workers revised and improved their methods. Now, large, high-resolution gels (46×30 cm) can be cast with a separation capacity of up to 10 000 polypeptide spots (Klose and Kobalz 1995). The non-equilibrium pH gradient electrophoresis (NEPHGE) technique offers an excellent resolution capacity and a relatively good separation of hydrophobic proteins (Wittmann-Liebold et al. 2006). But, with these techniques it is often difficult to obtain reproducible results for the less experienced researcher, due to pH gradient instability and batch-to-batch variation of the chemically synthesised carrier ampholytes. These difficulties were overcome with the establishment of immobilised pH gradients (Bjellqvist et al. 1982).

The pH gradients in these gels are produced by co-polymerisation of acrylamide monomers with acrylamide derivatives containing carboxylic and tertiary amino groups. Because the buffering groups are covalently linked to the acrylamide matrix, the pH gradient cannot drift. The immobilised pH gradient gels cast onto GelBond PAGfilm can be self-made in the laboratory (Görg et al. 2000), but easier to use are the commercially available IPG dry strips (GE Healthcare, Bio-Rad, Sigma-Aldrich, Serva). Before use, the dry IPG strips have to be rehydrated overnight with a rehydration buffer. Several types of dry IPG strips with different lengths are available (7, 11, 13, 18, 24 cm). Linear or non-linear IPG strips with a wide pH range (e.g. pH 3–11) are most suitable to obtain a global overview, whereas medium (e.g. pH 4–7, pH 6–9) and ultra-narrow IPG strips (pH 4.5–5.5) allow the application of higher protein loads and give better resolutions. By the application of overlapping narrow IPG strips covering one pH unit, Wildgruber et al. (2000) could reveal a total of 2286 yeast protein spots compared to 755 protein spots in a pH 3–10 gradient.

In principle, there are two ways of applying protein samples for IEF. IPG dry strips are either rehydrated with the sample, which has been mixed before with rehydration buffer, or the sample is applied during the IEF via cup-loading. The available temperature-controlled, integrated power supplies (BioRad, GE Healthcare, NextGen Sciences, Syngene etc.) for IEF can generate voltages of up to 10 000 V and allow a fast and relatively reproducible separation. After the IEF step, the IPG strips have to be equilibrated to allow the separated proteins to fully interact with SDS and to reduce and alkylate sulfhydryl groups (Görg et al. 2004). Subsequently the IPG strips are ready for the SDS-PAGE dimension.

D. Second Dimension: SDS-PAGE

SDS-PAGE can be performed on horizontal or vertical gel systems using a Tris-glycine running buffer based on the technique described by Laemmli (1970). In most cases the stacking gel is omitted, since the polypeptides within the IPG strips are already concentrated. High-throughput electrophoresis chambers, first developed by Anderson and Anderson (1978a), are available which allow a run of up to 12 gels in parallel. Usually, homogenous gels with a polyacrylamide concentration of 12.5% are used, which allow a good separation of polypeptides with molecular masses in the range of 14–100 kDa. Gradient gels can separate proteins over a wider range, but they are difficult to prepare without batch-to-batch variations. However, proteins with molecular masses below 5 kDa or above 200 kDa cannot be separated by the Laemmli system. For the resolution of proteins smaller than 30 kDa the Tricine-SDS-PAGE system first described by Schägger and Jagow (1987) is the preferred method (Schägger 2006). For a detailed description of the 2D gel electrophoresis technology with IPG strips, see also Görg et al. (2004) or Carrette et al. (2006).

E. Detection of Proteins

After 2D gel electrophoresis the proteins in the SDS gels have to be visualised. Universal detection methods with visible dyes (Zinc imidazol, Commassie blue, Silver staining), fluorescence dyes (Sypro ruby, Cy-dye-labelling) or labelling with radioactive isotopes is possible. All staining methods have their pros and cons. Coomassie brilliant blue is still the most widely used staining technique, since it is cheap, easy to reproduce, visible by eye and compatible to mass spectrometry. Colloidal Coomassie staining with Coomassie blue G 250 (Neuhoff et al. 1988) is the most often applied method for 2D-PAGE (Kang et al. 2002; Candiano et al. 2004; Westermeier 2006).

Silver staining is by far more sensitive than Coomassie staining (detection limit is around 0.1–1.0 ng protein/band). However, this staining technique is not well suited to quantitative analysis, since the linear dynamic range is reported to be low (Smejkal 2004). Furthermore, silver staining is more elaborative, not easy to reproduce and not as well compatible with mass spectrometry (MS) analysis as Coomassie staining (Rabilloud et al. 1988; Klose and Kobalz 1995; Swain and Ross 1995; Shevchenko et al. 1996; Yan et al. 2000; Winkler et al. 2007). Even more confident results in terms of sensitivity are obtained by detection methods based on the radiolabelling of proteins and subsequent detection with X-ray films or by phosphoimager. Isotopic labelling has often been applied for yeast using radioactive labelled amino acids, e.g. [^{35}S] methionine (Blomberg 2002).

Meanwhile fluorescence dyes have become a good alternative to silver staining and Sypro ruby has become one of the most popular fluorescent dyes for 2D-PAGE. This ruthenium metal chelate has a broad linear dynamic range and is compatible with MS analysis (Berggren et al. 2000, 2002). There are also alternatives to stains based on ruthenium metal chelates, such as a compound named Epicoconone from the marine fungus *Epicoccum nigrum* (Bell and Karuso 2003) or the recently introduced Flamingo fluorescence stain from Bio-Rad and fluorescein-derivatives such as C16-F (Kang et al. 2003).

Another approach is the labelling of proteins prior to electrophoresis. This novel experimental design was introduced by Alban et al. (2003) as the DIGE technique (two-dimensional difference gel electrophoresis). Samples are labelled with the spectrally resolvable fluorescent dyes Cy2, Cy3 and Cy5 mixed prior to IEF and separated on the same 2D gel.

The N-hydroxyl succinimidyl ester-derivatives of Cy dyes are covalently linked to lysine residues via an amide linkage. The dye:protein ratio is kept below 3% to ensure that only single lysine residues per polypeptide are labelled. Proteins are visualised using a laser scanner. The big advantage of this so-called multiplexing approach is the reduction of gel-to-gel variations, because two samples are compared in the same gel. For the comparison of several gels a pooled internal standard can be used, resulting in statistically more reliable results (Knowles et al. 2003).

More recently, a very sensitive type of cyanine dyes with the two fluorophores Cy3 and Cy5 has been developed. These dyes contain maleimide reactive groups, which are linked via thioethers to cysteine residues. All cysteine residues within a protein are labelled. This type of labelling is particularly useful in cases where sample amounts are scarce; 10 μg of protein is sufficient (Shaw et al. 2003). For a more comprehensive overview of protein stains, the reader is referred to Lilley and Friedman (2004), Smejkal (2004) and Miller et al. (2006).

F. Scanning and Image Analysis

Gels stained with Coomassie or silver are digitised by employing conventional scanners. For fluorescence dyes, laser-based scanners or charge-coupled device (CCD) camera systems are used. In general, fluorescence scanners are more sensitive and give pictures with a higher resolution, especially when large-format gels are scanned (Miura 2003). The image data are further processed by an image analysis software package. With the help of the software, changes in the abundance of proteins spots from an experimental sample in comparison to a control can be detected. Protein spots of interest can be excised and the protein digested and identified using MS. The general workflow for an image analysis software package is the following: (a) image normalisation and background subtraction, (b) spot detection and quantification, (c) matching of protein spots in different gels, (d) identification of protein spots differing significantly in protein abundance and (e) statistical analysis, data preparation and interpretation (Görg et al. 2004).

Several commercial 2D image analysis software packages are on the market, such as Delta 2-D, Melanie, PDQuest, Phoretix, Progenesis, Z3, Proteomeweaver, DeCyder, Gellab II+ and Dymension. They perform the steps mentioned above by using different spot detection and matching algorithms. Software has improved in recent years and each software package has its advantages and disadvantages. Only a few comparative studies have been published (Raman et al. 2002; Rosengren et al. 2003; Wheelock and Buckpitt 2005). Taken together, all software packages are still far from allowing totally automatic image analysis without the requirement for time-consuming user interventions.

G. Tryptic Digestion and Mass Spectrometry Identification

A protein spot cannot be unambiguously identified just according to its position in the gel. For this purpose, proteins of interest are usually analysed by mass spectrometry (MS). In most cases, spots showing significant changes in abundance after differential image analysis are designated for protein identification (Fig. 8.1). They are excised from the gel and undergo proteolysis to obtain smaller peptide fragments. The most commonly used enzyme for protein cleavage is the serine protease trypsin, but for special applications other enzymes are frequently used, e.g. chymotrypsin, endoproteinase or elastase. Digestion of the proteins is usually carried out in the excised gel plugs. This so-called in-gel digestion was originally introduced by Wilm et al. (1996) and Shevchenko et al. (1996). The method has recently been updated (Shevchenko et al. 2006).

Once extracted from the gel plug, the peptides can be analysed by MS. Depending on the chosen ionisation method, the extracted peptides are either applied directly or they are concentrated and unwanted contaminants are removed (Kussmann et al. 1997).

H. Protein Identification by Mass Spectrometry

MS is nowadays the most powerful tool for protein identification and, in return, the amino acid sequence determination by Edmann degradation has become less important. For the determination of the exact mass, the sample molecules have to be converted into ions in the gas phase, where they can be separated according to their mass:charge ratio (m/z). For the analysis of intact peptides, two soft ionisation techniques are applied which were developed in the late 1980s: (a) electrospray ionisation (ESI; Fenn et al. 1989) and (b) matrix-assisted laser desorption/ionisation (MALDI; Karas and Hillenkamp 1988). The development of these two techniques was a breakthrough for the mass spectrometric analysis of bigger biomolecules (Fig. 8.2). In 2002, the two pioneering scientist in this field,

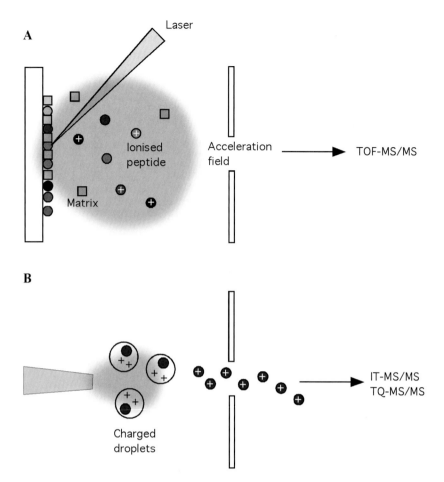

Fig. 8.2. Technical principles of matrix-assisted laser desorption/ionisation (MALDI) and electrospray ionisation (ESI). **A** MALDI. The ionisation is triggered by a laser beam (often nitrogen laser). A matrix supports vaporisation and ionisation. **B** ESI. The peptides are dissolved in a large amount of solvent. Droplets are formed, the solvent evaporates and multiple-charged ions are formed

John B. Fenn and Kochi Tanaka, were honoured with the Nobel Prize (Cho and Normile 2002).

During ESI, peptides dissolved in an acidic solvent are sprayed from a metal syringe needle, which is under the influence of a high electric field (developed by J. Fenn). Fine highly charged droplets are formed, from which the solvent quickly evaporates. The charged ions are then directed into the mass analyser. Typical for ESI is the formation of multiply charged ions, which reduces the observable mass (change in the m/z ratio). These ions are also more readily fragmented during tandem-MS (MS/MS) analysis. Very often ESI is coupled to a chromatography device, allowing the separation of a peptide mixture. The MALDI technique (invented by K. Tamaka and independently by M. Karas and F. Hillenkamp) is based on the principle of peptide ionisation by pulsed laser shots. The extracted peptides are mixed with a light-absorbing compound called the matrix and prepared on a sample plate known as the MALDI target. The peptides form co-crystals with a matrix, e.g. α-cyano-4-hydroxycinnamic acid, on the MALDI target. Short intensive pulses of the laser cause an ionisation of the matrix/peptide co-crystals and MALDI plumes are formed in the vacuum. The ions are then accelerated into the MS analyser (Fig. 8.2). Characteristic for MALDI is the predominant production of singly charged ions (Westermeier and Naven 2002; Aebersold and Mann 2003; Hufnagel and Rabus 2006).

The formed peptide ions are separated according to their m/z ratio by the mass analyser. There are five basic types of mass analysers used in proteomic research: quadrupole, ion trap, time-of-flight, Fourier-transform ion cyclotron resonance and orbitrap (see also Westermeier and Naven 2002;

Simpson 2003). They differ in sensitivity, resolution, mass accuracy, dynamic range and the ability to generate additional fragment ions from the analysed peptides (MS/MS spectra).

Several combinations of ionisation methods with the described mass analysers are possible. Usually, MALDI is coupled to TOF analysers to measure the mass of intact peptides, whereas ESI is most often combined with an ion trap or triple-quadrupole instrument. The latter MS devices are often used for the generation of MS/MS data. For MS/MS a selected peptide, the so-called precursor (parent) ion, is further cleaved into smaller peptide fragments, the product (daughter) ions (Aebersold and Goodlet 2001; Aebersold and Mann 2003; Domon and Aebersold 2006).

J. Peptide Mass Fingerprint and Tandem-Mass Spectrometry

The technique of peptide mass fingerprinting was introduced at the beginning of the 1990s (Westermeier and Naven 2002). Data are commonly obtained via MALDI-TOF analysis. This method comprises the following steps: (a) tryptic digestion of the protein, (b) determination of the peptide fragment masses that provide a fingerprint of the protein of interest and (c) identification of the protein by sequence database search algorithms. For identification, every protein in the database is theoretically digested (for trypsin: cleavage after each lysine and arginine), giving a huge amount of theoretically possible peptides. The experimentally obtained peptide masses are then compared with the theoretical peptide masses. A score is calculated, which ranks the quality of the matches. Peptide mass fingerprinting is generally restricted to proteins of organisms with fully sequenced genomes. A drawback of this method is that, in some cases, proteins cannot be identified because of ionisation effects that result in weak ion responses of some peptides in the presence of other peptides (Beavis and Chait 1990). Nevertheless, this method is sensitive, quick and therefore suitable for high-throughput approaches. It is very often the method of choice for the identification of proteins from 2D gels (Simpson 2003).

The MS/MS approach uses single peptides for protein identification and produces sequence-specific spectra (Aebersold and Goodlet 2001). As mentioned above, during tandem-mass spectrometry (MS/MS) a selected peptide is further cleaved into smaller product ions. Repeating ion selection and fragmentation leads to MS^n. The most widely used method for the introduction of energy into a precursor ion is collision with an inert gas, called collision-induced dissociation (CID). The collision causes fragmentation of the peptide at the amide bonds in two complementary ion series: the N-terminal ion series (or b-ion series) and the C-terminal ion series (or y-ion series; Hayes and Gross 1990). In addition, the neutral loss of amino acid side-chains produces further fragments. These fragment ions can be used to interpret the amino acid sequence. A number of search engines have been developed for database searching, such as Mascot (Matrix Science), MS-TAG (Prospector), Phenyx (Genebio), SEQUEST (Thermo Finnigan), X!Tandem (GPM organisation), etc. These search engines match obtained CID spectra against theoretical fragmentation patterns of all tryptic peptides derived from proteins in the database. In contrast to peptide mass fingerprinting, EST databases can also be used for searching. Detailed sequence information (i.e. collection of a significant number of contiguous fragments) enables de novo sequencing (Hunt et al. 1986) and allows the identification of proteins from organisms with unsequenced genomes by similarity searches (Shevchenko et al. 2001).

Besides protein identification, MS/MS analysis can be used for the characterisation of post-translational modifications. The analysis of post-translational modifications is, however, by far more complex than simple protein identification, since quite often only a minor fraction of a protein to be investigated has undergone modification. During MS/MS analysis the peptide that contains the modified residue needs to be isolated. In general, diagnostic ions, which are characteristic of the modifying group or of its loss during fragmentation, are used for the targeted analysis (Aebersold and Goodlet 2001). For a detailed description of the analysis of post-translational modifications, the reader is referred to Aebersold and Goodlett (2001), Mann and Jensen (2003), Sickmann et al. (2003a) and Reinders and Sickmann (2005).

IV. Gel-Independent Techniques

A. Liquid Chromatography-Based MS Techniques

The combination of protein separation by 2D-PAGE and subsequent identification by mass spectrometry is a powerful tool, but also has its drawbacks, such as a limitation in sensitivity and the failure to detect very hydrophobic, basic or large proteins. Hence, several methods have been established which by-pass 2D-PAGE and which are based on the separation of proteins by alternative methods [liquid chromatography (LC), SDS-PAGE]. Most of the "gel-free" methods employ LC coupled to MS. In such cases, peptides of a digested protein mixture are separated by a reversed phase chromatography and subsequently eluted into a mass spectrometer using ES as the ionisation method. With such an LC system it is possible to distinguish proteins in a complex mixture containing more than 50 different proteins (McCormack et al. 1997). The LC-separation of a complex mixture of tryptically digested peptides is called a "bottom-up approach" to distinguish it from top-down approaches, in which intact proteins are isolated before MS identification and characterisation takes place.

With the use of capillary columns this approach was further improved. Nano-LC systems have very low flow rates (25–50 nl/min) which allow for more time to analyse each peptide (peak width of 30 s). Consequently, increased sensitivity is attained (Wilm and Mann 1996). However, for complex peptide mixtures, the separation capacity of a single reversed-phase column is not sufficient. Co-elution of peptides is likely, so that MS would be unable to select all peptides for fragmentation. For this reason, the multi-dimensional protein identification technology (MuDPIT) was developed (Link et al. 1999). In this approach, complex peptide mixtures are loaded on a biphasic column in which two different stationary phases, a cation-exchange and a stationary phase material, are packed into a nanospray capillary. By applying this technology, more than 2000 differen proteins were identified in *Saccharomyces cerevisiae*, including proteins with three or more predicted membrane domains, protein kinases and transcription factors (Washburn et al. 2001; Wei et al. 2005; De Godoy et al. 2006).

Another strategy to reduce the complexity of a peptide mixture is the off-line coupling of chromatographic separation and MS analysis. So, liquid chromatography can be combined with MALDI-MS (LC MALDI-MS). In this case, the LC effluent is automatically mixed with a suitable matrix and deposited onto a MALDI plate. As in traditional chromatography, discrete spots correspond to a certain fraction. This method offers the possibility to reanalyse samples and to decouple MS experiments from MS/MS analysis (Mirgorodskaya et al. 2005).

Another off-line technique is a liquid-based 2D liquid column chromatographic separation system using a combination of chromatofocusing and non-porous RP column chromatography to separate intact proteins. This method provides pI information and, for this reason, a 2D liquid protein map of two different samples can be displayed for the identification of differently expressed proteins. This technique has recently been applied to characterize *C. albicans*-infected macrophages (Shin et al. 2006).

Altogether, the development of technical advances in LC-MS/MS has been impressively fast. LC-MS/MS can be easily automated and shows very high sensitivity. Nevertheless, it is still a challenge to analyse the whole proteome of an organism by LC-MS completely. The sensitivity of the method is limited by the sequencing speed and the effective dynamic range of the used instruments (de Godoy et al. 2006). Sequencing speed depends on the time the mass spectrometer needs to switch between MS and MS/MS modes. An effective sequencing is further impaired by repeated sequencing of the same peptide. Another problem is the dynamic range of the instruments, which is smaller than the dynamic range of most organisms (de Godoy et al. 2006). Furthermore, a reliable interpretation of MS/MS data yielded by shotgun proteomics approaches is often time-consuming. However, the identification of peptides by database searches alone often gives false-positive identifications rates of about 1% (Elias et al. 2005). To estimate the number of false-positives, searches can be performed against a database in which the sequences have been reversed or randomised, so-called "decoy" databases (Elias and Gygi 2007).

However, increased coverage of proteomes of eukaryotic micro-organisms and more reliable identification of peptides appear achievable in future by improved acquisition software and instrumentation. New techniques for the analysis of peptides will emerge; such as the ion mobility spectrometry, the development of which is still in its infancy, but which will probably have more impact in the future (McLean et al. 2005).

B. Quantitative Proteomics Using LC-MS

Mass spectrometry is not inherently quantitative, since different peptides have different mass spectrometric response. For the study of global protein expression changes, multi-dimensional LC-MS-MS can be combined with stable isotope tagging to enable quantitation of proteome samples. Several stable, non-radioactive isotopes like ^{2}H, ^{13}C, ^{15}N and ^{18}O have been used for this purpose. There are three strategies for introducing stable isotopes into peptides or proteins: (a) chemical labelling by binding an isotopic tag to a specific amino acid residue (Gygi et al. 1999; Thompson et al. 2003; Ross et al. 2004; Schmidt et al. 2005; Wiese et al. 2007), (b) enzymatic labelling by using a protease and oxygen-18 water (Myagi and Rao 2007) and (c) metabolic labelling by using isotopically enriched media. For the comparison of two proteomes, one sample is labelled with the isotopically light reagent, the other with the heavy counterpart. The ratios of signal intensities of differentially mass-tagged peptide pairs are compared after MS or MS/MS analysis (for a review, see Mann 2006).

In parallel to the development of labelling strategies for the quantitation of global changes in protein expression (for reviews, see Righetti et al. 2004; Ong and Mann 2005; Leitner and Lindner 2006), non-labelling approaches for quantitation of peptides have gained importance. Especially the improvement in reproducibility of chromatography systems, the high performance of new mass spectrometers and the development of suitable software have made label-free quantification feasible. In most cases, ion currents of the same peptides in different experiments are compared, but extremely reproducible sample preparation is required (see Wiener et al. 2004; Ishihama et al. 2005; Wang et al. 2005; Silva et al. 2006).

In summary, quantitative protein expression profiling is a crucial part of proteomics. Both LC- and gel-based technologies have their pros and cons. Kolkman et al. (2005) compared DIGE and metabolic stable isotope labelling for the comparison of nitrogen- with carbon-limited grown yeast cells. Surprisingly, the DIGE technique showed a better signal-to-noise-ratio for lower-concentrated proteins in comparison with stable isotope labelling.

V. Sample Fractionation to Identify Low-Abundant Proteins

The analysis of complex protein samples can be laborious; and often low-abundant proteins are not detected. Depending on the question to be elucidated, several strategies can be applied to enrich the proteins of interest.

A. Subcellular Fractionation

One of the big challenges in proteomic studies is the complexity and the huge dynamic range of proteins in a cell. One strategy to reduce the complexity is to carry out a prefractionation step prior to separation by 2D-PAGE or LC-MS/MS. By doing so, certain proteins including low-copy number proteins are enriched and, hence, are amenable to further investigations. Fractionation of protein samples can be accomplished by several methods: (a) isolation of organelles or cell compartments by fractional centrifugation, (b) application of classical chromatographic separation techniques, (c) separation of proteins or organelles by free-flow electrophoresis, (d) use of multi-compartment electrolysers based on the electrofocusing of proteins with soluble carrier ampholytes, (e) immunoprecipitation of protein complexes, (f) sequential extraction of proteins with differently powerful solubilising buffers and (g) depletion of high-abundant proteins (reviewed in Righetti et al. 2005).

Combining large-scale proteomics studies with classic cell biology techniques for the isolation of organelles has become a popular strategy. It gives both access to low-abundant proteins and additional information about the localisation of a protein (Yates et al. 2005; Andersen and Mann 2006). The mitochondrial proteome has been a matter of particular interest, because mitochondria play a central role in many cellular processes. Several groups have attempted to define the mitochondrial proteins of the model organism *S. cerevisiae* and up to 851 proteins have already been identified (Sickmann et al. 2003b; Prokisch et al. 2004; Reinders et al. 2006).

B. Membrane Proteomics

Detection of membrane or very hydrophobic proteins in 2D gels is limited; and mainly the hydrophilic, high abundant proteins of a protein extract are resolved. The under-representation of membrane proteins can be attributed to several factors: (a) they are low in abundance, (b) their pIs are generally above pH 7 and (c) they are poorly soluble in the buffers used for 2D-PAGE (Santoni et al. 2000).

Since membrane proteins play important roles in many cellular processes, such as signal transduction, cell adhesion and transport, several strategies have been developed to improve the detection rate for membrane proteins. They can be enriched by the preparation of membrane fractions and IPG strips with extended pH gradients are available, but the major challenge in membrane proteomics is the solubilisation in aqueous solutions. Several groups use strong detergents (Molloy 2000; Fountoulakis and Takacs et al. 2001; Henningsen et al. 2002; Luche et al. 2003) to improve solubilisation. Nevertheless, proteins with more than six trans-membrane domains are hardly detectable in 2D gels. For this reason, several other methodologies have been developed to overcome the under-representation of membrane proteins with a high number of trans-membrane domains. Delom et al. (2006) combined ion-exchange chromatography of solubilized membrane proteins with a one-dimensional lithium dodecyl-sulfate-PAGE procedure. Other 2DE approaches use a combination of a cationic acid PAGE with a subsequent anionic basic PAGE. In the first dimension, proteins are solubilised and separated in the presence of the cationic detergent cetyltrimethylammonium bromide (CTAB) or benzyldimethyl-n-hexadecylammonium chloride (16-BAC) under acidic buffer conditions (Macfarlane et al. 1989; Akins et al. 1992; Kramer 2006). By a subsequent second-dimension SDS-PAGE, a widely spread diagonal spot pattern occurs due to the different separation behaviours of membrane proteins in the two different dimensions (Zahedi et al. 2005). Membrane proteins can also be resolved by a double SDS-PAGE that employs 6 M urea SDS-PAGE, using a low acrylamide concentration in the first dimension followed by urea-free SDS-PAGE with a high acrylamide concentration in the second dimension (Rais et al. 2004; Williams et al. 2006). Furthermore, the combination of Blue-native PAGE and Tricine-SDS-PAGE can be used for the isolation of protein complexes from biological membranes (Wittig et al. 2006). Alternatively, the aforementioned gel-free separation systems such as MudPIT can be applied for the separation of membrane proteins. To enrich membrane fractions, detergents, organic solvents or organics acids compatible with subsequent proteolytic digestions/chemical cleavage have been used (Wu and Yates 2003). There is still no method which is able to resolve all the membrane proteins of an organism. The application of several techniques increases the number of identified membrane proteins significantly (Burré et al. 2006).

C. Phosphoproteomics

The reversible phosphorylation of proteins regulates many major cellular processes, like signal transduction. It is the most predominant post-translational modification in eukaryotic cells (Hunter 2000). Moreover, signalling pathways, which include phosphorylation steps, are involved in the infection process of human-pathogenic fungi (Brakhage and Liebmann 2005; Bahn et al. 2007).

Most common are phosphorylation of the amino acids serine, threonine and tyrosine (O-phosphorylation) and occasionally N-phosphorylation of histidine and lysine, S-phosphorylation of cysteine and acyl-phosphorylation of aspartic and glutamic acid (Scott and Lecomte 2005). Because of the importance of protein phosphorylation in cellular processes, a lot of new methods for the analysis of complex phosphoproteomes have been developed in recent years. Gel-based techniques are often applied, but mass spectrometry is becoming an increasingly important alternative to the traditional methods based on gel electrophoresis.

The detection of phosphorylated proteins is often hampered by the fact that many signalling proteins are not abundantly expressed and the proportion of phosporylated to unphosphorylated protein can be quite low. So prefractionation strategies for the reduction of the sample complexity are often applied (Morandell et al. 2006).

1. Detection of Phosphoproteins with 2D-PAGE

By phosphorylation, proteins change their position on 2D gels towards a more acidic isoelectric point. Nevertheless, this shift can also be caused by other post-translational modifications or by artificial modification of proteins during sample preparation.

Phosphoproteins can be unambiguously visualised on 2D gels by autoradiography or phosphorimaging after the incorporation of radioactive [^{32}P] or [^{33}P] into cellular proteins via treatment with labelled inorganic phosphate or ATP (Cobitz et al. 1989). This method is very sensitive and especially easy to apply for microbial cultures.

An alternative approach for the detection of phosphorylated proteins is the application of phosphospecific antibodies or phosphoprotein-specific stains, which enable the detection of protein phosphorylation in a steady state. After electrophoresis, the proteins are blotted onto a membrane and detected by the use of anti-phosphoserine, anti-phosphotyrosine or anti-phosphothreonine antibodies (Soskic et al. 1999; Kaufmann et al. 2001). Recently, phosphoprotein gel stains based on fluorescence-based detection technology were introduced (Schulenberg et al. 2004; Kinoshita et al. 2006). Another method, which does not require any radiolabelling or electroblotting, is the mapping of phosphorylated proteins in 2D gels by the application of a protein phosphatase, as shown by Yamagata et al. (2002). Prior to 2D-PAGE, protein extracts were divided into two aliquots. One aliquot was dephosphorylated by a recombinant λ-protein phosphatase, whereas the other aliquot was not treated. Phosphorylated proteins showed a shift to a more basic position in gels loaded with the phosphatase-treated sample in comparison with the non-treated control.

2. Enrichment of Phosphoproteins

All the methods described above are dealing with the problem that the proportion of phosphorylated to unphosphorylated proteins can be quite low. Many phosphorylated proteins are not detectable in 2D gels, or the amount of protein is insufficient for the identification by MS-analysis. Furthermore, detection of phosphopeptides by MS is often hampered by suppression effects due to the presence of unphosphorylated proteins. To increase the sensitivity of detection, several techniques have been developed to specifically enrich phosphorylated proteins. Antibodies against phosphorylated amino acids immobilised onto a column matrix can be used for the enrichment of tyrosine-phosphorylated proteins (Pandey et al. 2000). Antibodies against phosphothreonine and phosphoserine are less suitable for this approach since they are more unspecific. At present, immobilised metal affinity chromatography (IMAC) is one of the most popular techniques, introduced by Andersson and Porath (1986). It uses Fe^{3+} or Ga^{3+} metal ions, which interact electrostatically with the negatively charged phosphate groups. Alternative approaches with a higher selectivity are affinity chromatography methods with metal oxide/hydroxide (Sano and Nakmura 2004; Wolschin and Weckwerth 2005) or dinuclear zinc(II) complexes bound to an agarose matrix (Kinoshina-Kikuta 2006).

Further methods for the enrichment of phosphoproteins use chemical modifications of the phosphate groups. A variety of phosphopeptide/protein derivatizations have been developed, which are also summarised by Reinders and Sickmann (2005).

3. Chromatographic Methods for Phosphoproteomics

As described, the recent developments in MS technology have made it increasingly feasible to analyse very complex peptide mixtures by LC-MS/MS without prior gel separation. Enrichment strategies for phosphopeptides can be coupled in-line or off-line to liquid chromatography, e.g. Ficarro et al. (2002) digested a whole cell lysate from *S. cerevisiae* with trypsin. The resulting peptides were converted to methyl esters and enriched for phosphopeptides by IMAC chromatography. The methylation of the digested peptides was carried out to avoid binding of non-phosphorylated peptides to the IMAC column. More than 1000 phosphorylated peptides were detected by neutral loss ion scanning, using an LCQ-ion trap mass spectrometer. In combination with stable-isotope labelling, LC-MS/MS methods allow also the quantification of phosphorylated proteins (Bonenfant et al. 2003; Gruhler et al. 2005).

VI. Standard Development

The field of proteomics has grown at a considerable rate and a diverse set of new technologies has been applied to investigate the change of protein expression on a global scale. The progress is impressive. Nevertheless, inadequate attention has often been paid to a proper experimental design of proteomics studies, which allows producing statistically significant data. In many cases, the degree of stringency required in proteomic data generation and analysis has been underestimated (Wilkins et al. 2006). In the light of these perceived problems, a lot of papers have been published lately which discussed several important points such as the correct use of bioinformatics tools, the impact of technical and biological variations on the statistical importance of the results, the right criteria for the correct identification of proteins, etc. (Hunt et al. 2005; Biron et al. 2006; Urfer et al. 2006). At the same time, proteomic journals have initiated the development of a set of minimum guidelines for the publication of proteomics research (Wilkins et al. 2006). Furthermore, a proteomics standards initiative (PSI) was founded within the Human Proteome Organisation (HUPO) in 2002. The PSI work group aims at defining community standards for data representation in proteomics to facilitate systematic data capture, comparison, exchange and verification (Hermjakob 2006). Following the "minimum information about a microarray experiment" (MIAME) guidelines (Brazma et al. 2001), the corresponding "minimum information about a proteomics experiment" (MIAPE) requirements will be deployed in future (Taylor 2006). Besides the development of MIAPE reporting requirement, the PSI is working on platform-independent data exchange formats, standardisation of terminology and the promotion of publicly available proteomics databases (Hermjakob 2006).

VII. Proteomics of Human-Pathogenic Fungi

A. Sample Preparation

Sample preparation is a critical step and is essential for reproducible results. The cell wall makes up the majority of the cell mass in fungi and it is exceptionally robust (Ruiz-Herrera 1992). Thus, cell lysis is a difficult step in sample preparation of fungi for 2D-PAGE. For *C. albicans*, cell disruption has often been achieved by sonication or disruption with glass beads. In many cases Tris- or phosphate buffers containing supplements, such as protease inhibitors, EDTA, or reducing agents, were used. Later, samples (sometimes concentrated) were resuspended in lysis buffer containing CHAPS, urea and DTT (Manning and Mitchell 1980b; Niimi et al. 1996; Choi et al. 2003; Hernandez et al. 2004).

For filamentous fungi such as *A. fumigatus*, it is recommendable to remove lipids, pigments and polysaccharides. The classic TCA/acetone precipitation has already been applied for the preparation of extracts from fungi (Harder et al. 1999; Nandakumar 2003). Kniemeyer et al. (2006) carried out a comparison of a commercial kit for protein precipitation (GE Healthcare Bio-Sciences) with a standard TCA/acetone precipitation procedure. Additionally, direct precipitation of fungal proteins after cell disruption was compared with precipitation of SDS-extracted proteins. Direct lysis of ground mycelia resulted in irreproducible gels with streaks, a strong background and faint spots. The TCA/acetone precipitation of proteins from *A. fumigatus* crude extracts resulted in 690 detectable spots on a mini-gel. Direct treatment of samples with a commercially available 2D clean-up kit showed gels of similar quality, but fewer spots were detected which, however, showed less vertical streaking. Furthermore, the protein yield was higher with the commercial kit. An SDS extraction of proteins with a subsequent precipitation of proteins to remove interfering compounds, including SDS, did not produce superior results. Therefore the impact of methods for the precipitation of fungal proteins was clearly proven. Another strategy was pursued by Shimizu and Wariishi (2005) for the preparation of samples suitable for 2D-PAGE analysis. They utilised fungal protoplasts, prepared from the brown-rot basidiomycete *Tyromyces palustris*, for protein extraction. Proteome maps of the protoplasts indicated more spots with a wider range of molecular masses. In addition, a good subcellular fractionation into cytosolic and membrane proteins was possible from fungal protoplasts. In another report, Nandakumar and Marten (2002) compared four different lysis methods for *A. oryzae*: boiling in 0.2 M NaOH, boiling in SDS lysis buffer, chemical lysis in commercially available Y-PER reagent and mechanical lysis via agitation with glass beads. Mechanical disruption was found to be the most efficient extraction method, which is in good agreement with the results from Kniemeyer et al. (2006).

Another critical step during 2D-PAGE is the solubility of proteins during IEF. Many proteins tend to aggregate or precipitate at their isoelectric point. Kniemeyer et al. (2006) evaluated five lysis buffers for the solubilisation of proteins from the filamentous fungus *A. fumigatus*. The application of buffers containing the detergents CHAPS, CHAPS and Zwittergent 3–10 and CHAPS, Zwittergent 3–10 and ASB 14 resulted in gels of almost the same quality and with the approximately same number of spots. More spots were detected when not only CHAPS was used, but additionally another sulfobetaine detergent such as Zwittergent 3–10. The 2D-PAGE protocol optimised by Kniemeyer et al. (2006) was evaluated. *A. fumigatus* was grown on glucose and ethanol and the different protein maps for each growth condition were compared. As expected, spots representing the enzymes of gluconeogenesis, glyoxylate cycle and ethanol degradation pathway were differently regulated and all key enzymes of gluconeogenic growth were detected.

B. Proteomics Studies of Fungal Physiology

A lot of proteomic studies investigated the morphological transition of *C. albicans*, since the yeast-to-mycelium conversion plays a role in the pathogenesis of candidiasis. A pioneering work of Manning and Mitchell (1980b) analysed cultures of yeast and true hyphal cells by 2D gel electrophoresis. Two strains were selected for this study, of which one was only able to produce yeast cells. Pulse chase experiments with [^{35}S]-sulfate revealed the biosynthesis of new proteins after morphological transition, but no differences in the major cytoplasmic proteins of the yeast- or mycelial-phase extracts were identified. The same was observed in a subsequent study (Niimi et al. 1996). Besides 2D-PAGE, the authors applied heparin-agarose affinity chromatography to detect low abundant polypeptides, such as DNA-binding proteins. A small fraction of polypeptides were bound to the column that were preferentially synthesised during the early stage of transition between yeast and hyphal form, but they were not characterised further. In another 2D-PAGE study with immobilized pH gradients three proteins increased during hyphal differentiation: pH-regulated antigen (Pra1p), thiol specific antioxidant (Tsa1p) and pH response protein (Phr1p). Pra1p is a glycosylated surface protein and deletion of the corresponding gene led to a temperature-dependent defect in hyphal formation (Sentandreu 1998). Interestingly, a role in adhesion to leukocytes was later shown by Soloviev et al. (2007). Also Tsa1p was later characterised in more detail. Tsa1p, a thioredoxin peroxidase, conferred resistance against oxidative stress but, in addition, it was involved in the correct composition of hyphal cell walls. Therefore, Tsa1p has a bifunctional role and a correlation of its functions with its respective localisation was supposed, due to the fact that Tsa1p was only present in the cell wall of true hyphae, but not in yeast cells (Shin et al. 2005a; Urban et al. 2005). The increase of Phr1p levels in hyphae was also confirmed by Northern blot analysis on the transcript level (Choi et al. 2003). Phr1p is regulated in response to the pH of the culture medium and belongs to a family of fungal glucanosyltransferases. Mutants lacking *PHR1* exhibited a pH-dependent morphogenic defect and were less virulent in a mouse model of systemic candidiasis (Ghannoum et al. 1995; Saporito-Irwin et al. 1995). The different levels of Phr1p in yeast- and hyphal cells of *C. albicans* were quantified more precisely by LC-MALDI-MS/MS, using a triple-quadrupole as mass analyser (Melanson et al. 2006). Selected reaction monitoring (SRM) was employed for quantitation. In an SRM experiment the analyte ion of interest was selected in the first quadrupole (Q1), fragmented in the second quadrupole (Q2) and a selected fragment ion was monitored in the third quadrupole (Q3). Five Phr1p peptides were quantified and an up-regulation of 7.7- to 8.7-fold for hyphae was measured.

The influence of the metabolic status of *C. albicans* cells on morphology was recently studied by mitochondrial proteomics using DIGE (Vellucci et al. 2007). An implication of the pyruvate dehydrogenase complex protein X (Pdx1p) in filamentation was postulated. The level of this protein increased significantly in a mutant which was impaired in the mitochondrial electron transport chain and was defective in filamentation. Presumably, the observed upregulation of Pdx1p compensated for the respiratory malfunction. Due to high respiratory activity in the mutant its yeast morphology was maintained. In this way, a postulated association of a high level of respiration with the yeast morphology of *C. albicans* was confirmed.

An elaborate study of proteins found in hyphal cells of *C. albicans* was carried out by Hernández et al. (2004). A 2D gel reference map was created with 66 different proteins from 106 spots. Most of the identified proteins were involved in metabolism, transcription and protein synthesis, as well as cell rescue and defence.

Besides global proteomic studies and the creation of proteome maps, specific protein complexes or cell membrane structures of *C. albicans* have also been analysed by 2D-PAGE or SDS-PAGE. The 20S core particle of the proteasome of *C. albicans* was purified and the subunits were separated by 2D-PAGE, which revealed 14 polypeptides (Murray et al. 2000). Proteins in *C. albicans* asssociated with lipid rafts (microdomains) were investigated by Insenser et al. (2006). Lipid rafts are membrane domains with higher amounts of sphingolipids and sterols. The polarisation of lipid rafts are expected to contribute to hyphal growth in *C. albicans*. In addition, these microdomains are presumbly involved in the control of signalling pathways (Martin and Konopka 2004; Wachtler and Balasubramanian 2006). The protocol developed by Insenser et al. (2006) was based on the fact that lipid rafts are insoluble in non-ionic detergents, such as Triton X-100, at low temperature. In total, 29 lipid-raft associated proteins were identified.

The question how *C. albicans* adapts to environmental changes has been addressed by several proteomics surveys based on 2D-PAGE. In *C. albicans*, starvation for amino acids induces the synthesis of the bZIP transcriptional activator GCN_{4p}. GCN_{4p} activates the transcription of amino acid biosynthetic genes via the general control response elements (GCRE) in the promoter region of target genes (Tripathi et al. 2002). Yin et al. (2004) characterised the general amino acid control (GCN) response of *C. albicans* at the level of the proteome and compared it with data from *S. cerevisiae*. To elucidate Gcn4p-dependent changes, the proteome of wild-type cells was compared with a Δ*gcn4/gcn4* mutant under amino acid starvation conditions. Most aspects of the GCN response of *C. albicans* resembled the response of *S. cerevisiae*: amino acid biosynthesis enzymes and proteins of the carbon metabolism were induced. However, purine biosynthetic enzymes were upregulated in *S. cerevisiae*, but not in *C. albicans*.

The pathogenicity of *C. albicans* also depends upon its ability to cope with host defence. During the infection process the oxidative burst of immune effector cells is suggested to play a significant role. Kusch et al. (2007a) analysed the oxidative stress response of *C. albicans* after treatment with hydrogen peroxide or diamide. Among the induced proteins were proteins with antioxidant functions, such as catalase A, glutathion reductase, glutathion peroxidase, thioredoxin reductase and a set of oxidoreductases, as well as heat-shock proteins. The targets of the key regulator Cap1p, a homologue of the yeast transcription factor Yap1p that is involved in the adaptive response to oxidative stress, were also elucidated. The proteome of wild-type cells were compared with a mutant exhibiting a hyperactive allele of *CAP1*. Twelve proteins were controlled by Cap1p, e.g., catalase A, glutathione reductase, thioredoxin reductase, NADPH dehydrogenase and several dehydrogenases.

The most comprehensive protein map of cytoplasmic proteins in *C. albicans* yeast cells was recently published (Kusch et al. 2007b). Altogether, 746 protein spots representing 360 different proteins were identified. In this study, a proteome signature of exponentially-grown cells and cells from the stationary phase were yielded. During exponential growth, proteins required for the synthesis of proteins and nucleotides were present in large amounts. During the stationary phase, a switch to pathways for the utilisation of alternative carbon sources other than glucose was observed: proteins of the glyoxylate cycle, gluconeogenesis and glutamate degradation were up-regulated.

In comparison with *C. albicans*, only a few studies were carried out with filamentous fungi in the first decades after the introduction of the 2D-PAGE technique; and the number of publication increased just recently. In most cases, the model organism *A. nidulans*, but not *A. fumigatus*, was object for studies using 2D-PAGE as a method. Sheir-Neiss et al. (1978) resolved the α- and β-tubulins of a wild-type strain and a *benA* mutant strain by 2D-PAGE. Later, Weatherbee et al. (1985) did the same. They proved the involvement of β3-tubulin in conidial development of *A. nidulans*.

More global proteome studies of filamentous fungi have been started recently. Kim et al. (2007) studied the osmoadaptation of *A. nidulans* by 2D-PAGE. They compared the protein expression pattern of *A. nidulans* during growth with or without the osmolyte KCl (0.6 M). No significant changes in growth rate or morphology in the presence of 0.6 M KCl was observed, but 90 spots were differently regulated, of which 30 were identified by MALDI-TOF. One of the proteins that were up-regulated in osmoadapted *A. nidulans* mycelium was the glycolysis enzyme glyceraldehyde-3-phosphate dehydrogenase. Together with a reduced expression of enolase and an increased expression of aldehyde dehydrogenase, the authors supposed an accumulation of 3-phosphoglycerate and an increased production of glycerol due to osmotic changes. In addition, an increased protein turnover presumably takes place during osmoadaptation: heat-shock proteins and the proteins involved in protein degradation via the ubiquitin/proteasome pathway were significantly up-regulated. Besides *A. nidulans*, proteomics approaches were also applied to *A. fumigatus*. By 2D-PAGE and subsequent MALDI-MS, 50 cytosolic proteins were identified. The majority of identified proteins were involved in primary metabolic processes such as glycolysis, citric acid cycle, pentose phosphate pathway, fatty acid metabolism and oxidative phosphorylation. Furthermore, structural proteins, chaperones and signalling proteins were also detected (Carberry et al. 2006). The same authors purified glutathione-binding proteins by affinity chromatography, and subsequently separated them by 2D-PAGE. They identified and characterised the elongation factor eEF1Bγ proteins, ElfA and ElfB, as glutathione-binding proteins. A glutathione transferase activity was proven for ElfA. Hence, the authors speculated about a putative role of ElfA in controlling translation in response to stress.

Immune effector cells are strongly suggested to kill fungal pathogens by the secretion of reactive oxygen intermediates. For this reason the response of *A. fumigatus* to reactive oxygen intermediates, such as hydrogen peroxide, was investigated by 2D-PAGE (Lessing et al. 2007). Primarily, proteins required for antioxidant defense, heat-shock proteins, the protein translation apparatus, proteins of the central metabolic pathways (glycolysis, TCA cycle, pentose phosphate cycle) and proteins of the trehalose metabolism as well as those of the cytoskeleton were effected in this study. Since several of the identified proteins and their corresponding genes, respectively, were apparently regulated by a putative *S. cerevisiae* homolog Yap1p in *A. fumigatus* the corresponding gene was deleted. The proteome of the Δ*yap1* strain was characterised and compared with the wild type. Over 20 proteins were directly or indirectly controlled by the Yap1p homologue, including catalase 2.

C. Drug-Induced Changes

As discussed before, the development of new antifungal drugs is of immense importance. So far, only a few compounds are available and there is only a small number of known drug targets. Proteomic approaches are suitable for gaining a deeper insight into the response of fungi towards antibiotic compounds. De Backer et al. (2000a) characterised a single allele knock-out of the *C. albicans* gene *CGT1* coding for an mRNA capping enzyme. The level of *CGT1* mRNA was reduced 2- to 5-fold compared with the parental strain. The disrupted strain was significantly more resistant to hygromycin B and heat stress. To elucidate possible mechanisms behind this observation, the authors looked at the changes in the proteome by 2D-PAGE. The effect of hygromycin on the translational fidelity was in agreement with the finding that the translational elongation factor Ef-1αp and the ribosomal protein Rps5p were up-regulated in the mutant. Besides, the cell wall-associated heat-shock protein Ssa2p showed a higher abundance in the *cgt1/CGT1* heterozygote. Bruneau et al. (2003) used a proteomic approach to classify antifungal compounds according to their mechanism of action. They investigated the drug-induced proteome changes in *C. albicans* caused by triazoles and antibiotics of the echinocandin class. These two classes of antifungal drugs have different targets: Azoles inhibit the ergosterol biosynthesis pathway enzyme lanosterol 14α demethylase, whereas echinocandins impair the synthesis of the fungal cell wall component β-1,3 glucan. The different modes of action were reflected by different polypeptide patterns which may help to classify and to study antifungal compounds with an unknown mechanism. A more detailed proteomic study on the specific mechanism of acquired resistance against fluconazole revealed a metabolic shift in a fluconazole-resistant *C. albicans* strain. On the one hand, proteins involved in oxidative phosphorylation showed a decreased expression level; on the other hand enzymes involved in glycolysis and the glyoxylate cycle were up-regulated (Yan et al. 2007). Hooshdaran et al. (2004) found quite similar results by a comparison of a fluconazole-sensitive and a fluconazole-resistant *C. albicans* isolate. Protein Erg10p, a protein involved in ergosterol biosynthesis, was up-regulated in the azole-resistant strain, as were several alcohol dehydrogenases.

Melin et al. (2002) analysed the effect of the antibiotic concanamycin A, produced by a *Streptomyces* species, in the filamentous fungus *A. nidulans*. Twenty proteins were differently regulated. Remarkably, one of the down-regulated proteins, CpcB, is known to be involved in the initiation of sexual development and the cross-pathway control.

D. Biofilm Formation

The rise in *C. albicans* infections is related to the increased application of various invasive indwelling medical devices (IMD), including central venous catheters, cardiovascular devices, dialysis tubings, etc. *C. albicans* cells are able to adhere to the IMD surface by forming biofilms. In biofilms, cells of *C. albicans* are attached to a surface and are encased in a self-produced organic polymeric matrix. The ability to form biofilms is believed to contribute to the invasiveness and virulence of *C. albicans*. Growing as a biofilm correlates with a higher resistance to commonly used antifungal drugs (Mukherjee et al. 2005). To get a deeper understanding of *C. albicans* biofilm formation, several groups investigated the expression of biofilm-specific genes, but some proteomic studies have been carried out as well. Thomas et al. (2006a) identified eight proteins with a different level of expression in biofilm-grown cells in comparison with planktonic cells. Especially, glycolytic enzymes, the alcohol dehydrogenase 1 (Adh1p) and stress proteins were up-regulated, whereas another alcohol dehydrogenase (Adh2p) was down-regulated.

A chitinase (Cht3p) as well as mannoproteins (Mp65p, Mp58p) represented abundant proteins in the biofilm extracellular matrix. Nevertheless, the polypeptide pattern of biofilm and planktonic cultures showed an astonishingly high similarity. Vediyappan and Chaffin (2006) came to the same conclusion, but found a surface-dependent protein expression. Mukherjee et al. (2006) demonstrated that alcohol dehydrogenase 1 (Adh1p) is down-regulated in cell walls in the early phase of *C. albicans* biofilms. The authors concluded that the production of ethanol inhibits the formation of biofilms. This ethanol-based regulation was confirmed by a *ADH1*-deleted mutant strain of *C. albicans*, which formed more and thicker biofilms.

In contrast to *C. albicans*, little is known about biofilm formation in *A. fumigatus*. It was recently shown that *A. fumigatus* is able to form an extracellular hydrophobic matrix when it is grown under static, aerial conditions. The extracellular matrix is composed of α-1,3 glucans, monosaccharides and polyols, melanin and proteins, such as catalase B, dipeptidylpeptidase V and ribotoxin (ASPF1), which are known as secreted antigens (Beauvais et al. 2007).

E. Proteomics of the Cell Wall

The cell wall of fungi is a complex multi-layered structure composed of polysaccharides, proteins and lipids. Since the fungal cell wall is the outermost cellular structure, it is the first cellular component in contact with the host. β-Glucans, cell wall proteins and chitin are interconnected through covalent bonds and non-covalent interactions. A glycosylphosphatidylinositol (GPI)-anchoring machinery is involved in the attachment of proteins to the fungal cell wall.

GPI proteins are anchored to the plasma membrane at first. A few GPI proteins reside permanently at the plasma membrane, whereas most others are integrated into the cell wall by covalent attachment to cell wall glucans (de Groot et al. 2005). GPI proteins have several functions, such as mediating adhesion to surfaces or remodelling the cell wall (Sundstrom 2002; Albrecht et al. 2006).

For this reason, interest has grown for proteomic approaches which enable scientists to study the cell envelope of pathogenic fungi. For *C. albicans* several different fractionation techniques have been published (see also Sohn et al. 2006). Masuoka et al. (1998) subjected yeast cell walls to limited enzymatic digestion with β-1,3 glucanase. The released proteins were separated by a preparative electrofocusing system with soluble carrier ampholytes (Rotofor, Bio Rad). Each fraction was subsequently run on SDS-PAGE gels. The authors identified a hydrophobic protein which reacted with monoclonal antibodies of mice immunised with *C. albicans* cell wall proteins. The corresponding gene was later cloned and designated as *CHS1* (cell surface hydrophobicity; Singleton et al. 2001). Another report showed that cell surface proteins were released by treatment of intact cells with ammonium carbonate buffer containing β-mercaptoethanol (Vediyappan et al. 2000). Pitarch et al. (2002) developed a more systematic approach for the analysis of the cell wall proteome of *C. albicans* by sequential extraction. Proteins attached to the cell wall by non-covalent interactions or disulfide bonds were extracted with hot SDS and dithiothreitol treatment. Proteins such as chaperones, glycolytic proteins and elongation factors were found in this fraction. Proteins directly linked to β-1,3 glucan through their O-glycosyl side-chains, such as Pir proteins (family of proteins with internal repeats), were released by incubation with 30 mM sodium hydroxide. Mannoproteins indirectly attached to β-1,3 glucan by a phosphodiester bridge (GPI linkage) and other cell wall proteins anchored to β-1,3 glucan or chitin were extracted by digestion with β-1,3 glucanase (Quantazyme) and exochitinase, respectively. The authors improved the MS detection rate of extracted cell wall proteins by enzymatic deglycosylation prior to tryptic digestion. Obviously, an increased rate of glycolysation is typical for cell wall associated proteins, but also other post-translational modifications have been reported (Sepulveda et al. 1996).

Besides enzymatic treatment, chemical methods can also be applied to release cell wall proteins of *C. albicans* (de Groot et al. 2004). The use of NaOH and HF-pyridine led to the identification of 12 GPI proteins and two mild alkali-sensitive proteins. The proteins detected included proteins involved in carbohydrate metabolism, adhesion and stress response (superoxide dismutase). The same group established a method for the analysis of cell wall proteins of *S. cerevisiae* without a prior protein solubilisation step (Yin et al. 2005). SDS-treated cell walls were directly incubated with trypsin or the proteinase Glu-C and the solubilised peptides were analysed by LC-MS/MS. The same method was also applied in a comprehensive study of the cell wall proteome of yeast and hyphal forms of *C. albicans* (Ebanks et al. 2006). Seventeen proteins were only found in hyphal protein extracts including heat shock proteins and a fatty-acyl-CoA synthase.

Labelling of *C. albicans* cell surface-associated proteins with a membrane-impermeable biotin derivative and subsequent affinity chromatography purification is an alternative technique for the investigation of cell surface proteins (Urban et al. 2003).

Only little information is currently available on the cell wall proteome of *A. fumigatus*. The mycelial cell wall of *A. fumigatus* is very different from the yeast *S. cerevisiae*, e.g. it lacks β-1,6 glucans (Latgé et al. 2005). Mouyna et al. (2000) showed that a GPI-anchored glucanosyltransferase in *A. fumigatus* plays an important role in fungal cell wall biosynthesis. The same group identified other GPI-anchored proteins by a proteomic approach (Bruneau et al. 2001). A membrane fraction was treated with the detergent β-*n*-octylglucopyranoside and GPI-anchored proteins were released by the activity of an endogenous GPI phospholipase. Solubilised proteins were purified by liquid chromatography and subsequently separated by 2D-PAGE. Nine GPI-anchored proteins were identified by MALDI-TOF or ESI-MS/MS. Five of them showed high sequence similarities to putatively GPI-anchored yeast proteins and the authors concluded that the identified proteins are involved in *A. fumigatus* cell wall organisation.

F. Immunoproteomics

For a better diagnosis and management of fungal infections it is important to characterise the antigens reactive to serum from patients. For instance, specific antigens could be related to specific allergic diseases caused by *A. fumigatus* and could be used for immunodiagnosis (*Aspergillus* asthma, extrinsic allergic alveolitis, allergic bronchopulmonary aspergillosis). The same is also true for *C. albicans*, since systemic candidiasis rarely produces specific clinical symptoms and microbiological blood cultures require several days for diagnosis. Hence, culture-independent clinical tests are of great interest (Pitarch et al. 2006c). Beside the improvement of diagnosis, there are attempts to develop a vaccination against invasive mycoses with protection-inducing fungal antigens (Bellochio et al. 2005).

An approach often applied to study the antibody response to fungal proteins is the combination of 2D-PAGE with immunoblotting. One of the first studies in *C. albicans* was the comparison of the expression profiles of yeast and hyphae-grown cells (Manning and Mitchell 1980a). Rabbits were immunised with cytoplasmic fractions of two different strains to produce anti-yeast-phase and anti-mycelium-phase immune sera. One strain produced exclusively yeast cells, whereas the other strain was able to undergo yeast-to-hyphae transition. After 2D-PAGE followed by Western blotting 168 cytoplasmic antigen spots were detected, but none of them was mycelium-specific.

Later, several studies followed which aimed at characterising the "immunome" of *C. albicans*. The technique was improved by using polyvinylidene difluoride membranes. Restaining of the blotted proteins with Coomassie after antibody detection made an identification of antigens in a complex mixture easier (Shen et al. 1990) and new chemiluminescence techniques with anti-human IgG antibodies coupled to horse radish peroxidase decreased the detection limit (Barea et al. 1999). Also, different protein preparations were tested. Pitarch et al. (1999) used cytoplasmic extracts, protoplast lysates and proteins secreted by regenerating protoplasts for immunodetection of *C. albicans* antigens. During protoplast regeneration many cell wall components are secreted into the medium. By this means, samples enriched in cell wall proteins can be obtained. Western blotting with sera from patients with systemic candidiasis allowed the detection of several antigens, e.g. enolase, glyceraldehyde-3-phosphate dehydrogenase, aconitase, pyruvate kinase, phosphoglycerate mutase and methionine synthase. Protoplast lysates allowed easier immunodetection in comparison with the cytoplasmic extracts and one antigen of acidic character was only detected in fractions obtained from regenerating protoplasts (Pitarch et al. 1999; Pardo et al. 2000). In patients with haematological malignancies about 85 immunoreactive proteins were detected, including 42 different house-keeping enzymes (heat-shock, metabolic, ribosomal proteins). Interestingly, there was a correlation between changes in the level of certain anti-*C.albicans* antibodies and the patient's outcome. The rise of high anti-enolase antibody concentrations and the falling of anti-phosphoglycerate kinase and anti-methionine synthase antibody levels were related to the recovery from systemic candidiasis in three patients (Pitarch et al. 2004). Later, the same group analysed the serum levels of IgG antibodies against *C. albicans* cell wall-associated proteins by 2D-PAGE in serum of systemic candidiasis patients and control groups (Pitarch et al. 2006a). Statistical tools such as hierarchical clustering and principal component analysis were applied to group samples of the study population. High levels of IgG antibodies against the β-1,3 glucosyltransferase (Bgl2p) and phosphoglycerate kinase were the only independent predictors for systemic candidiasis and were therefore potential diagnostic biomarkers. Antibodies against enolase were an indicator for a lower risk of dying of a systemic *C. albicans* infection. Hence, previous results were confirmed. Interestingly, only IgG antibodies against the cell wall-associated form of enolase seemed to confer protection.

Similar antigens as described for human serum samples were also obtained during the analysis of the serological response to systemic *C. albicans* infections in different mice strains, which indicated the usefulness of mouse models for studying the antibody response in systemic candidiasis (Pitarch et al. 2001).

Mice were also used as animal models to investigate the role of protective antibodies during *C. albicans* infections (Fernández-Arenas et al. 2004a, b). Low-virulent *C. albicans* mutant strains were used as vaccines to induce the formation of antibodies and protection against a subsequent infection with wild-type *C. albicans* strains. A mutant strain deleted in the MAP-kinase gene *HOG1* and therefore attenuated in virulence was able to confer protection in mice (60–70% survival). Hog1p is essential for the oxidative stress and hyperosmolarity response in *C. albicans*. Using 2D-PAGE followed by Western blotting, several immunogenic proteins were detected. Furthermore, antibody patterns in sera of mice infected with non-protective low-virulent *C. albicans* strains were distinguishable from the profile detected in sera from surviving animals (vaccinated with the mutant deleted in *HOG1*). The protective sera shared antibodies (IgG2a isotype) against the following proteins: enolase, pyruvate kinase, pyruvate decarboxylase, 40S ribosomal subunit Bel1p, triosephosphate isomerase, DL-glycerol phosphatase, fructose-bisphosphate aldolase, IMP dehydrogense and acetyl-CoA synthetase (Fernández-Arenas et al. 2004b). Thomas et al. (2006c) used β-mercaptoethanol extracts from *C. albicans* cell wall preparations to vaccinate mice. A protection against disseminated candidiasis was observed in 75% of vaccinated mice. The spectrum of the antibody response in sera was characterised in more detail by ELISA: IgG1, IgG2a, IgG2b, IgG3, IgA and IgM. IgG1, IgG2b and IgG2a antibodies were highly abundant, but IgM reactivity was also high. This suggested a mixed Th1/Th2 immune response. Twenty distinct immunogenic proteins were identified, of which several had already been found in previous studies, e.g. enolase, fructose-bisphosphate aldolase, triosephosphate isomerase and pyruvate kinase (Thomas et al. 2006c).

Likewise, the immunoproteome of *A. fumigatus* has been investigated by immunoproteomics. De Repentigny et al. (1991) used a rabbit model to study the serologic response to *A. fumigatus* antigens. They separated mycelial proteins, which had been purified by concanavalin–sepharose chromatography before, by SDS-PAGE and 2D-PAGE. They detected three antigens (41, 54, 71 kDa), which were derived from one single protein. López-Medrano et al. (1995) separated cytosolic fraction of mycelial extracts from the three fungi *A. nidulans*, *A. flavus* and *A. fumigatus* by SDS-PAGE and 2D-PAGE. The proteins were transferred to nitrocellulose membranes and were incubated with serum samples from aspergilloma patients. In 90% of all cases four main antigenic bands (designated p90, p60, p40, p37) were recognised in cytosolic fractions from *A. fumigatus*, but not in fractions from *A. nidulans* or *A. flavus*. All four bands consisted of several N-glycosylated isoforms, but they were not characterised further. Chou et al. (1999) identified an alkaline serine proteinase from *A. flavus* as major IgE-binding allergen in sera from asthmatic patients. IgE cross-reactivity could be found for extracts from *A. flavus*, *A. fumigatus* and *Penicillium citrinum*. Additionally, the authors concluded that the results should be useful in clinical fungal allergy analyses. The same group also found a vacuolar serine proteinase as a major allergen of *A. fumigatus* (Shen et al. 2001). Again, cross-reactivity was also detected for proteins of other fungi. The N-terminal sequence of the characterised protease was not identical to the previous antigen from *A. flavus*. Presumably the number of antigenic active serine proteases is quite diverse, since Nigam et al. (2003) found another allergenic serine protease from *A. fumigatus* by lectin affinity chromatography. Lai et al. (2002) discovered that an enolase, a cytosolic enzyme active in glycolysis, is an allergen from *P. citrinum* and *A. fumigatus*. They used sera from asthmatic patients for IgE-binding assays. Asif et al. (2006) used conidia, but not mycelium for the identification of potential allergens and vaccine candidates, since conidia mediate the first contact with the immune response of the host. They extracted conidial surface proteins by treatment of the conidia with β-1,3 glucanase. After TCA precipitation, the extracted conidial surface proteins were separated by 2D-PAGE. Overall, 26 different conidial proteins were identified and 12 proteins contained a signal peptide for secretion. Besides hypothetical proteins, aspartic endopeptidase, surface protein rodlet A, an extracellular lipase, a putative disulfide isomerase and the allergen Aspf3 were identified. So far, the identified proteins have not been tested for allergenic activity by using sera from *Aspergillus*-infected patients yet.

G. Secretome

Secreted proteins of pathogenic fungi have gained a lot of attention since they are potential virulence factors and potential diagnostic markers (for proteinases in *C. albicans*, see Naglik et al. 2004), but only a few studies of extracellular proteins of filamentous fungi have been carried out by proteomic approaches.

The aspartyl proteinase Sap2p was shown to be responsible for the digestion of the gastrointestinal mucosa around *C. albicans* colonies in the host. This enzyme was found in culture filtrates and was recognised by immunoblotting with an anti-secretory aspartyl proteinase (Colina et al. 1996).

In *A. flavus* the extracellular enzymes secreted during growth on the flavonoid rutin were compared with those enzymes secreted when the fungus was grown in common rich medium (Medina et al. 2004). This is of interest, since some flavonoids have antimicrobial properties, particularly against filamentous fungi. In this study, after lyophilisation of the supernatant and subsequent TCA precipitation, 15 rutin-induced proteins and seven non-induced proteins were identified from 2D gels. Nevertheless, 90 spots remained unidentified by MALDI-TOF. So, in a following study, the authors analysed the secreted proteins of *A. flavus* by nano-LC MS/MS. A total of 51 unique extracellular proteins were identified, of which 18 were only secreted during growth on rutin. For instance, the protein quercetin 2,3-dioxygenase involved in flavonoid degradation and a chitinase were specifically induced.

Oda et al. (2006) carried out a comparative proteome analysis of extracellular proteins in solid-state and liquid cultures of *A. oryzae*. This filamentous fungus has received increasing attention as a host for heterologous protein production. Secreted proteins were precipitated by ammonium sulfate precipitation and analysed by 2D-PAGE afterwards. The secreted proteins could be classified into four groups: Group 1 consisted of proteins specifically produced in solid-state growth conditions, such as glucoamlyase B and alanyl dipeptidyl peptidase. Group 2 was represented by proteins specifically produced during liquid cultivation, e.g. glucoamylase A and xylanase G2. Xylanase G1 was produced under both conditions (group 3) and group 4 consisted of proteins that were secreted to the medium during solid-state cultivation, but trapped in the cell wall in the submerged conditions, e.g. α-amylase and β-glucosidase. Interestingly, the secretion of proteins was regulated at the post-transcriptional level at many cases.

No detailed 2D-PAGE study of extracellular proteins of *A. fumigatus* has been carried out so far. Schwienbacher et al. (2005) characterised the major *in vitro* secreted proteins of *A. fumigatus* by SDS-PAGE and Edman degradation or MS analysis. The three major proteins secreted during growth were ribotoxin, mitogillin and a chitosanase. Antichitosanase and antimitogillin antibodies were detected in the sera of patients suffering from invasive aspergillosis, but not in the control sera of healthy subjects.

H. Virulence Factors

Proteomic techniques allow us to characterise and to elucidate putative virulence determinants of pathogenic fungi. A prerequisite for invasive candidiasis is the ability of *C. albicans* to penetrate the mucosal barriers. Many pathogenic micro-organisms produce receptors for plasminogen, which can be converted into the serine protease plasmin by mammalian plasminogen activators. This activated form of plasminogen can degrade extracellular matrix proteins and promote migration across the human extracellular matrix. Both yeast cells and isolated cell wall proteins of *C. albicans* were able to bind plasminogen via C-terminal lysine residues (Crowe et al. 2003). Proteomic analysis identified eight major plasminogen-binding proteins in a cell wall fraction obtained by digestion with β-1,3 glucanase: phosphoglycerate mutase, alcohol dehydrogenase, thioredoxin peroxidase, catalase, transcription elongation factor, glyceraldehyde-3-phosphate dehydrogenase, phosphoglycerate kinase and fructose bisphosphate aldolase. Bound plasminogen degraded the blood-clotting protein fibrin, but a role for plasminogen in enhancing the virulence of *C. albicans* was not shown (Crowe et al. 2003).

An important molecule involved in the morphogenetic switch between yeast and hyphal form is the transcription factor Efg1p (Stoldt et al. 1997). *Candida* strains deleted in *EFG1* are avirulent, since they are unable to form true hyphae under most conditions. Deletion of the general regulator Efg1p may also have effects other than missing filamentation which contribute to the avirulence of Δ*efg1*/Δ*efg1* mutants. Additionally, adherence to host cells may be effected by a *EFG1* deletion. For this reason Saville et al. (2006) investigated the role of Efg1p in regulating the interaction between *Candida* cells and extracellular matrix. The Δ*efg1*/Δ*efg1* mutant strain showed a reduced ability to interact with extracellular matrices. Proteomic analyses of the surface-associated proteins released by treatment with β-mercaptoethanol in the Δ*efg1*/Δ*efg1* mutant and the parental strain led to the identification of 30 proteins, whose expression levels varied. The identified proteins were glycolytic enzymes, chaperones and proteins with plasminogen-binding capacity found previously by Crowe et al. (2003). Downstream effectors of Efg1p and Cph1p, another important transcription factor for yeast-to-hyphae transition, were also identified by Sohn et al. (2005). Two isoforms of the protein Cor33p were strongly down-regulated in 2D gels of a Δ*cph1*/Δ*efg1* double mutant in comparison with wild-type cells. Cor33p shared homology with phenylcoumarin benzylic ether reductases and isoflavone reductase and conferred tolerance towards oxidative stress due to the fact, that Δ*cor33* deletion mutants showed increased sensitivity to hydrogen peroxide.

J. Host–Pathogen Interaction

It is particularly important to study the interaction of immune effector cells with fungi to discover which mechanisms are responsible for the killing of the pathogens. The innate immune system is the most important line of defence against *Candida* and *Aspergillus* infections. Both macrophages and neutrophilic granulocytes contribute to the killing of *C. albicans* and *A. fumigatus* by their phagocytic and microbicidal activity. Several studies have been carried out to investigate the induced changes of the proteome in macrophages when they are in contact with *C. albicans*. Martinez-Solano et al. (2006) added *C. albicans* yeast cells to a mouse macrophages cell line. After 45 min of co-incubation, macrophage proteins were extracted with detergents and separated by 2DE. Several processes were affected in the macrophages upon interaction with *C. albicans* cells: cytoskeletal organisation, oxidative response mediated by annexin 1 and Rho-GTPases signalling, protein biosynthesis and refolding. Shin and coworkers (2005b) used a quite similar approach based on 2DE. They incubated *C. albicans* with macrophages for 3 h and examined the changes of protein expression by 2DE. Enzymes involved in glucose metabolism were significantly down-regulated. The authors assumed that energy depletion might be one of the key consequences of *C. albicans* infections. In addition, proteins involved in the maintenance of cell integrity were up-regulated, whereas the level of enzymes involved in NO production were reduced. The authors suggested that the observed effects might contribute to the death of *C. albicans*-infected macrophages. In a later publication, the same authors reported a suppression of galectin-3 in macrophages using the 2D LC separation system Proteome Lab PF2D for a better separation of basic proteins. Galectin-3 is a relatively high abundant lectin in macrophages and contributes to phagocytosis (Shin et al. 2006).

Not only macrophages display a significant change of the proteome after confrontation with *C. albicans*. Also, *C. albicans* responds to circumvent phagocytosis. Fernández-Arenas et al. (2007) analysed the response of *C. albicans* after encountering macrophages by 2D-PAGE. A protocol for the separation of yeast cells from macrophages could be established and 132 differentially expressed yeast protein species were found. A high percentage of metabolic proteins were affected indicating a response to nutrient starvation. Furthermore, the levels of stress-related proteins (chaperones, detoxification system) as well as proteins belonging to the proteasome increased. After an integrative analysis with transcriptomic data, the authors speculated that different cell death pathways mediated by mitochondria are activated in *C. albicans* yeast cells after contact with macrophages.

VIII. Conclusion: Towards Integrative Analysis–Systems Biology

Many proteomic studies on fungal organisms have used the yeast *S. cerevisiae*. The application of established methods of proteomics to human-pathogenic fungi such as *C. albicans* and *A. fumigatus* has enormous potential. It is safe to say that the development of new techniques in the field of proteomics and other "omics" technologies (transcriptomics, metabolomics) will even accelerate the accumulation of knowledge about the biology and pathogenicity of *C. albicans* and *A. fumigatus* in the future. The emergence of large-scale studies has come along with a change in biology from a classical, reductionistic approach, where individual genes or proteins are studied, to an integrative, holistic approach. Systems biology is now an emerging field in life science. It describes and analyses quantitatively the interaction among all the individual components of a cell. The final aim of this approach is to develop computational models which can describe and predict biological systems (Ideker et al. 2001; Ge et al. 2003). To reach this ambitious target, an interdisciplinary collaboration of biologists, technologists and computational scientists is needed.

Up to now, the tools for systems biology are still in their infancies. There are only a few examples of the application of systems approaches to the biology of human-pathogenic fungi. Gene regulatory network models have been described for *A. fumigatus* and *A. nidulans* (David et al. 2006; Guthke et al. 2007), but proteomic data have not been processed and analysed by mathematical models so far (Albrecht et al. 2008). Hence, also integrative data analyses of transcriptome and proteome data of *C. albicans* and *A. fumigatus* are missing. Besides the development of computational tools for the analysis of data sets, the set-up and development of databases and data warehouses play a crucial part on the way towards an integrative analysis of the biology of *C. albicans* and *A. fumigatus* (Albrecht et al. 2007b). So far, no databases are available which incorporate experimental data from different cellular levels (transcriptome, proteome, metabolome) of *C. albicans* and *A. fumigatus*.

References

Aebersold R, Goodlett DR (2001) Mass spectrometry in proteomics. Chem Rev 101:269–295

Aebersold R, Mann M (2003) Mass spectrometry-based proteomics. Nature 422:198–207

Akins RE, Levin PM, Tuan RS (1992) Cetyltrimethylammonium bromide discontinuous gel electrophoresis: M_r-based separation of proteins with retention of enzymatic activity. Anal Biochem 202:172–178

Alban A, David SO, Bjorkesten L, Andersson C, Sloge E, Lewis S, Currie I (2003) A novel experimental design for comparative two-dimensional gel analysis: two dimensional difference gel electrophoresis incorporating a pooled internal standard. Proteomics 3:36–44

Albrecht A, Felk A, Pichova I, Naglik JR, Schaller M, de Groot P, Maccallum D, Odds FC, Schäfer W, Klis F, Monod M, Hube B (2006) Glycosylphosphatidylinositol-anchored proteases of *Candida albicans* target proteins necessary for both cellular processes and host–pathogen interactions. J Biol Chem 281:688–694

Albrecht D, Kniemeyer O, Brakhage AA, Berth M, Guthke R (2007) Integration of transcriptome and proteome data from human-pathogenic fungi by using a data warehouse. J Integr Bioinf 4:52

Albrecht D, Guthke R, Kniemeyer O, Brakhage AA (2008) Systems biology of human-pathogenic fungi. In: Daskalaki A (ed) Handbook of research on systems biology applications in medicine. Information Science Reference, Hershey (in press)

Andersen JS, Mann M (2006) Organellar proteomics: turning inventories into insights. EMBO Rep 7:874–879

Anderson N, Anderson N (1978a) Analytical techniques for cell fractions. XXII. Two-dimensional analysis of serum and tissue proteins: multiple gradient-slab gel electrophoresis. Anal Biochem 85:341–354

Anderson N, Anderson N (1978b) Analytical techniques for cell fractions. XXI. Two-dimensional analysis of serum and tissue proteins: multiple isoelectric focusing 1. Anal Biochem 85:331–340

Andersson L, Porath J (1986) Isolation of phosphoproteins by immobilized metal (Fe^{3+}) affinity chromatography. Anal Biochem 154:250–254

Asif AR, Oellerich M, Amstrong VW, Riemenschneider B, Monod M, Reichard U (2006) Proteome of conidial surface associated proteins of *Aspergillus fumigatus* reflecting potential vaccine candidates and allergens. J Proteome Res 5:954–962

Bader T, Schröppel K, Bentink S, Agabian N, Köhler G, Morschhäuser J (2006) Role of calcineurin in stress resistance, morphogenesis, and virulence of a *Candida albicans* wild-type strain. Infect Immun 74:4366–4369

Bahn YS, Xue C, Idnurm A, Rutherford JC, Heitman J, Cardenas ME (2007) Sensing the environment: lessons from fungi. Nat Rev Microbiol 5:57–69

Barea PL, Calvo E, Rodriguez JA, Rementeria A, Calcedo R, Sevilla MJ, Ponton J, Hernando FL (1999) Characterization of *Candida albicans* antigenic determinants by two-dimensional polyacrylamide gel electrophoresis and enhanced chemiluminescence. FEMS Immunol Med Microbiol 23:343–354

Beauvais A, Schmidt C, Guadagnini S, Roux P, Perret E, Henry C, Paris S, Mallet A, Prévost MC, Latgé JP (2007) An extracellular matrix glues together the aerial-grown hyphae of *Aspergillus fumigatus*. Cell Microbiol 9:1588–1600

Beavis RC, Chait BT (1990) Rapid, sensitive analysis of protein mixtures by mass spectrometry. Proc Natl Acad Sci USA 87:6873–6877

Bell PJ, Karuso P (2003) Epicocconone, a novel fluorescent compound from the fungus *Epicoccum nigrim*. J Am Chem Soc 125:9304–9305

Bellocchio S, Bozza S, Montagnoli C, Perruccio K, Gaziano R, Pitzurra L, Romani L (2005) Immunity to *Aspergillus fumigatus*: the basis for immunotherapy and vaccination. Med Mycol 43: S181–S188

Berggren KN, Chernokalskaya E, Steinberg TH, Kemper C, Lopez MF, Diwu Z, Haugland RP, Patton WF (2000) Background-free, high sensitivity staining of proteins in one- and two-dimensional sodium dodecyl sulfate-polyacrylamide gels using luminescent ruthenium complex. Electrophoresis 21:2509–2521

Berggren KN, Schulenberg B, Lopez MF, Steinberg TH, Bogdanova A, Smejkal G, Wang A, Patton WF (2002) An improved formulation of SYPRO Ruby protein gel stain: comparison with the original formulation and with a ruthenium II tris (bathophenanthroline disulfonate) formulation. Proteomics 2:486–498

Biron DG, Brun C, Lefevre T, Lebarbenchon C, Loxdale HD, Chevenet F, Brizard JP, Thomas F (2006) The pitfalls of proteomics experiments without the correct use of bioinformatics tools. Proteomics 6:5577–5596

Bjellqvist B, Ek K, Righetti PG, Gianazza E, Görg A, Westermeier R, Postel W (1982) High-resolution two-dimensional electrophoresis with isoelectric focusing in immobilized pH gradients. J Biochem Biophys Methods 6:317–339

Blomberg A (2002) Use of two-dimensional gels in yeast proteomics. Methods Enzymol 350:559–584

Bonenfant D, Schmelzle T, Jacinto E, Crespo JL, Mini T, Hall MN, Jenoe P (2003) Quantitation of changes in protein phosphorylation: a simple method based on stable isotope labeling and mass spectrometry. Proc Natl Acad Sci USA:100:880–885

Bradford MM (1976) A rapid and sensitive method for the quantitation of microgram quantities of protein utilizing the principle of protein–dye binding. Anal Biochem 7:248–254

Brakhage AA (2005) Systemic fungal infections caused by *Aspergillus* species: Epidemiology, infection process and virulence determinants. Curr Drug Targets 6:875–886

Brakhage AA, Langfelder K (2002) Menacing mold: the molecular biology of *Aspergillus fumigatus*. Annu Rev Microbiol 56:433–455

Brakhage AA, Liebmann B (2005) *Aspergillus fumigatus* conidial pigment and cAMP signal transduction: significance for virulence. Med Mycol 43:S75–S82

Braun BR et al (2005) A human-curated annotation of the *Candida albicans* genome. PloS Genet 1(1):e1

Brazma A, Hingamp P, Quackenbush J, Sherlock G, Spellman P, Stoeckert C, Aach J, Ansorge W, Ball CA, Causton HC, Gaasterland T, Glenisson P, Holstege FC, Kim IF, Markowitz V, Matese JC, Parkinson H, Robinson A, Sarkans U, Schulze-Kremer S, Stewart J, Tayloer R, Vilo J, Vingron M (2001) Minimum information about a microarray experiment (MIAME)-toward standards for microarray data. Nat Genet 29:365–371

Bruneau JM, Magnin T, Tagat E, Legrand R, Bernard M, Diaquin M, Fudali C, Latgé J-P (2001) Proteome analysis of *Aspergillus fumigatus* identifies glycosylphosphatidylinositol-anchored proteins associated to the cell wall biosynthesis. Electrophoresis 22:2812–2823

Bruneau JM, Maillet I, Tagat E, Legrand R, Supatto F, Fudali C, Le Caer JP, Labas V, Lecaque D, Hodgson J (2003) Drug induced proteome changes in *Candida albicans*: comparison of the effect of α(1,3) glucan synthase inhibitors and two triazoles, fluconazole and itraconazole. Proteomics 3:325–336

Burré J, Beckhaus T, Schägger H, Corvey C, Hofmann S, Karas M, Zimmermann H, Volknandt W (2006) Analysis of the synaptic vesicle proteome using three gel-based protein separation techniques. Proteomics 6:6250–6262

Candiano G, Bruschi M, Musante L, Santuci L, Ghiggeri GM, Carnemolla B, Orecchia P, Zardi L, Righetti PG (2004) Blue silver: a very sensitive colloidal Coomassie G-250 staining for proteome analysis. Electrophoresis 25:1327–1333

Carberry S, Neville CM, Kavanagh KA, Doyle S (2006) Analysis of major intracellular proteins of *Aspergillus fumigatus* by MALDI mass spectrometry: Identification and characterisation of an elongation factor 1B protein with glutathione transferase activity. Biochem Biophys Res Commun 341:1096–1104

Carrette O, Burkhard PR, Sanchez J-C, Hochstrasser DF (2006) State-of-the-art two-dimensional gel electrophoresis: a key tool of proteomics research. Nat Protoc 1:812–823

Cho A, Normile D (2002) Nobel prize in chemistry: mastering macromolecules. Science 298:527–528

Choi W, Yoo YJ, Kim M, Shin D, Jeon HB, Choi W (2003) Identification of proteins highly expressed in the hyphae of *Candida albicans* by two-dimensional electrophoresis. Yeast 20:1053–1060

Chou H, Lin WL, Tam MF, Wang SR, Han SH, Shen HD (1999) Alkaline serine proteinase is a major allergen of Aspergillus flavus, a prevalent airborne *Aspergillus* species in the Taipei area. Int Arch Allergy Immunol 119:282–290

Cobitz AR, Yim EH, Brown WR, Perou CM, Tamanoi F (1989) Phosphorylation of RAS1 and RAS2 proteins in *Saccharomyces cerevisiae*. Proc Natl Acad Sci 86:858–862

Colina AR, Aumont F, Deslauriers N, Belhumeur P, De Repentigny L (1996) Evidence for degradation of gastrointestinal mucin by *Candida albicans* secretory aspartyl proteinase. Infect Immun 64:4514–4519

Crowe JD, Sievwright IK, Auld GC, Moore NR, Gow NAR, Booth NA (2003) *Candida albicans* binds human plasminogen: identification of eight plasminogen-binding proteins. Mol Microbiol 47:1637–1651

Damerval C, Devienne D, Zivy M, Thiellement H (1986) Technical improvements in two-dimensional electrophoresis increase the level of genetic variation detected in wheat-seedling proteins. Electrophoresis 7:52–54

Da Silva Ferreira ME, Kress MR, Savoldi M, Goldman MH, Hartl A, Heinekamp T, Brakhage AA, Goldman GH (2006a) The akuB(KU80) mutant deficient for nonhomologous end joining is a powerful tool for analyzing pathogenicity in *Aspergillus fumigatus*. Eukaryot Cell 5:207–211

Da Silva Ferreira ME, Malavazi I, Savoldi M, Brakhage AA, Goldman MH, Kim HS, Nierman WC, Goldman GH (2006b) Transcriptome analysis of *Aspergillus fumigatus* exposed to voriconazole. Curr Genet 50:32–44

David H, Hofmann G, Oliveira AP, Jarmer H, Nielsen J (2006) Metabolic network driven analysis of genome-wide transcription data from *Aspergillus nidulans*. Genome Biol 7:R108

De Backer J, De Hoogt RA, Froyen G, Odds FC, Simons F, Contreras R, Luyten WHM (2000a) Single allele knock-out of *Candida albicans* CGT1 leads to unexpected resistance to hygromycin B and elevated temperature. Microbiology 146:353–365

De Backer MD, Magee PT, Pla J (2000b) Recent developments in molecular genetics of *Candida albicans*. Annu Rev Microbiol 54:463–498

De Godoy LFM, Olsen JV, De Souza GA, Li G, Mortensen P, Mann M (2006) Status of complete proteome analysis by mass spectrometry: SILAC labeled yeast as model system. Genome Biol 7:R50

De Groot PWJ, De Boer AD, Cunningham J, Dekker HL, De Jong L, Hellingwerf KJ, De Koster C, Klis FM (2004) Proteomic analysis of *Candida albicans* cell walls reveals covalently bound carbohydrate-active enzymes and adhesins. Eukaryot Cell 3:955–965

De Groot PW, Ram AF, Klis FM (2005) Features and functions of covalently linked proteins in fungal cell walls. Fungal Genet Biol 42:657–675

De Hoog CL, Mann M (2004) Proteomics. Annu Rev Genomics Hum Genet 5:267–293

Delom F, Szponarski W, Sommerei N, Boyer JC, Bruneau JM, Rossignol M, Gibrat R (2006) The plasma membrane proteome of *Saccharomyces cerevisiae* and its response to the antifungal calcofluor. Proteomics 6:3029–3039

De Repentigny L, Kilanowski E, Pedneault L, Boushira M (1991) Immunoblot analyses of the serologic response to *Aspergillus fumigatus* antigens in experimental invasive aspergillosis. J Infect Dis 163:1305–1311

Domon B, Aebersold R (2006) Mass spectrometry and protein analysis. Science 312:212–217

Ebanks RO, Chisholm K, McKinnon S, Whiteway M, Pinto DM (2006) Proteomic analysis of *Candida albicans* yeast and hyphal cell wall and associated proteins. Proteomics 6:2147–2156

Elias JE, Gygi SP (2007) Target-decoy search strategy for increased confidence in large-scale protein identification by mass spectrometry. Nat Methods 4:207–214

Elias JE, Hass W, Faherty BK, Gygi SP (2005) Comparative evaluation of mass spectrometry platforms used in large-scale proteomics investigations. Nat Methods 2:667–675

Fenn JB, Mann M, Meng CK, Wong SF, Whitehouse CM (1989) Electrospray ionisation for the mass spectrometry of large biomolecules. Science 246:64–71

Fernández-Arenas E, Molero G, Nombela C, Diez-Orejas R, Gil C (2004a) Contribution of the antibodies response induced by a low virulent *Candida albicans* strain in protection against systemic candidiasis. Proteomics 4:1204–1215

Fernández-Arenas E, Molero G, Nombela C, Diez-Orejas R, Gil C (2004b) Low virulent strains of *Candida albicans*: unravelling the antigens for a future vaccine. Proteomics 4:3007–3020

Fernández-Arenas E, Cabezón V, Bermejo C, Arroyo J, Nombela C, Diez-Orejas R, Gil C (2007) Integrated proteomic and genomic strategies bring new insight into *Candida albicans* response upon macrophage interaction. Mol Cell Proteomics 6:460–478

Ficarro SB, McCleland ML, Stukenberg PT, Burke DJ, Ross MM, Shabanowitz J, Hunt DF, White FM (2002) Phosphoproteome analysis by mass spectrometry and its application to *S. cerevisiae*. Nat Biotechnol 20:301–305

Fountoulakis M, Takács B (2001) Effect of strong detergents and chaotropes on the detection of proteins in two-dimensional gels. Electrophoresis 22:1593–1602

Futcher B, Latter GI, Monardo P, McLaughlin CS, Garrels JI (1999) A sampling of the yeast proteome. Mol Cell Biol 19:7357–7368

Galagan JE et al (2005) Sequencing of *Aspergillus nidulans* and comparative analysis with *A. fumigatus* and *A. oryzae*. Nature 438:1105–1115

Ge H, Walhout AJM, Vidal M (2003) Integrating 'omic' information: a bridge between genomics and systems biology. Trends Genet 19:551–560

Ghannoum MA, Spellberg B, Saporito-Irwin SM, Fonzi WA (1995) Reduced virulence of *Candida albicans* PHR1 mutants. Infect Immun 63:4528–4530

Görg A, Obermaier C, Boguth G, Harder A (2000) The current state of two-dimensional electrophoresis with immobilized pH gradients. Electrophoresis 21:1037–1053

Görg A, Weiss W, Dunn MJ (2004) Current two-dimensional electrophoresis technology for proteomics. Proteomics 4:3665–3685

Griffin TJ, Gygi SP, Ideker T, Rist B, Eng J, Hood L, Aebersold R (2002) Complementary profiling of gene expression at the transcriptome and proteome levels in *Saccharomyces cerevisiae*. Mol Cell Proteomics 1:323–333

Gruhler A, Olsen JV, Mohammed S, Mortensen P, Faergeman NJ, Mann M, Jensen ON (2005) Quantitative phosphoproteomics applied to the yeast pheromone signaling pathway. Mol Cell Proteomics 4:310–327

Guthke R, Kniemeyer O, Albrecht D, Brakhage AA, Moeller U (2007) Discovery of gene regulatory networks in *Aspergillus fumigatus*. Lect N Bioinf 4366:22–41

Gygi SP, Rist B, Gerber SA, Turecek F, Gelb MH, Aebersold R (1999) Quantitative analysis of complex protein mixtures using isotope-coded affinity tags. Nat Biotechnol 17:994–999

Harder A, Wildgruber R, Nawrocki A, Fey SJ, Larsen PM, Görg A (1999) Comparison of yeast cell protein solubilisation procedures for two-dimensional electrophoresis. Electrophoresis 20:826–829

Hayes RN, Gross ML (1990) Collision-induced dissociation. Methods Enzymol 193:237–263

Henningsen R, Gale BL, Straub KM, DeNagel DC (2002) Application of zwitterionic detergents to the solubilization of integral membrane proteins for two-dimensional gel electrophoresis and mass spectrometry. Proteomics 2:1479–1488

Hermjakob H (2006) The HUPO proteomics standards initiative – overcoming the fragmentation of proteomics data. Pract Proteomics 1/2:34–38

Hernández R, Nombela C, Diez-Orejas R, Gil C (2004) Two-dimensional reference map of *Candida albicans* hyphal forms. Proteomics 4:374–382

Hooshdaran MZ, Barker KS, Hilliard GM, Kusch H, Morschhäuser J, Rogers PD (2004) Proteomic analysis of azole resistance in *Candida albicans* clinical isolates. Antimicrob Agents Chemother 48:2733–2735

Hsiang T, Baillie DL (2006) Issues in comparative fungal genomics. In: Arora DK, Berka R, Singh GB (eds) Bioinformatics. (Applied mycology and biotechnology, vol 6) Elsevier, Amsterdam, pp 1–26

Hube B (2006) Infection-associated genes of *Candida albicans*. Future Microbiol 1:209–218

Hufnagel P, Rabus R (2006) Mass spectrometric identification of proteins in complex post-genomic projects. J Mol Microbiol Biotechnol 11:53–81

Hunt DF, Yates JR III, Shabanowitz J, Winston S, Hauer CR (1986) Protein sequencing by tandem mass spectrometry. Proc Natl Acad Sci USA 83:6233–6237

Hunt SMN, Thomas MR, Sebastian LT, Pedersen SK, Harcourt RL, Sloane AJ, Wilkins MR (2005) Optimal replication and the importance of experimental design for gel-based quantitative proteomics. J Proteome Res 4:809–819

Hunter T (2000) Signaling-2000 and beyond. Cell 100:113–127

Ideker T, Galitski T, Hood L (2001) A new approach to decoding life: systems biology. Annu Rev Genomics Hum Genet 2:343–372

Insenser M, Nombela C, Molero G, Gil C (2006) Proteomic analysis of detergent-resistant membranes from *Candida albicans*. Proteomics 6:S74–S81

Isaacson T, Damasceno CMB, Saravanan RS, He Y, Catalá C, Saladié M, Rose JKC (2006) Sample extraction techniques for enhanced proteomic analysis of plant tissues. Nat Protoc 1:769–774

Ishihama Y, Oda Y, Tabata T, Sato T, Nagasu T, Rappsilber J, Mann M (2005) Exponentially modified protein abundance index (emPAI) for estimation of absolute protein amount in proteomics by the number of sequenced peptides per protein. Mol Cell Proteomics 4:1265–1272

Jones T, Federspiel NA, Chibana H, Dungan J, Kalman S, Magee BB, Newport G, Thorstenson YR, Agabian N, Magee PT, Davis RW, Scherer S (2004) The diploid genome sequence of *Candida albicans*. Proc Natl Acad Sci 101:7329–7334

Kang C, Kim HJ, Kang D, Jung DY, Suh M (2003) Highly sensitive and simple fluorescence staining of proteins in sodium dodecyl sulfate–polyacrylamide-based gels by using hydrophobic tail-mediated enhancement of fluorescein luminescence. Electrophoresis 24:3297–3304

Kang D, Gho YS, Su M, Kang C (2002) Highly sensitive and fast protein detection with Coomassie brilliant blue in sodium dodecyl sulfate–polyacrylamide gel electrophoresis. 23:1511–1512

Karas M, Hillenkamp F (1988) Laser desorption ionisation of proteins with molecular mass exceeding 100,00 daltons. Anal Chem 60:2299–2301

Kaufmann H, Bailey JE, Fussenegger M (2001) Use of antibodies for detection of phosphorylated proteins separated by two-dimensional gel electrophoresis. Proteomics 1:194–199

Kim Y, Nandakumar MP, Marten MR (2007) Proteome map of *Aspergillus nidulans* during osmoadaptation. Fungal Genet Biol 44:886–895

Kinoshita E, Kinoshita-Kikuta E, Takiyama K, Koike T (2006) Phosphate-binding tag, a new tool to visualise phosphorylated proteins. Mol Cell Proteomics 5:749–757

Kinoshita-Kikuta E, Kinoshita E, Yamada A, Endo M, Koike T (2006) Enrichment of phosphorylated proteins from cell lysate using a novel phosphate-affinity chromatography at physiological pH. Proteomics 6:5088–5095

Klose J (1975) Protein mapping by combined isoelectric focusing and electrophoresis of mouse tissue: a novel approach to testing for induced point mutations in mammals. Humangenetik 26:231–243

Klose J, Kobalz U (1995) Two-dimensional electrophoresis of proteins: an updated protocol and implications for a functional analysis of the genome. Electrophoresis 16:1034–1059

Kniemeyer O, Lessing F, Scheibner O, Hertweck C, Brakhage AA (2006) Optimisation of a 2-D gel electrophoresis protocol for the human-pathogenic fungus *Aspergillus fumigatus*. Curr Genet 49:178–189

Knowles MR, Cervino S, Skynner HA et al (2003) Multiplex proteomics analysis by two-dimensional differential in-gel electrophoresis. Proteomics 3:1162–1171

Kolkman A, Dirksen EHC, Slijper M, Heck AJR (2005) Double standards in quantitative proteomics. Mol. Cell Proteomics 4:255–266

Kontoyiannis DP, Lewis RE (2002) Antifungal drug resistance of pathogenic fungi. Lancet 359:1135–1144

Kramer ML (2006) A new multiphasic buffer system for benzyldimethyl-*n*-hexadecylammonium chloride polyacrylamide gel electrophoresis of proteins providing efficient stacking. Electrophoresis 27:347–356

Krappmann S, Sasse C, Braus GH (2006) Gene targeting in *Aspergillus fumigatus* by homologous recombination is facilitated in a nonhomologous end-joining-deficient genetic background. Eukaryot Cell 5:212–215

Kusch H, Engelmann S, Albrecht D, Morschhäuser J, Hecker M (2007a) Proteomic analysis of the oxidative stress response in *Candida albicans*. Proteomics 7:686–697

Kusch H, Engelmann S, Bode R, Albrecht D, Morschhäuser J, Hecker M (2007b) A proteomic view of *Candida albicans* yeast cell metabolism in exponential and stationary growth phases. Int J Med Microbiol 298:291–318

Kussmann M, Nordhoff E, Rahbek-Nielsen H, Haebel S, Rossel-Larsen M, Jakobsen L, Gobom J, Mirgorodskaya E, Kroll Kristensen A, Palm L, Roepstorff (1997) MALDI-MS sample preparation techniques designed for various peptide and protein analytes. J Mass Spectrom 32:593–601

Laemmli UK (1970) Cleavage of structural proteins during assembly of the head of bacteriophage T4. Nature 227:680–685

Lai HY, Tam MF, Tang RB, Chou H, Chang CY, Tsai JJ, Shen HD (2002) cDNA cloning and immunological characterization of a newly identified enolase allergen from

Penicillium citrinum and *Aspergillus fumigatus*. Int Arch Allergy Immunol 127:181–190

Langfelder K, Jahn B, Gehringer H, Schmidt A, Wanner G, Brakhage AA (1998) Identification of a polyketide synthase gene (*pksP*) of *Aspergillus fumigatus* involved in conidial pigment biosynthesis and virulence. Med Microbiol Immunol 187:79–89

Latgé JP (2001) The pathobiology of *Aspergillus fumigatus*. Trends Microbiol 9:382–389

Latgé JP, Mouyna I, Tekaia F, Beauvais A, Debeaupuis JP, Nierman W (2005) Specific molecular features in the organisation and biosynthesis of the cell wall of *Aspergillus fumigatus*. Med Mycol 43:S15–S22

Leitner A, Lindner W (2006) Chemistry meets proteomics: The use of chemical tagging reactions for MS-based proteomics. Proteomics 6:5418–5434

Lessing F, Kniemeyer O, Wozniok I, Loeffler J, Kurzai O, Haertl A, Brakhage AA (2007) The *Aspergillus fumigatus* transcriptional regulator AfYap1 represents the major regulator for defense against reactive oxygen intermediates but is dispensable for pathogenicity in an intranasal mouse model. Eukaryot Cell 6:2290–2302

Lilley KS, Friedman (2004) All about DIGE: quantification technology for differential-display 2D-gel proteomics. Expert Rev Proteomics 1:401–409

Link AJ, Eng J, Schieltz DM, Carmack E, Mize GJ, Morris DR, Garvik BM, Yates JR III (1999) Direct analysis of protein complexes using mass spectrometry. Nat Biotechnol 17:676–682

López-Medrano R, Ovejero MC, Calera JA, Puente P, Leal F (1995) *Aspergillus fumigatus* antigens. Microbiology 141:2699–2704

Luche S, Santoni V, Rabilloud T (2003) Evaluation of nonionic and zwitterionic detergents as membrane protein solubilizers in two-dimensional electrophoresis. Proteomics 3:249–253

Macfarlane DE (1989) Two dimensional benzyldimethyl-n-hexadecylammonium chloride → sodium dodecyl sulfate preparative polyacrylamide gel electrophoresis: a high capacity high resolution technique for the purification of proteins from complex mixtures Anal Biochem 176:457–463

Mackintosh JA, Choi HY, Bae SH et al (2003) A fluorescent natural product for ultrasensitive detection of proteins in one-dimensional and two-dimensional gel electrophoresis. Proteomics 3:2273–2288

Magee BB, Magee PT (2005) Recent advances in the genomic analysis of *Candida albicans*. Rev Iberoam Micol 22:187–193

Mann M (2006) Functional and quantitative proteomics using SILAC. Nat Rev Mol Cell Biol 7:952–958

Mann M, Jensen ON (2003) Proteomic analysis of post-translational modifications. Nat Biotechnol 21:255–261

Manning M, Mitchell TG (1980a) Anaylsis of cytoplasmic antigens of the yeast and mycelial phases of *Candida albicans* by two-dimensional electrophoresis. Infect Immun 30:484–495

Manning M, Mitchell TG (1980b) Morphogenesis of *Candida albicans* and cytoplasmic proteins associated with differences in morphology, strain, or temperature. J Bacteriol 144:258–273

Marr KA, Carter RA, Crippa F, Wald A, Corey L (2002) Epidemiology and outcome of mould infections in hematopoietic stem cell transplant recipients. Clin Infect Dis 34:909–917

Martin SW, Konopka JB (2004) Lipid raft polarization contributes to hyphal growth in *Candida albicans*. Eukaryot Cell 3:675–684

Martinez-Solano L, Nombela C, Molero G, Gil C (2006) Differential protein expression of murine macrophages upon interaction with *Candida albicans*. Proteomics 6:S133–S144

Masuoka J, Glee PM, Hazen KC (1998) Preparative isoelectric focusing and preparative electrophoresis of hydrophobic *Candida albicans* cell wall proteins with in-line transfer to polyvinylidene difluoride membranes for sequencing. Electrophoresis 19:675–678

McCormack AL, Schieltz DM, Goode B, Yang S, Barnes G, Drubin D, Yates JR III (1997) Direct analysis and identification of proteins in mixtures by LC/MS/MS and database searching at the low-femtomole level. Anal Chem 69:767–776

McLachlin DT, Chait BT (2001) Analyis of phosphorylated proteins and peptides by mass spectrometry. Curr Opin Chem Biol 5:591–602

McLean JA, Ruotolo BT, Gillig KJ, Russel DH (2005) Ion mobility-mass spectrometry: a new paradigm for proteomics. Int J Mass Spectrom 240:301–315

Medina ML, Kiernan UA, Francisco WA (2004) Proteomic analysis of rutin-induced secreted proteins from *Aspergillus flavus*. Fungal Genet Biol 41:327–335

Melanson JE, Chisholm KA, Pinto DM (2006) Targeted comparative proteomics by liquid chromatography/matrix-assisted laser desorption/ionization triple-quadrupole mass spectrometry. Rapid Commun Mass Spectrom 20:904–910

Melin P, Schnürer J, Wagner EG (2002) Proteome analysis of *Aspergillus nidulans* reveals proteins associated with the response to the antibiotic concanamycin A, produced by *Streptomyces* species. Mol Genet Genomics 267:695–702

Miller I, Crawford J, Gianazza E (2006) Protein stains for proteomic applications: which, when, why? Proteomics 6:5385–5408

Mirgorodskaya E, Braeuer C, Fucini P, Lehrach H, Gobom J (2005) Nanoflow liquid chromatography coupled to matrix-assisted laser desorption/ionization mass spectrometry: sample preparation, data analysis, and application to the analysis of complex peptide mixtures. Proteomics 5:399–408

Miura K (2003) Imaging technologies for the detection of multiple stains in proteomics. Proteomics 3:1097–1108

Miyagi M, Rao KC (2007) Proteolytic ^{18}O-Labelling strategies for quantitative proteomics. Mass Spectrom Rev 26:121–126

Molloy MP (2000) Two-dimensional electrophoresis of membrane proteins using immobilized pH gradients. Anal Biochem 280:1–10

Morandell S, Stasyk T, Grosstessner-Hain K, Roitinger E, Mechtler K, Bonn GK, Huber LA (2006) Phosphoproteomics strategies for the functional analysis of signal transduction. Proteomics 6:4047–4056

Mouyna I, Fontaine T, Vai M, Monod M, Fonzi WA, Diaquin M, Popolo L, Hartland RP, Latgé J-P (2000) Glycosylphosphatidylinositol-anchored glucanosyltransferases play an active role in the biosynthesis of the fungal cell wall. J Biol Chem 275:14882–14889

Mukherjee PK, Zhou G, Munyon R, Ghannoum MA (2005) Candida biofilm: a well-designed protected environment. Med Mycol 43:191–208

Mukherjee PK, Mohamed S, Chandra J, Kuhn D, Liu S, Antar OS, Munyon R, Mitchell AP, Andes D, Chance MR, Rouabhia M, Ghannoum MA (2006) Alcohol dehydrogenase restricts the ability of the pathogen Candida albicans to form a biofilm on catheter surfaces through an ethanol-based mechanism. Infect Immun 74:3804–3816

Murry PF, Biscoglio MJ, Passeron S (2000) Purification and characterization of 20S proteasome: identification of four proteasomal subunits. Arch Biochem Biophys 375:211–219

Naglik J, Albrecht A, Bader O, Hube B (2004) Candida albicans proteinases and host/pathogen interactions. Cell Microbiol 6:915–926

Nandakumar MP, Marten MR (2002) Comparison of lysis methods and preparation protocols for one- and two-dimensional electrophoresis of Aspergillus oryzae intracellular proteins. Electrophoresis 23:2216–2222

Nandakumar MP, Shen J, Raman B, Marten MR (2003) Solubilization of TCA precipitated microbial proteins via NaOH for two-dimensional gel electrophoresis. J Proteome Res 2:89–93

Navarro-García F, Sánchez M, Nombela C, Pla J (2001) Virulence genes in the pathogenic yeast Candida albicans. FEMS Microbiol Rev 25:245–268

Neuhoff V, Arold N, Taube D, Ehrhardt W (1988) Improved staining of proteins in polyacrylamide gels including isoelectric focusing gels with clear background at nanogram sensitivity using Coomassie brilliant blue G-250 and R-250. Electrophoresis 9:255–262

Nierman WC, Pain A, Anderson MJ, Wortman JR, Kim HS, Arroyo J, Berriman M, Abe K, Archer DB, Bermejo C, Bennett J, Bowyer P, Chen D, Collins M, Coulsen R, Davies R, Dyer PS, Farman M, Fedorova N, Fedorova N, Feldblyum TV, Fischer R, Fosker N, Fraser A, García JL, García MJ, Goble A, Goldman GH, Gomi K, Griffith-Jones S, Gwilliam R, Haas B, Haas H, Harris D, Horiuchi H, Huang J, Humphray S, Jiménez J, Keller N, Khouri H, Kitamoto K, Kobayashi T, Konzack S, Kulkarni R, Kumagai T, Lafon A, Latgé JP, Li W, Lord A, Lu C, Majoros WH, May GS, Miller BL, Mohamoud Y, Molina M, Monod M, Mouyna I, Mulligan S, Murphy L, O'Neil S, Paulsen I, Peñalva MA, Pertea M, Price C, Pritchard BL, Quail MA, Rabbinowitsch E, Rawlins N, Rajandream MA, Reichard U, Renauld H, Robson GD, Rodriguez de Córdoba S, Rodríguez-Peña JM, Ronning CM, Rutter S, Salzberg SL, Sanchez M, Sánchez-Ferrero JC, Saunders D, Seeger K, Squares R, Squares S, Takeuchi M, Tekaia F, Turner G, Vazquez de Aldana CR, Weidman J, White O, Woodward J, Yu JH, Fraser C, Galagan JE, Asai K, Machida M, Hall N, Barrell B, Denning DW (2005) Genomic sequence of the pathogenic and allergenic filamentous fungus Aspergillus fumigatus. Nature 438:1151–1156

Nigam S, Ghosh PC, Sarma PU (2002) A new glycoprotein allergen/antigen with the protease activity from Aspergillus fumigatus. Int Arch Allergy Immunol 132:124–131

Niimi M, Shepherd MG, Monk BC (1996) Differential profiles of soluble proteins during the initiation of morphogenesis in Candida albicans. Arch Microbiol 166:260–268

Niimi M, Cannon RD, Monk BC (1999) Candida albicans pathogenicity: a proteomic perspective. Electrophoresis 20:2299–2308

O'Farrel PH (1975) High resolution two dimensional gel electrophoresis of proteins. J Biol Chem 250:4007–4021

Oda K, Kakizono D Yamada O, Iefuji H, Akita O, Iwashita K (2006) Proteomic analysis of extracellular proteins from Aspergillus oryzae grown under submerged and solid-state culture conditions. Appl Environ Microbiol 72:3448–3457

Odds FC, Brown AJP, Gow NAR (2003) Antifungal agents: mechanisms of action. Trends Microbiol 11:272–279

Odds FC, Gow NAR, Brown AJP (2001) Fungal virulence studies come of age. Genome Biol 2:1009.1–1009.4

Odds FC, Gow NAR, Brown AJP (2006) Towards a molecular understanding of Candida albicans virulence. In: Heitman J et al (eds) Molecular principles of fungal pathogenesis, ASM Press, Washington, D.C., pp 305–321

Ong S-E, Mann M (2005) Mass spectrometry-based proteomics turns quantitative. Nat Chem Biol 1:252–262

Pandey A, Podtelejnikov AV, Blagoev B, Bustelo XR et al (2000) Analysis of receptor signaling pathways by mass spectrometry: identification of vav-2 as a substrate of the epidermal and platelet-derived growth factor receptors. Proc Natl Acad Sci USA 97:179–184

Pardo M, Ward M, Pitarch A, Sánchez M, Nombela C, Blackstock W, Gil C (2000) Cross-species identification of novel Candida albicans immunogenic proteins by combination of two-dimensional polyacrylamide gel electrophoresis and mass spectrometry. Electrophoresis 21:2651–2659

Patterson SD (2004) How much of the proteome do we see with discovery-based proteomics methods and how much do we need to see? Curr Proteomics 1:3–12

Patterson SD, Aebersold RH (2003) Proteomics: the first decade and beyond. Nat Genet 33:311–323

Perlroth J, Choi B, Spellberg B (2007) Nosocomial fungal infections: epidemiology, diagnosis, and treatment. Med Mycol 45:321–346

Pfaller MA, Diekema DJ (2004) Rare and emerging opportunistic fungal pathogens: concern for resistance beyond Candida albicans and Aspergillus fumigatus. J Clin Microbiol 42:4419–4431

Phizicky E, Bastiaens PIH, Zhu H, Snyder M, Fields S (2003) Protein analysis on a proteomic scale. Nature 422:208–215

Pitarch A, Abian J, Carrascal M, Sánchez M, Nombela C, Gil C (2004) Proteomics-based identification of novel Candida albicans antigens for diagnosis of systemic candidiasis in patients with underlying hematological malignancies. Proteomics 4:3084–3106

Pitarch A, Diez-Orejas R, Molero G, Pardo M, Sánchez M, Gil C, Nombela C (2001) Analysis of the serologic response to systemic *Candida albicans* infection in a murine model. Proteomics 1:550–559

Pitarch A, Jiménez A, César Nombela, Gil C (2006a) Decoding serological response to *Candida* cell wall immunome into novel diagnostic, prognostic, and therapeutic candidates for systemic candidiasis by proteomic and bioinformatic analyses. Mol Cell Proteomics 5:79–96

Pitarch A, Molero G, Monteoliva L, Thomas DP, Lopez-Ribot JL, Nombela C, Gil C (2007) Proteomics in *Candida* species. In: d'Enfert C, Hube B (ed) *Candida*: comparative and functional genomics. Caister Academic, London, pp 169–194

Pitarch A, Nombela C, Gil C (2006b) *Candida* biology and pathogenicity: insights from proteomics. In: Hecker M (ed) Microbial proteomics: functional biology of whole organisms. Wiley, Hoboken, pp 285–330

Pitarch A, Nombela C, Gil C (2006c) Contribution of proteomics to diagnosis, treatment, and prevention of candidiasis. In: Hecker M (ed) Microbial proteomics: functional biology of whole organisms. Wiley, Hoboken, pp 285–330

Pitarch A, Pardo M, Jiménez A, Pla J, Gil C, Sánchez M, Nombela C (1999) Two-dimensional gel electrophoresis as analytical tool for identifying *Candida albicans* immunogenic proteins. Electrophoresis 20:1001–1010

Pitarch A, Sánchez M, Nombela C, Gil C (2002) Sequential fractionation and two-dimensional gel analysis unravels the complexity of the dimorphic fungus *Candida albicans* cell wall proteome. Mol Cell Proteomics 1:967–982

Pitarch A, Sanchez M, Nombela C, Gil C (2003) Analysis of the Candida albicans proteome I. Strategies and applications. J Chromatogr B 787:101–128

Prokisch H, Scharfe C, Camp DG 2nd, Wenzhong X, David L, Andreoli C, Monroe ME, Moore RJ, Gritsenko MA, Kozany C, Hixson KK, Mottaz HM, Zischka H, Ueffing M, Herman ZS, Davis RW, Meitinger T, Oefner PJ, Smith RD, Steinmetz LM (2004) Integrative analysis of the mitochondrial proteome in yeast. PloS Biol 2:795–804

Rabilloud T (1996) Solubilization of proteins for electrophoretic analyses. Electrophoresis 17:813–829

Rabilloud T, Carpentier G, Tarrox P (1988) Improvement and simplification of low-background silver staining of proteins by using sodium dithionite. Electrophoresis 9:288–291

Rais I, Karas M, Schägger H (2004) Two-dimensional electrophoresis for the isolation of integral membrane proteins and mass spectrometric identification. Proteomics 4:2567–2571

Raman B, Cheung A, Marten MR (2002) Quantitative comparison and evaluation of two commercially available, two-dimensional electrophoresis image analysis software packages, Z3 and Melanie. Electrophoresis 23:2194–2202

Reinders J, Sickmann A (2005) State-of-the-art in phosphoproteomics. Proteomics 5:4052–4061

Reinders J, Zahedi RP, Pfanner N, Meisinger C, Sickmann A (2006) Toward the complete yeast mitochondrial proteome: multidimensional separation techniques for mitochondrial proteomics. J Proteome Res 5:1543–1554

Rementeria A, López-Molina N, Ludwig A, Vivanco AB, Bikandi J, Pontón J, Garaizar J (2005) Genes and molecules involved in *Aspergillus fumigatus* virulence. Rev Iberoam Micol 22:1–23

Rhodes JC, Brakhage AA (2006) Molecular determinants of virulence in *Aspergillus fumigatus*. In: Heitman J et al (eds) Molecular principles of fungal pathogenesis. ASM Press, Washington, D.C., pp 333–345

Richardson MD (2005) Changing patterns and trends in systemic fungal infections. J Antimicrob Chemother 56[Suppl S1]:i5–i11

Richardson MD, Warnock DW (2003) Fungal infections: diagnosis and management. Blackwell, Oxford

Righetti PG, Campostrini N, Pascali J, Hamdan M, Astner H (2004) Quantitative proteomics: a review of different methodologies. Eur J Mass Spectrom 10:335–348

Righetti PG, Castagna A, Antonioli P, Boschetti E (2005) Prefractionation techniques in proteome analysis: The mining tools of the third millenium. Electrophoresis 26:297–319

Rosengren AT, Salmi JM, Aittokallio T, Westerholm J, Lahesmaa R, Nyman TA, Nevalainen OS (2003) Comparisonn of PDQuest and Progenesis software packages in the analysis of two-dimensional electrophoresis gels. Proteomics 3:1936–1946

Ross PL, Huang YN, Marchese JN, Williamson B, Parker K, Hattan S, Khainovski N, Pillai S, Dey S, Daniels S, Purkayastha S, Juhasz P, Martin S, Bartlet-Jones M, He F, Jacobson A, Pappin DJ (2004) Multiplexed protein quantitation in *Saccharomyces cerevisiae* using amine-reactive isobaric tagging reagents. Mol Cell Proteomics 3:1154–1169

Ruiz-Herrera J (1992) Fungal cell wall: structure, synthesis and assembly. CRC Press, Boca Raton

Rupp S (2004) Proteomics on its way to study host-pathogen interaction in *Candida albicans*. Curr Opin Microbiol 7:330–335

Sano A, Nakamura H (2004) Titania as a chemo-affinity support for the column-switching HPLC analysis of phosphopeptides: application to the characterization of phosphorylation sites in proteins by combination with protease digestion and electrospray ionization mass spectrometry. Anal Sci 20:861–864

Santoni V, Molloy M, Rabilloud T (2000) Membrane proteins and proteomics: un amour impossibile? Electrophoresis 21:1054–1070

Saporito-Irwin SM, Birse CE, Sypherd PS, Fonzi WA (1995) PHR1, a pH-regulated gene of *Candida albicans*, is required for morphogenesis. Mol Cell Biol 15:601–613

Saville SP, Thomas DP, López Ribot JL (2006) A role for Efg1p in *Candida albicans* interactions with extracellular matrices. FEMS Microbiol Lett 256:151–158

Schägger H (2006) Tricine-SDS-PAGE. Nat Protoc 1:16–22

Schägger H, Von Jagow G (1987) Tricine–sodium dodecyl sulfate polyacrylamide gel electrophoresis for the separation of proteins in the range from 1–100 kDalton. Anal Biochem 166:368–379

Schmidt A, Kellermann J, Lottspeich F (2005) A novel strategy for quantitative proteomics using isotope-coded protein labels. Proteomics 5:4–15

Schrettl M, Bignell E, Kragl C, Joechl C, Rogers T, Arst HN Jr, Haynes K, Hass H (2004) Siderophore biosynthesis but not reductive iron assimilation is essential for *Aspergillus fumigatus* virulence J Exp Med 200:1213–1219

Schulenberg B, Goodman TN, Aggeler R, Capaldi RA, Patton WF (2004) Characterization of dynamic and steady-state protein phosphorylation using a fluorescent phosphorylation gel stain and mass spectrometry. Electrophoresis 25:2526–2532

Schwienbacher M, Weig M, Thies S, Regula JT, Heesemann J, Ebel F (2005) Analysis of the major proteins secreted by the human opportunistic pathogen *Aspergillus fumigatus* under in vitro conditions. Med Mycol 43:623–630

Scott NL, Lecomte JTJ (2005) Protein structure: unusual covalent bonds. In: Nature encyclopedia of life sciences. Wiley, Chichester. http://www.els.net/ doi:10.1038/npg.els.0003015

Sentandreu M, Elorza MV, Sentandreu R, Fonzi WA (1998) Cloning and characterization of PRA1, a gene encoding a novel pH-regulated antigen of *Candida albicans*. J Bacteriol 180:282–289

Sepulveda P, Lopez-Ribot JL, Gozalbo D, Cervera A, Martinez JP, Chaffin WL (1996) Ubiquitin-like epitopes associated with *Candida albicans* cell surface receptors. Infect Immun 64:4406–4408

Shaw MM, Riederer BM, (2003) Sample preparation for two-dimensional gel electrophoresis. Proteomics 3:1408–1427

Shaw J, Rowlinson R, Nickson J, Stone T, Sweet A, Williams K, Tonge R (2003) Evaluation of saturation labelling two-dimensional difference gel electrophoresis fluorescent dyes. Proteomics 3:1181–1195

Sheir-Ness G, Lai MG, Morris NR (1978) Identification of a gene for β-tubulin in *Aspergillus nidulans*. Cell 15:639–647

Shen HD, Choo KB, Lin WL, Lin RY, Han SH (1990) An improved scheme for the identification of antigens recognized by specific antibodies in two-dimensional gel electrophoresis and immunoblotting. Electrophoresis 11:878–882

Shen HD, Lin WL, Tam FM, Chou H, Wang CW, Tsai JJ, Wang SR, Han SH (2001) Identification of vacuolar serine proteinase as a major allergen of *Aspergillus fumigatus* by immunoblotting and N-terminal amino acid sequence analysis. Clin Exp Allergy 31:295–302

Shevchenko A, Wilm M, Vorm O, Mann M (1996) Mass spectrometric sequencing of proteins silver-stained polyacrylamide gels. Anal Chem 68:850–858

Shevchenko A, Sunyaev S, Loboda A, Shevchenko A, Bork P, Ens W, Standing KW (2001) Charting the proteomes of organisms with unsequenced genomes by MALDI-quadrupole time-of-flight mass spectrometry and BLAST homology searching. Anal Chem 73:1917–1926

Shevchenko A, Tomas H, Havlis J, Olsen JV, Mann M (2006) In-gel digestion for mass spectrometric characterization of proteins and proteomes. Nat Protoc 1:2856–2860

Shimizu M, Wariishi H (2005) Development of a sample preparation method for fungal proteomics. FEMS Microbiol Lett 247:17–22

Shin DH, Jung S, Park SJ, Kim YJ, Ahn JM, Kim W, Choi W (2005a) Characterization of thiol-specific antioxidant 1 (TSA1) of *Candida albicans*. Yeast 22:907–918

Shin YK, Kim KY, Paik YK (2005b) Alterations of protein expression in macrophages in response to *Candida albicans* infection. Mol Cells 20:271–279

Shin YK, Lee HJ, Lee JS, Paik YK (2006) Proteomic analysis of mammalian basic proteins by liquid-based two-dimensional column chromatography. Proteomics 6:1143–1150

Sickmann A, Mreyen M, Meyer HE (2003a) Mass spectrometry – a key technology in proteom research. Adv Biochem Biotechnol 83:141–176

Sickmann A, Reinders J, Wagner Y, Joppich C, Zahedi R, Meyer HE, Schönfisch B, Perschil I, Chacinska A, Guiard B, Rehling P, Pfanner N (2003b) The proteome of *Saccharomyces cerevisiae* mitochondria. Proc Natl Acad Sci USA 100:13207–13212

Silva JC, Gorenstein MV, Li G-Z, Vissers JPC, Geromanos SJ (2006) Absolute quantification of proteins by LCMSE. Mol Cell Proteomics 5:144–156

Simpson RJ (2003) Proteins and proteomics – a laboratory manual. Cold Spring Harbor Laboratory, Cold Spring Harbor, N.Y.

Simpson DC, Smith RD (2005) Combining capillary electrophoresis with mass spectrometry for applications in proteomics. Electrophoresis 26:1291–1305

Singleton DR, Masuoka J, Hazen KC (2001) Cloning and analysis of a *Candida albicans* gene that affects cell surface hydrophobicity. J Bacteriol 183:3582–3588

Smejkal GB (2004) The Coomassie chronicles: past, present and future perspectives in polyacrylamide gel staining. Expert Rev Proteomics 1:381–387

Sohn K, Roehm M, Urban C, Saunders N, Rothenstein D, Lottspeich F, Schröppel K, Brunner H, Rupp S (2005) Identification and characterization of Cor33p, a novel protein implicated in tolerance towards oxidative stress in *Candida albicans*. Eukaryot Cell 4:2160–2169

Sohn K, Schwenk J, Urban C, Lechner J, Schweikert M, Rupp S (2006) Getting in touch with *Candida albicans*: the cell wall of a fungal pathogen. Curr Drug Targets 7:505–512

Soloviev DA, Fonzi WA, Sentandreu R, Pluskota E, Forsyth CB, Yadav S, Plow EF (2007) Identification of pH-regulated antigen 1 released from *Candida albicans* as the major ligand for leukocyte integrin alphaMbeta2. J Immunol 178:2038–2046

Soskic V, Gorlach M, Poznanovic S, Boehmer FD, Godovac-Zimmermann J (1999) Functional proteomics analysis of signal transduction pathways of the platelet-derived growth factor beta receptor. Biochemistry 38:1757–1764

Stasyk T, Hellman U, Souchelnytskyi (2001) Optimising sample preparation for 2-D electrophoresis. Life Sci News 9:9–12

Stensballe A, Andersen S, Jensen ON (2001) Characterization of phosphoproteins from electrophoretic gels by nanoscale Fe(III) affinity chromatography with off-line mass spectrometry analysis. Proteomics 1:207–222

Stoldt VR, Sonneborn A, Leuker CE, Ernst JF (1997) Efg1p, an essential regulator of morphogenesis of the human pathogen *Candida albicans*, is a member of a conserved class of bHLH proteins regulating morphogenetic processes in fungi. EMBO J 16:1982–1991

Sundstrom P (2002) Adhesion in *Candida* spp. Cell Microbiol 4:461–469

Sundstrom P (2006) *Candida albicans* hypha formation and virulence. In: Heitman J et al (ed) Molecular principles of fungal pathogenesis, ASM Press, Washington, D.C., pp 45–48

Swain M, Ross NW (1995) A silver stain protocol for proteins yielding high resolution and transparent background in sodium dodecyl sulfate–polyacrylamide gels. Electrophoresis 16:948–951

Taylor CF (2006) Minimum reporting requirements for proteomics: a MIAPE primer. Pract Proteomics 1/2:39–44

Tekaia F, Latgé JP (2005) Aspergillus fumigatus: saprophyte or pathogen? Curr Opin Microbiol 8:385–392

Thomas DP, Bachmann SP, Lopez-Ribot J (2006a) Proteomics for the analysis of the *Candida albicans* biofilm lifestyle. Proteomics 6:5795–5804

Thomas DP, Pitarch A, Monteoliva L, Gil C, Lopez-Ribot JL (2006b) Proteomics to study *Candida albicans* biology and pathogenicity. Infect Disord Drug Targets 6:335–341

Thomas DP, Viudes A, Monteagudo C, Lazzell AL, Saville SP, López-Ribot JL (2006c) A proteomic-based approach for the identification of *Candida albicans* protein components present in a subunit vaccine that protects agains disseminated candidiasis. Proteomics 6:6033–6041

Thompson A, Schäfer J, Kuhn K, Kienle S, Schwarz J, Schmidt G, Neumann T, Hamon C (2003) Tandem mass tags: a novel quantification strategy for comparative analysis of complex protein mixtures by MS/MS. Anal Chem 75:1895–1904

Tripathi G, Wiltshire C, Macaskill S, Tournu H, Budge S, Brown AJ (2002) Gcn4 co-ordinates morphogenetic and metabolic responses to amino acid starvation in Candida albicans. EMBO J 21:5448–5465

Urban C, Sohn K, Lottspeich F, Brunner H, Rupp S (2003) Identification of cell surface determinants in *Candida albicans* reveals Tsa1p, a protein differentially localized in the cell. FEBS Lett 533:228–235

Urban C, Xiong X, Sohn K, Schröppel K. Brunner H, Ruff S (2005) The moonlighting protein Tsa1p is implicated in oxidative stress response and in cell wall biogenesis in *Candida albicans*. Mol Microbiol 57:1318–1341

Urfer W, Grzegorczyk M, Jung K (2006) Statistics for proteomics: a review of tools for analyzing expermental data. Pract Proteomics 1/2:48–55

Vandewoude KH, Vogelaers D, Blot SI (2006) Aspergillosis in the ICU – the new 21st century problem? Med Mycol 44:S71–S76

Vediyappan G, Chaffin WL (2006) Non-glucan attached proteins of *Candida albicans* biofilm formed on various surfaces. Mycopathologia 161:3–10

Vediyappan G, Bikandi J, Braley R, Chaffin WL (2000) Cell surface proteins of *Candida albicans*: preparation of extracts and improved detection. Electrophoresis 21:956–961

Vellucci VF, Gygax SE, Hostetter MK (2007) Involvement of *Candida albicans* pyruvate dehydrogenase complex protein X (Pdx1) in filamentation. Fungal Genet Biol 44:979–990

Wachtler V, Balasubramanian MK (2006) Yeast lipid rafts?- an emerging view. Trends Cell Biol 16:1–4

Wang G, Wells WW, Zeng W, Chou C-L, Shen R-F (2006) Label-free protein quantification using LC-coupled ion trap or FT mass spectrometry: reproducibility, linearity, and application with complex proteomes. J Proteome Res 5:1214–1223

Wang W, Scali M, Vignani R, Spadafora A, Sensi E, Mazzuca S, Cresti M (2003) Protein extraction for two-dimensional electrophoresis from olive leaf, a plant tissue containing high levels of interfering compounds. Electrophoresis 24:2369–2375

Washburn MP, Wolters D, Yates JR III (2001) Large-scale analysis of the yeast proteome by multidimensional protein identification technology. Nat Biotechnol 19:242–247

Weatherbee JA, May GS, Gambino J, Morris NR (1985) Involvement of a particular species of beta-tubulin (beta3) in conidial development in *Aspergillus nidulans*. J Cell Biol 101:706–711

Wei J, Sun J, Yu W, Jones A, Oeller P, Keller M, Woodnutt G, Short JM (2005) Global proteome discovery using an online three-dimensional LC-MS/MS. J Proteome Res 4:801–808

Wessel D, Flügge UI (1984) A method for the quantitative recovery of protein in dilute solution in the presence of detergents and lipids. Anal Biochem 138:141–143

Westermeier R (2006) Sensitive, quantitative, and fast modifications for Coomassie blue staining of polyacrylamide gels. Proteomics 6[Suppl 2]:61–64

Westermeier R, Naven T (2002) Proteomics in practice. Wiley-VCH, Weinheim

Wheelock AM, Buckpitt AR (2005) Software-induced variance in two-dimensional gel electrophoresis image analysis. Electrophoresis 26:4508–4520

Wiener MC, Sachs JR, Deyanova EG, Yates NA (2004) Differential mass spectrometry: a label-free LC-MS method for finding significant differences in complex peptide and protein mixtures. Anal Chem 76:6085–6096

Wiese S, Reidegeld KA, Meyer HE, Warscheid B (2007) Protein labeling by iTRAQ: A new tool for quantitative mass spectrometry in proteome research. Proteomics 7:340–350

Wildgruber R, Harder A, Obermaier C, Boguth G, Weiss W, Fey SJ, Larsen PM, Görg A (2000) Towards higher resolution: two-dimensional electrophoresis of *Saccharomyces cerevisiae* proteins using overlapping narrow immobilized pH gradients. Electrophoresis 21:2610–2616

Wildgruber R, Reil G, Drews O, Parlar H, Görg A (2002) Web-based two-dimensional database of *Saccharomyces cerevisiae* proteins using immobilized pH gradients from pH 6 to pH 12 and matrix-assisted laser desorption/ionization-time of flight mass spectrometry. Proteomics 2:727–732

Wilkins MR, Sanchez JR, Gooley AA, Appel RD, Humphery-Smith I, Hochstrasser DF, Williams KL (1995) Progress with proteome projects, why all proteins expressed by a genome should be identified and how to do it. Biotechnol Gen Eng Rev 13:19–50

Wilkins MR, Appel RD, Van Eyk JE, Chung MCM, Görg A, Hecker M. Huber LA, Langen H, Link AJ, Paik YK, Patterson SD, Pennington SR, Rabilloud T, Simpson RJ, Weiss W, Dunn MJ (2006) Guidelines for the next 10 years of proteomics. Proteomics 6:4–8

Williams TI, Combs JC, Thakur AP, Strobel HJ, Lynn BC (2006) A novel Bicine running buffer system for doubled sodium dodecyl sulfate-polyacrylamide gel electrophoresis of membrane proteins. Electrophoresis 27:2984–2995

Wilm MS, Mann M (1996) Analytical properties of the nanoelectrospray ion source. Anal Chem 68:1–8

Wilm MS, Shevchenko A, Houthaeve T, Breit S, Schweigerer L, Fotsis T, Mann M (1996) Femtomole sequencing of proteins from polyacrylamide gels by nano-electrospray mass spectrometry. Nature 379:466–469

Winkler C, Denker K Wortelkamp S, Sickmann A (2007) Silver- and Coomassie-staining protocols: detection limits and compatibility with ESI MS. Electrophoresis 28:2095–2099

Wittig I, Braun HP, Schägger H (2006) Blue native PAGE. Nat Protoc 1:418–428

Wittmann-Liebold B, Graack H-R, Pohl T (2006) Two-dimensional gel electrophoresis as tool for proteomics studies in combination with protein identification by mass spectrometry. Proteomics 6:4688–4703

Wolschin F, Weckwerth W (2005) Combining metal oxide affinity chromatography (MOAC) and selective mass spectrometry for robust identification of *in vivo* protein phosphorylation. Plant Methods 1:9

Wu CC, Yates JR 3rd (2003) The application of mass spectrometry to membrane proteomics. Nat Biotechnol 21:262–267

Yamagata A, Kristensen DB, Takeda Y, Miyamoto Y, Okada K, Inamatsu M, Yoshizato K (2002) Mapping of phosphorylated proteins on two-dimensional polyacrylamide gels using protein phosphatase. Proteomics 2:1267–1276

Yan JX, Wait R, Berkelman T, Harr RA, Westbrook JA, Wheeler CH, Dunn MJ (2000) A modified silver staining protocol for visualization of proteins compatible with matrix-assisted laser desorption/ionization and electrospray ionization–mass spectrometry. Electrophoresis 21:3666–3672

Yan L, Zhang JD, Cao YB, Gao PH, Jiang YY (2007) Proteomic analysis reveals a metabolism shift in a laboratory fluconazole-resistant *Candida albicans* strain. J Proteome Res 6:2248–2256

Yates JR III, Gilchrist A, Howell KE, Bergeron JJM (2005) Proteomics of organelles and large cellular structures. Nat Rev Mol Cell Biol 6:702–714

Yeo SF, Wong B (2002) Current status of nonculture methods for diagnosis of invasive fungal infections. Clin Microbiol Rev 15:465–484

Yin Z, Stead D, Selway L, Walker J, Riba-Garcia I, McInerney T, Gaskell S, Oliver SG, Cash P, Brown AJ (2004) Proteomic response to amino acid starvation in *Candida albicans* and *Saccharomyces cerevisiae*. Proteomics 4:2425–2436

Yin QY, Groot PWJ de, Dekker HL, Jong L de, Klis FM, Koster CG d e (2005) Comprehensive proteomic analysis of *Saccharomyces cerevisiae* cell walls. J Biol Chem 280:20894–20901

Zahedi RP, Meisinger C, Sickmann A (2005) Two-dimensional benzyldimethyl-*n*-hexadecylammonium chloride/SDS-PAGE for membrane proteomics. Proteomics 5:3581–3588

Zhu H, Bilgin M, Snyder M (2003) Proteomics. Annu Rev Biochem 72:783–812

9 Transcriptomics of the Fungal Pathogens, Focusing on *Candida albicans*

Steffen Rupp[1]

CONTENTS

I. Introduction.........................	187
A. Prerequisite for Transcriptomics: Genomic Sequences................	188
II. Transcriptomics of Fungal Pathogens....	193
A. Transcriptomics of Primary Fungal Pathogens..................	193
B. Transcriptomics of Opportunistic Fungal Pathogens..................	193
III. Transcriptomics of *Candida albicans*	194
A. Resistance Mechanisms to Antimycotics	195
1. Response of *C. albicans* to Antimycotics.................	195
2. Transcriptional Profiling of Clinical Isolates Resistant to Antimycotics.................	198
3. Experimental Induction of Resistance	200
4. Antimycotics for Topical Applications.....................	201
B. Stress Response	202
1. General Stress Response in *C. albicans*	202
2. Response to NO..................	203
3. pH Regulation	204
C. Polymorphism of *C. albicans*.........	205
1. Yeast to Hyphae Transition	206
2. The APSES Proteins Efg1 and Efh1 in *C. albicans*............	207
3. cAMP Signalling	208
4. Repression of Transcription as Key for Morphogenesis	210
5. Phenotypic Switching and Mating.....................	211
D. Host–Pathogen Interaction	212
1. Neutrophils	212
2. Macrophages	213
3. Blood..........................	214
4. Epithelial Surfaces................	215
E. Biofilm Formation..................	216
IV. Conclusions	217
References...........................	218

I. Introduction

The past century brought the availability of vaccines and antibiotics, leading to a dramatic fall in mortalities caused by infectious diseases. This led to the assumption that infectious disease has been defeated by medicine. In 1969 the United States Surgeon General actually claimed that "we can close the book on infectious diseases". However, today we know that this assumption was naïve, not taking into account that evolution is a constant motor in adapting the existing organisms to changing environmental conditions, including the adaptation of pathogens to changes in the host. Today nearly 25% of the annual deaths world-wide are directly related to pathogens (Morens et al. 2004). This can be attributed to the appearance of new diseases, like HIV, SARS or West Nile Virus, but also to an increase of resistance to antibiotics in pathogens thought to be defeated, like *Mycobacterium tuberculosis* or *Staphylococcus* and *Enterococcus* strains. In addition the progress in medical care results in a large proportion of immune-deficient patients and consequently in an increase in opportunistic infections. Especially fungi have gained an infamous reputation during recent decades as being highly detrimental to patients with haematologic–oncologic diseases, neutropenia or after organ transplantation. A review of the current literature identified 1415 species as known to be pathogenic to humans, including 538 bacteria and 307 fungi (Cleaveland et al. 2001). The fungi are a large group of diverse eukaryotic organisms. Only about 74 000 to 120 000 of the estimated 1.5×10^6 existing species of fungi have been described. Of the approximately 300 fungal species that are known to cause human infections, the most commonly observed live threatening systemic infections are caused by opportunistic infections of *Candida* species or *Aspergilli*. Therefore the major scientific interest with regard to fungal pathomechanisms has focused on these organisms in the past decade.

[1] Fraunhofer IGB, Nobelstrasse 12, 70569 Stuttgart, Germany; e-mail: rupp@igb.fhg.de

The early availability of the genome sequence of *Candida albicans* (the first assembly of the genome sequence was publicly available in 2000, at http://www-sequence.stanford.edu/group/candida), the availability of molecular tools and the use of *Saccharomyces cerevisiae* as a model for many characteristics of *C. albicans* relevant for pathogenesis (including morphogenesis and signalling pathways involved in stress response) resulted in a major body of work concerning this opportunistic fungal pathogen (Berman and Sudbery 2002; Braun et al. 2005; Jones et al. 2004). Until 2004 *C. albicans* was actually the only fungal pathogen on which genome-wide transcriptomics using arrays had been published. The sequences of other pathogenic fungi or the tools required for genome-wide transcriptomics had not been available to the public until then. However, a major body of molecular work has been performed on other opportunistic fungal pathogens, including *C. glabrata*, *C. parapsylosis*, *C. tropicalis*, *Cryptococcus neoformans* and *Aspergillus* species, where *Aspergillus fumigatus* is leading the clinically relevant species, as well as on the genera of primary fungal pathogens, including *Blastomyces*, *Coccidioides*, *Histoplasma* and *Paracoccidioides*. Due to the advancement of *Candida albicans* transcriptomics this chapter mainly focuses on this organism and only briefly touches the current work on other fungal pathogens.

A. Prerequisite for Transcriptomics: Genomic Sequences

The availability of complete genomic sequences and new methods to use this knowledge was key for the development and use of genome-wide technologies, including array technologies. Within the past decade sequencing of entire genomes has been a major effort both in academic as well as in commercial research. The first sequenced eukaryotic genome was the genome of *Sac. cerevisiae* (the second completely sequenced genome at all) in 1996. Only five years later a first draft of the human genomic sequence was published (Lander et al. 2001; Venter et al. 2001). Now the genome sequences of significantly more than 1000 organisms, including all kingdoms and viral genomes as well as the genomes of almost all the major pathogenic microbes, can be found in various databases, e.g. at NCBI Entrez Genomes, in different stages of their emergence, e.g. as completed annotated sequences, sequences in assembly or progress (http://www.ncbi.nlm.nih.gov/entrez/query.fcgi?CMD=Details&DB=genome). As a consequence, a new discipline has arisen, which has been named "pathogenomics." As the name implies, pathogenomics is the analysis at the genomic level of the processes involved in pathogenesis caused by the interaction of pathogenic microbes and their hosts (for a review, see Pompe et al. 2005).

A recent review by Galagan summarizes all the fungal genome sequencing projects that are publicly available at any of the stages, from nominated candidates to fully assembled and annotated genomes (Galagan et al. 2005). Galagan focuses on comparative genomics. Currently 85 fungal genome sequences are listed at the NCBI of which nine are referenced as completed and 44 are at the assembly stage (as of 4 August 2006; Table 9.1). A total of 16 links to publications of fungal genomic sequences are given, including annotations of the respective genome (http://www.ncbi.nlm.nih.gov/genomes/leuks.cgi). Besides model organism like *Sac. cerevisiae* or *Schizosaccharomyces pombe*, this list contains several of the human pathogenic fungi, including *C. albicans, C. glabrata, C. tropicalis, Cryptococcus neoformans* var. *neoformans, A. fumigatus, Histoplasma capsulatum* and *Coccidioides immitis* among others (Dujon et al. 2004; Jones et al. 2004; Loftus et al. 2005; Nierman et al. 2005). The availability of the increasing wealth of fungal sequences is largely due to initiatives for fungal genomics applying a kingdom-wide approach like the Fungal Genome Initiative of the Broad Institute (http://www.broad.mit.edu/annotation/fungi/fgi/index.html) or the Genolevures consortium (http://cbi.labri.fr/Genolevures/). Both initiatives selected a well defined collection of fungi (rather than choosing individual fungi in isolation) that maximizes the overall value for comparative genomics, evolutionary studies, eukaryotic biology and medical studies.

Based on these complete genomic sequences, DNA-microarray technology has been widely used for expression profiling, to monitor changes of transcriptional activity of every known or annotated gene of the respective fungi in a single experiment. For *Sac. cerevisiae* the first genome-wide transcriptional analyses appeared shortly after completion of the genomic sequence (DeRisi et al. 1997; Hauser et al. 1998; Wodicka et al. 1997). This set the start for genome-wide analysis of an organism based on the knowledge of its genome. Thus *Sac.*

Table 9.1. Status of fungal genome sequencing projects as of August 2006

Organism name	Organism subgroup	Size (Mb)	Status	Depth	Sequence release date (month/day/year)	Center/Consortium
Ajellomyces capsulatus G186AR	Ascomycetes	24.0	In progress	2×		Washington University (WashU)
Aje. capsulatus G217B	Ascomycetes	24.0	In progress	8×		Washington University (WashU)
Aje. capsulatus NAm1 NAm I	Ascomycetes	28.0	Assembly	4×	09/21/2005	Broad Institute
Aje. dermatitidis ATCC 26199	Ascomycetes	28.0	In progress	3×		Washington University (WashU)
Antonospora locustae	Other Fungi	2.9	In progress	3×		Marine Biological Laboratory
Ascosphaera apis USDA-ARSEF 7405	Ascomycetes	24.0	Assembly	4×	07.06.2006	Baylor College of Medicine
Aspergillus clavatus NRRL 1	Ascomycetes	35.0	Assembly	11.4×	09.09.2005	TIGR
Asp. flavus NRRL3357	Ascomycetes	36.0	Assembly	5×	08.01.2005	TIGR
Asp. fumigatus Af293	Ascomycetes	30.0	Assembly	10×	06.01.2005	TIGR/Sanger Institute
Asp. nidulans FGSC A4	Ascomycetes	31.0	Assembly	13×	04.07.2003	Broad Institute
Asp. parasiticus	Ascomycetes		In progress			University of Oklahoma
Asp. terreus ATCC 20542	Ascomycetes	35.0	Assembly		01/22/2003	Microbia
Asp. terreus NIH2624	Ascomycetes	35.0	Assembly	11.05×	08/30/2005	Broad Institute
Batrachochytrium dendrobatidis	Other Fungi	20.0	In progress	10×		Broad Institute
Botryotinia fuckeliana	Ascomycetes	38.0	In progress	10×		Genoscope/Bayer CropScience
Bot. fuckeliana B05.10	Ascomycetes	38.0	Assembly	5.4×	10/20/2005	Syngenta Biotech., Inc./Broad Institute
Candida albicans 1161	Ascomycetes		In progress			Welcome Trust Sanger Institute
Can. albicans SC5314	Ascomycetes	16.0	Assembly		02/24/2001	Stanford University
Can. albicans WO-1	Ascomycetes	14.0	Assembly	10×	03/28/2006	Broad Institute
Can. glabrata CBS 138	Ascomycetes	12.2	Complete	8×	07.02.2004	Genolevures Consortium
Can. tropicalis CBS 94	Ascomycetes	15.0	In progress	0.2×		Genolevures Consortium
Can. tropicalis MYA-3404	Ascomycetes	15.0	Assembly	10×	03/16/2005	Broad Institute
Chaetomium globosum CBS 148.51	Ascomycetes	36.0	Assembly	7×	03/14/2005	Broad Institute
Clavispora lusitaniae ATCC 42720	Ascomycetes	16.0	Assembly	9×	03/16/2005	Broad Institute
Coccidioides immitis H538.4	Ascomycetes	29.0	In progress	3×		Broad Institute
Coc. immitis RS	Ascomycetes	28.7	Assembly	10×	10.04.2004	Broad Institute

(continued)

Table 9.1. (continued)

Organism name	Organism subgroup	Size (Mb)	Status	Depth	Sequence release date (month/day/year)	Center/Consortium
Coc. posadasii C735	Ascomycetes	29.0	In progress			TIGR
Coprinopsis cinerea okayama7#130	Basidiomycetes	37.5	Assembly	10×	07/30/2003	Broad Institute
Cryptococcus neoformans R265	Basidiomycetes	20.0	Assembly	6×	03/17/2005	Broad Institute
Cry. neoformans WM276	Basidiomycetes	18.3	In progress	6×		Genome Sciences Centre/Univ. of British Columbia
Cry. neoformans var. grubii H99	Basidiomycetes	20.0	Assembly	11×	06/24/2003	Broad Institute/Duke University
Cry. neoformans var. neoformans B-3501A	Basidiomycetes	18.5	Assembly		07/13/2004	Stanford Univ.
Cry. neoformans var. neoformans JEC21	Basidiomycetes	19.1	Complete	12.5×	01.07.2005	TIGR/Stanford University
Debaryomyces hansenii CBS767	Ascomycetes	12.2	Complete	9.7×	07.02.2004	Genolevures Consortium
Encephalitozoon cuniculi GB-M1	Other Fungi	2.5	Complete		11/24/2001	Genoscope|Universite Blaise Pascal
Eremothecium gossypii ATCC 10895	Ascomycetes	8.74	Complete		03.06.2004	Zool. Institut der Univ. Basel, Switzerland
Gibberella moniliformis 7600	Ascomycetes	46.0	Assembly	4.2×	10/20/2005	Syngenta Biotech., Inc./Broad Institute
G. zeae PH-1	Ascomycetes	40.0	Assembly	10×	05.09.2003	International Gibberella zeae Genomics Consortium/Broad Institute
Kazachstania exigua CBS 379	Ascomycetes	18.0	In progress	0.2×		Genolevures Consortium
Kluyveromyces lactis NRRL Y-1140	Ascomycetes	10.7	Complete	11.4×	07.02.2004	Genolevures Consortium
Klu. marxianus CBS 712	Ascomycetes	14.0	In progress	0.2×		Genolevures Consortium
Klu. thermotolerans CBS 6340	Ascomycetes	10.6	In progress	0.2×		Genolevures Consortium
Klu. waltii NCYC 2644	Ascomycetes	10.9	Assembly	8×	03.09.2004	Broad Institute
Lodderomyces elongisporus NRLL YB-4239	Ascomycetes	16.0	In progress	8×		Broad Institute
Magnaporthe grisea 70-15	Ascomycetes	40.0	Assembly	7×	10/31/2003	International Rice Blast Genome Consortium/Broad Institute/Fungal Genomics Lab., North Carolina State Univ.
M. grisea 70-15	Ascomycetes	40.0	Assembly		01/30/2006	North Carolina State University (NCSU)

(continued)

Table 9.1. (continued)

Organism name	Organism subgroup	Size (Mb)	Status	Depth	Sequence release date (month/day/year)	Center/Consortium
Nectria haematococca MPVI	Ascomycetes	40.0	In progress			DOE Joint Genome Institute
Neosartorya fischeri NRRL 181	Ascomycetes	35.0	Assembly		09.09.2005	TIGR
Neurospora crassa OR74A	Ascomycetes	43.0	Assembly	10×	04/25/2003	Broad Institute
Phaeosphaeria nodorum SN15	Ascomycetes		Assembly		04.04.2005	Broad Institute
Phakopsora meibomiae	Basidiomycetes		In progress	8×		DOE Joint Genome Institute
Phk. pachyrhizi	Basidiomycetes	50.0	In progress	8×		DOE Joint Genome Institute
Phanerochaete chrysosporium RP-78	Basidiomycetes	30.0	Assembly		05.04.2004	DOE Joint Genome Institute
Pichia angusta CBS 4732	Ascomycetes		In progress	0.5×		Genolevures Consortium
Pic. angusta RB11	Ascomycetes	9.5	In progress	8×		Qiagen
Pic. farinosa CBS 7064	Ascomycetes	13.9	In progress	0.4×		Genolevures Consortium
Pic. guilliermondii ATCC 6260	Ascomycetes	12.0	Assembly	12×	03/17/2005	Broad Institute
Pic. stipitis CBS 6054	Ascomycetes	15.4	In progress			DOE Joint Genome Inst./Stanford Univ.
Pneumocystis carinii	Ascomycetes	8.0	In progress			University of Cincinnati
Podospora anserina	Ascomycetes	34.0	In progress	7×		Broad Institute
Pod. anserina S mat+	Ascomycetes	34.0	In progress	10×		Genoscope/CGM at the CNRS/Orsay Univ./Univ. Bordeaux II/Wageningen Univ.
Rhizopus oryzae RA 99-880	Other Fungi	40.0	Assembly	10×	03/17/2005	Broad Institute
Saccharomyces bayanus 623-6C	Ascomycetes	11.5	Assembly	2.9×	05/31/2003	Washington University (WashU)
Sac. bayanus MCYC 623	Ascomycetes	11.5	Assembly	6.4×	05/16/2003	Broad Institute
Sac. castellii NRRL Y-12630	Ascomycetes		Assembly	3.9×	05/31/2003	Washington University (WashU)
Sac. cerevisiae S288c	Ascomycetes	12.1	Complete		10/25/1996	Sanger Institute/European Yeast Genome Sequencing Network (EYGSN)/McGill Univ./Stanford Univ./Tsukuba Life Science Center/Washington Univ. (WashU)
Sac. cerevisiae RM11-1a	Ascomycetes	12.0	Assembly	10×	03/16/2005	Broad Institute
Sac. cerevisiae YJM789	Ascomycetes	16.0	Assembly	10×	01.06.2005	Stanford University

(continued)

Table 9.1. (continued)

Organism name	Organism subgroup	Size (Mb)	Status	Depth	Sequence release date (month/day/year)	Center/Consortium
Sac. kluyveri NRRL Y-12651	Ascomycetes	12.6	Assembly	3.6×	05/31/2003	Washington University (WashU)
Sac. kluyveri NRRL Y-12651	Ascomycetes		In progress	0.2×		Genolevures Consortium
Sac. kudriavzevii IFO 1802	Ascomycetes		Assembly	3.4×	05/31/2003	Washington University (WashU)
Sac. mikatae IFO 1815	Ascomycetes	12.1	Assembly	5.9×	05/16/2003	Broad Institute
Sac. mikatae IFO 1815	Ascomycetes	12.1	Assembly	2.8×	05/31/2003	Washington University (WashU)
Sac. paradoxus NRRL Y-17217	Ascomycetes	11.8	Assembly	7.7×	05/16/2003	Broad Institute
Sac. servazzii CBS 4311	Ascomycetes	12.3	In progress	0.2×		Genolevures Consortium
Sac. uvarum CLIB 533	Ascomycetes		In progress	0.4×		Genolevures Consortium
Schizosaccharomyces japonicus	Ascomycetes	14.0	In progress	7×		Broad Institute
Sch. octosporus	Ascomycetes	14.0	In progress	7×		Broad Institute
Sch. pombe 972h⁻	Ascomycetes	12.5	Complete	8×	02/21/2002	Sch. pombe European Sequencing Consortium (EUPOM)/Sanger Institute/Cold Spring Harbor Lab
Sclerotinia sclerotiorum 1980	Ascomycetes	38.0	Assembly	8×	09/20/2005	Broad Institute
Trichoderma reesei QM9414	Ascomycetes	33.0	Assembly		08.04.2005	DOE Joint Genome Institute/Los Alamos National Laboratory
Uncinocarpus reesii 1704	Ascomycetes	30.0	Assembly	5×	09/13/2005	Broad Institute
Ustilago maydis 521	Basidiomycetes	20.0	Assembly	10×	07/29/2003	Broad Institute
Yarrowia lipolytica CLIB122	Ascomycetes	20.5	Complete	10×	07.02.2004	Genolevures Consortium
Zygosaccharomyces rouxii CBS 732	Ascomycetes	12.8	In progress	0.4×		Genolevures Consortium

cerevisiae was also key in developing both the biochemical and bioinformatic methods necessary for transcriptional profiling. Today thousands of transcription profiles have been generated from Sac. cerevisiae, which also can be of use for comparison to the biology of other fungi. At the Saccharomyces Genome Database (SGD, http://www.yeastgenome.org/), links to publicly available data sets have been set and many results thereof were integrated into the functional descriptions of genes (see SGD at http://www.yeastgenome.org/).

This amount of data shows that, besides the knowledge of the genome, ways to generate and analyse the data are a prerequisite for transcriptomics. Microarray fabrication and data analysis, however, are beyond the scope of this chapter and cannot be reviewed. For recent reviews, see Barbulovic-Nad et al. (2006) and Eisenstein (2006).

Transcriptomics can be used in several ways. The most straightforward way to use transcriptomics is to use it as a screening tool to study gene function or to identify individual genes required for adaptation to certain environmental conditions. This strategy mimics conventional mutational or promoter activation screens. Especially for pathogenic fungi with cryptic or absent sexual cycles – excluding the use of classic genetics – this technology greatly facilitates research. DNA-microarrays have the advantage that the mRNA levels of all genes represented on the array can be monitored in a single experiment avoiding tedious screening procedures which may be developed using conventional methods if possible at all. However, genes which have not been included in the arrays, e.g. due to exclusion from annotation because of size (usually the limit is at 100 bp), sequencing and annotation problems or other reasons, are not included in the analysis. A more comprehensive approach is the analysis of large sets of transcriptome data in order to create a picture of interconnected networks of signalling pathways. This can take account of the fact that signalling pathways in general are not operating in an isolated fashion, but rather form an entangled web of numerous possible signalling avenues and feedback loops (Carter et al. 2006; Ihmels et al. 2005).

There are also pitfalls one should keep in mind if working on a genome-wide level. For many of the genes in a genome defined by an open reading frame and annotated by homology to a gene from another organism, still no experimental evidence has been generated for its function, localisation or other parameters. Although the predictive models for gene/protein function are getting better and better, they are based on present knowledge. This might lead to incorrect conclusions based on incorrect assignment of functionalities. Furthermore, completely new functionalities may not be recognized in transcriptional profiling experiments based on the predictions included in the gene annotations. In addition regulatory mechanisms which do not require changes in mRNA levels, e.g. activating of signalling molecules by post-translational modifications, can not be detected using transcriptomics. Although transcriptional profiling is an excellent tool for unravelling complex interaction of cellular pathways, it should also be seen as a tool with certain limitations.

Nevertheless transcriptomics have been used to create a comprehensive picture of changes in gene expression of pathogenic fungi during host–pathogen interaction, stress response or other environmental challenges which ultimately will contribute to the better understanding of the mechanisms of infection and thus foster the development of new diagnostics, therapeutics and vaccines.

II. Transcriptomics of Fungal Pathogens

A. Transcriptomics of Primary Fungal Pathogens

Primary fungal pathogens are able to cause disease in individuals who have no immune deficiencies and are in good health. Dermatophytes are the most important group causing primary infections in humans, including athletes foot, tinea and other skin infections. Responsible for these infections are dermatophytes of the genera *Trichophyton*, *Microsporum* and *Epidermophyton*. Although these are probably the most common fungal infections worldwide, research on dermatophytes is under-represented. Only one publicly available genome project for dermatophytes (*T. rubum*) is currently at the planning stage (http://www.broad.mit.edu/annotation/fungi/fgi/nominated.html). Therefore no genome-wide transcriptional profiling experiments have been published to date.

Another group of primary fungal pathogens has been investigated in more detail. These encompass genera of dimorphic fungi including *Blastomyces*, *Coccidioides*, *Histoplasma* and *Paracoccidioides*. These fungi cause endemic mycoses in subtropical or tropical areas of the Americas or Africa. Sequencing projects for *Coccoides imitis* strains and *Histoplasma capsulatum* strains are well advanced (see Table 9.1). For *H. capsulatum* transcriptional studies using a shotgun array covering approximately one-third of the genome have been published already (Gebhart et al. 2006; Hwang et al. 2003b). For *C. imitis* a partial array containing 1000 putative ORFs has just been employed (Johannesson et al. 2006). *Blastomyces dermatitidis* and *Paracoccidioides brasiliensis* are on the list of nominated candidates for sequencing (http://www.broad.mit.edu/annotation/fungi/fgi/nominated.html). A significant body of work based on an EST-based transcriptomic approach has been reported recently for *P. brasiliensis* (Andrade et al. 2005).

B. Transcriptomics of Opportunistic Fungal Pathogens

Opportunistic fungal pathogens normally are not able to infect immunocompetent individuals. Some of them, e.g. *Candida* species like *C. albicans* are even part of the normal microflora of a large part of the population. Other fungi like *A. fumigatus* are normally saprophytic and *Cryptococcus neoformans* is associated with birds. These fungi may cause mycoses of various grades, including a high rate of fatal cases, in immunocompromised hosts like HIV, cancer and diabetes patients among others. For several of the opportunistic fungal pathogens the genomes have been completed, including sufficient annotation to enable the design of genome-wide arrays for transcriptional profiling.

Currently, genome-wide transcriptional profiling experiments using arrays based on the genomic sequence of the organism have been published from four species of opportunistic fungi, *Candida albicans*, *C. glabrata* (Vermitsky et al. 2006), *Cryptococcus neoformans* (Cramer et al. 2006; Kraus et al. 2004) and *A. fumigatus* (da Silva Ferreira et al. 2006; Nierman et al. 2005). By far the largest body of data has been created from *Candida albicans*. The Candida Genome database (Arnaud et al. 2005; Costanzo et al. 2006) lists 81 hits for genome-wide analysis publications (August 2006; http://www.candidagenome.org/cache/genome-wide-analysis.html) of which 66 are using transcriptome analysis. For the other organisms in total less than ten are found in PubMed (as of August 2006). Therefore, this review focuses on the transcriptomics of *C. albicans*.

III. Transcriptomics of *Candida albicans*

C. albicans is mostly found as a commensal organism which colonizes the gastrointestinal/urogenital tract in a large part of the population without causing any symptoms. However, *C. albicans* is able to switch between a commensal and a pathogen. This is in contrast to other opportunistic fungal pathogens like *Aspergilli* which are in general not part of the human microbial flora. Thus *C. albicans* has not only developed mechanisms to colonize, infect and invade into the host but must have developed mechanisms to persist in the host in large numbers without causing any damage. This switch between commensal and pathogen is central to *C. albicans* and has resulted in new definitions of pathogens (Casadevall and Pirofski 2003). Currently, no appropriate model system for commensalisms of fungi exists which restricts the work performed to study pathogenesis to more or less artificial model systems. Therefore, the symptom-free interplay between the host and a commensal fungus, which often is a prerequisite for infection, so far cannot been studied. The majority of the studies undertaken to date have been focusing on fundamental mechanisms generally thought to be required for pathogenesis, e.g. stress response, response to antimycotics, switching between different morphologies (yeast-to-hyphae transition) or the function of specific virulence genes by comparing mutant and wild-type or revertant strains in vitro. Furthermore, simple model systems mimicking host–pathogen interaction, e.g. *C. albicans* encountering the host defense, like macrophages or neutrophils or adhering and penetrating into tissue using different reconstituted tissue models derived from cell lines or primary cells, have been used to shed light into early stages of candidiasis. Also first attempts to directly investigate gene expression profiles of *C. albicans* isolated from patients have been reported.

For the *C. albicans* genome, several sequencing projects have been carried out, mostly commercially driven. A publicly available sequence of the genome was completed and assembled in 2000 by the Stanford Technology Centre (http://www-sequence.stanford.edu/group/candida). However, due to the diploid character of the organism, even a ten-fold coverage of the genomic sequence could only be assembled in more than 1200 contigs, indicating significant ambiguities (assembly 4). The following assemblies were significantly improved. For the recent assembly 19, published in May 2002, the ploidy of the organism was taken into account and a diploid genome of *C. albicans* was assembled (266 contigs over eight chromosomes with a total of 14.88 Mb of sequence). Assembly 20 has been released recently; however, in this case two different strain backgrounds have been inappropriately mixed to fill-in sequence gaps, resulting in small sequence stretches which might be inaccuarate (Arnaud et al. 2007; Nantel 2006). Nevertheless, even with the early preliminary assembled sequences and many uncertainties in gene annotation, several comprehensive databases were initiated, describing predicted genes with respect to homologies to other organisms, especially *Sac. cerevisiae* (e.g. http://genolist.pasteur.fr/CandidaDB/, http://wwwsequence.stanford.edu/group/candida or http://alces.med.umn.edu:80/Candida.html). While early on over 9000 ORFs were found in an automated ORF prediction procedure at the Stanford Technology Centre, between 6000 and 7200 ORFs were annotated in non-redundant gene sets in databases of different groups. These efforts resulted in several recent publications of the diploid sequence of *C. albicans*, its annotated ORFs and functional assignments (Braun et al. 2005; Costanzo et al. 2006; d'Enfert et al. 2005; Jones et al. 2004).

The publicly available sequence of the *C. albicans* genome by the Stanford Technology Centre was the basis for most of the transcriptome studies published to date. Early gene expression data, performed on DNA-arrays comprising subsets of about 300 up to 2000 genes of the genome as well as arrays containing groups of functionally related genes, e.g. like genes encoding for cell wall proteins, have been published by several groups (Bensen et al. 2002; Braun et al. 2001; Lane et al. 2001; Lotz et al. 2004; Murad et al. 2001; Sohn et al. 2003). The first microarray data predicting to cover almost the complete genome, however, derived from a commercial source of about 6600 cDNA library sequences, investigating *C. albicans* response to drug treatment (De Backer et al. 2001). Since then several groups have been developing genome-wide arrays based on the public available sequence. Arrays used for the studies described here include arrays generated by Incyte Genomics (De Backer et al. 2001), the high-density oligonucleotide GeneChip manufactured by Affymetrix (Santa Clara, Calif.; Lan et al. 2002), arrays based on spotted PCR products by the Biotechnology Research Institute (NRC, Montreal, Canada; http://www.irb-bri.cnrc.gc.ca/microarraylab/; Nantel et al. 2002), by Eurogentec SA (Ivoz-Ramet, Belgium) in collaboration with the European Galar Fungail Consortium (www.pasteur.fr/recherche/unites/Galar_Fungail/; Fradin et al. 2003) and by Fraunhofer-IGB (Stuttgart, Germany; Sohn et al. 2003, 2006) among others. A summary of the arrays including their characteristics is given in Table 9.2. These arrays have been used to study *C. albicans* with regard to response to antimycotics and development of resistance, environmental stresses, hyphal development, host–pathogen interaction, biofilm formation and cell wall biogenesis as well as switching and mating among other studies (Fig. 9.1). Some of the results found in these studies are reviewed below. For a comprehensive review of the biology of *C. albicans*, see Calderone (2002).

Although most of these arrays are derived from the same genomic sequence, significant differences both in early and late established arrays as well as in the definition of ORFs resulted in microarrays which differ in their extent of gene specific probes predicted to be detectable. Furthermore, the different array facilities have designed distinct oligo sets or PCR-products for detection of the individual transcripts, which may result in differences in the detection of these transcripts. This is certainly one of the reasons for differences in the expression level and even the presence of individual genes between the individual arrays, which in some cases are evident. Nevertheless, the general picture of transcriptional profiling in general is comparable also when the different arrays developed in various laboratories have been used for similar experiments.

A. Resistance Mechanisms to Antimycotics

One of the most intensively studied topics in *C. albicans* is its defence mechanisms against antimycotics and the development of resistance. Several approaches to identify resistance mechanisms using transcriptomics are reported in the literature. One way is to directly confront *C. albicans* with the respective drug and monitor the change in gene expression. The first published transcriptomics study for *C. albicans* used this approach (De Backer et al. 2001). Another approach is to look at changes of gene expression in resistant clinical isolates or series of isolates with increasing resistance, e.g. (Rogers and Barker 2003). A third approach found in the literature is to follow adaptation of *C. albicans* in experimental microbial populations to (sub)-inhibitory concentrations of an antimicrobial drug. This experiment mimicked the evolution of drug resistance and identified genetic changes which were accompanied by changes in gene expression that persisted in the absence of the drug, resulting in new constitutive patterns of drug resistance (Cowen et al. 2002).

1. Response of *C. albicans* to Antimycotics

The first publication on transcriptomics of *C. albicans* was on its response to itraconazole (De Backer et al. 2001). The microarray was generated by Incyte Genomics containing 6600 ORFs, which were identified from genomic DNA sequences and cDNA sequences. In this study CAI-4, a *URA3* deficient strain generally used in the community was used (Fonzi and Irwin, 1993). Among other

Table 9.2. Characteristics of genome-wide microarrays used for *C. albicans* transcriptomics reviewed in this chapter. The denotation within brackets in column 1 (*Consortium/fabrication*) is used within the text to identify the DNA-microarray used for the respective study

Consortium/fabrication	Type	References/links
Jansen Research Foundation/Incyte Genomics, USA (Incyte)	Spotted PCR-products: 6600 ORFs derived from genomic sequence and cDNA sequences	De Backer et al. (2001)
European Galar Fungail Consortium/Eurogentec SA (Ivoz-Ramet, Belgium) (Eurogentec)	Spotted PCR-products: 6039 putative ORFs derived from assembly 6 of *C. albicans* genome ~300 bp PCR-product of each ORF	http://www.pasteur.fr/recherche/unites/Galar_Fungail/
Biotechnology Research Institute, National Research Council, Montreal, Canada (BRI)	Spotted PCR-products: several versions, starting with assembly 4 (6580 ORFs greater 250 bp) up to assembly 19 (6002 ORFs). Latest version: 70mer oligonucleotide probes 6354 potential ORFs	http://www.irb-bri.cnrc.gc.ca/microarray-lab/
Rupp Lab, Fraunhofer IGB, Stuttgart, Germany (IGB)	Spotted PCR-products: 7200 orfs (<100 base pairs). ORFs derived from assembly 6 of *C. albicans* genome, 300–600 bp PCR-product for each ORF	http://www.igb.fraunhofer.de/
Agabian Lab, UCSF/Affymetrix, USA (Affymetrix)	Custom high-density oligonucleotide Genechip: 13 025 probe sets; these probe sets reflect 7116 large ORFs (<100 amino acids), 247 structural RNA targets, 4208 unannotated small ORFs	Lan et al. (2002)
Fink and Johnson Labs, Whitehead Institute and UCSF, USA (UCSF/WI)	Spotted PCR-products: Primer design and ORF amplification were performed in collaboration between A. Johnson, UCSF and G. Fink, Whitehead Institute, giving 10 000 PCR products representing ~6300 ORFs, with many ORFs represented by more than one spot	Bennett et al. (2003)
Consortium for Candida DNA Microarray Facilities, USA (CCDMF)	Spotted PCR-products: PCR-products of 6175 unique *C. albicans* ORFs derived from assembly 6	Bensen et al. (2004)
Operon (Operon)	70mer oligonucleotide	http://www.operon.com/arrays/oligosets_yeasts_overview.php

genes, this study revealed a global up-regulation of the *ERG* genes, in agreement with studies showing that the target of the azoles is the ergosterol biosynthetic pathway. In addition changes were observed in cell wall maintenance genes, lipid biosynthesis and gene products involved in vesicular transport. Interestingly induction of multi-drug transporters was not observed under the conditions used (24 h incubation with 10 µM itraconazol in synthetic glucose medium). Over all, treatment of cells with 10 µM itraconazole resulted in 296 responsive genes. For 116 genes transcript levels were decreased at least 2.5-fold, while for 180 genes transcript levels were similarly increased. The ERG genes *ERG11* and *ERG5* were found to be up-regulated approximately 12-fold. In addition, a significant up-regulation was observed for *ERG6*, *ERG1*, *ERG3*, *ERG4*, *ERG10*, *ERG9*, *ERG26*, *ERG25*, *ERG2*, *IDI1*, *HMGS*, *NCP1* and *FEN2*, all of which are genes known to be involved in ergosterol biosynthesis.

Liu et al. (2005) examined changes in the gene expression profile of *C. albicans* following exposure to representatives of the four currently available classes of antifungal agents used in the treatment of systemic fungal infections, the azoles, polyenes, echinocandines and nucleotide analoga. The most remarkable finding of this study is that none of the differentially regulated genes found exhibited similar changes in expression for all

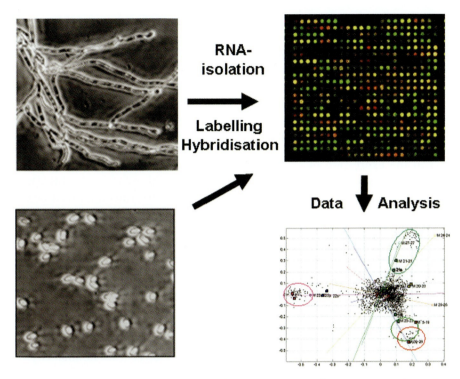

Fig. 9.1. Principle of transcriptomics. Cells are grown under two conditions of interest, e.g. yeast and hyphal growth conditions. RNA is isolated from cells grown under both conditions, labelled and co-hybridized to DNA-microarrays (for Affymetrix GenChips each labeled mRNA population is hybridized separately). Data analysis can be performed using clustering algorithms, e.g. correspondence analysis as shown here (Fellenberg et al. 2006)

four drugs. Therefore, the response to the individual drug seems to be highly specific. This result could give an explanation for the observed benefit of combinatorial treatments due to the additive effect of multiple cellular defects induced by the combination of drugs. The experiments were conducted using microarrays manufactured by Eurogentec. SC5314 was grown in synthetic complete dextrose medium at a drug concentration corresponding to the IC_{50}. All genes deviating by a factor of at least 1.5-fold from SC5314 incubated without the respective drug were defined as differentially expressed genes.

For each of the conditions used the authors identified a number of differentially regulated genes, ranging from 82 for incubation with ketoconazole (KTZ), 256 for incubation with Amphotericin B (AMB), 439 for incubation with 5-flucytosine (5-FC), up to 480 for incubation with Caspofungin (CPF). Only two of the genes responded similar to AMB, KTZ and CPF. These were *DDR48*, a gene known to be responsive to stress, and *FET33* encoding for a ferroxidase required for high affinity iron uptake. Approximately 60 genes responded similar to AMB and CPF. This included the down-regulation of genes involved in iron regulation (*CFL1, CFL2, FTR1, FTR2, CTR1, FTH1*), chromatin/chromosome structure (*HHF21, HHT21, HHF22, HTA1, HHT3, HTB1, NHP6A*), and lipid, fatty acid and sterol metabolism (*ERG25, ERG251, SAH1, ERG3, OPI3, ERG13, FAS1, ERG9*). The authors speculated that these commonly differentially expressed genes reflect a response to a compromised cellular integrity, as both AMB and CPF are fungicidal agents. Other combinations found were marginal in number.

The response to the individual drugs was also described in detail. KTZ exposure increased the expression of genes involved in lipid, fatty acid and sterol metabolism, including *NCP1, MCR1, CYB5, ERG2, ERG3, ERG10, ERG25, ERG251* and *ERG11* (the azole target). KTZ also increased expression of several genes associated with azole resistance, including *CDR1, CDR2, IFD4* (*CSH1*), *DDR48* and *RTA3*. In agreement with the study by De Backer et al. (2001) several genes were found in both studies including *ERG2, ERG3, POT14* (*ERG10*), *ERG11, ERG25, NCP1, CYB5, SAH1, DDR48, CWH8* and *CTR1*. Differences in the results of these studies were explained by the authors by differences between the drugs used and the experimental designs employed (in the study by De Backer, *C. albicans* was exposed to itraconazole for 24 h). Interestingly, there seems to be no *UPC2* regulon in *C. albicans* as observed in *Sac. cerevisiae* (Agarwal et al. 2003). In *Sac. cerevisiae* *UPC2* is the transcriptional regulator for sterol uptake under anaerobic conditions (Vik and Rine 2001).

AMB produced changes in the expression of genes involved in small-molecule transport, especially up-regulation of ion transporters, including calcium (*IPF11550*, *IPF11560*), potassium (*IPF9136*), sodium (*ENA21*), zinc (*ZRT2*), and sulfate (*SUL1*) transport, indicative of ion loss across the plasma membrane. This is consistent with the role of AMB in disruption of the plasma membrane. Furthermore, genes involved in stress response were identified (*YHB1*, *CTA1*, *AOX1*, *SOD2*, *GSH1*), consistent with the role of AMP in causing oxidative stress (Sokol-Anderson et al. 1988). Also observed was a decreased expression of genes involved in ergosterol biosynthesis, including *ERG26*, *ERG16*, *ERG11*, *ERG9*, *ERG25*, *ERG13*, *ERG251* and *ERG3*, the fatty acid biosynthesis genes *FEN12*, *FAS1*, *FAS2* and *ACB1*, and the phospholipid biosynthesis genes *CHO2* and *OPI3*, indicative of the need of the cell to avoid the presence of ergosterol in the plasma membrane. In contrast to the study of Barker et al. (2004) who investigated in vitro generated strains resistant to AMB overexpression of *ERG5*, *ERG6* and *ERG25* -a- ll facilitating the alternate production of sterols – was not observed.

CPF targets the cell wall by inhibition of β-1,3-glucan synthase. Therefore, consistent with its mechanism of action, it induced changes in expression of genes encoding cell wall maintenance proteins, including the β-1,3-glucan synthase subunit *GSL22*, as well as other cell wall maintenance genes including *PHR1*, *ECM21*, *ECM33* and *FEN12*. The genes responsive to both CPF and AMB have been mentioned above. As these also include genes involved in lipid, fatty acid and sterol metabolism additional modes of action for CPF might be possible.

5-FC increased the expression of genes involved in purine and pyrimidine biosynthesis, including *YNK1*, *FUR1* and its target *CDC21*. The changes in gene expression observed in response to 5-FC exposure are consistent with a need for the cell to increase RNA, DNA, and protein synthesis, including uracil phosphoribosyltransferase and thymidylate synthetase.

The relationship between antimicrobial pharmacodynamics and gene expression was examined by Lepak et al. (2006) in order to gain insight into the mechanism of persisting fluconazole (FCZ) effects following drug exposure. The authors examined a *C. albicans* culture at two time-points (1 h, 3 h) during drug exposure (4xMIC) and at two time-points after drug exposure (6 h, 9 h) using microarrays (BRI-Arrays). During FCZ exposure 126 genes were found to be up-regulated and 148 to be down-regulated (threshold 1.5-fold). During recovery from FCZ treatment, a much larger number of genes were up-regulated (1055) and 35 were found to be down-regulated. Among the genes with known function that were up-regulated during exposure, most were related to plasma membrane/cell wall synthesis (18%) including nine genes of the ergosterol pathway, stress responses (7%), and metabolism (6%). The categories of down-regulated genes during exposure included protein synthesis (15%), DNA synthesis/repair (7%) and transport (7%) genes. The majority of genes identified at the postexposure time-points again were from the protein (17%) and DNA (7%) synthesis categories. Three genes (*CDR1*, *CDR2*, *ERG11*) were examined in greater detail following FCZ exposure in vitro and in vivo using a neutropenic mouse model of disseminated candidiasis. Expression levels from the in vitro and in vivo studies for these three genes were congruent. *CDR1* and *CDR2* transcripts were reduced during in vitro FCZ exposure and during supra-MIC exposure in vivo. In the post-exposure period, the mRNA abundance of both pumps increased. For *ERG11* the opposite effect was observed, expression increased during exposure and fell in the post-exposure period. The expression of the three genes responded in a dose-dependent manner. Post-antifungal effects could be observed neither in vitro nor in vivo in this study, in contrast to previous studies (Andes and van Ogtrop 1999). This study may define targets to exceed the drug effects or block recovery from the drugs.

2. Transcriptional Profiling of Clinical Isolates Resistant to Antimycotics

Rogers et al. (Rogers and Barker 2003) used a set of four isolates, selected from a set of 17 isogenic previously characterized clinical isolates, to investigate FCZ resistance (Lyons and White 2000; Pfaller et al. 1994; Redding et al. 1994; White

1997a, b). In a preceding study two of the four isolates were investigated using a DNA-Microarray prepared by Incyte Genomics (Rogers and Barker 2002). In the second, more comprehensive study, a set of four matched clinical isolates (including the two previously studied) were investigated using a microarray manufactured by Eurogentec (Rogers and Barker 2003).

These studies identified groups of genes that are co-ordinately expressed with either *CDR1* and *CDR2* or *MDR1*, or with genetic changes in the *ERG11* gene across the set of isolates. These findings led to the suggestion that these sets of co-ordinately regulated genes may be controlled by common regulatory systems. Indeed, Coste et al. (2004) could show in a later study that *TAC1* is a transcription factor representing a common factor for *CDR1* and *CDR2* (see below). Rogers and Baker identified a total of 32 genes to be up-regulated and 14 genes to be down-regulated between isolates 2–79 (sensitive to FCZ MIC 2 µg/ml) and 12–99 (resistant ≥64 µg/ml). These included genes previously shown to be up-regulated (*CDR1, CDR2, MDR1, ERG2, GPX1, RTA3, IFD5, IPF5987, CRD2*) and down-regulated (*FET34*) in this series. Besides changes in membrane metabolism and efflux pumps, this study could confirm a connection of azole resistance to iron metabolism (via *FET34*) and to oxidative stress response [via *SOD5* (*IPF1222*) and *CRD2*]. Genes coordinately regulated with *MDR1* encompass 14 genes up-regulated and 10 genes down-regulated, including the up-regulation of *IFD1, IFD4, IFD5, IFD7, GRP2, DPP1, CRD2* and *INO1* and the down-regulation of *FET34, OPI3* and *IPF1222*. Interestingly, the expression of several of the ERG genes (*ERG5, ERG13, ERG25*) was not correlated with the expression of either *CDR1* or *CDR2* nor with *MDR1* in this set of clinical isolates.

The co-regulation of genes with respect to *CDR1/2* or *MDR1* expression as identified in clinical isolates (see above; Rogers and Barker 2003) asks for identification of the respective regulons. In order to define the regulons of these efflux pumps Karababa and co-workers (2004) directly confronted *C. albicans* strains with drugs specifically up-regulating either *CDR1/2* or *MDR1* and comparing the results to expression profiles of clinical isolates overexpressing either *CDR1/2* or *MDR1*. For this purpose they compared the transcript profiles (Eurogentec array) of a laboratory strain (CAF2-1) exposed to fluphenazine, resulting in specific up-regulation of *CDR1* and *CDR2* (de Micheli et al. 2002) or to benomyl, resulting in specific *MDR1* up-regulation (Gupta et al. 1998), with those of two matched pairs of azole-susceptible and -resistant strains overexpressing *CDR1* and *CDR2* (CDR isolates) or *MDR1* (MDR isolates). The clinical isolates were incubated without drugs (YPD at 30 °C) and CAF2-1 was incubated without or with exposure either for 20 min to 10 mg/l of fluphenazine or for 30 min to 25 mg/l of benomyl (YPD, 30 °C). In each of the four experiments between 200 and 300 genes were differentially regulated by at least 2-fold. However, only a small portion of all genes was found to be commonly regulated, indicating that in vitro drug-induced gene expression only partially mimics expression profiles observed in azole-resistant clinical strains.

Between fluphenazine-exposed cells and CDR isolates 42 commonly regulated genes (8.6% of all regulated genes 2-fold; 19 genes 4-fold) could be identified. Most strongly induced besides *CDR1* and *CDR2* were *IFU5, RTA3* (encoding putative membrane proteins), *HSP12* (encoding heat-shock protein) and *IPF4065* (potentially involved in stress response). Four of these six genes, but not *HSP12* and *IPF4065*, contain a putative *cis*-acting drug responsive element (DRE) in their promoters. Interestingly, five out of nine genes specifically induced by fluphenazine (4-fold) were potentially involved in stress response (*CFL2, IPF6629, GRP2, IPF17282, SAS3*), whereas three of the four genes specific for the clinical isolates were part of the ERG pathway (*ERG3, ERG6, ERG 251*).

Commonly regulated between benomyl-exposed cells and MDR isolates were 57 genes (11.5% of all regulated genes 2-fold; 23 genes 4-fold). The most up-regulated besides *MDR1* were genes with oxido-reductive functions such as IFD genes, *IPF5987, GRP2* (belonging to the aldo-keto reductase family), *IPF7817* [NAD(P)H oxidoreductase] and *IPF17186*. This study revealed that for the benomyl specific gene induction 21 of the 29 genes identified (4-fold) contained a potential Cap1 binding site, a transcription factor known to be involved in oxidative stress response (Zhang et al. 2000). Seven of these genes (*IPF2897, IPF11105, PST2, IPF3264, SOD1, TTR1, TRX1*) are potentially implicated in oxidative stress response. Five other genes included in this cluster play a role in other stress responses. Among the 17 commonly regulated genes, nine are also involved in oxidative stress response (*PYC2, GPX1, GRP2, IPF7817, IFD1, IFD4, IFD5, IFD7*).

Karababa et al. compared their results also to the studies from Rogers et al. (Rogers and Barker 2003) and Cowen et al. (2002; see above). They found only very limited overlap between the individual results. Between the clinical isolates up-regulating the CDRs, five commonly identified genes were found *CDR1*, *CDR2*, *RTA3*, *IFU5* and *GPX1*. These genes are also among those that are also commonly regulated between the fluphenazine and the CDR experiment of this study. Since four of these five commonly regulated genes contained a drug response element (DRE) in their promoter, these results strongly suggest the existence of a common transcriptional pathway important for their regulation. This transcription factor could later be identified as *TAC1* (Coste et al. 2004).

When comparing the set of genes commonly regulated in isolates up-regulating *MDR1* only eight genes were commonly up-regulated in the three studies: *MDR1*, *GRP2*, *IFD1*, *IFD5*, *IPF5987*, *GDH3*, *ARO8* and *SNZ1*. Besides the antifungal drug resistance function of *MDR1*, the other genes have oxido-reductive functions (*GRP2*, *GDH3*, *IFD1/5*) or are potentially involved in pyridoxine (vitamin B_6) synthesis (*IPF5987*, *SNZ1*).

Since *MDR1* can be also induced by H_2O_2, Karababa and co-workers compared their *MDR1*-related results with a study published by Enjalbert et al. (2003) investigating the response of *C. albicans* to H_2O_2. They could show further parallels between *MDR1* up-regulation, benomyl and oxidative stress response. *IFD1*, *GRP2*, *IFD5*, *IFD4* and *IFD7* were induced more than 4-fold under these three conditions, whereas *GRP4*, *IPF12303*, *IFR2*, *TTR1*, *IPF13081*, *PST2* and *IPF20104* were found induced only in benomyl and H_2O_2 response. Again, most of these genes are implicated in response to stress or have oxido-reductive functions. Among these, *GRP2*, *GRP4*, *IPF12303*, *TTR1*, *PST2* and *IPF20104* contain a putative Cap1 binding site in their promoters. Therefore, a functional linkage involving Cap1 probably exists between benomyl and H_2O_2 exposure.

The transcription factor responsible for regulation of the *CDR1/2* regulon, *TAC1*, was identified by Coste and co-workers (2004) based on homology search in the genome for zinc-finger transcription factors. Tac1 was shown to bind to the DRE element located upstream of *CDR1* and *CDR2* as well as other genes induced by antifungals. Both, strains deleted for *TAC1* as well as clinical isolates containing dominant alleles of *TAC1* were investigated by DNA-microarrays (Eurogentec) in order to identify genes regulated by Tac1. The authors compared genes up-regulated in a clinical isolate (DSY296, azole-resistant) against its progeny (DSY294, azole-susceptible) as well as in a strain deleted for *TAC1* carrying either an activated *TAC1-2* allele (derived from DSY296) or *TAC1*. Furthermore, the response of the strain CAF2-1 to fluphenazine was monitored in the presence or absence of *TAC1*. Only four genes were commonly regulated in the microarray experiments: *CDR1*, *CDR2* and *RTA3* (probable transmembrane protein similar to *Sac. cerevisiae YOR049c*), all containing a consensus DRE in their promoters, and *HSP12* (heat-shock protein). In addition *IFU5*, which was identified previously as a DRE-containing gene (de Micheli et al. 2002), was confirmed as regulated by *TAC1* using Northern blotting in this study.

These results are in accordance with the studies by Karababa et al. (2004), Rogers et al. (Rogers and Barker 2003) and Cowen et al. (2002).

3. Experimental Induction of Resistance

Cowen et al. (2002) have used DNA-microarrays (BRI Array with 5,000 ORFs) to examine changes in gene expression during experimental acquisition of resistance to FCZ. The changes in gene expression were followed in four replicate populations (D8, D9, D11, D12) during 330 generations of evolution characterized previously (Cowen et al. 2000). The MIC for FCZ could be induced from 0.25 µg/ml to 4 µg/ml (D8, D12) or 64 µg/ml (D9, D11) respectively for two of the populations. Interestingly reversion from high resistance (64 µg/ml) at generation 260 to reduced resistance at generation 330 (4 µg/ml) was observed for one of the cultures (D12). The final culture (D12-330) showed a higher fitness in the presence of the drug than its ancestor. In the absence of the drug (YPD, 30 °C) 301 genes whose expression was at last 1.5-fold different over all populations were identified. Cluster analysis showed that in the final outcome three of the four populations (D9, 11, 12) grouped together whereas a forth population (D8) evolved differently. A third cluster reflected the early stage of adaptation. For D8 in total eight different ORFs could be identified as up-regulated, of which *CDR2* is known to be implicated in drug resistance (*CDR1* was not present on the array). For D9, D11 and D12 at generation 330 a significantly larger number of genes were deregulated (up to 124 from

301 genes). All of them showed overexpession of *MDR1*. The nine genes showing the largest change in expression in the microarray experiments were *ADH4*, *MDR1*, *YPL88*, *YPX98*, *YPR127W*, *GRE99*, *YNL229C*, *HYR1* and *HSP12*. The expression of these genes including *CDR2* were screened in 30 clinical samples with MICs of FCZ >4µg/ml and could be clustered in the same three patterns of gene expression. More than half of the isolates (17 of 30) clustered with D8, nine clinical isolates clustered with the control population that evolved without the drug and four samples of which three derived from one patient are similar to the early- and late-stage pattern of D9, D11 and D12.

Besides *MDR1* and *CDR2*, a set of genes (*YPL88*, *YPX98*, *YPR127w*, *ADH4*) implicated in oxidative stress was found. The authors speculate that they may contribute to drug resistance because the azoles sensitize fungal cells to oxidative metabolites through inhibition of the target (a cytochrome P450 enzyme) in the ergosterol biosynthesis pathway (White et al. 1998). In contrast to other studies in fungi (Bammert and Fostel 2000; De Backer et al. 2001), only a moderate transcriptional modulation of the ERG pathway was observed in this study (for *ERG1*, *ERG3*, *ERG11*, *ERG13*).

Resistance of *C. albicans* to AMB was generated in a study by Barker et al. (2004) by growing SC5314 in the presence of increasing concentrations of the drug. Interestingly, cross-resistance to FCZ in the resistant strains generated was observed, as reported previously for a clinical isolate (Kelly et al. 1997). The MICs for the resistant isolate created were >32 mg/l for AMB and >256 mg/l FCZ. However, the doubling time for the resistant isolate SC5314-AR was 181±19 min as compared with 97±3 min for isolate SC5314, indicating a significant loss of fitness (as was confirmed by the loss of stability of the resistance after 28 doublings growth in the absence of AMB). The changes in gene expression profile associated with the experimentally induced resistance to AMB, were identified in microarray experiments (Eurogentec arrays). They identified 133 genes that were differentially expressed, by at least 2-fold in SC5314-AR, with 27 genes up-regulated and 106 genes down-regulated. These experiments revealed overexpression of *ERG5*, *ERG6* and *ERG25*, all facilitating the alternate production of sterols. Down-regulation of ERG genes was not observed. The sterol content of the resistant strain revealed that ergosterol was basically missing in the membrane fraction and replaced mainly by lanosterol and euburicol (24-methylene lanosterol), explaining the resistance against AMB. The synthetic pathway to lanosterol involves *ERG1* and *ERG7* but not *ERG11*, explaining the cross-resistance against Fluconazole. Again stress response genes like *DDR48* and iron transporters like *FTR1* and *FET34* were found to be induced. The majority of genes down-regulated represent genes involved in protein synthesis, especially genes encoding for ribosomal proteins, which is in agreement with its slow-growth phenotype.

4. Antimycotics for Topical Applications

Besides antifungals used against systemic infections the reaction of *C. albicans* to ciclopirox olamine, a drug used for treatment of superficial mycoses was investigated by two groups using microarrays (both from Eurogentec). Ciclopirox is a topical antifungal agent of the hydroxypyridone class. Sigle et al. (2005) investigated the reponse of *C. albicans* in Sabouraud glucose medium at subinhibitory concentrations of ciclopirox (0.6 mg/l). Only 25 genes were found to be induced by more than 2-fold and 21 genes were repressed by more than 2-fold. The vast majority of the up-regulated genes (15 genes) were involved in iron metabolism. These included known genes encoding iron reductases (*CFL1*), iron permeases and transporters (*FTR1*, *FTR2*, *FTH1*) which were previously found (Niewerth et al. 2003) and genes possibly involved in iron metabolism that have not yet been described in *C. albicans* (*CFL2*, *CFL12*, *FET5*, *FET32*, *FET33*, *FET34*, *FRE5*, *FRE31*, *FRE32*). Furthermore, a number of genes were identified encoding proteins similar to the GPI protein Rbt5 (*RBT5*, *RBT2*, *IPF12101*, *CSA1*) which has been reported to be involved in utilization of haemin and haemoglobin as iron sources (Weissman and Kornitzer 2004). These results in combination with microbiological assays led to the conclusion that ciclopirox acts as an iron chelator. Consequently, addition of iron ions strongly reduces the inhibitory effect of ciclopirox. Additional experiments showed that cells with induced oxidative stress proteins or grown in the absence of glucose were less susceptible to ciclopirox. This study indicates that metabolic activity, oxygen accessibility and iron levels are critical parameters in the mode of action of ciclopirox olamine.

A parallel study of Lee et al. (2005) came to similar conclusions. Lee et al. used the same strain, *C. albicans* SC5314, but a different medium (synthetic dextrose) and ciclopirox olamine concentration were used [equivalent to the IC_{50} (0.24 mg/l) for 3 h]. A total of 49 genes were found to be responsive to ciclopirox olamine (cut off: 2-fold), including 36 up-regulated and 13 down-regulated genes. These included genes involved in small molecule transport (*HGT11, HXT5, ENA22, PHO84, CDR4*), iron uptake (*FRE30, FET34, FTR1, FTR2, SIT1*) and cell stress (*SOD1, SOD22, CDR1, DDR48*).

Comparing the induced genes found in both studies it is apparent that only nine of 25 or 36 up-regulated genes have been found in both studies. These are *RBT5, FET34.3eoc, FTR1, FTR2, FTH1, CCC2, CFL1, CSA1* and *IPF12101*. All of these genes are involved in iron metabolism or potential cell surface proteins. The differences between both studies may be explained by different conditions, especially since media composition has been reported to have a strong effect on the activity of ciclopirox against *C. albicans* (Sigle et al. 2005).

B. Stress Response

An appropriate stress response is thought to be one of the key elements for a pathogen to successfully colonize different niches of a host and to escape the host's defence mechanisms. Several stresses have been imposed to *C. albicans* in vitro in order to study the mechanisms of stress response and compare it with other non-pathogenic fungi. These stresses include peroxides and nitric oxide, as these compounds are encountered by *C. albicans* during cellular defence mechanisms of the host. Furthermore, heat shock, NaCl/osmotic stress and heavy metal exposure have been analysed by transcriptional profiling. In addition, adaptation to different pH is a key feature for *C. albicans* inhabiting environments like the gut, the skin and the vaginal tract with strongly differing pH values.

One of the main questions investigated was whether *C. albicans*, like *Sac. cerevisiae* and *Sch. pombe*, has a general stress response resulting in cross-protection to various stresses or whether the stress response is regulated by direct response to the individual stress.

1. General Stress Response in *C. albicans*

Two publications by Enjalbert et al. (2003, 2006) address this question. In total four different stresses were analysed. They include heat shock, oxidative stress, osmotic stress and heavy metal stress (cadmium). In the first publication heat-shock response was triggered by a temperature shift from 23 °C to 37 °C, osmotic stress by adding 0.3 M NaCl and oxidative stress by the addition of H_2O_2 to a final concentration of 0.4 mM. The second publication also triggered osmotic stress by 0.3 M NaCl, however they used 5 mM H_2O_2 to induce oxidative stress and 0.5 mM $CdSO_4$ for heavy metal stress. Furthermore, they focus more on the role of Hog1p, including profiling of mutants deficient in *HOG1*. The different experiments conducted were analysed on two distinct arrays (BRI, Eurogentec). After analysis of the data from these experiments, conducted under partially different conditions, distinct conclusions were drawn with regard to the existence of a general stress response in *C. albicans*. Whereas the first study (Enjalbert et al. 2003) clearly states that no general stress response is present in *C. albicans*, the second study (Enjalbert et al. 2006) comes to the conclusion that a limited core stress response exists as well in this organism. However, it was confirmed that *C. albicans* has diverged from corresponding stress response networks in other yeasts (the model yeasts *Sac. cerevisiae* and *Sch. pombe*) and that in *C. albicans* several pathways function in parallel to regulate the core transcriptional response to stress. One of the main reasons for this divergent conclusions in these two studies is that *C. albicans* seems to be rather resistant to some stresses in comparison to *Sac. cerevisiae* and *Sch. pombe*. The response to 5 mM and 0.4 mM H_2O_2 respectively are significantly different in *C. albicans*, indicating that 0.4 mM H_2O_2 may not be sufficient for triggering oxidative stress response. Whereas 347 genes showed expression that was modified specifically in response to 5 mM H_2O_2, only 265 genes show altered expression at 0.4 mM H_2O_2. Genes involved in the detoxification of peroxide stress were induced generally under both conditions (*CAP1, CTA1, GPX1, GST3, TRR1, TRX1*). However, other subsets of genes were differentially induced, for example, genes involved in carbohydrate metabolism were only induced in response to high levels of peroxide stress (*ICL1, GPM2, GSY1, MLS1, NTH1, PCK1*), whereas the DNA-damage response appeared to be evoked specifically at low levels of H_2O_2 (*HNT2, IPF4708, IPF4356, RGA2*).

Comparing the results from the high H_2O_2 concentration, Cd stress, osmotic stress and heat-shock experiments from both studies, a set of nine induced core stress response factors was identified in the data set (induced 1.5-fold or more). These include *ECM41*, *GLK1*, *GRP2*, *HSP12*, *HXT61*, *HSP31*, *orf19.675*, *AHP1* and *orf19.7085*. The proteins encoded by these genes have known or putative functions in carbohydrate metabolism (*GLK1*, *HXT61*), cell wall (*ECM41*, *orf19.675*), redox processes (*GRP2*, *AHP1*) and as chaperones (*HSP12*, *HSP31*).

There is considerably more overlap between the lists of stress-repressed genes in these experiments than within the stress-induced genes (34 vs 38 genes in both studies). This is not surprising because all the stresses have the common feature of reducing growth rates. The nature of the repressed common genes therefore reflects in general a reduction in growth. A significant proportion of these genes are involved in protein synthesis and RNA processing (e.g. *IPF966*, *IPF3709*, *NOP4*, *NMD3*, *MRPL3*, *RCL1*).

The role of *HOG1* was investigated in more detail (Enjalbert et al. 2006). Inactivation of *HOG1* significantly attenuates transcriptional response to osmotic and Cd-induced stresses. However, a less dramatic effect on the transcriptional response to oxidative stress was observed (actually none of the 46 *HOG1* dependent genes also induced by H_2O_2 have a known antioxidant function). This is also reflected in the result showing that *hog1* mutants are highly sensitive to osmotic stress but only sensitive to peroxides at high concentrations. In *C. albicans CAP1* was shown already to be responsible for the resistance to oxidative stress at significantly lower levels of H_2O_2 (Alarco and Raymond 1999; Zhang et al. 2000). Hence, *CAP1* is required for *C. albicans* to survive both low and high doses of peroxide stress, whereas *HOG1* seems to be required only for the response to high levels of oxidative stress. This is consistent with previous reports that Hog1 is activated only in response to high levels of peroxide (Smith et al. 2004). In agreement with these findings *CAP1* was found to be crucial for regulating genes involved in oxidative stress response like *CTA1*, *IPF20104* (both core stress response genes) and *IFR2* (responding to osmotic and oxidative stress) independently of *HOG1*.

Considering only three stresses, Cd, oxidative and osmotic/NaCl stress, 24 genes were identified in *C. albicans* as core stress response genes. When compared to the available transcriptional data on the response of *Sac. cerevisiae* and *Sch. pombe* to these stresses it was found that in *Sac. cerevisiae* 5-fold and in *Sch. pombe* 7-fold more genes constitute the core stress response. This is also reflected in cross-protection experiments. In *C. albicans* cross-protection experiments show a 2-fold increase of resistance after a mild heat stress followed by a strong oxidative stress and no improvement in survival in the case of the mild oxidative stress or hyperosmotic stress followed by a strong heat shock (Enjalbert et al. 2003). Thus, the acquired resistance seems to be weak in *C. albicans* (maximum 2-fold) compared with the more than 100-fold increase in *Sac. cerevisiae* survival (Lewis et al. 1995).

Therefore, although there is a high degree of functional overlap in the global oxidative, osmotic and heavy metal stress response in the three yeasts, there has been significant divergence between the stress responses in these fungi. Especially with regard to a core stress response the available data indicate that it is rather limited in *C. albicans* and does not result in comparable cross-protection, as observed for *Sac. cerevisisae*. Interestingly, the functions of Msn2- and Msn4-related proteins, key elements of the core stress response in *Sac. cerevisisae* (O'Rourke and Herskowitz 2004), also appear to have been reassigned in *C. albicans* (Nicholls et al. 2004), giving further evidence for the divergence of stress response mechanisms in these organisms.

Activation of Hog1 seems also to be distinct in *C. albicans* when compared to *Sac. cerevisiae* in which Hog1 is activated by the Ssk1 response regulator. Chauhan et al. (2003) found a divergent function for Ssk1 in *C. albicans*. Microarray studies showed that *C. albicans* utilizes the Ssk1 response regulator protein to adapt cells to oxidative stress, while its role in the adaptation to osmotic stress is less certain. Further, *SSK1* appears to have a regulatory function in some aspects of cell wall biosynthesis. In *C. albicans* a deletion of *SSK1* is not sensitive to osmotic stress imposed by the addition of sorbitol; however, significant sensitivity against oxidative stress was observed. Du and co-workers (2005) showed that an *ssk1* mutant strain was more susceptible to killing by neutrophils than the wild type. Besides the high sensitivity to oxidative stress it was shown that the sensitivity of a Δ*ssk1* strain to human defensin-1, one of the non-oxidative antimicrobial peptides of PMNs, was also greater than that of the wild type, demonstrating that non-oxidative killing in PMNs may contribute to the increased susceptibility of the *ssk1* mutant.

2. Response to NO

Nitric oxide is a key antimicrobial compound produced by the innate immune system (Fang 2004).

Hromatka et al. (2005) investigated the reaction of *C. albicans* to this compound. The transcriptional response of *C. albicans* to 1.0 mM DPTA NONOate (a chemical agent that releases NO in a pH-dependent manner) in YPD was monitored in a time-course experiment over 120 min (UCSF/WI array). They identified in total a number of 131 genes differentially regulated by a factor of at least two. Most of these genes are only transiently induced. All of the ~65 repressed genes returned to normal levels of expression by the 40 min timepoint. Only a group of nine genes, mostly involved in ion transport or redox processes, remains highly expressed throughout the 2 h time-course, including *YHB1, AOX2, SSU1, YOL075c, YMR209c, CTR2, RBT5* and *AOX1*. *YHB1*, a flavohemoglobin, was most strongly induced by NO and its role in resistance to NO in vitro could be confirmed by deletion studies (see also Ullmann et al. 2004). Furthermore, deletion of *YHB1* resulted in prolonged induction of genes only transiently induced in the wild type and a set of additional genes involved partially in DNA damage repair, indicative of a main protective role of *YHB1* for detoxification of NO. In addition, it was shown that deletion of *YHB1* inappropriately activates parts of the filamentous growth pathway, as eight genes known to be hyphal specific were induced in YPD more than 6-fold, including *HWP1, ALS3, RBT1* and *ECE1*. This is in agreement with the hyperfilamentous phenotype of the ∆*yhb1* mutant strain. Interestingly, deletion of *YHB1* resulted only in a moderate reduction of virulence in the tail-vein model of systemic infections in mice (Ullmann et al. 2004). More importantly it could be shown that mice lacking iNOS2, the main gene responsible for production of NO in the innate immune system, show no difference in susceptibility to *C. albicans* wild type or *yhb1* mutant strains, indicating that NO has no significant role in host defence mechanisms in a tail vein model of systemic infection in mice (Hromatka et al. 2005). Thus, the observed virulence defect for ∆*yhb1* strains may not be attributed to increased sensitivity to NO.

Comparing the genes identified in this study as induced for at least 2 h in the presence of NO in wild type or ∆*yhb1* strains with the genes defined as core stress-response genes by Enjalbert and co-workers (2006; *ECM41, GLK1, GRP2, HSP12, HXT61, HSP31, orf19.675, AHP1, orf19.7085*) it is apparent that none of the genes is part of the gene set describing the core stress response. However, the transcriptional response to DPTA NONOate of *Sac. cerevisiae* is reported to bear significant homologies to the response of *C. albicans*, including the induction of *YHB1* and *SSU1* (Sarver and DeRisi 2005).

3. pH Regulation

C. albicans encounters a multitude of different pH ranges in the different host niches it colonizes, starting from the oral cavity (in which highly fluctuating pH values occur due to nutritional uptake), to extremely acidic in the stomach (pH 2), less acidic in the duodenum (pH 5), to alkaline in the intestine (pH 7.7) or acidic in the vaginal tract (pH 4). To counteract this pH stress, *C. albicans* has developed a system in which Rim101, a pH-responsive transcription factor, plays a central role (Davis et al. 2000; Ramon et al. 1999). Bensen et al. (2004) investigated how Rim101 governs gene expression at pH 4 and pH 8 (M199 medium, 37 °C) by comparing a *rim101* deletion mutant with the wild type using whole genome microarrays (CCDMF). Comparing the transcriptome of the wild type at pH 4 and pH 8 identified differential regulation of 514 ORFs from 4715 detectable transcripts (<2-fold). About half were down-regulated and the other half up-regulated at pH 8/pH 4 respectively. Besides the known Rim-regulated genes like *PHR2, RIM8, PHR1, PRA1* and *RIM101* itself, a pH-dependent bias was found for genes involved in hyphal growth, ion transport, protein synthesis and electron transport. Hyphal growth-specific genes were expected, since growth in M199 at 37 °C, pH 8, promotes hyphal cell growth and growth at 37 °C, pH 4, promotes yeast cell growth. The known hyphal-specific genes *CSA1/WAP1, ECE1, HWP1, HYR1, IHD1, RBT1, SAP4* and *SAP6* were expressed 2- to 31-fold higher at pH 8 compared with pH 4. Seventeen ORFs up-regulated at pH 8 were classified as ion transporters, of which ten are predicted to function in iron transport, indicating that alkaline pH induces iron starvation. These include two high-affinity iron permeases (*FTR1/2, FTH1*), one multicopper oxidase (*FET34*), five ferric reductases (*CFL1, FRE2, FRE7, FRE9, FRP2*), one vacuolar iron transporter (*SMF3*) and one iron-siderophore binding protein (*ARN1/SIT1*) as well as *CTR1*, a copper transporter required for iron assimilation.

The contribution of Rim101 to this pH-dependent gene regulation was determined by comparing the transcriptional profile of a *RIM101* deletion mutant with the wild type at pH 8 or pH 4 (M199 medium, 37 °C). At pH 4 only eight genes showed a more than 2-fold difference in the *RIM101* deletion mutant, compared with the wild type. This indicates that Rim101 has no active role at pH 4 under the conditions tested, which is in agreement with its inactive state and low expression level (Davis et al. 2000; El Barkani et al. 2000; Porta et al. 2001; Ramon et al. 1999). At pH 8, 186 genes showed a more than 2-fold difference in the *rim101* deletion mutant, compared with the wild type. Of these 186 genes 70 were not identified as pH regulated genes in wild type, 49 are alkaline repressed genes and 67 were alkaline induced genes. Thus, about a quarter of the genes deregulated in M199, 37 °C at pH 8 by a *rim101* deletion are not primarily pH-dependent but may serve other functions at alkaline pH. Rim101 is responsible for both repression and activation of genes at alkaline pH. Rim101 dependent alkaline induced genes include the hyphal-specific genes *CSA1/WAP1*, *ECE1*, *HWP1*, *HYR1*, *IHD1* and *RBT1*, as well as genes required for ion transport, especially for iron metabolism (*ARN1*, *CTR1*, *ENA2*, *FET34*, *FRE2*, *FRE5*, *FRP2*). Consequently, *rim101* mutants are defective in hyphal development at M199, pH 8, 37 °C and show high sensitivity to iron starvation under these conditions. Thus the authors conclude that one important new aspect of the Rim101p-dependent alkaline pH response is to adapt to iron starvation conditions (for a more detailed study on iron metabolism, see Lan et al. 2004).

In *Sac. cerevisiae* it is known that Rim101 partially acts through Nrg1 (Lamb and Mitchell 2003). Bensen and co-workers investigated the relation between Nrg1 and Rim101 in *C. albicans*. In epistatic experiments strains deleted for *RIM101*, *NRG1* or both were tested for hyphal growth on M199, pH 8 plates (5 days, 37 °C). The authors conclude from their results that Rim101 does not act through Nrg1 as the case of *Sac. cerevisiae*.

In a publication focusing on Rim101 binding sites Ramon et al. (Ramon and Fonzi 2003) described only 20 genes identified as deregulated in a *rim101* mutant strain that was incubated in M199 medium (pH 7.5, 28 °C, OD_{600} = 0.6–0.7) and compared with the wild type using genome-wide microarray analysis (BRI arrays, Candida Chips 5.2). The genes found overlap with the genes identified by Bensen et al. (2004). As described by Bensen et al., based on the microarray analysis performed, *NRG1* expression was not influenced by *RIM101* under the conditions used in this study.

However, Lotz et al. (2004) found that that Rim101 and Nrg1 do interact. In their initial analysis a limited microarray focused on genes encoding for putative cell wall proteins was used to investigate the Rim101 mediated effect on the cell wall. Using both, a deletion mutant of *rim101* as well as the dominant active allele *RIM101-1426*, it could be shown that the level of Rim101 activity inversely correlates with the level of *NRG1* transcript (α-MEM medium, pH 4.5 or 7.4, 25 °C or 37 °C). Dominant active Rim101 results in a strong decrease in *NRG1* mRNA under all conditions tested, most strongly at low temperatures and low pH values where *NRG1* levels are high in the wild type. In contrast, deletion of *RIM101* resulted in an increase in *NRG1* transcript levels when compared with the wild type, most strongly at pH 7.4. In parallel to Rim101 activity, transcript levels of *HWP1* and *RBT1* are induced. For *RBR1/PGA20*, *RBR2/PGA21* and *RBR3* the opposite pattern was observed, namely activated Rim101 blocks their expression, whereas *NRG1* is required for their expression. Thus Rim101-activity, directly or indirectly, regulates *NRG1* transcript levels under the conditions used in this study. Thus the balanced regulation of *RIM101* and *NRG1* expression contributes to the control of the hyphal specific genes investigated.

Comparing the list of hyphal induced genes regulated by Rim101 with the genes identified as repressed by Nrg1 or Nrg1 and Tup1 (Kadosh and Johnson 2005) highlights the genes *ECE1*, *HYR1*, *HWP1*, *IHD1* and *RBT1* as co-regulated by these repressors and Rim101. In addition, *PHR1* also was identified as repressed by Nrg1 and Tup1, as well as *RNH1*, *PGA58*, *orf6.5146*, *DDR48*, *HIS1*, *PGA13* and *ARG1* which are also regulated by Rim101. This indicates that for expression of hyphae-specific (cell wall) genes, there seems to be an interaction between Rim101 and Nrg1. For other functions described for Nrg1 and Rim101 this seems to be not the case. A direct comparison of the transcriptomes of Rim101, Nrg1 and Rim101/Nrg1 deletion strains might further clarify the relation of these transcription factors in *C. albicans*.

C. Polymorphism of *C. albicans*

C. albicans is a polymorphic organism occurring in several distinct morphologies, including yeast-form cells or blastospores, chlamydospores, pseudohyphal growth forms and true hyphae (Sudbery et al. 2004). Depending on the environmental conditions *C. albicans* is able to change from one to the other growth form. The regulation of morphogenesis is governed by a multitude of signalling pathways, most of which have been reported to be of relevance for virulence (Liu 2001, 2002; Whiteway and Oberholzer 2004). The polymorphism of *C. albicans* is one of the key features of this organism and has been shown to be critical for pathogenesis. Mutants which are predominantly in the yeast form or in a pseudohyphal/hyphal morphology have been show to be strongly attenuated in virulence (Braun and Johnson 1997; Lo et al. 1997). Therefore, the molecular mechanisms underlying morphogenesis are of major interest. Several aspects have been selected for this review, focusing on the basic transcriptional changes during the yeast-to-hyphal transition, some of the transcription factors involved, the cAMP pathway and white–opaque switching and mating.

1. Yeast to Hyphae Transition

One of the first studies using microarrays containing an almost complete set of predicted ORFs (BRI, 5668 ORFs, based on assembly 4) was presented by Nantel et al. (2002). A goal of this study was to determine the genes which are differentially regulated during transition from the yeast to hyphal growth form and the contribution of Efg1p and Cph1p, two transcription factors relevant for morphogenesis and virulence (Lo et al. 1997). Efg1 was shown to be a central regulator of virulence and morphogenesis, which is regulated via the cAMP pathway (Bockmuhl and Ernst 2001), whereas Cph1 was shown to be regulated by a MAPK pathway involved in morphogenesis and mating (Lane et al. 2001; Liu 2002).

Hyphal induction was performed by the addition of serum to rich medium (YPD 30 °C, to YPD 37 °C + fetal calf serum; FCS) and by a shift of Lee's-medium from 25 °C to 37 °C. In addition, the effect of temperature induction alone (in YPD) or addition of serum at 25 °C (where no hyphal development is observed) was monitored to exclude genes not involved in morphogenesis but responding to a shift in temperature or serum itself. Data analysis revealed that 18 genes were consistently induced by at least 2-fold (additional 56 genes by a factor of 1.5), whereas 46 genes were consistently down-regulated (1.5-fold) 6 h after the shift to the hyphal conditions employed. Besides previously identified hyphal specific genes, like *HWP1*, *ECE1*, *SAP4-6* or *RBT1* among others, this study identified genes connected to actin remodelling *PFY1* (profilin) and *RDI1* (inhibitor of Rho-GTPases) the secretory pathway (*SEC24* and *YBL060w*), *SOD5* a previously undescribed superoxide dismutase (see also Fradin et al. 2005) as well as ORFs without homologies to known proteins. Interestingly, *RBT1* and *ECE1* respond to serum at low temperatures under conditions when *C. albicans* is not in a hyphal growth form, thus excluding them from the strictly morphogenesis-related genes.

The genes most strongly down-regulated were genes of unknown function (*RHD1,2,3* repressed by hyphal development), cell surface proteins (*FLO1*, *CSP27*), a set of genes involved in lipid metabolism (*YER73*, *YKR70*, *DAK2*, *SOU1*, *PLB1*), DNA-binding proteins (*NRG1*, *GIS2*, *CBF1*, *YDR73*, *TYE7*, *CUP9*) as well as other functions: *HSP12*, *CHT2*, *YHB1*, *RHR2*, *YLR63* and *PCK1*. For genes like *CHT2* and *RHD2*, a down-regulation by a temperature shift from 30 °C to 37 °C without addition of serum was observed, identifying them as temperature-regulated genes rather than morphogenetic genes.

In a time-course experiment looking at the time-points 30 min and 60 min additional genes were identified which were only transiently expressed and therefore potentially relevant for the initiation of germ tubes. In total 232 genes were reported to show significant variation at the time-points investigated. These include chaperones encoded by *WOS2*, *RAD14*, *YNP115* and *CYP2*, as well as proteins like a *BEM2* homolog, a Rho1-GAP involved in cell wall maintenance and Rho3p, a small GTPase involved in cell polarity. The majority of the genes transiently down-regulated encode proteins involved in translation.

The role of *EFG1* and *CPH1* in morphogenesis was addressed by transcriptional profiling of the respective mutants deleted for theses transcription factors. Transcriptional profiling of mutants in *EFG1* and *CPH1* revealed a change in the expression of 74 genes (30 genes induced, 44 genes repressed) in YPD at 30 °C. Some of the repressed genes have been shown in *Sac. cerevisiae* to be involved in stress response, including *HSP12*, *GLK1*, *SNO1*, *ECM4* and *GRE2*. The transcriptional profiles of the *efg1cph1* mutant strain indicate that most of the hyphal induced genes do not respond to induction with serum at 37 °C. Instead, the transcriptional profile resembles the profile of the wild type after adaptation to 37 °C (without serum), indicating that Efg1p and Cph1p are responsible for initiating the transcriptional response to serum (including hyphal development). In addition, it was found that Cph1p does not have a significant *EFG1*-independent role in yeast morphology under the conditions tested.

A similar result was found in a study investigating specifically cell wall biogenesis during the yeast to hyphal transition in dependence of *EFG1* and *CPH1*. Sohn and co-workers (2003) used two hyphae-inducing conditions (YPD, 37 °C + serum, or α-MEM, 37 °C) and two yeast form inducing conditions (YPD, 30 °C, or α-MEM, 25 °C) to investigate the transcriptional profile of 117 genes involved in cell wall biogenesis in SC5314 as well as in strains deleted for *CPH1*, *EFG1*, or both transcription factors. As reported by Nantel et al. (2002) Cph1p did not have a significant *EFG1*-independent role. In addition it was found that Efg1p is a major regulator of cell wall biogenesis. This study revealed a high variability of the cell wall transcriptome under the conditions tested and could identify both yeast-form-specific and hyphae-specific transcripts of potential cell wall genes. About 60% of all genes present on the array were changed by more than 2-fold under at least one of the conditions investigated. As hyphal specific genes up-regulated under the hyphal inducing conditions used *HWP1*, *RBT1*, *RBT4*, *HYR1*, *HWP2* (*ORF6.2933*), *CHS4* and *ORF6.2071* were identified. Deletion of *EFG1* resulted both in the repression as well as in the induction of potential cell wall genes, indicating a dual function of Efg1 as repressor and activator of gene expression. This was shown in detail for the newly described genes *HWP2* and *YWP1*, whose expression depended on the presence of *EFG1* and for *RBE1* which is only expressed significantly in the absence of Efg1. The major changes of the cell wall composition observed in an *EFG1* mutant, as implicated from this study, is well in agreement with the reduced adhesion observed for strains deficient in Efg1 (Dieterich et al. 2002) and helps to rationalize the reduced pathogenicity of these strains observed in mouse models of systemic infections (Lo et al. 1997).

2. The APSES Proteins Efg1 and Efh1 in *C. albicans*

In *C. albicans* two genes encoding APSES proteins (named after the members of the family all encoding fungal transcription factors involved in morphogenesis) are present: *EFG1* and *EFH1*. The function of both genes was studied in detail using DNA-microarrays (Eurogentec) initially also focusing on morphogenesis (Doedt et al. 2004). In contrast to Nantel et al. (2002) in this study the early stage of hyphal induction (30 min) and its dependence on the APSES proteins was investigated.

The function of *EFG1* and *EFH1* were investigated both in rich medium and under hyphae-inducing conditions by comparing the respective deletion strains to the wild type. Genome-wide transcriptional profiling revealed that *EFG1* and *EFH1* regulate partially overlapping sets of genes associated with filament formation. Most interestingly, Efg1p not only regulates genes involved in morphogenesis but also strongly influences the expression of metabolic genes, inducing glycolytic genes and repressing genes essential for oxidative metabolism. By using one- and two-hybrid assays, it was furthermore demonstrated that Efg1p acts as a repressor of transcription, whereas Efh1p acts as an activator of gene expression.

In rich medium (YPD, 30 °C) *EFG1* was found to modulate the expression of 283 genes by a factor of at least 1.5, with 100 being up- and 183 down-regulated. Most interestingly deletion of *EFG1* had a major impact on genes known to regulate carbon metabolism. From all genes deregulated in an *efg1* mutant strain 27% could be assigned to metabolism. Almost all glycolytic enzymes, like the key regulators *FBA1* or *PFK1*, as well as genes required for the accumulation of reserve carbohydrate (*TPS2*, *TPS3* among others) were repressed in the absence of Efg1, whereas genes encoding for enzymes of the TCA cycle were induced. Thus, the presence of Efg1p favours fermentative and represses oxidative growth. Consequently, *efg1* mutant strains are more sensitive to antimycin A, a drug blocking ATP synthesis via the respiratory chain. This is also in agreement with results by Lan et al. (2002) who showed that white cells had enhanced expression of glycolytic genes compared with the opaque cells favouring oxidative metabolism. Opaque cells require very low levels of Efg1p, whereas high expression levels of Efg1p induce white cells (Sonneborn et al. 1999; see below).

In contrast to *EFG1*, only nine genes (eight up- and one down-regulated) were found to be regulated by *EFH1* in YPD, 30 °C, indicating a minor role for Efh1 under these conditions. This is reflected in the low transcript level of *EFH1* which is about 10-fold less than the transcript level of *EFG1*.

The simultaneous deletion of *EFG1* and *EFH1* showed a transcriptional pattern that only partially overlapped with that of the *efg1* single mutant (49 from 283 genes). The majority of the genes (233 genes) were deregulated only by *efg1* but not in the *efg1 efh1* double mutant, however, a new subset of 63 genes was affected in the *efg1 efh1* double mutant that was not detected in the *efg1* or *efh1* single mutant. This result is indicative of synthetic interactions between *EFG1* and *EFH1*. Synthetic phenotypes of the *efg1efh1* mutant were observed for embedding and microaerophilic conditions. Basically, additional deletion of *efh1* in an *efg1* mutant strain resulted in reversion of the phenotype back to the wild type, consistent with the results from the transcriptional profiling experiments.

The effect of *EFG1* and *EFH1* on morphogenesis was also investigated in this study. In contrast to Nantel et al. (2002) hyphal induction was initiated by adding cells to YP + 10% horse serum for 30 min without the addition of glucose to the medium. Under this regimen 243 genes were affected in the *efg1* strain (factor 1.5) and 39 in the *efh1* strain, again showing the predominat effect of Efg1p. In a similar way as in YPD, 30 °C, the double mutant *efg1efh1* revealed an additional set of 58 genes but also only 47 which were affected by deletion of *EFG1* alone. Again glycolytic enzymes were expressed at a lower level in cells lacking Efg1, but in this case the TCA cycle was not up-regulated, most likely due to the fact that no glucose was present in the medium (10% serum only). The induction of hyphal associated cell wall genes was observed in a similar way as reported (Nantel et al. 2002; Sohn et al. 2003). Differences in these data to the data generated by Nantel and co-workers may be due to the different experimental conditions used.

The effect of overexpression of *EFG1* and *EFH1* in *C. albicans* was also investigated. The use of the *PCK1*-promoter required the use of SSAC medium, a synthetic medium containing no glucose (Leuker et al. 1997). Overexpression of *EFG1* resulted in the down-regulation of 53 and up-regulation of only 32 genes, indicating that Efg1p predominantly acts as a repressor as indicated in a previous study (Sohn et al. 2003). Interestingly, only 14 genes identified in this experiment were also identified in strains deleted for *EFG1* (as oppositely regulated). This may also be due to the significantly different growth conditions used for both experiments including the change of the main carbon source. Among these 14 genes, cell wall genes associated with hyphal development, *HWP1*, *ALS10*, *RBT5*, *ECE1* and *PHR1*, as well as genes encoding stress-response proteins like *DDR48* and *SOD5* were identified. In contrast to *EFG1* overexpression, *EFH1* overexpression resulted in 53 up-regulated genes (including *HWP1*, *ALS10*, *ECE1*, *DDR48*) and 28 down-regulated genes, indicative of a transcriptional activator. *EFH1* overexpression, like *EFG1* overexpression, results in the formation of pseudohyphae, blocks true hyphae formation and triggers opaque to white switching, both of which requires *EFG1*.

Promoter activation studies in *C. albicans* using LexA-Efg1 and LexA-Efh1 fusions confirmed these results. Interestingly, in *Sac. cerevisiae* using a Gal4 DNA-binding domain, fusion to *EFG1* did not result in repression of the corresponding promoter, indicating that additional cofactors from *C. albicans* are required to exert the repressing function of Efg1.

These results indicate that Efh1 supports the regulatory functions of the primary regulator, Efg1, supporting a dual role for these APSES proteins in the regulation of fungal morphogenesis and metabolism.

Cao et al. (2006) could show that *FLO8* controls a subset of the genes controlled by *EFG1*. The Δ*flo8* mutant was shown to be avirulent in a mouse model of systemic infection, similar to a Δ*efg1* mutant. Genome-wide transcription profiling of Δ*efg1* and Δ*flo8* using a *C. albicans* DNA microarray (70mer set by QIAGEN Operon) suggested that Flo8 controls subsets of Efg1-regulated genes. Most of these genes are hyphae-specific, including *HWP1*, *HYR1*, *ALS3*, *ALS10*, *RBT1*, *HGC1* and *IHD1*. Most interestingly, all genes identified to be regulated by *FLO8* are also regulated by *EFG1* in a similar way but not vice versa. Consistent with this finding, it was shown by in vivo immune-precipitation that Flo8 interacts with Efg1 in yeast and hyphal cells. Similar to Δ*efg1* and Δ*cdc35* (adenylate cyclase) strains, Δ*flo8* strains shows enhanced hyphal growth under an embedded growth condition. These results suggest that Flo8 may function downstream of the cAMP/PKA pathway and together with Efg1 regulate the expression of hyphae-specific genes in *C. albicans*.

3. cAMP Signalling

cAMP is a signalling molecule activating one of the major protein kinases, PKA, in fungi. Harcus and co-workers (2004) investigated the consequences of the absence of adenylyl cyclase (*CDC35*), *RAS1* and *EFG1* using transcription profiling (BRI array, 6002 ORFs). Cdc35 is the only known enzyme responsible for cAMP production in *C. albicans* which is activated in part by Ras1 (Rocha et al. 2001). Efg1, as describe above, has been proposed to be one of the key transcription factors regulated by the cAMP-PKA pathway (Bockmuhl and Ernst 2001). To investigate conditions inducing yeast and hyphal growth forms the respective deletion mutants were grown in YPD, 30 °C, or YPD + serum, 37 °C. Genes modulated by at least 1.4-fold were selected and data from Nantel et al. (2002) and Lee et al. (2004) were analysed together with the data generated in this study. A collection of 1168 genes was identified as significantly modulated (1.4-fold) under at least one of the conditions used. Under all conditions examined, the profiles of the *ras1* and *cdc35* mutants were similar to each other (in the same dendrogram sub-branch), whereas the profile of the *efg1* mutant was shown to be different. Morphologically, the three mutants were distinct: when growing in YPD, 37 °C + serum the *efg1* and *cdc35* mutants both remained nonhyphal whereas *ras1* was still able to form hyphae. Thus, transcription profiling provides a different picture of the relationships among the elements than did the cellular morphology.

Comparison of the *cdc35* mutant with the wild type resulted in the largest differences (600 transcripts in yeast form, 800 transcripts in hyphal growth conditions). The profiles from yeast and hyphal growth conditions correlated significantly. The majority of genes encoding for ribosomal proteins or for subunits of the RNA polymerase holoenzymes were repressed in the *cdc35* mutant. Similarly, the loss of *CDC35* was associated with repression of metabolic pathways such as the TCA cycle, pyrimidine metabolism and the synthesis of heme and sterol. This reflects the reduced growth rate exhibited by the *cdc35* mutant. Besides a large number of genes without known homologs, a notable group of the transcripts elevated in the absence of cAMP encode proteins involved in the formation and function of the cell wall. In accordance with this finding the *cdc35* mutant tends to aggregate and was shown to be significantly more resistant than wild type cells to calcofluor white, which binds to chitin, as well as to zymolyase, which is primarily a β-1,3-glucanase. In addition a significant correlation was observed between the *cdc35* profile and the profile observed in osmotically shocked cells (Enjalbert et al. 2003). Cdc35 cells also exhibited an increased sensitivity to osmotic stress. Further results showed that, during the yeast-to-hyphal transition, almost all of the genes that were modulated in wild-type cells, including classic hyphae-induced genes such as *ECE1*, *HWP1* and *SAP4*, are no longer responsive to the serum and heat signals in the *cdc35* mutant. This suggests that most of the response to a shift from 30 °C to 37 °C + FCS in *C. albicans* is mediated by the cAMP pathway. However, a few transcripts, including *CHA2*, *GAP4*, *HMO1*, *RHD1*, *RHD3*, *SNZ1* and *orf19.7531*, still respond as they did in the wild type, suggesting that a cAMP-independent pathway may contribute to morphogenesis.

The loss of Ras1p function was less severe than the loss of adenylyl cyclase. Only 72 transcripts were significantly more abundant in *RAS1*-deleted cells (YPD, 30 °C) than in the wild type, whereas four transcripts were less abundant (100 during hyphal induction conditions). The majority of the Ras1p-influenced transcripts are a subset of those that are modulated by cAMP (both for hyphal and yeast-form growth). Besides many of the genes of unknown function, a number of cell wall genes showed similar behaviour in *ras1* and *cdc35* cells. As observed for strains deleted for *CDC35*, *ras1* mutants showed increased resistance to zymolyase and calcofluor white. However, the *ras1* mutant is not responsive to osmotic stress (both phenotypically and at the transcriptional level). The transcript levels of some of the hyphal-specific genes, such as *ECE1*, *RBT1* and *HWP1*, were clearly reduced when *ras1* was compared with the wild type in hyphal conditions. However, they were still partially responsive in the *ras1* whereas in the *cdc35* mutant they were totally unresponsive.

Interestingly, Harcus and co-workers found that the majority of *EFG1*-modulated genes were distinct from those modulated by *RAS1* or *CDC35*. Efg1 had the strongest effect on gene expression during the yeast to hyphal transition (200 genes modulated), whereas during yeast growth only 85 genes were identified as modulated. Doedt et al. (2004) found a significantly larger number of genes modulated by Efg1, partially under the same conditions (283 genes in YPD, 30 °C; 243 genes in YP + serum, 37 °C, 30 min). Both studies identified the genes required for glycolysis and gluconeogenesis modulated by Efg1, as well as a correlation with profiles of the white to opaque switch (Lan et al. 2002). For hyphal growth conditions, the small number of modulated transcripts that were commonly influenced in the *efg1*, *cdc35* and *ras1* mutants identified most of the highly modulated, hyphae-specific genes that were initially used to define this signalling pathway (e.g. *HWP1*, *ECE1*). The factors of induction or repression in this study, however, were in general lower than in other studies (e.g. Doedt et al. 2004; Kadosh and Johnson 2005). A reason for this has been proposed by Kadosh et al. (see below).

4. Repression of Transcription as Key for Morphogenesis

Two publications described the global effect of transcriptional repressors on gene expression for the yeast to hyphal switch in *C. albicans*. Kadosh focused on the three transcriptional repressors Tup1, Nrg1 and Rfg1 (Kadosh and Johnson 2005), whereas Garcia-Sanchez focused on Ssn6, Tup1 and Nrg1 (Garcia-Sanchez et al. 2005).

Kadosh and Johnson compared the transcriptional changes of mutants deficient in Rfg1, Nrg1 and Tup1 to the wild type in yeast and hyphal growth conditions (UCSF/WI array; YPD, 30 °C or 37 °C, YPD + serum, 30 °C or 37 °C, similar to Nantel et al. 2002). It was pointed out that the status of the culture to inoculate the main culture was critical for the experiment. An overnight culture had to have OD_{600} >13 if a high rate (close to 100%) of filament formation was to be obtained. Otherwise only partial filamentation was observed. Consequently, analysis of these mixed cultures resulted in apparently reduced induction of filament-specific genes, whereas cultures with high levels of filamentation resulted in much higher induction rates of hyphae-specific gene expression.

DNA-microarray analysis identified 61 genes that are significantly induced (≥2-fold) during the yeast-to-hyphae transition [YPD + serum, 37 °C, compared with the same time-points (1, 2, 3, 5 h time-points) in YPD, 30 °C]. Approximately one-third of these genes are induced ≥10-fold after 1 h (including *ECE1*, *HYR1*, *ALS10*, *ALS3*, *HWP1*, *SAP5*, *SAP4*, *PRY4*, *orf19.3698*, *POP4*, *IHD1*, *PHR1*, *USO6*, *orf19.1120*, *SOD5*). Of the 61 genes identified in this study, ten correspond to the 18 genes (induced ≥2-fold) identified by Nantel et al. (*ECE1*, *SAP4,5*, *HWP1*, *SOD5*, *RBT1*, *DDR48*, *PHR1*, *YBL060w*, *IHD1*). Most of them are part of the genes induced ≥10-fold. Approximately half of the 61 genes are transcriptionally repressed in the yeast-form state by at least one of the three transcriptional repressors: Rfg1, Nrg1, and Tup1. From these results, the authors conclude that the relief of transcriptional repression plays a key role in activating the *C. albicans* filamentous growth programme. Intriguingly, several of the highly induced genes found in this study, including *ALS10*, *HWP1*, *ECE1*, *RBT1*, *SOD5*, *DDR48* and *PHR1*, are among the genes induced by overexpression of Efg1p (Doedt et al. 2004).

Garcia-Sanchez and co-workers (2005) characterized the regulons of the transcriptional repressors Ssn6, Nrg1, and Tup1 (Eurogentec Array). In *Sac. cerevisiae* Tup1-Ssn6 constitute a well defined co-repressor which is conserved from yeast to man (Redd et al. 1997). Overlapping sets of genes are a sign for co-regulation and therefore would indicate interaction between these factors. In contrast to *Sac. cerevisiae*, transcriptional profiling in *C. albicans* revealed only a small overlap between Tup1 and Ssn6. This is in agreement with the distinct phenotypes of a deletion in *TUP1* or *SSN6*. Deletion of *SSN6* promotes morphological events reminiscent of morphological switching rather than filamentous growth, whereas deletion of *TUP1*, as well as deletion of *NRG1*, results in constitutive filamentation. Of 224 genes which were up-regulated in a Δ*tup1* strain, only 38 genes were coregulated in the Δ*ssn6* strain, indicating that for the repression of 186 Tup1-regulated genes *SSN6* is not necessary. Looking at down-regulated genes the discrepancy is even larger (only five of 117 *TUP1*-regulated genes are co-regulated with *SSN6*). The overlap with Δ*nrg1* strain was similar, with a higher proportion of Nrg1 and Tup1 co-regulation than Nrg1 and Ssn1 co-regulation. Consistent with the study of Kadosh and Johnson (2005) the hyphae-specific genes *HWP1*, *ECE1*, *RBT1* and *RBT5*, as well as genes like *DDR48* and *ALS10* were identified as regulated both by *NRG1* and *TUP1*. Other hyphae-specific genes like *HYR1* were identified as *TUP1*-regulated. All of these genes were not found to be derepressed in the Δ*ssn6* strain under the conditions used (YPD, 30 °C). [Hwang reported partial derepression of *HWP1* and *ECE1* in YPD, 37 °C (Hwang et al. 2003a).] Genes co-regulated by Tup1 and Ssn6 were related to amino acid or carbon metabolism, like the key gluconeogenetic genes *PCK1* and *FBP1* as well as the glyoxalate cycle gene *MLS1*. Phenotypic switching is also associated with changes in carbon metabolism and has been shown to depend on Efg1 (see above and Doedt et al. 2004; Lan et al. 2002). In addition the white-phase-specific gene *WH11* was shown (by Northern blot) to be repressed by Ssn6 but not by Tup1 or Nrg1.

These results show that Ssn6 and Tup1 in general play distinct roles in *C. albicans*. Nevertheless, both Ssn6 and Tup1 were required for Nrg1 mediated repression of an artificial *NRE* (Nrg1 Response Element) promoter, indicating that in some cases a Tup1-Ssn6 co-repressor exists, as in *Sac. cerevisiae*.

5. Phenotypic Switching and Mating

White–opaque switching in the human fungal pathogen *C. albicans* is an alternation between two quasi-stable, heritable transcriptional states observed in a few clinical isolates. This has been most extensively described in the patient isolate WO-1 (Slutsky et al. 1987). WO-1 alternates between white hemispherical colonies, designated white (W), and grey flat colonies, designated opaque (O). W/O phenotypic switching affects the shape and size of cells, their ability to form hyphae, their surface properties (e.g. adhesion, permeability), membrane composition, range of secretory products, sensitivity to neutrophils and oxidants, antigenicity and drug susceptibility (Soll 1997). Recently, it was shown that white–opaque switching and mating are both controlled by the mating type locus homeodomain proteins (Miller and Johnson 2002). The majority of *C. albicans* strains are heterozygous for the mating type locus *MTL* (a/α) and cannot undergo white–opaque switching (Lockhart et al. 2002). However, when these cells undergo homozygosis at the mating type locus (i.e. become a/a or α/α), they can switch, and they have to switch in order to mate efficiently (WO-1 is *MTL* α/α). Opaque cells were shown to mate approximately 10^6 times more efficiently than white cells. These results showed that opaque cells are a mating-competent form of *C. albicans* and that this pathogen may undergo a white-to-opaque switch as a critical step in the mating process. As white cells are generally more robust than are opaque cells, this strategy may allow *C. albicans* to survive the harsh environments within a mammalian host, but still retain the ability to generate mating-competent cells.

For the analysis of switching Lan and co-workers (2002) analysed the transcriptome of both cell types (in WO-1) at four time-points (12 h, 18 h, 24 h, 48 h; Lee's medium; Affymetrix GeneChip). A total of 373 ORFs demonstrated a greater than 2-fold difference in expression level between the switch phenotypes (in at least three time points); 221 ORFs were expressed at a higher level in opaque cells than in white cells; and 152 were more highly expressed in white cells. Affected genes represent functions as diverse as metabolism, adhesion, cell surface composition, stress response, signalling, mating type and virulence. Approximately one-third of the differences between cell types were shown to be related to metabolic pathways. Most interestingly, opaque cells were expressing a transcriptional profile consistent with oxidative metabolism and white cells were expressing a fermentative metabolism. This bias was obtained regardless of carbon source, suggesting a connection between phenotypic switching and metabolic flexibility. Efg1 seems to be involved in these events as it was shown previously that the expression level of Efg1 determines the phase of the cells (high Efg1 levels induce white cells; Sonneborn et al. 1999) and regulate metabolism (Doedt et al. 2004). *EFG1* is primarily expressed in white cells which accordingly have a fermentative metabolism. In addition it was found that W and O cells differentially express genes presumed to function in mating-type differentiation and cell-type control. The α-pheromone encoded by *ORF6.4306*, a homologue of the *Sac. cerevisiae* a-factor pheromone receptor (*STE3*) and a putative mating-type regulatory protein encoded by *MTLα1* are all more highly expressed in O cells, consistent with WO-1 being equivalent to *MTL*α/α.

In contrast to mating in *Sac. cerevisiae* (Herskowitz 1989), Tsong and co-workers could show that in *C. albicans* each of the two mating type alleles (a, α) contributes a positive regulator of its respective mating type (a2, α1; Tsong et al. 2003). Additionally, each allele contributes one-half of a heterodimer that negatively regulates mating competency (a1, α2). Each half of the heterodimer – on its own – has no regulatory activity. To identify genes regulated by the *C. albicans* a1, a2, α1 and α2 proteins strains carrying the 16 possible combinations of the four *MTL* genes described above were compared by transcriptional profiling (UCSF/WI array). All 16 strains were analyzed in the white phase (in SC or YPD media); for the 12 strains competent to switch from white to opaque, also the transcriptional profiles of the opaque forms were analyzed. In all experiments, a reproducible 2-fold change was considered significant.

For the 16 white strains of each *MTL* configuration, seven genes were found that were reproducibly repressed under multiple conditions 2- to 8-fold by a1 and α2 working together in the white phase (in *Sac. cerevisiae* 20–30 genes are repressed by a1-α2; Galitski et al. 1999). This gene cluster is comprised of *CAG1* (homologous to *Sac. cerevisiae GPA1*), *FUS3*, *FAR1*, *STE2*, *YEL003w* and two ORFS with no homology to any known genes. These seven genes are controlled by the a1-α2 heterodimer, independent of the white-to-opaque transition. No significant differences between the expression patterns of a and α white cells were observed.

In addition white and opaque versions of strains carrying the 12 *MTL* configurations permissive for white-opaque switching were compared to a white control strain carrying an intact *MTL*. Irrespective of the *MTL* configuration, 237 genes were identified that are up-regulated and 197 genes that are down-regulated in the opaque phase. Transcripts up-regulated in the opaque phase included some likely to be involved in mating in both a and α cells, such as *STE4* and *FUS3*. This set of white and opaque-specific transcripts overlaps with genes identified in a comparison of the white and opaque phases of the genetically different strain WO-1 (Lan et al. 2002).

From the combination of strains investigated two α-specific genes were identified. *STE3* and *MFα1* are highly induced in opaque strains relative to white strains (300- and 1000-fold, respectively), but only in those that carry an intact *MTLα1* gene (which are able to mate as α-cells). In this data set no a-specific genes were identified in opaque-phase a-type cells. However, addition of α-factor to cells containing a2 (in the absence of *MTLα*) resulted in the induction of 12 genes, including *STE6*, *RAM2*, *ECE1*, *HWP1*, *FIG1*, *RBT1* and *CEK1*. These genes of course could also be pheromone-induced genes rather than a-specific genes (a publication describing a-factor in *C. albicans* has been just been accepted during finalisation of this article; Dignard et al. 2007).

By comparing the mating circuits of *Sac. cerevisiae* and *C. albicans* several major differences were identified. One major example of divergence between *C. albicans* and *Sac. cerevisiae* in mating type regulation is that *C. albicans* has retained a positive regulator of a-type mating from a common ancestor, while *Sac. cerevisiae* has lost this regulator. The second major example of divergence is the interposition of an additional layer of transcriptional control in the *C. albicans* mating type circuit. As described above *C. albicans* a

and α cells must undergo a "phenotypic switch" from the white phase to the opaque phase before they are competent to mate (Lockhart et al. 2002; Miller and Johnson 2002). This switch is governed by the mating type locus: two homeodomain proteins (a1, α2) – one from the a locus and one from the α locus – co-operate to repress the switching, thereby assuring that a/a and α/α, but not a/α cells, can mate. *C. albicans* a1 and α2 proteins repress a few genes directly, but control many more indirectly by governing white-opaque switching. This indirect regulation of mating competency by a1-α2 in *C. albicans* constitutes an additional layer of transcriptional regulation, absent in *Sac. cerevisiae*, which ensures that mating only occurs in specific environments. Recently, a transcriptional regulator, *WOR1/TOS9*, acting as a master switch regulator, was identified by several groups which is required for establishment of the opaque phase (Huang et al. 2006; Srikantha et al. 2006; Zordan et al. 2006).

D. Host–Pathogen Interaction

One of the most interesting events in *C. albicans* biology is the direct interaction with the host. The switch between commensalisms and pathogen until now could not be studied due to the lack of appropriate models. Experiments focusing on vaginal candidosis have been performed using human volunteers, however, not on a genome-wide platform (Fidel 2007). Mouse models of systemic infection have been used frequently and first attempts to isolate *C. albicans* for genome-wide profiling have been reported (Andes et al. 2005; Fradin et al. 2003). Currently, simple model systems mimicking host-pathogen interaction, e.g. *C. albicans* encountering the host defence, like macrophages and neutrophils or adhering and penetrating into tissue using different reconstituted tissue models derived from cell lines or primary cells, have been used to shed light into *C. albicans* pathogenesis.

1. Neutrophils

Rubin-Bejerano et al. (2003) compared the transcriptional response of *Sac. cerevisiae* and *C. albicans* engulfed by neutrophils (UCSF/WI array). In addition the uptake of *Sac. cerevisiae* in monocytes was investigated. After phagocytosis by neutrophils, both *Sac. cerevisiae* and *C. albicans* respond by inducing genes of the methionine and arginine pathways. Neither of these pathways is induced upon phagocytosis by monocytes. Both fungi show a similar induction of these pathways when transferred from amino acid-rich medium to amino acid-deficient medium. From these data the authors conclude that the internal phagosome of the neutrophil is an amino acid-deficient environment. In contrast to engulfment by macrophages, *Sac. cerevisiae* and *C. albicans* were killed 3 h after phagocytosis in neutrophils. For *Sac. cerevisiae* induction of the methionine genes by deprivation was found to be independent of Gcn4, whereas induction of the arginine genes was dependent on Gcn4. In *C. albicans GCN4* and *PCL5*, encoding a Gcn4-stabilizing protein, were induced upon exposure to neutrophils. A stronger oxidative stress response of *C. albicans* than of *Sac. cerevisiae* to neutrophils was observed. *SOD1*, *CCP1-1*, *CTA1-1*, *CTA1-2*, *GPX3-1* and *GPX3-2* were induced between 10- and 40-fold. Interestingly, in contrast to monocytes/macrophages, *C. albicans* is not able to form filaments within neutrophils.

2. Macrophages

Lorenz and co-workers (2004) analysed the global transcriptional response of *C. albicans* upon internalization by mouse macrophage line J774A (UCSF/WI array). They could show that phagocytosis stimulates an immediate transcriptional response (1 h time-point). The early pattern is characterized by a dramatic up-regulation of the gluconeogenesis/glyoxylate pathways and down-regulation of glycolysis and the genes encoding the translation apparatus. Genes of the TCA cycle which are not part of the glyoxalate cycle are not affected by macrophages. Isocitrate lyase (*ICL1*) and malate synthase (*MLS1*) have been found in a previous study to be up-regulated in *Sac. cerevisiae* and *C. albicans* (Lorenz and Fink 2001). Consequently, *C. albicans* mutants lacking *ICL1* are markedly less virulent in mice than the wild type. The glyoxalate cycle is also up-regulated in other pathogens like *Cryptococcus neoformans*, however, deletion of *ICL1* does not affect pathogenesis in this organism. In addition to gluconeogenesis and the glyoxylate pathway, β-oxidation is activated in *Candida albicans*, indicating a flow of carbon from fatty acids to glucose. Almost all the genes encoding for ribosomal proteins as well as many genes required for the translation apparatus are strongly down-regulated. Mitochondrial translation, however, is reported to be unaffected by phagocytosis (both the glyoxalate cycle and β-oxidation require mitochondria). A specific nonmetabolic response distinct from filamentation embedded in the early pattern that responds to stresses presented by macrophage contact was also identified, including machinery for DNA damage repair, oxidative stress responses, peptide uptake systems and arginine biosynthesis. Filamentation-specific genes have not been identified since the control cultures in RPMI + serum induced hyphae at the same rate as in the presence of macrophages. Thus macrophages seem to not specific induce a filamentation response. This early transcription profile switches to a later profile which is highly similar to the profile of the cells grown without macrophages. This basically reflects the observation that *C. albicans* has escaped form the macrophages and grows now in tissue culture medium (RPMI + serum), like the control culture. The non-filamentous *cph1 efg1* mutant (Lo et al. 1997), which is internalized but cannot escape, remains frozen in the early pattern. Interestingly, the avirulent *cph1efg1* mutant is not killed within the macrophages (J774A cell line), it actually is able to double within the 6 h time-course experiment conducted, indicating that the macrophages used in this study do not play a role in the host's defence mechanism against *C. albicans* (J774A is able to kill *Sac. cerevisiae*).

In order to confirm that the metabolic reprogramming after phagocytosis is due to lack of nutrients, similar starvation conditions were mimicked in vitro by omitting C-, or N-sources or both in synthetic growth media. From these data it was concluded that the similarities between phagocytosed cells and starvation were only to cells deprived of a carbon source. Nitrogen-depleted cells show a significantly different pattern of gene expression, with little to no overlap with ingested cells. A cluster of 227 genes specific for response to macrophages could be identified from the data. Only one metabolic pathway specific to macrophage phagocytosis was found by comparison with the in vitro starvation conditions. This was the arginine biosynthetic pathway with all but one of the ten genes strongly up-regulated. The reason for this is unknown. In addition a set of 117 genes including transporters required mostly for N-source uptake and vacuolar proteases have been identified. Only a few genes responsible for oxidative stress defence (including *YHB1*, *GPX3* and *CCP1*) or metal homeostasis (including *FRE3*, *FRE7* and *CTR1* required for iron homeostasis) and DNA repair were mentioned as macrophage-specific.

Most interestingly the early response described in this study is basically absent in the *Sac. cerevisiae*. In contrast to *C. albicans* where 545 genes respond to internalisation, only 53 respond in *Sac. cerevisiae* (Lorenz and Fink 2001), underscoring the pathogen/non-pathogen differences and revealing a highly co-ordinated system in *C. albicans* for immune evasion.

3. Blood

In two studies, Fradin et al. (2003, 2005) focused on the transcriptional response of *C. albicans* to human blood and the individual components of blood (Eurogentec array). This environment is interesting because survival in blood and escape from blood vessels into tissues are essential steps for the pathogen to cause systemic infections. Whole blood induced genes that are involved in general and oxidative stress response, the glyoxylate cycle and protein biosynthesis. Subsets of these genes were also detected from *C. albicans* isolated after tail-vein infection of mice (Fradin et al. 2003). To determine how different blood components affect the *C. albicans* gene expression profile, the blood was separated into five fractions: enriched in erythrocytes (EC), polymorphonuclear leukocytes (PMN) and mononuclear leukocytes (MNC) or depleted of neutrophils or all blood cells (i.e. plasma; Fradin et al. 2005). *C. albicans* exposed to PMNs rapidly loses viability, whereas exposure to MNCs, plasma or erythrocytes has no effect on its viability. These fractions were inoculated for 30 min with *C. albicans* (5×10^6 cells/ml) and compared with *C. albicans* cells incubated with erythrocytes. Of the ORFs represented on the microarray, 25% (1518 genes) were shown to be modulated (1.5-fold changes) under at least one of the five conditions tested. These studies revealed that the transcript profiles of *C. albicans* exposed to EC, MNC and plasma are highly similar, whereas the transcriptional profiles of PMN and whole blood are similar to each other, indicating that PMN, consisting (to 90%) of neutrophils, are the cellular component responsible for the strong reaction of *C. albicans* to blood. This was confirmed by depleting whole blood from neutrophils using an antibody directed against CD15 (present on neutrophils and eosinophils). Upon exposure to EC, MNC, plasma or blood lacking neutrophils *C. albicans* rapidly switched to filamentous growth. The presence of neutrophils blocked hyphal development and resulted in growth arrest (see also Rubin-Bejerano et al. 2003). Consequently, most of the known hyphae-specific genes, such as *SAP4–SAP6*, *HYR1*, *ECE1* and *ALS3*, were repressed in the PMN fractions. Growth inhibition is reflected in the dramatic transcript reduction of genes involved in protein synthesis, including genes coding for ribosomal proteins (*RPS10*, *RPL12*), translation elongation factors (*EFB1*, *EFT3*), or translation initiators (*GCD7*, *GCN3*). This is paralleled by the induction of genes required for the response to nitrogen and carbohydrate starvation, like genes involved in the arginine, leucine, lysine and methionine biosynthesis pathways, *GCN4* (transcriptional activator of amino acid biosynthesis), several genes encoding amino acid transporters involved in general nitrogen metabolism and the ammonium permeases Mep2 and Mep3. In addition genes encoding the vacuolar proteases Prb1, Prb2, Apr1, Prc1/Cpy1 and Prc2 were also expressed at higher levels in the presence of neutrophils. The availability of carbohydrates to *C. albicans* also seemed to be reduced in the presence of neutrophils, as the genes encoding the key enzymes of the glyoxylate cycle (*MLS1*, *ICL1*, *ACS1*) were strongly up-regulated (not reported by Rubin-Bejerano et al. 2003). Thus, growth inhibition by PMNs might be due to nutrient starvation. Fungal cells incubated with MNC also expressed to higher level genes involved in nitrogen metabolism, the glyoxylate cycle and the antioxidative response. However, the expression of these genes was reported to be significantly lower than for *C. albicans* cells exposed to neutrophils (see also Lorenz et al. 2004).

C. albicans genes encoding the cytoplasmic and the surface superoxide dismutase (*SOD1*, *SOD5*), the catalase (*CTA1*), the glutathione peroxidase/glutathione reductase complex and the thioredoxin peroxidase/thioredoxin reductase complex were up-regulated strongly in the presence of neutrophils. Three out of the 18 antioxidant genes identified in this study were also found to be induced by *C. albicans* in response to neutrophils by Rubin-Bejerano et al. (2003): *SOD1*, *CTA1* and *GPX3*. In addition, several of these genes (including *TTR1*, *TRX1*, *CTA1*, *CAP1*) were identified by Enjalbert et al. (2003) who investigated the transcriptional profile of *C. albicans* exposed to oxidative stress.

Most interestingly, only 38% of all fungal cells were phagocytosed by and 57.5% attached to neutrophils in the PMN fraction, suggesting that the observed effects resulted from both intra- and extracellular activities of neutrophils. Most of these cells showed *SOD5* expression (as monitored by GFP fluorescence) and were arrested in the yeast form. Studies by Urban et al. (2006) indeed showed that neutophils are able to act not only by phagocytosis. The supernatant of PMNs does not result in inhibition of hyphal growth, suggesting that contact between *C. albicans* and neutrophils is crucial for inhibition of the yeast to hyphal transition.

4. Epithelial Surfaces

Adhesion to mammalian epithelia is one of the prerequisites for colonisation and invasion of *C. albicans* in the host. *C. albicans* is able to adhere to a plethora of different host niches consisting of different cell types providing individual micro-environments for colonization (Fig. 9.2).

Sandovsky-Losica investigated the transcriptional response of *C. albicans* to HEp2 epithelial cells (Sandovsky-Losica et al. 2006). Changes in gene transcription of *C. albicans* were determined following infection of HEp2 cells compared to control cultures grown in the absence of HEp2 cells. Among the approximately 300 genes which were identified as differentially regulated for at least 2-fold following

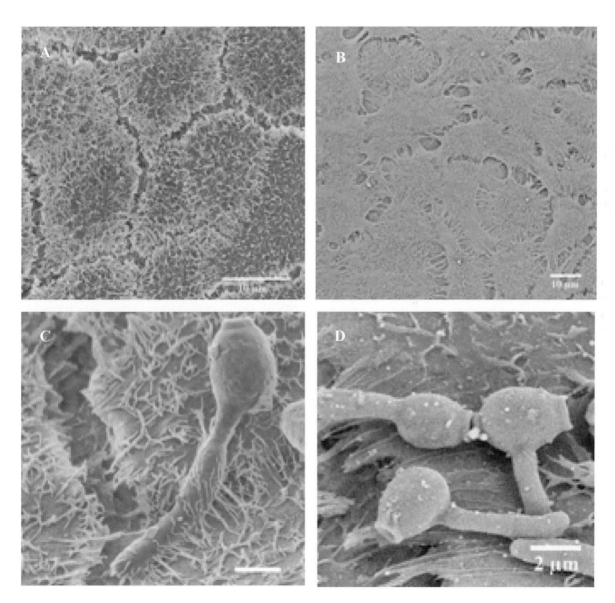

Fig. 9.2. Scanning electron microscopy of in vitro-infected epithelia. The human colorectal adenocarcinoma cell line Caco-2 (**A**) and the human epidermoid cell line A-431(**B**) are used as model systems for early events of adhesion and invasion (Sohn 2006). *C. albicans* was incubated for 2 h on Caco-2 (**C**) or A-431 cells (**D**). The different surface structure of each epithelia influences the interaction between *C. albicans* and the epithelial surface (e.g. see interaction between *C. albicans* and microvilli-like structures on the Caco-2 cells in (**C**))

3 h incubation with HEp2 *ALS2* and *ALS5* were identified as up-regulated. Both genes encode proteins that provide an adherence function for *C. albicans*.

To study the early response of *C. albicans* adhering to different surfaces on the transcriptional level Sohn et al. (2006) have established an in vitro adhesion assay exploiting confluent monolayers of the human colorectal carcinoma cell line Caco-2 or epidermoid vulvo-vaginal A-431 cells. *C. albicans* very efficiently adheres to these epithelia growing as hyphae within 1–3 h. Transcriptional profiles (IGB array) of *C. albicans* adhering to Caco-2 or to A-431 cells, although very similar, still significantly differ from those of Candida cells adhering to plastic surfaces or grown in suspension. Correspondingly from the 260 genes differentially regulated, several cell surface genes were identified, including *PRA1*, *PGA23*, *PGA7* and *HWP1*, showing either a cell-type or adhesion-dependent induction of transcription. Especially the kinetics of hyphal induction were much faster when *C. albicans* was grown on a surface, plastic or epithelia, as shown morphologically and molecularly by a more rapid and stronger induction of hyphae-specific genes like *HWP1*. Obviously, *C. albicans* is able to respond specifically to very subtle differences in the environment during adhesion to various growth substrates.

E. Biofilm Formation

Biofilm formation by *C. albicans* is a complex process with significant consequences for human health. It contributes to implanted medical device-associated infections and results in resistance, especially to azoles. Initial studies of *C. albicans* biofilm visualized yeast cells, pseudohyphae and hyphae embedded in an extracellular matrix using scanning electron microscopy. This and subsequent studies (Hawser and Douglas 1994; Baillie and Douglas 1999; Chandra et al. 2001; Ramage et al. 2001; Douglas 2003) showed that biofilm formation in vitro can be broken down into three basic stages: (a) attachment and colonization of yeast cells to a surface, (b) growth and proliferation of yeast cells to allow the formation of a basal layer of anchoring cells and (c) growth of pseudohyphae and extensive hyphae together with the production of extracellular matrix material. For recent reviews, see Nett and Andes (2006) and Nobile and Mitchell (2006). Several models for biofilm formation have been developed and transcriptional profiling has been performed on early and late stages of biofilm formation as well as on the impact of farnesol on biofilm formation (Cao et al. 2005; Garcia-Sanchez et al. 2004; Murillo et al. 2005).

Garcia-Sanchez and co-workers (2004) compared biofilm and planktonic cultures produced under different conditions of nutrient flow, aerobiosis or glucose concentration by overall gene expression correlation (partial macroarray containing 2002 ORFs). Correlation was much higher between biofilms than planktonic populations irrespective of the growth conditions, indicating that biofilm populations formed in different environments display very similar and specific transcript profiles. The authors found over-representation of amino acid biosynthesis genes in biofilms. Consequently, Gcn4p, a regulator of amino acid metabolism, was shown to be required for normal biofilm growth. Hyphal formation has been thought to be required for biofilm formation; however, a biofilm-like structure formed by the *efg1cph1* mutant strain locked in the yeast/pseudohyphal growth forms lacks hyphae entirely (Garcia-Sanchez et al. 2004). Still the majority of genes discovered to be involved in biofilm formation are also required for hyphal formation indicating the importance of hyphae for biofilm formation (Richard et al. 2005). Consequently, farnesol, a quorum-sensing molecule which inhibits hyphal morphogenesis results in inhibition of biofilm formation (Ramage et al. 2002b). Addition of farnesol alters the expression of 274 genes in a biofilm, including a significant number of hyphal-associated gene expression in biofilms (Cao et al. 2005). This again suggested that hyphal formation is a key factor for biofilm formation.

Cell surface contact is a key requirement for biofilm formation. At just 30 min after *C. albicans* yeast cells contact a polystyrene surface, a gene expression programme is initiated that is distinct from that of planktonic cells grown under otherwise similar conditions (Murillo et al. 2005; Sohn et al. 2006). This includes the development of drug resistance. Mateus et al. (2004) have observed that *CDR1* and *MDR1* promoter activities increase within 15–30 min after the adherence of cells to a glass slide, using promoter fusions to a GFP reporter gene (not observed in profiling studies by Murillo et al. 2005). The importance of *CDR1/2* and *MDR1* in resistance of a biofilm to FCZ was confirmed by mutational studies (Mateus et al.

2004). Interestingly at later time-points during biofilm formation the efflux pumps have no further effect on drug resistance in a biofilm (Mukherjee et al. 2003), suggesting additional mechanisms for resistance.

Noteworthy is the up-regulation of several methionine and cysteine biosynthetic genes, whose transcript levels remain elevated for many hours (Garcia-Sanchez et al. 2004; Murillo et al. 2005). The methionine and cysteine gene set is up-regulated during biofilm development under diverse conditions, even in the biofilm formed by the *efg1cph1* mutant that lacks hyphae (Garcia-Sanchez et al. 2004). Although the functional role of methionine and cysteine biosynthetic genes in biofilm development has yet to be determined, the rapid up-regulation of these and other genes suggests that *C. albicans* may sense cell surface contact or perhaps the presence of neighbouring cells.

One of the key regulators required specifically for biofilm formation (and not for hyphal development in general) is Bcr1, which in turn is regulated by *TEC1* (Nobile and Mitchell 2005). Comparison of transcriptional profiling (Operon array) of a *brc1* mutant with its complemented strain (Δ*brc1* + *pBRC1*; in suspension culture, Spider medium, 37 °C) revealed among the 22 most severely altered genes, 11 specifying cell surface- or cell wall-modifiying proteins, including *HYR1*, *ECE1*, *RBT5*, *ECM331*, *HWP1*, *ALS1*, *ALS3* and *ALS9*. Using the *brc1* mutant strain made it possible to separate the circuits required for hyphal morphogenesis and biofilm formation. The failure of the *brc1* mutant to create a biofilm indicates that the hyphal surface proteins are required for biofilm formation. Recent mutational approaches focusing on *ALS1*, *ALS3*, *HWP1* and *ECE1* could confirm their importance for biofilm formation (Nobile et al. 2006a, b; Zhao et al. 2006).

IV. Conclusions

Transcriptomics has been used to create a comprehensive picture of changes in gene expression of *C. albicans* during host–pathogen interaction, stress response or other environmental challenges. Individual genes could be identified as central components in several pathways, making them especially important for understanding pathogenesis.

In addition these studies also reveal that besides individual virulence factors, the appropriate regulation of metabolic pathways are essential to adapt to the host and survive in it. Indeed the expression of virulence factors and metabolic pathways are tightly coupled, e.g. via Efg1. This manifests the complexity and interaction between different pathways reflecting a dense biological network rather than linear pathways.

This review also shows that similar experiments using microarrays based on the same genome sequence might result in only very limited overlap with regard to the genes found to be differentially regulated. One of the reasons certainly is that apparently slightly different conditions in the experimental setup might lead to significant changes in the transcriptome. Furthermore, significant differences both in early and late established arrays, the definition of ORFs and the design of various distinct oligo sets or PCR-products for detection of the individual transcripts result in differences in the detection of the individual transcripts. However, verification of the quality of transcriptome data as well as their analysis is crucial for their interpretation. Especially, selecting the "good" signals from the "bad" in a complex series of experiments, including all necessary controls, is still a field with a lot of opportunities for advancement. Therefore, results from transcriptome studies should in general not be considered as complete genome-wide datasets but rather a dataset of significantly expressed genes which are modulated under the specific conditions applied. Although there are differences in the present studies, the general concusion drawn is usually identical. This reflects the huge benefits which can be gained from this technology.

The availability of the genome sequence was key for the realisation of genome-wide arrays as tools to perform these studies. The increasing wealth of genomic sequences from both pathogenic and non-pathogenic fungi will lead to increasing numbers of genome-wide transcriptional data from multiple species. As a result, we will be able to compare pathogenic fungi not only on the genome level, but on a transcriptional level and thereby further advance our understanding of pathogenesis. This will ultimately contribute to the better understanding of the mechanisms of infection and thus foster the development of new diagnostics, therapeutics and vaccines.

Acknowledgments. I would like to thank Rosa Hernadez-Barbado, Kai Sohn and Martin Zavrel for suggestions and critical reading of the manuscript. This chapter has been made possible by the DFG (Ru608/4) and the EU (Marie Curie Research Training Network, CanTrain).

References

Agarwal AK, Rogers PD, Baerson SR, Jacob MR, Barker KS, Cleary JD, Walker LA, Nagle DG, Clark AM (2003) Genome-wide expression profiling of the response to polyene, pyrimidine, azole, and echinocandin antifungal agents in *Saccharomyces cerevisiae*. J Biol Chem 278:34998–35015

Alarco AM, Raymond M (1999) The bZip transcription factor Cap1p is involved in multidrug resistance and oxidative stress response in *Candida albicans*. J Bacteriol 181:700–708

Andes D, Ogtrop M van (1999) Characterization and quantitation of the pharmacodynamics of fluconazole in a neutropenic murine disseminated candidiasis infection model. Antimicrob Agents Chemother 43:2116–2120

Andes D, Lepak A, Pitula A, Marchillo K, Clark J (2005) A simple approach for estimating gene expression in *Candida albicans* directly from a systemic infection site. J Infect Dis 192:893–900

Andrade RV, Da Silva SP, Torres FA, Pocas-Fonseca MJ, Silva-Pereira I, Maranhao AQ, Campos EG, Moraes LM, Jesuino RS, Pereira M et al (2005) Overview and perspectives the transcriptome of *Paracoccidioides brasiliensis*. Rev Iberoam Micol 22:203–212

Arnaud MB, Costanzo MC, Skrzypek MS, Binkley G, Lane C, Miyasato SR, Sherlock G (2005) The *Candida* genome database (CGD), a community resource for *Candida albicans* gene and protein information. Nucleic Acids Res 33:D358–D363

Arnaud MB, Costanzo MC, Skrzypek MS, Shah P, Binkley G, Lane C, Miyasato SR, Sherlock G (2007) Sequence resources at the *Candida* genome database. Nucleic Acids Res 35:D452–D456

Bammert GF, Fostel JM (2000) Genome-wide expression patterns in *Saccharomyces cerevisiae*: comparison of drug treatments and genetic alterations affecting biosynthesis of ergosterol. Antimicrob Agents Chemother 44:1255–1265

Barbulovic-Nad I, Lucente M, Sun Y, Zhang M, Wheeler AR, Bussmann M (2006) Bio-microarray fabrication techniques – a review. Crit Rev Biotechnol 26:237–259

Barker KS, Crisp S, Wiederhold N, Lewis RE, Bareither B, Eckstein J, Barbuch R, Bard M, Rogers PD (2004) Genome-wide expression profiling reveals genes associated with amphotericin B and fluconazole resistance in experimentally induced antifungal resistant isolates of *Candida albicans*. J Antimicrob Chemother 54:376–385

Bennett RJ, Uhl MA, Miller MG, Johnson AD (2003) Identification and characterization of a *Candida albicans* mating pheromone. Mol Cell Biol 23:8189–8201

Bensen ES, Filler SG, Berman J (2002) A forkhead transcription factor is important for true hyphal as well as yeast morphogenesis in *Candida albicans*. Eukaryot Cell 1:787–798

Bensen ES, Martin SJ, Li M, Berman J, Davis DA (2004) Transcriptional profiling in *Candida albicans* reveals new adaptive responses to extracellular pH and functions for Rim101p. Mol Microbiol 54:1335–1351

Berman J, Sudbery PE (2002) *Candida albicans*: a molecular revolution built on lessons from budding yeast. Nat Rev Genet 3:918–930

Bockmuhl DP, Ernst JF (2001) A potential phosphorylation site for an A-type kinase in the Efg1 regulator protein contributes to hyphal morphogenesis of *Candida albicans*. Genetics 157:1523–1530

Braun BR, Johnson AD (1997) Control of filament formation in *Candida albicans* by the transcriptional repressor TUP1 [see comments]. Science 277:105–109

Braun BR, Kadosh D, Johnson AD (2001) NRG1, a repressor of filamentous growth in *C. albicans*, is down-regulated during filament induction. EMBO J 20:4753–4761

Braun BR, Het Hoog M van, d'Enfert C, Martchenko M, Dungan J, Kuo A, Inglis DO, Uhl MA, Hogues H, Berriman M et al (2005) A human-curated annotation of the *Candida albicans* genome. PLoS Genet 1:36–57

Calderone RA (2002) *Candida* and candidiasis. ASM Press, Washington, D.C.

Cao F, Lane S, Raniga PP, Lu Y, Zhou Z, Ramon K, Chen J, Liu H (2006) The Flo8 transcription factor is essential for hyphal development and virulence in *Candida albicans*. Mol Biol Cell 17:295–307

Cao YY, Cao YB, Xu Z, Ying K, Li Y, Xie Y, Zhu ZY, Chen WS, Jiang YY (2005) cDNA microarray analysis of differential gene expression in *Candida albicans* biofilm exposed to farnesol. Antimicrob Agents Chemother 49:584–589

Carter GW, Rupp S, Fink GR, Galitski T (2006) Disentangling information flow in the Ras-cAMP signaling network. Genome Res 16:520–526

Casadevall A, Pirofski LA (2003) The damage-response framework of microbial pathogenesis. Nat Rev Microbiol 1:17–24

Chauhan N, Inglis D, Roman E, Pla J, Li D, Calera JA, Calderone R (2003) *Candida albicans* response regulator gene SSK1 regulates a subset of genes whose functions are associated with cell wall biosynthesis and adaptation to oxidative stress. Eukaryot Cell 2:1018–1024

Cleaveland S, Laurenson MK, Taylor LH (2001) Diseases of humans and their domestic mammals: pathogen characteristics, host range and the risk of emergence. Philos Trans R Soc Lond B Biol Sci 356:991–999

Costanzo MC, Arnaud MB, Skrzypek MS, Binkley G, Lane C, Miyasato SR, Sherlock G (2006) The *Candida* genome database: facilitating research on *Candida albicans* molecular biology. FEMS Yeast Res 6:671–684

Coste AT, Karababa M, Ischer F, Bille J, Sanglard D (2004) TAC1, transcriptional activator of CDR genes, is a new transcription factor involved in the regulation of *Candida albicans* ABC transporters CDR1 and CDR2. Eukaryot Cell 3:1639–1652

Cowen LE, Sanglard D, Calabrese D, Sirjusingh C, Anderson JB, Kohn LM (2000) Evolution of drug resistance in experimental populations of *Candida albicans*. J Bacteriol 182:1515–1522

Cowen LE, Nantel A, Whiteway MS, Thomas DY, Tessier DC, Kohn LM, Anderson JB (2002) Population genomics of drug resistance in *Candida albicans*. Proc Natl Acad Sci USA 99:9284–9289

Cramer KL, Gerrald QD, Nichols CB, Price MS, Alspaugh JA (2006) Transcription factor Nrg1 mediates capsule formation, stress response, and pathogenesis in *Cryptococcus neoformans*. Eukaryot Cell 5:1147–1156

Davis D, Wilson RB, Mitchell AP (2000) RIM101-dependent and -independent pathways govern pH responses in *Candida albicans*. Mol Cell Biol 20:971–978

De Backer MD, Ilyina T, Ma XJ, Vandoninck S, Luyten WH, Vanden Bossche H (2001) Genomic profiling of the response of *Candida albicans* to itraconazole treatment using a DNA microarray. Antimicrob Agents Chemother 45:1660–1670

d'Enfert C, Goyard S, Rodriguez-Arnaveilhe S, Frangeul L, Jones L, Tekaia F, Bader O, Albrecht A, Castillo L, Dominguez A et al (2005) CandidaDB: a genome database for *Candida albicans* pathogenomics. Nucleic Acids Res 33:D353–D357

DeRisi JL, Iyer VR, Brown PO (1997) Exploring the metabolic and genetic control of gene expression on a genomic scale. Science 278:680–686

Dieterich C, Schandar M, Noll M, Johannes FJ, Brunner H, Graeve T, Rupp S (2002) In vitro reconstructed human epithelia reveal contributions of *Candida albicans* EFG1 and CPH1 to adhesion and invasion. Microbiology 148:497–506

Dignard D, El-Naggar AL, Logue ME, Butler G, Whiteway M (2007) Identification and characterization of MFA1; the gene encoding *Candida albicans* a-factor pheromone. Eukaryot Cell (in press)

Doedt T, Krishnamurthy S, Bockmuhl DP, Tebarth B, Stempel C, Russell CL, Brown AJ, Ernst JF (2004) APSES proteins regulate morphogenesis and metabolism in *Candida albicans*. Mol Biol Cell 15:3167–3180

Du C, Calderone R, Richert J, Li D (2005) Deletion of the SSK1 response regulator gene in *Candida albicans* contributes to enhanced killing by human polymorphonuclear neutrophils. Infect Immun 73:865–871

Dujon B, Sherman D, Fischer G, Durrens P, Casaregola S, Lafontaine I, De Montigny J, Marck C, Neuveglise C, Talla E et al (2004) Genome evolution in yeasts. Nature 430:35–44

Eisenstein M (2006) Microarrays: quality control. Nature 442:1067–1070

El Barkani A, Kurzai O, Fonzi WA, Ramon A, Porta A, Frosch M, Muhlschlegel FA (2000) Dominant active alleles of RIM101 (PRR2) bypass the pH restriction on filamentation of *Candida albicans*. Mol Cell Biol 20:4635–4647

Enjalbert B, Nantel A, Whiteway M (2003) Stress-induced gene expression in *Candida albicans*: absence of a general stress response. Mol Biol Cell 14:1460–1467

Enjalbert B, Smith DA, Cornell MJ, Alam I, Nicholls S, Brown AJ, Quinn J (2006) Role of the Hog1 stress-activated protein kinase in the global transcriptional response to stress in the fungal pathogen *Candida albicans*. Mol Biol Cell 17:1018–1032

Fang FC (2004) Antimicrobial reactive oxygen and nitrogen species: concepts and controversies. Nat Rev Microbiol 2: 820–832

Fellenberg K, Busold CH, Witt O, Bauer A, Beckmann B, Hauser NC, Frohme M, Winter S, Dippon J, Hoheisel JD (2006) Systematic interpretation of microarray data using experiment annotations. BMC Genomics 7:319

Fidel PL Jr (2007) History and update on host defense against vaginal candidiasis. Am J Reprod Immunol 57:2–12

Fonzi WA, Irwin MY (1993) Isogenic strain construction and gene mapping in *Candida albicans*. Genetics 134:717–728

Fradin C, Kretschmar M, Nichterlein T, Gaillardin C, d'Enfert C, Hube B (2003) Stage-specific gene expression of *Candida albicans* in human blood. Mol Microbiol 47:1523–1543

Fradin C, De Groot P, MacCallum D, Schaller M, Klis F, Odds FC, Hube B (2005) Granulocytes govern the transcriptional response, morphology and proliferation of *Candida albicans* in human blood. Mol Microbiol 56:397–415

Galagan JE, Henn MR, Ma LJ, Cuomo CA, Birren B (2005) Genomics of the fungal kingdom: insights into eukaryotic biology. Genome Res 15:1620–1631

Galitski T, Saldanha AJ, Styles CA, Lander ES, Fink GR (1999) Ploidy regulation of gene expression. Science 285:251–254

Garcia-Sanchez S, Aubert S, Iraqui I, Janbon G, Ghigo JM, d'Enfert C (2004) *Candida albicans* biofilms: a developmental state associated with specific and stable gene expression patterns. Eukaryot Cell 3:536–545

Garcia-Sanchez S, Mavor AL, Russell CL, Argimon S, Dennison P, Enjalbert B, Brown AJ (2005) Global roles of Ssn6 in Tup1- and Nrg1-dependent gene regulation in the fungal pathogen, *Candida albicans*. Mol Biol Cell 16:2913–2925

Gebhart D, Bahrami AK, Sil A (2006) Identification of a copper-inducible promoter for use in ectopic expression in the fungal pathogen *Histoplasma capsulatum*. Eukaryot Cell 5:935–944

Gupta V, Kohli A, Krishnamurthy S, Puri N, Aalamgeer SA, Panwar S, Prasad R (1998) Identification of polymorphic mutant alleles of CaMDR1, a major facilitator of *Candida albicans* which confers multidrug resistance, and its in vitro transcriptional activation. Curr Genet 34:192–199

Harcus D, Nantel A, Marcil A, Rigby T, Whiteway M (2004) Transcription profiling of cyclic AMP signaling in *Candida albicans*. Mol Biol Cell 15:4490–4499

Hauser NC, Vingron M, Scheideler M, Krems B, Hellmuth K, Entian KD, Hoheisel JD (1998) Transcriptional profiling on all open reading frames of *Saccharomyces cerevisiae*. Yeast 14:1209–1221

Herskowitz I (1989) A regulatory hierarchy for cell specialization in yeast. Nature 342:749–757

Hromatka BS, Noble SM, Johnson AD (2005) Transcriptional response of *Candida albicans* to nitric oxide and the role of the YHB1 gene in nitrosative stress and virulence. Mol Biol Cell 16:4814–4826

Huang G, Wang H, Chou S, Nie X, Chen J, Liu H (2006) Bistable expression of WOR1, a master regulator of white-opaque switching in *Candida albicans*. Proc Natl Acad Sci USA 103:12813–12818

Hwang CS, Oh JH, Huh WK, Yim HS, Kang SO (2003a) Ssn6, an important factor of morphological conversion and virulence in *Candida albicans*. Mol Microbiol 47:1029–1043

Hwang L, Hocking-Murray D, Bahrami AK, Andersson M, Rine J, Sil A (2003b) Identifying phase-specific genes in the fungal pathogen *Histoplasma capsulatum* using a genomic shotgun microarray. Mol Biol Cell 14:2314–2326

Ihmels J, Bergmann S, Berman J, Barkai N (2005) Comparative gene expression analysis by differential clustering approach: application to the *Candida albicans* transcription program. PLoS Genet 1:e39

Johannesson H, Kasuga T, Schaller RA, Good B, Gardner MJ, Townsend JP, Cole GT, Taylor JW (2006) Phase-specific gene expression underlying morphological adaptations of the dimorphic human pathogenic fungus, *Coccidioides posadasii*. Fungal Genet Biol 43:545–559

Jones T, Federspiel NA, Chibana H, Dungan J, Kalman S, Magee BB, Newport G, Thorstenson YR, Agabian N, Magee PT et al (2004) The diploid genome sequence of *Candida albicans*. Proc Natl Acad Sci USA 101:7329–7334

Kadosh D, Johnson AD (2005) Induction of the *Candida albicans* filamentous growth program by relief of transcriptional repression: a genome-wide analysis. Mol Biol Cell 16:2903–2912

Karababa M, Coste AT, Rognon B, Bille J, Sanglard D (2004) Comparison of gene expression profiles of *Candida albicans* azole-resistant clinical isolates and laboratory strains exposed to drugs inducing multidrug transporters. Antimicrob Agents Chemother 48:3064–3079

Kelly SL, Lamb DC, Kelly DE, Manning NJ, Loeffler J, Hebart H, Schumacher U, Einsele H (1997) Resistance to fluconazole and cross-resistance to amphotericin B in *Candida albicans* from AIDS patients caused by defective sterol delta5,6-desaturation. FEBS Lett 400:80–82

Kraus PR, Boily MJ, Giles SS, Stajich JE, Allen A, Cox GM, Dietrich FS, Perfect JR, Heitman J (2004) Identification of *Cryptococcus neoformans* temperature-regulated genes with a genomic-DNA microarray. Eukaryot Cell 3:1249–1260

Lamb TM, Mitchell AP (2003) The transcription factor Rim101p governs ion tolerance and cell differentiation by direct repression of the regulatory genes NRG1 and SMP1 in *Saccharomyces cerevisiae*. Mol Cell Biol 23:677–686

Lan CY, Newport G, Murillo LA, Jones T, Scherer S, Davis RW, Agabian N (2002) Metabolic specialization associated with phenotypic switching in *Candida albicans*. Proc Natl Acad Sci USA 99:14907–14912

Lan CY, Rodarte G, Murillo LA, Jones T, Davis RW, Dungan J, Newport G, Agabian N (2004) Regulatory networks affected by iron availability in *Candida albicans*. Mol Microbiol 53:1451–1469

Lander ES, Linton LM, Birren B, Nusbaum C, Zody MC, Baldwin J, Devon K, Dewar K, Doyle M, FitzHugh W et al (2001) Initial sequencing and analysis of the human genome. Nature 409:860–921

Lane S, Birse C, Zhou S, Matson R, Liu H (2001) DNA array studies demonstrate convergent regulation of virulence factors by Cph1, Cph2, and Efg1 in *Candida albicans*. J Biol Chem 276:48988–48996

Lee CM, Nantel A, Jiang L, Whiteway M, Shen SH (2004) The serine/threonine protein phosphatase SIT4 modulates yeast-to-hypha morphogenesis and virulence in *Candida albicans*. Mol Microbiol 51:691–709

Lee RE, Liu TT, Barker KS, Rogers PD (2005) Genome-wide expression profiling of the response to ciclopirox olamine in *Candida albicans*. J Antimicrob Chemother 55:655–662

Lepak A, Nett J, Lincoln L, Marchillo K, Andes D (2006) Time course of microbiologic outcome and gene expression in *Candida albicans* during and following in vitro and in vivo exposure to fluconazole. Antimicrob Agents Chemother 50:1311–1319

Leuker CE, Sonneborn A, Delbruck S, Ernst JF (1997) Sequence and promoter regulation of the PCK1 gene encoding phosphoenolpyruvate carboxykinase of the fungal pathogen *Candida albicans*. Gene 192:235–240

Lewis JG, Learmonth RP, Watson K (1995) Induction of heat, freezing and salt tolerance by heat and salt shock in *Saccharomyces cerevisiae*. Microbiology 141:687–694

Liu H (2001) Transcriptional control of dimorphism in *Candida albicans*. Curr Opin Microbiol 4:728–735

Liu H (2002) Co-regulation of pathogenesis with dimorphism and phenotypic switching in *Candida albicans*, a commensal and a pathogen. Int J Med Microbiol 292:299–311

Liu TT, Lee RE, Barker KS, Wei L, Homayouni R, Rogers PD (2005) Genome-wide expression profiling of the response to azole, polyene, echinocandin, and pyrimidine antifungal agents in *Candida albicans*. Antimicrob Agents Chemother 49:2226–2236

Lo HJ, Kohler JR, Di Domenico B, Loebenberg D, Cacciapuoti A, Fink GR (1997) Nonfilamentous *C. albicans* mutants are avirulent. Cell 90:939–949

Lockhart SR, Pujol C, Daniels KJ, Miller MG, Johnson AD, Pfaller MA, Soll DR (2002) In *Candida albicans*, white-opaque switchers are homozygous for mating type. Genetics 162:737–745

Loftus BJ, Fung E, Roncaglia P, Rowley D, Amedeo P, Bruno D, Vamathevan J, Miranda M, Anderson IJ, Fraser JA et al (2005) The genome of the basidiomycetous yeast and human pathogen *Cryptococcus neoformans*. Science 307:1321–1324

Lorenz MC, Fink GR (2001) The glyoxylate cycle is required for fungal virulence. Nature 412:83–86

Lorenz MC, Bender JA, Fink GR (2004) Transcriptional response of *Candida albicans* upon internalization by macrophages. Eukaryot Cell 3:1076–1087

Lotz H, Sohn K, Brunner H, Muhlschlegel FA, Rupp S (2004) RBR1, a novel pH-regulated cell wall gene of *Candida albicans*, is repressed by RIM101 and activated by NRG1. Eukaryot Cell 3:776–784

Lyons CN, White TC (2000) Transcriptional analyses of antifungal drug resistance in *Candida albicans*. Antimicrob Agents Chemother 44:2296–2303

Mateus C, Crow SA Jr, Ahearn DG (2004) Adherence of *Candida albicans* to silicone induces immediate enhanced tolerance to fluconazole. Antimicrob Agents Chemother 48:3358–3366

Micheli M de, Bille J, Schueller C, Sanglard D (2002) A common drug-responsive element mediates the upregulation of the *Candida albicans* ABC transporters CDR1 and CDR2, two genes involved in antifungal drug resistance. Mol Microbiol 43:1197–1214

Miller MG, Johnson AD (2002) White-opaque switching in *Candida albicans* is controlled by mating-type locus homeodomain proteins and allows efficient mating. Cell 110:293–302

Morens DM, Folkers GK, Fauci AS (2004) The challenge of emerging and re-emerging infectious diseases. Nature 430:242–249

Mukherjee PK, Chandra J, Kuhn DM, Ghannoum MA (2003) Mechanism of fluconazole resistance in *Candida albicans* biofilms: phase-specific role of efflux pumps and membrane sterols. Infect Immun 71:4333–4340

Murad AM, d'Enfert C, Gaillardin C, Tournu H, Tekaia F, Talibi D, Marechal D, Marchais V, Cottin J, Brown AJ (2001) Transcript profiling in *Candida albicans* reveals new cellular functions for the transcriptional repressors CaTup1, CaMig1 and CaNrg1. Mol Microbiol 42:981–993

Murillo LA, Newport G, Lan CY, Habelitz S, Dungan J, Agabian NM (2005) Genome-wide transcription profiling of the early phase of biofilm formation by *Candida albicans*. Eukaryot Cell 4:1562–1573

Nantel A (2006) The long hard road to a completed *Candida albicans* genome. Fungal Genet Biol 43:311–315

Nantel A, Dignard D, Bachewich C, Harcus D, Marcil A, Bouin AP, Sensen CW, Hogues H, Hoog M van het, Gordon P et al (2002) Transcription profiling of *Candida albicans* cells undergoing the yeast-to-hyphal transition. Mol Biol Cell 13:3452–3465

Nett J, Andes D (2006) *Candida albicans* biofilm development, modeling a host-pathogen interaction. Curr Opin Microbiol 9:340–345

Nicholls S, Straffon M, Enjalbert B, Nantel A, Macaskill S, Whiteway M, Brown AJ (2004) Msn2- and Msn4-like transcription factors play no obvious roles in the stress responses of the fungal pathogen *Candida albicans*. Eukaryot Cell 3:1111–1123

Nierman WC, Pain A, Anderson MJ, Wortman JR, Kim HS, Arroyo J, Berriman M, Abe K, Archer DB, Bermejo C et al (2005) Genomic sequence of the pathogenic and allergenic filamentous fungus *Aspergillus fumigatus*. Nature 438:1151–1156

Niewerth M, Kunze D, Seibold M, Schaller M, Korting HC, Hube B (2003) Ciclopirox olamine treatment affects the expression pattern of *Candida albicans* genes encoding virulence factors, iron metabolism proteins, and drug resistance factors. Antimicrob Agents Chemother 47:1805–1817

Nobile CJ, Mitchell AP (2005) Regulation of cell-surface genes and biofilm formation by the *C. albicans* transcription factor Bcr1p. Curr Biol 15:1150–1155

Nobile CJ, Mitchell AP (2006) Genetics and genomics of *Candida albicans* biofilm formation. Cell Microbiol 8:1382–1391

Nobile CJ, Andes DR, Nett JE, Smith FJ, Yue F, Phan QT, Edwards JE, Filler SG, Mitchell AP (2006a) Critical role of Bcr1-dependent adhesins in *C. albicans* biofilm formation in vitro and in vivo. PLoS Pathog 2:e63

Nobile CJ, Nett JE, Andes DR, Mitchell AP (2006b) Function of *Candida albicans* adhesin Hwp1 in biofilm formation. Eukaryot Cell 5:1604–1610

O'Rourke SM, Herskowitz I (2004) Unique and redundant roles for HOG MAPK pathway components as revealed by whole-genome expression analysis. Mol Biol Cell 15:532–542

Pfaller MA, Rhine-Chalberg J, Redding SW, Smith J, Farinacci G, Fothergill AW, Rinaldi MG (1994) Variations in fluconazole susceptibility and electrophoretic karyotype among oral isolates of *Candida albicans* from patients with AIDS and oral candidiasis. J Clin Microbiol 32:59–64

Pompe S, Simon J, Wiedemann PM, Tannert C (2005) Future trends and challenges in pathogenomics. A Foresight study. EMBO Rep 6:600–605

Porta A, Wang Z, Ramon A, Muhlschlegel FA, Fonzi WA (2001) Spontaneous second-site suppressors of the filamentation defect of prr1Delta mutants define a critical domain of Rim101p in *Candida albicans*. Mol Genet Genomics 266:624–631

Ramon AM, Fonzi WA (2003) Diverged binding specificity of Rim101p, the *Candida albicans* ortholog of PacC. Eukaryot Cell 2:718–728

Ramon AM, Porta A, Fonzi WA (1999) Effect of environmental pH on morphological development of *Candida albicans* is mediated via the PacC-related transcription factor encoded by PRR2. J Bacteriol 181:7524–7530

Redd MJ, Arnaud MB, Johnson AD (1997) A complex composed of tup1 and ssn6 represses transcription in vitro. J Biol Chem 272:11193–11197

Redding S, Smith J, Farinacci G, Rinaldi M, Fothergill A, Rhine-Chalberg J, Pfaller M (1994) Resistance of *Candida albicans* to fluconazole during treatment of oropharyngeal candidiasis in a patient with AIDS: documentation by in vitro susceptibility testing and DNA subtype analysis. Clin Infect Dis 18:240–242

Richard ML, Nobile CJ, Bruno VM, Mitchell AP (2005) *Candida albicans* biofilm-defective mutants. Eukaryot Cell 4:1493–1502

Rocha CR, Schroppel K, Harcus D, Marcil A, Dignard D, Taylor BN, Thomas DY, Whiteway M, Leberer E (2001) Signaling through adenylyl cyclase is essential for hyphal growth and virulence in the pathogenic fungus *Candida albicans*. Mol Biol Cell 12:3631–3643

Rogers PD, Barker KS (2002) Evaluation of differential gene expression in fluconazole-susceptible and -resistant isolates of *Candida albicans* by cDNA microarray analysis. Antimicrob Agents Chemother 46:3412–3417

Rogers PD, Barker KS (2003) Genome-wide expression profile analysis reveals coordinately regulated genes associated with stepwise acquisition of azole resistance in *Candida albicans* clinical isolates. Antimicrob Agents Chemother 47:1220–1227

Rubin-Bejerano I, Fraser I, Grisafi P, Fink GR (2003) Phagocytosis by neutrophils induces an amino acid deprivation response in *Saccharomyces cerevisiae* and *Candida albicans*. Proc Natl Acad Sci USA 100:11007–11012

Sandovsky-Losica H, Chauhan N, Calderone R, Segal E (2006) Gene transcription studies of *Candida albicans* following infection of HEp2 epithelial cells. Med Mycol 44:329–334

Sarver A, DeRisi J (2005) Fzf1p regulates an inducible response to nitrosative stress in *Saccharomyces cerevisiae*. Mol Biol Cell 16:4781–4791

Sigle HC, Thewes S, Niewerth M, Korting HC, Schafer-Korting M, Hube B (2005) Oxygen accessibility and iron levels are critical factors for the antifungal action of ciclopirox against *Candida albicans*. J Antimicrob Chemother 55:663–673

Silva Ferreira ME da, Malavazi I, Savoldi M, Brakhage AA, Goldman MH, Kim HS, Nierman WC, Goldman GH (2006) Transcriptome analysis of *Aspergillus fumigatus* exposed to voriconazole. Curr Genet 50:32–44

Slutsky B, Staebell M, Anderson J, Risen L, Pfaller M, Soll DR (1987) "White-opaque transition": a second high-frequency switching system in *Candida albicans*. J Bacteriol 169:189–197

Smith DA, Nicholls S, Morgan BA, Brown AJ, Quinn J (2004) A conserved stress-activated protein kinase regulates a core stress response in the human pathogen *Candida albicans*. Mol Biol Cell 15:4179–4190

Sohn K, Senyurek I, Fertey J, Konigsdorfer A, Joffroy C, Hauser N, Zelt G, Brunner H, Rupp S (2006) An in vitro assay to study the transcriptional response during adherence of *Candida albicans* to different human epithelia. FEMS Yeast Res 6:1085–1093

Sohn K, Urban C, Brunner H, Rupp S (2003) EFG1 is a major regulator of cell wall dynamics in *Candida albicans* as revealed by DNA microarrays. Mol Microbiol 47:89–102

Sokol-Anderson M, Sligh JE Jr, Elberg S, Brajtburg J, Kobayashi GS, Medoff G (1988) Role of cell defense against oxidative damage in the resistance of *Candida albicans* to the killing effect of amphotericin B. Antimicrob Agents Chemother 32:702–705

Soll DR (1997) Gene regulation during high-frequency switching in *Candida albicans*. Microbiology 143:279–288

Sonneborn A, Tebarth B, Ernst JF (1999) Control of white-opaque phenotypic switching in *Candida albicans* by the Efg1p morphogenetic regulator. Infect Immun 67:4655–4660

Srikantha T, Borneman AR, Daniels KJ, Pujol C, Wu W, Seringhaus MR, Gerstein M, Yi S, Snyder M, Soll DR (2006) TOS9 regulates white-opaque switching in *Candida albicans*. Eukaryot Cell 5:1674–1687

Sudbery P, Gow N, Berman J (2004) The distinct morphogenic states of *Candida albicans*. Trends Microbiol 12:317–324

Tsong AE, Miller MG, Raisner RM, Johnson AD (2003) Evolution of a combinatorial transcriptional circuit: a case study in yeasts. Cell 115:389–399

Ullmann BD, Myers H, Chiranand W, Lazzell AL, Zhao Q, Vega LA, Lopez-Ribot JL, Gardner PR, Gustin MC (2004) Inducible defense mechanism against nitric oxide in *Candida albicans*. Eukaryot Cell 3:715–723

Urban CF, Reichard U, Brinkmann V, Zychlinsky A (2006) Neutrophil extracellular traps capture and kill *Candida albicans* yeast and hyphal forms. Cell Microbiol 8:668–676

Venter JC, Adams MD, Myers EW, Li PW, Mural RJ, Sutton GG, Smith HO, Yandell M, Evans CA, Holt RA et al (2001) The sequence of the human genome. Science 291:1304–1351

Vermitsky JP, Earhart KD, Smith WL, Homayouni R, Edlind TD, Rogers PD (2006) Pdr1 regulates multidrug resistance in *Candida glabrata*: gene disruption and genome-wide expression studies. Mol Microbiol 61:704–722

Vik A, Rine J (2001) Upc2p and Ecm22p, dual regulators of sterol biosynthesis in *Saccharomyces cerevisiae*. Mol Cell Biol 21:6395–6405

Weissman Z, Kornitzer D (2004) A family of *Candida* cell surface haem-binding proteins involved in haemin and haemoglobin-iron utilization. Mol Microbiol 53:1209–1220

White TC (1997a) Increased mRNA levels of ERG16, CDR, and MDR1 correlate with increases in azole resistance in *Candida albicans* isolates from a patient infected with human immunodeficiency virus. Antimicrob Agents Chemother 41:1482–1487

White TC (1997b) The presence of an R467K amino acid substitution and loss of allelic variation correlate with an azole-resistant lanosterol 14alpha demethylase in *Candida albicans*. Antimicrob Agents Chemother 41:1488–1494

White TC, Marr KA, Bowden RA (1998) Clinical, cellular, and molecular factors that contribute to antifungal drug resistance. Clin Microbiol Rev 11:382–402

Whiteway M, Oberholzer U (2004) *Candida* morphogenesis and host–pathogen interactions. Curr Opin Microbiol 7:350–357

Wodicka L, Dong H, Mittmann M, Ho MH, Lockhart DJ (1997) Genome-wide expression monitoring in *Saccharomyces cerevisiae*. Nat Biotechnol 15:1359–1367

Zhang X, De Micheli M, Coleman ST, Sanglard D, Moye-Rowley WS (2000) Analysis of the oxidative stress regulation of the *Candida albicans* transcription factor, Cap1p. Mol Microbiol 36:618–629

Zhao X, Daniels KJ, Oh SH, Green CB, Yeater KM, Soll DR, Hoyer LL (2006) *Candida albicans* Als3p is required for wild-type biofilm formation on silicone elastomer surfaces. Microbiology 152:2287–2299

Zordan RE, Galgoczy DJ, Johnson AD (2006) Epigenetic properties of white-opaque switching in *Candida albicans* are based on a self-sustaining transcriptional feedback loop. Proc Natl Acad Sci USA, 103:12807–12812

Host

10 Yeast Infections in Immunocompromised Hosts

Emmanuel Rollides[1], Thomas J. Walsh[1]

CONTENTS

I. Introduction 225
II. Candidiasis 225
    A. Mucosal Candidiasis 225
    B. Deeply Invasive Candidiasis 226
III. Cryptococcosis 227
IV. Infections due to *Trichosporon* 228
V. Infections due to *Blastoschizomyces capitatus* 229
VI. Infections due to *Malassezia* spp. 230
VII. Infections due to Dematiaceous yeasts ... 230
VIII. Conclusions 230
    References 231

I. Introduction

A number of yeast fungi are pathogenic, but the two genera that contain the most important animal and human pathogens are *Candida* and *Cryptococcus*. In addition, there are a number of other yeasts that have been, more rarely, implicated in disease.

II. Candidiasis

Members of the genus *Candida* cause superficial infections of the skin, nails and mucosal membranes of the gastrointestinal tract and vagina. They can also invade the tissues and cause fungemia and deep-seated infection of the body.

A. Mucosal Candidiasis

Candida albicans is a normal component of the mucocutaneous flora of humans. Other *Candida* species, such as *C. tropicalis, C. parapsilosis,*

---

[1] Immunocompromised Host Section, Pediatric Oncology Branch, National Cancer Institute, Building 10, CRC 1-5750, Bethesda, MD 20892, USA; e-mail: walsht@mail.nih.gov

C. krusei, C. glabrata and *C. lusitaniae*, are occasionally isolated but are more frequently recovered in immuncompromised patients and in those receiving antifungal therapy. Among the host defenses against mucosal candidiasis, endogenous bacterial flora are an important factor in suppressing proliferation of *Candida* spp. on mucocutaneous surfaces. Cell-mediated immunity (CMI) and mucosal immunity are closely related to inhibit proliferation and germination on mucosal surfaces (Fidel 2005). Epithelial cells of oral cavity have direct anti-*Candida* activity. In addition, during mucosal infection with *Candida*, a large number of pro-inflammatory and immunoregulatory cytokines are generated by epithelial cells. These cytokines may stimulate chemotaxis, phagocytosis and intracellular killing of infiltrating neutrophils as well as functions of CD4+ and CD8+ T-cells (Chauhan et al. 2006). Phagocytes can recognize *Candida* blastoconidia and hyphae by Toll-like receptors (van der Graaf et al. 2005). A potential link between lower levels of certain pro-inflammatory cytokines and susceptibility to oral *C. albicans* infection has been found suggesting involvement of such cytokines in protection (Dongari-Bagtzoglou and Fidel 2005). Antibody of the IgA class also may assume a role in mucosal immunity to *Candida* infections. Mucus and an intact epithelial cell surface appear to provide an additional line of defense to local *Candida* invasion of mucosal surfaces. Abrogation of these mucosal host defenses against *Candida* may lead to proliferation, local invasion and blood-borne dissemination.

The clinical spectrum of mucosal candidiasis includes oropharyngeal, esophageal, epiglottic and vaginal candidiasis (Pankhurst 2005; Spence 2005). Broad-spectrum antibacterial antibiotics, especially third generation cephalosporins and carbapenems (Maraki et al. 1999; Samonis et al. 2006), may lead to mucosal candidiasis by reducing the normal competing bacterial flora. Mucosal candidiasis following the administration of antibiotics is often manifested as oral candidiasis or vaginal candidiasis. However, esophageal candidiasis and severe gastrointestinal candidiasis can develop in immunocompromised patients following the administration of broad-spectrum antibiotics.

The natural history of HIV infection has underscored the critical role of intact CMI in contributing to mucosal host defense against Candida (Ohmit et al. 2003). Mucosal candidiasis evolves as an early and frequent manifestation of HIV infection (Klein et al. 1984). As a reflection of impaired CMI, oropharyngeal candidiasis developing in an HIV-positive patient carries an ominous prognosis of developing advanced complications of acquired immune deficiency syndrome (AIDS), such as Pneumocystis jirovecii pneumonia. Vaginal candidiasis also may be a recurrent debilitating infection in HIV-positive women. As a logical extension of these mucocutaneous manifestations, development of esophageal candidiasis without other predisposing events in a previously asymptomatic HIV-positive patient has been deemed as an AIDS-defining illness (Laine 1994).

Chronic mucocutaneous candidiasis (CMC) is yet another example of the importance of systemic CMI to mucosal host defense against Candida (Kirkpatrick 2001). Chronic mucocutaneous candidiasis is a disease identified most frequently in children. Patients with this disease may have severe recurrent episodes of mucocutaneous candidiasis related to impaired CMI recognition, processing, or response to Candida antigens. Recently, altered patterns of cytokine production in response to Candida spp. with decreased production of some Th1 cytokines and increased levels of interleukin-10 were found (Lilic 2002). The underlying genetic defect remains unknown but studies are in progress addressing the putative role of dendritic cells and pattern recognition receptors in directing cytokine responses.

Systemic and inhalational corticosteroid therapy also may lead to mucosal candidiasis in patients (Buhl 2006). Studies of experimental mucosal candidiasis reveal that parenterally administered corticosteroid therapy leads to a marked increase in esophageal and gastrointestinal candidiasis in comparison to saline-treated controls. Rabbits treated with parenteral corticosteroid have profound depletion of gut-associated lymphoid tissue (GALT; Roy and Walsh 1992). Lymphoid domes and follicles in such animals are considerably reduced in size. The dome epithelial layer is markedly depleted of M cells and lymphocytes, while the follicular B cell and T cell regions are severely involuted, thus indicating the potentially profound effect of systemic corticoseroids on mucosal immunity (Walsh and Pizzo 1992).

Disruption of an intact epithelium is an important component of locally invasive candidiasis, particularly in those patients who are receiving cytotoxic chemotherapy for cancer. This disruption of mucosal integrity permits invasion of Candida into the submucosal regions of the alimentary tract, invasion of blood vessels and systemic dissemination.

Mucosal candidiasis often can be treated by topical therapy, such as with nystatin, clotrimazole, or miconazole (Odds 1992). More severe forms of mucosal candidiasis can be treated with fluconazole (Pons et al. 1993), itraconazole (Saag et al. 1999), voriconazole, amphotericin B formulations or echinocandins, caspofungin, micafungin and anidulafungin. The latter drugs are usually more potent in vitro and have a broader spectrum of activity, including activity against fluconazole-resistant Candida species and may be used in the management of refractory mucosal candidiasis (Vazquez 2003).

B. Deeply Invasive Candidiasis

Neutrophils, peripheral blood monocytes and macrophages maintain a critical role in host defense against deeply invasive Candida infections. Neutrophils and peripheral blood monocytes phagocytose Candida blastoconidia and damage the cell walls and cell membranes of blastoconidia, pseudohyphae and hyphae. Macrophages in liver and spleen clear circulating blastoconidia (Kappe et al. 1992). A number of Th1 and Th2 cytokines as well as hemopoietic growth factors modulate the effector functions of these innate immune cells in response to Candida spp. (Roilides and Walsh 2004).

Patients who are neutropenic due to cytotoxic chemotherapy or aplastic anemia have a high risk of invasive candidiasis, particularly in the setting of severe mucosal disruption (Maksymiuk et al. 1984). Additional corticosteroid-mediated suppression of phagocytosis of Candida by macrophages further increases the risk of deeply invasive candidiasis in neutropenic patients. Preterm neonates are also susceptible to colonization and infection by *Candida* spp. (Leibovitz 2002).

While the alimentary tract is the putative portal of entry in many patients with deeply invasive candidiasis, vascular catheters afford another site of entry for *Candida*. By passing mucosal host defenses, *Candida* may be introduced directly through the lumen of the catheter into the blood stream. Consistent with this mechanism of entry are the observations of various epidemiological studies which impute central venous catheters as an independent risk factor for fungemia (Lecciones et al. 1992). It has become an important problem in the intensive care unit followed by increased mortality (Ostrosky-Zeichner and Pappas 2006).

Deeply invasive candidiasis may be classified as fungemia, acute disseminated candidiasis, chronic disseminated candidiasis and single-organ candidiasis. Fungemia may be classified as transient or persistent. Acute disseminated candidiasis is characterized by the development of fungemia and tissue-proven candidiasis. Patients with this acute disseminated candidiasis may have hemodynamic instability and septic shock. By comparison, patients with chronic disseminated candidiasis, other-wise known as hepatosplenic candidiasis, present with a more indolent process of infection of the liver, spleen and other tissues. Single-organ infection is usually the result of disseminated candidiasis that becomes clinically overt at a single organ site; e.g., *Candida* osteomyelitis, meningitis, renal candidiasis and endophthalmitis. *Candida* spp. can form biofilms on catheters and other foreign bodies and become extremely resistant to the antifungal action of both drugs and host immune cells. Echinocandins may be more effective than azoles in eradicating *Candida* biofilms from foreign bodies (d'Enfert 2006).

Treatment of deeply invasive candidiasis depends upon the host and severity of infection (Boucher et al. 2004). Uncomplicated fungemia due to susceptible organisms may be treated with fluconazole in non-neutropenic patients and possibly in neutropenic patients. Other patients with fungemia may be treated with conventional or lipid amphotericin B formulations or with echinocandins caspofungin, micafungin and anidulafungin. Removal of central venous catheters is recommended in patients with candidemia if feasible (Lecciones et al. 1992). Chronic disseminated candidiasis can be treated through a variety of strategies utilizing amphotericin B, lipid formulations of amphotericin B, and alternatively with fluconazole (Pappas et al. 2004).

III. Cryptococcosis

In its most commonly encountered form, cryptococcosis is a chronic, wasting, frequently fatal disease, if untreated. It is characterized by a pronounced predilection for the central nervous system (CNS) and is caused by the basidiomycetous yeast *Cryptococcus neoformans*.

There are two varieties of *C. neoformans* causing disease to humans: *C. neoformans* var. *neoformans* and *C. neoformans* var. *gattii*. The two varieties have different geographic distributions. *C. neoformans* var. *gattii* is most often found in tropical or subtropical regions. In addition, the two varieties have different ecological niches: *C. neoformans* var. *neoformans* is found in association with avian habitats, especially pigeons, while the only natural source of *C. neoformans* var. *gattii* so far identified is debris from eucalyptus trees (Kwon-Chung and Bennett 1984).

Impairment of CMI) is the central immunological deficit leading to increased risk of cryptococcosis. Patients, such as those with HIV infection, those receiving corticosteroids and those with lymphoma, have a particularly increased risk due to impaired CMI. Central to host defense against these infections are T-lymphocytes, particularly T-helper cells, which are markedly depleted or functionally altered in patients with HIV infection. Monocytes, activated macrophages and NK cells also have been identified as playing a role in conferring protection against cryptococcosis.

As clinical evidence of the critical role of CMI in host defense against cryptococcosis, meningoencephalitis due to *Cryptococcus neoformans* occurs in approximately 6–13% of HIV-infected adults and approximately 1% of HIV-infected children. By comparison, cryptococcal infections are rarely seen in neutropenic patients or in immunologically competent hosts. With the advent of HAART the incidence of cryptococcal infection in HIV-infected patients has been dramatically reduced (Ruhnke 2004).

Meningoencephalitis, pulmonary infection, fungemia and disseminated infection are the most common patterns of infection due to *Cryptococcus neoformans* (Chuck and Sande 1989; Panther and Sande 1990; Leggiadro et al. 1991; Gonzalez et al. 1996). Cryptococcal meningoencephalitis in HIV-infected patients often has few clinically overt signs early in the course of infection but may present in some patients with meningismus, photophobia and seizures (Viviani 1992). Fever, headache and altered mental status are the most common manifestations in cryptococcal meningoencephalitis. These symptoms are usually indolent, often evolving over the course of weeks to months. Unlike some other CNS mycoses, such as aspergillosis, cryptococcal meningoencephalitis seldom presents with focal neurological deficits (Walsh et al. 1985). Patients with altered mental status, evidence of increased intracranial pressure (e.g., papilledema), seizures and focal deficits are considered to be at particularly high risk for sudden death due to CNS cryptococcosis. Cutaneous lesions mimicking molluscum contagiosum may develop as a manifestation of disseminated cryptococcosis.

Diagnosis of cryptococcal meningitis in HIV infection can usually be established from cerebrospinal fluid (CSF) by a combination of direct examination on a wet mount, CSF culture and cryptococcal capsular polysaccharide antigen detection in CSF. The organism usually appears as an encapsulated budding yeast. However, infections by some "capsule-deficient" strains have been reported in patients with HIV infection (Bottone and Wormser 1985). These strains may be misdiagnosed upon direct exam as other yeasts or as contaminating particles. The CSF cell count, glucose and protein in patients with HIV infection may be virtually normal, due apparently to the paucity of an effective inflammatory response. Although a CT scan is usually non-specific in most cases of CNS cryptococcosis, the CT scan may reveal hydrocephalus or cryptococcomas. When stained with periodic acid Schiff (PAS) biopsy specimens of suspicious skin lesions may reveal encapsulated budding yeast cells. Mucicarmine or alcian blue stains can be used to specifically stain the mucopolysaccharide capsule.

Several features are more distinctive in cryptococcal meningitis in HIV infection in comparison to other immunocompromised populations, such as those with cancer or organ transplants. The CSF antigen in HIV-infected patients with meningeal involvement tends to be substantially higher, often exceeding 1:1024 in seriously ill patients. Consistent with these serological findings, the concentration of organisms in CSF of patients with HIV infection tends to be substantially higher than that of patients with cryptococcal meningitis who do not have HIV infection; India ink preparation is usually positive in patients with HIV infection and cryptococcal meningitis.

Conventional amphotericin B, $0.5-1.0\,\text{mg}\,\text{kg}^{-1}\,\text{day}^{-1}$ or lipid amphotericin B formulations with or without 5-FC for 4–8 weeks is the preferred regimen for the initial treatment of cryptococcal meningitis (Larsen et al. 1990; van der Horst et al. 1997). The role of flucytosine in combination with amphotericin B in HIV infection is controversial, due to dose-dependent suppression of hematopoiesis (Francis and Walsh). Serum concentrations of 5-FC are monitored and maintained at approximately $40\,\propto\text{g}\,\text{ml}^{-1}$ in order to avoid this complication. Titers of cryptococcal antigen should decline in the CSF during the course of therapy. The optimal duration of amphotericin B therapy is unclear, but due to toxicity at 2 weeks it should be switched to fluconazole. Maintenance therapy for prevention of recurrence of CNS cryptococcosis is necessary in HIV-infected and other chronically ill patients (Zuger et al. 1986; Dismukes 1993). A controlled trial found that fluconazole (200 mg day^{-1}, PO) was clearly superior to amphotericin B (1 mg kg^{-1} week^{-1}, IV) in preventing relapse of cryptococcal meningitis in HIV-infected adults (Powderly et al. 1992). Newer therapies such as interferon-gamma and antibody directed against cryptococcal polysaccharide have been tested in phase I–II trials.

IV. Infections due to *Trichosporon*

Trichosporon spp. during the past two decades have emerged as an infrequent but often lethal opportunistic pathogen in granulocytopenic and corticosteroid-treated patients (Walsh et al. 1990, 1993; Kontoyiannis et al. 2004). *Trichosporon* spp. cause a wide spectrum of conditions, which may be classified as summer-type hypersensitivity pneumonitis, white piedra, mucosal infection and deeply invasive infection, including fungemia, single organ infection and disseminated infection (Walsh et al. 1993).

Trichosporon beigelii was previously thought to be the only species. However, the genus *Trichosporon* has undergone extensive taxonomic reevaluation during last decade. Based on morphological, biochemical and most importantly ultrastructural and DNA characteristics *T. beigelii* was split into a number of distinct species, including *T. asahii*, *T. asteroides*, *T. cutaneum*, *T. inkin*, *T. jirovecii*, *T. mucoides* and *T. ovoides* (Gueho et al. 1994). *Trichosporon asahii* appears to be much more common in cases of systemic infections, while other *Trichosporon* species are involved in superficial skin lesions. This section reviews only the role of *Trichosporon* as an opportunistic fungal pathogen causing deep infection, which may be acute (most commonly recognized) or chronic in nature. Fungemia and tissue-proven disseminated infection are the two most frequently encountered patterns of infection due to *T. asahii*, particularly in neutropenic patients and organ transplant recipients.

The most frequent clinical manifestations of acute disseminated *Trichosporon* infection include persistent fever, cutaneous lesions, fungemia, renal dysfunction and pulmonary infiltrates. These clinical manifestations often develop despite administration of empirical amphotericin B in granulocytopenic and other immunosuppressed patients. Biopsy of these cutaneous lesions usually demonstrates hyaline hyphae, blastoconidia and arthroconidia within the dermis. The presence of arthroconidial forms of *Trichosporon* in tissue with the other two morphological forms serves to distinguish *Trichosporon* from other opportunistic yeasts. Cultures of cutaneous biopsy specimens usually yield *T. asahii*. Renal infection is manifest as hematuria, proteinuria, acute renal failure or glomerulonephritis with red blood cell casts. Histopathologically, there is infiltration by hyphae and arthroconidia of the glomeruli and renal tubules. *Trichosporon* pneumonitis is due either to hematogenous involvement or aspiration.

Trichosporon is a basidiomycetous yeast, which expresses cell wall antigens that cross-react with glucuronoxylomannan (GXM) capsular polysaccharide antigens of *C. neoformans* (Walsh et al. 1992; Lyman et al. 1995). The commercially available latex agglutination or enzyme immunoassay tests for *C. neoformans* may support the diagnosis of disseminated *Trichosporon* infection.

Isolates of *T. asahii* may be inhibited but not killed by safely achievable serum concentrations of amphotericin B ($2\,\mu g\,ml^{-1}$; Walsh et al. 1990). Amphotericin B is fungicidal only at concentrations greatly exceeding those attainable therapeutically in serum. This resistance to amphotericin B has been associated with the frequently persistent fungemia and fatal outcome of disseminated *Trichosporon* infection in neutropenic patients. In vivo and clinical data now support the use of fluconazole and voriconazole in treatment of *Trichosporon* infections (Walsh et al. 1992). Neutropenic patients with *Trichosporon* infection have been successfully treated with a combination of amphotericin B plus fluconazole. Non-neutropenic patients are treated with either fluconazole or amphotericin B, or both, depending upon severity of infection. GMCS and other cytokines reverse the GXM-induced immunosuppression of neutrophils and monocytes (Lyman et al. 1995). These therapeutic interventions have greatly improved the outcome from a frequently fatal infection to a treatable and survivable one.

V. Infections due to *Blastoschizomyces capitatus*

Cases of *Blastoschizomyces capitatus* (formerly *Trichosporon capitatum*) previously described in North America appeared to be indistinguishable from those of *T. asahii*. However, a review of 12 cases of infection due to *B. capitatus* from the University La Sapienza in Rome demonstrated patterns of infection that are distinct from those of *T. beigelii*. Specifically, four of seven patients with pulmonary infection had mycetoma-like cavitations; eight patients manifested clinical and radiological features of focal hepatic lesions similar to those of hepatic candidiasis; and three patients had clinically evident cerebritis and brain abscesses confirmed postmortem. Nevertheless, persistent fungemia, maculopapular cutaneous lesions and renal impairment similar to those of *Trichosporon* spp. also were observed. Mortality in these patients was approximately 60% (Martino et al. 1990). Another 26 cases were more recently reviewed from Spain (Martino et al. 2004). The outcome for neutropenic patients with *B. capitatus* infection was poor. Rapid removal of the central venous catheter was required for treatment of this rare infection. The preponderance of reported cases of *B. capitatus* have been reported from Western Europe, while most cases of *T. asahii* have been reported from the United States. These differences in geographic distribution and virulence between *T. asahii* and *B. capitatus* remain to be explored.

VI. Infections due to *Malassezia* spp.

Malassezia furfur, a lipophilic yeast, causes tinea versicolor, folliculitis and catheter-associated fungemia (Gueho et al. 1998). It has also been implicated as cause of seborrheic dermatitis. Tinea versicolor is an asymptomatic variably pigmented macular cutaneous lesion, usually distributed along the neck, chest and shoulders. Folliculitis due to *M. furfur* presents as a pruritic, papular to papulosquamous eruption distributed most prominently on the facial and neck areas. In immunocompromised patients, *M. furfur* folliculitis may simulate the lesions of acute disseminated candidiasis. Fungemia due to *M. furfur* has occurred in patients receiving lipid-supplemented total parenteral nutrition (TPN) via central venous catheters (Devlin 2006). The lipid component apparently provides a nutritional medium for the organism to proliferate in the host. A syndrome of acute respiratory failure and thrombocytopenia has been described in infants with *M. furfur* fungemia as a complication of lipid-supplemented TPN (Redline et al. 1985). This respiratory failure is related to the sequestration of *M. furfur* yeasts and lipids in the subendothelial regions of the pulmonary capillary bed. Dissemination to other organ sites, however, is seldom reported. *M. pachydermatis* also has been reported as a cause of fungemia; however, this species does not have the obligatory nutritional requirements for C_{12}-C_{24} lipids that characterize *M. furfur*.

Tinea versicolor may be identified by skin scrapings as a characteristic cluster of blastoconidia and hyphae. Detection of *M. furfur* from blood requires supplementation of agar plates with olive oil or another C_{12}-C_{24} oil to promote growth of this lipophilic yeast. This requirement for lipids may permit *M. furfur* to elude detection in blood cultures in the clinical microbiology laboratory, unless lipids are added to suspicious sub-cultures of blood. Management of this *M. furfur* fungemia requires discontinuing parenteral lipids and removing the catheter. Amphotericin B is the antifungal of choice for suspected cases of fungal infection. Immunocompromised children with *M. furfur* fungemia also may benefit from a course of an antifungal azole, such as fluconazole, itraconazole or voriconazole; however, resistant strains have been reported.

VII. Infections due to Dematiaceous yeasts

Among the dematiaceous yeasts causing infections in humans, the model *Wangiella dermatitidis* has a high propensity for infections of the CNS (Dixon and Polak-Wyss 1991). While some patients with infections due to *W. dermatitidis* have clinically overt immunodeficiencies, many have no apparent immune impairment that would suggest an increased risk for invasive fungal infection. Further implicating the intrinsic virulence of *W. dermatitidis*, immunocompetent mice may be infected with this dematiaceous yeast without administration of immunosuppressive agents. Traumatic inoculation of the skin and soft tissues may be the portal of entry for *W. dermatitidis*. The lungs are seldom infected.

Patients most commonly present with chronic granulomatous cutaneous lesions and/or focal neurologic deficits. Mortality appears to be host-dependent. For example, one study found that younger patients (<20 years) had significantly higher mortality than did older patients. Treatment of infections due to *W. dermatitidis* consists of complete resection of the lesion, where possible, and administration of antifungal chemotherapy. Itraconazole is considered the drug of choice in the treatment of *Wangiella* infections. Voriconazole and posaconazole have consistent in vitro activities. Due to a high propensity for recurrence, long-term follow-up of patients treated for *Wangiella* infections is important (Brandt and Warnock 2003; Revankar 2004).

VIII. Conclusions

Yeast infections are those caused by unicellular fungi that reproduce by budding (blastoconidia) or by arthroconidial formation. There are many such fungi that cause disease to humans but the two genera that contain the largest number of pathogens are *Candida* and *Cryptococcus*.

Candidiasis caused by *Candida* spp. consists of both superficial infections of the skin, nails and mucosal membranes of the gastrointestinal tract and vagina as well as invasive infections such as fungemia and deep-seated infection of the body. *Candida* spp. are commensals of their hosts, and infections are customarily acquired endogenously.

The appearance of the fungus in tissue is that of yeast cells and pseudohyphae, though the hyphal character is absent in *C. glabrata* and much reduced in *C. guilliermondii*.

Cryptococcosis in its most commonly diagnosed clinical form involves the CNS. The etiologic agent *C. neoformans* is an encapsulated yeast that occurs in two varieties: *C. neoformans* var. *neoformans* and *C. neoformans* var. *gatti*. Polysaccharides that comprise the capsule may occur apart from the blastoconidia and detection of them in body fluids is a valuable diagnostic procedure.

There are a number of other yeasts that may cause infections, including *Trichosporon* spp., *Malassezia* spp., *B. capitatis* and the dematiaceous yeasts, such as *W. dermatitidis*. They may cause serious infections in both immunocompetent hosts but especially in immunocompromised patients.

References

Bottone EJ, Wormser GP (1985) Capsule-deficient cryptococci in AIDS. Lancet 2:553

Boucher HW, Groll AH, Chiou CC, Walsh TJ (2004) Newer systemic antifungal agents: pharmacokinetics, safety and efficacy. Drugs 64:1997–2020

Brandt ME, Warnock DW (2003) Epidemiology, clinical manifestations, and therapy of infections caused by dematiaceous fungi. J Chemother 15[Suppl 2]:36–47

Buhl R (2006) Local oropharyngeal side effects of inhaled corticosteroids in patients with asthma. Allergy 61:518–526

Chauhan N, Latge JP, Calderone R (2006) Signalling and oxidant adaptation in *Candida albicans* and *Aspergillus fumigatus*. Nat Rev Microbiol 4:435–444

Chiou CC, Seibel NL, Derito FA, Bulas D, Walsh TJ, Groll AH (2006) Concomitant *Candida epiglottitis* and disseminated *Varicella zoster* virus infection associated with lymphoblastic leukemia. J Ped Hem Oncol 28:757–759

Chuck SL, Sande MA (1989) Infections with *Cryptococcus neoformans* in the acquired immunodeficiency syndrome. N Engl J Med 321:794–799

d'Enfert C (2006) Biofilms and their role in the resistance of pathogenic Candida to antifungal agents. Curr Drug Targets 7:465–470

Devlin RK (2006) Invasive fungal infections caused by *Candida* and *Malassezia* species in the neonatal intensive care unit. Adv Neonatal Care 6:68–77

Dismukes WE (1993) Management of cryptococcosis. Clin Infect Dis 17[Suppl 2]:S507–S512

Dixon DM, Polak-Wyss A (1991) The medically important dematiaceous fungi and their identification. Mycoses 34:1–18

Dongari-Bagtzoglou A, Fidel PL Jr (2005) The host cytokine responses and protective immunity in oropharyngeal candidiasis. J Dent Res 84:966–977

Fidel PL Jr (2005) Immunity in vaginal candidiasis. Curr Opin Infect Dis 18:107–111

Francis P, Walsh TJ (1992) Evolving role of flucytosine in immunocompromised patients: new insights into safety, pharmacokinetics, and antifungal therapy. Clin Invest Dis 15:1003–1018

Gonzalez CE, Shetty D, Lewis LL, Mueller BU, Pizzo PA, Walsh TJ (1996) Cryptococcosis in human immunodeficiency virus-infected children. Pediatr Infect Dis J 15:796–800

Graaf CA van der, Netea MG, Verschueren I, Meer JW, Kullberg BJ (2005) Differential cytokine production and Toll-like receptor signaling pathways by *Candida albicans* blastoconidia and hyphae. Infect Immun 73:7458–7464

Gueho E, Improvisi L, Hoog GS de, Dupont B (1994) Trichosporon on humans: a practical account. Mycoses 37:3–10

Gueho E, Boekhout T, Ashbee HR, Guillot J, Van Belkum A, Faergemann J (1998) The role of *Malassezia* species in the ecology of human skin and as pathogens. Med Mycol 36[Suppl 1]:220–229

Horst CM van der, Saag MS, Cloud GA, Hamill RJ, Graybill JR, Sobel JD, Johnson PC, Tuazon CU, Kerkering T, Moskovitz BL, Powderly WG, Dismukes WE (1997) Treatment of cryptococcal meningitis associated with the acquired immunodeficiency syndrome. (National institute of allergy and infectious diseases mycoses study group and AIDS clinical trials group) N Engl J Med 337:15–21

Kappe R, Levitz SM, Cassone A, Washburn RG (1992) Mechanisms of host defence against fungal infection. J Med Vet Mycol 30[Suppl 1]:167–177

Kirkpatrick CH (2001) Chronic mucocutaneous candidiasis. Pediatr Infect Dis J 20:197–206

Klein RS, Harris CA, Small CB, Moll B, Lesser M, Friedland GH (1984) Oral candidiasis in high-risk patients as the initial manifestation of the acquired immunodeficiency syndrome. N Engl J Med 311:354–358

Kontoyiannis DP, Torres HA, Chagua M, Hachem R, Tarrand JJ, Bodey GP, Raad, II (2004) Trichosporonosis in a tertiary care cancer center: risk factors, changing spectrum and determinants of outcome. Scand J Infect Dis 36:564–569

Kwon-Chung KJ, Bennett JE (1984) Epidemiologic differences between the two varieties of Cryptococcus neoformans. Am J Epidemiol 120:123–130

Laine L (1994) The natural history of esophageal candidiasis after successful treatment in patients with AIDS. Gastroenterology 107:744–746

Larsen RA, Leal MAE, Chan LS (1990) Fluconazole compared with amphotericin B plus flucytosine for cryptococcal meningitis in AIDS. Ann Intern Med 113:183–187

Lecciones JA, Lee JW, Navarro EE, Witebsky FG, Marshall D, Steinberg SM, Pizzo PA, Walsh TJ (1992) Vascular catheter-associated fungemia in patients with cancer: analysis of 155 episodes. Clin Infect Dis 14:875–883

Leggiadro RJ, Kline MW, Hughes WT (1991) Extrapulmonary cryptococcosis in children with acquired immunodeficiency syndrome. Pediatr Infect Dis J 10:658–662

Leibovitz E (2002) Neonatal candidosis: clinical picture, management controversies and consensus, and new therapeutic options. J Antimicrob Chemother 49[Suppl 1]:69–73

Lilic D (2002) New perspectives on the immunology of chronic mucocutaneous candidiasis. Curr Opin Infect Dis 15:143–147

Lyman CA, Devi SJ, Nathanson J, Frasch CE, Pizzo, Walsh TJ (1995) Detection and quantitation of the glycuronoxylomannan-like polysaccharide antigen from clinical and nonclinical isolates of Tricosporon beigelii and implications for pathogenicity. J Clin Microbiol 33:126–130

Maksymiuk AW, Thongprasert S, Hopfer R, Luna M, Fainstein V, Bodey GP (1984) Systemic candidiasis in cancer patients. Am J Med 77:20–27

Maraki S, Hajiioannou I, Anatoliotakis N, Plataki M, Chatzinikolaou I, Zoras O, Tselentis Y, Samonis G (1999) Ceftriaxone and dexamethasone affecting yeast gut flora in experimental mice. J Chemother 11:363–366

Martino P, Venditti M, Micozzi A, Morace G, Polonelli L, Mantovani MP, Petti MC, Burgio VL, Santini C, Serra P et al (1990) Blastoschizomyces capitatus: an emerging cause of invasive fungal disease in leukemia patients. Rev Infect Dis 12:570–582

Martino R, Salavert M, Parody R, Tomas JF, Camara R de la, Vazquez L, Jarque I, Prieto E, Sastre JL, Gadea I, Peman J, Sierra J (2004) Blastoschizomyces capitatus infection in patients with leukemia: report of 26 cases. Clin Infect Dis 38:335–341

Odds FC (1992) Candida infections in AIDS patients. Int J STD AIDS 3:157–160

Ohmit SE, Sobel JD, Schuman P, Duerr A, Mayer K, Rompalo A, Klein RS (2003) Longitudinal study of mucosal Candida species colonization and candidiasis among human immunodeficiency virus (HIV)-seropositive and at-risk HIV-seronegative women. J Infect Dis 188:118–127

Ostrosky-Zeichner L, Pappas PG (2006) Invasive candidiasis in the intensive care unit. Crit Care Med 34:857–863

Pankhurst C (2005) Candidiasis (oropharyngeal). Clin Evid:1701–1716

Panther LA, Sande MA (1990) Cryptococcal meningitis in the acquired immunodeficiency syndrome. Semin Respir Infect 5:138–145

Pappas PG, Rex JH, Sobel JD, Filler SG, Dismukes WE, Walsh TJ, Edwards JE (2004) Guidelines for treatment of candidiasis. Clin Infect Dis 38:161–189

Pons V, Greenspan D, Debruin M (1993) Therapy for oropharyngeal candidiasis in HIV-infected patients: a randomized, prospective multicenter study of oral fluconazole versus clotrimazole troches. The Multicenter Study Group. J Acquir Immun Defic Syndr 6:1311–1316

Powderly WG, Saag MS, Cloud GA, Robinson P, Meyer RD, Jacobson JM, Graybill JR, Sugar AM, McAuliffe VJ, Follansbee SE et al (1992) A controlled trial of fluconazole or amphotericin B to prevent relapse of cryptococcal meningitis in patients with the acquired immunodeficiency syndrome (the NIAID AIDS clinical trials group and mycoses study group). N Engl J Med 326:793–798

Redline RW, Redline SS, Boxerbaum B, Dahms BB (1985) Systemic Malassezia furfur infections in patients receiving intralipid therapy. Hum Pathol 16:815–822

Revankar SG (2004) Dematiaceous fungi. Semin Respir Crit Care Med 25:183–189

Roilides E, Walsh TJ (2004) Recombinant cytokines in augmentation and immunomodulation of host defenses against Candida spp. Med Mycol 42:113

Roy MJ, Walsh TJ (1992) Histopathologic and immunohistochemical changes in gut-associated lymphoid tissues after treatment of rabbits with dexamethasone. Lab Invest 66:437–443

Ruhnke M (2004) Mucosal and systemic fungal infections in patients with AIDS: prophylaxis and treatment. Drugs 64:1163–1180

Saag MS, Fessel WJ, Kaufman CA, Merrill KW, Ward DJ, Moskovitz BL, Thomas C, Oleka N, Guarnieri JA, Lee J, Brenner-Gati L, Klausner M (1999) Treatment of fluconazole-refractory oropharyngeal candidiasis with itraconazole oral solution in HIV-positive patients. AIDS Res Hum Retroviruses 15:1413–1417

Samonis G, Maraki S, Leventakos K, Spanaki AM, Kateifidis A, Galanakis E, Tselentis Y, Falagas ME, Mantadakis E (2006) Comparative effects of ertapenem, imipenem, and meropenem on the colonization of the gastrointestinal tract of mice by Candida albicans. Med Mycol 44:233–235

Spence D (2005) Candidiasis (vulvovaginal). Clin Evid:2200–2215

Vazquez JA (2003) Invasive oesophageal candidiasis: current and developing treatment options. Drugs 63:971–989

Viviani MA (1992) Opportunistic fungal infections in patients with acquired immune deficiency syndrome. Chemotherapy 38[Suppl 1]:35–42

Walsh TJ, Pizzo PA (1992) Experimental gastrointestinal and disseminated candidiasis in immunocompromised animals. Eur J Epidemiol 8:477–483

Walsh TJ, Hier DB, Caplan LR (1985) Fungal infections of the central nervous system: comparative analysis of risk factors and clinical signs in 57 patients. Neurology 35:1654–1657

Walsh TJ, Melcher GP, Rinaldi MG, Lecciones J, McGough DA, Kelly P, Lee J, Callender D, Rubin M, Pizzo PA (1990) Trichosporon beigelii, an emerging pathogen resistant to amphotericin B. J Clin Microbiol 28:1616–1622

Walsh TJ, Lee JW, Melcher GP, Navarro E, Bacher J, Callendar D, Reed KD, Wu T, Lopez- Berenstein G, Pizzo PA (1992) Experimental Trichosporon infection in persistently granulocytopenic rabbits: implications for pathogenesis, diagnosis, and treatment of an emerging opportunistic mycosis. J Infect Dis66:121–133

Walsh TJ, Melcher GP, Lee JW, Pizzo PA (1993) Infections due to Trichosporon species: new concepts in mycology, pathogenesis, diagnosis and treatment. Curr Top Med Mycol 5:79–113

Zuger A, Louie E, Holzman RS, Simberkoff MS, Rahal JJ (1986) Cryptococcal disease in patients with the acquired immunodeficiency syndrome. Diagnostic features and outcome of treatment. Ann Intern Med 104:234–240

11 The Host Innate Immune Response to Pathogenic *Candida albicans* and Other Fungal Pathogens

Peter F. Zipfel[1,2], Katharina Gropp[1], Michael Reuter[1], Susan Schindler[1], Christine Skerka[1]

CONTENTS

I. Introduction	233
II. Host Innate Immune Response	234
1. Complement	234
2. Binding of Host Complement Regulators	236
3. C3 Receptor	237
4. Antifungal Activity of C3a and C5a	237
III. Coagulation Cascade and Plasminogen Binding to Pathogenic Yeast	238
1. Acquisition of Host Coagulation Components in the Form of Plasminogen	238
IV. Macrophage Response to Pathogenic Fungi	238
1. Candida and the Innate Immune Response	238
V. Conclusions	240
References	241

I. Introduction

Infection with human pathogenic yeast and fungi is an increasing health problem, particularly for immunocompromised people. *Candida albicans*, as well as other fungi which are ubiquitously distributed and live as saprophytic commensals, colonizes the human mucosa and in addition can cause systemic, life-threatening infections (Mavor et al. 2005; Richardson 2005). By specific, but mostly unknown environmental changes, the yeast gains pathogenic potential and colonizes the host. During infection the pathogen is in contact with body fluids and host cells and is attacked by the host immune system. Apparently pathogenic fungi such as *C. albicans*, like other pathogens, has developed sophisticated means to inactivate and combat the host immune response. This immune evasion is considered due to the action of specific virulence factors. Although multiple virulence factors of *C. albicans* have been characterized in the past, the exact mechanisms how this pathogenic yeast modulates host innate immune responses are still unclear.

Infection is a complex process, as a vertebrate host exploits multiple, efficient and highly toxic immune reactions, which attack an invading pathogen on multiple levels. In general the host immune response acts on two major levels which are separated based on the time of initiation after the initial contact with the pathogen (Table 11.1). The *innate immune system*, which acts immediately and directly, includes physical barriers, enzymatic cascades in form of the complement and the coagulation cascade; and these are assisted by immune effector cells such as macrophages. The *adaptive immune system* (which is mediated by antibody-secreting B-lymphocytes and antigen receptor-specific T-lymphocytes) requires – after the first encounter with the pathogen – several days to generate a protective antigen-specific response. The course of infection of human pathogenic fungi, e.g. *C. albicans* in a human host, was covered by recent reviews (Calderone and Fonzi 2001; Mavor et al. 2005). Following infection of an immunocompetent host, *Candida* is exposed and attacked by the host immune system. In order to control this efficient and damaging immune responses pathogenic fungi (and likely all pathogens) actively inactivate the toxic immune response and prevent host immune activation. As a consequence, the pathogen survives the first phases of the immune response, resulting in infection and host disease (Zipfel et al. 2007a). In this review we characterize the initial immune response that is induced by the innate immune system acting immediately and directly against the pathogenic yeast. The later-acting adaptive response, which is mediated by the T- and B-lymphocytes is also covered by a recent excellent review (Zelante et al. 2007).

---

[1] Department of Infection Biology, Leibniz Institute for Natural Product Research and Infection Biology, Hans Knöll Institute, Beutenbergstrasse 11a, 07745 Jena, Germany; e-mail: peter.zipfel@hki-jena.de
[2] Friedrich Schiller University, Jena, Germany

Human and Animal Relationships, 2nd Edition
The Mycota VI
A.A. Brakhage and P.F. Zipfel (Eds.)
© Springer-Verlag Berlin Heidelberg 2008

Table 11.1. General characteristics of the immune system

	Innate	Adaptive
Recognition	Unique chemical patterns	Antigen-specific epitopes
Activation upon *first* contact	Rapid and immediate (seconds to hours)	Slow (1–2 weeks), antigen-driven activation
Activation upon *second* contact	Rapid and immediate (seconds to hours); direct recognition and response	Direct (hours), maturation
Effectors	Complement System, macrophages	T-Lymphoycytes, B-lymphocytes
Presentation Receptors	Toll-like receptors, complement receptors on macrophages	Antigen-specific receptors, T-cell receptor
Mediators	C3a and C5a, pore formation, cytokines, chemokines	Cytokines, chemokines, cytotoxic granules immunoglobulin
Memory	None, repetitive response	*Memory*, long (up to lifetime)

II. Host Innate Immune Response

The host innate immune system is comprised of pattern-recognition receptors which are able to identify unique structures on the surface of pathogens. The human complement system, particularly when activated via the alternative pathway, represents a pattern-recognition system (Zipfel et al. 2007b). Additional enzymatic systems like the coagulation system and particular cellular responses, such as the macrophage Toll-like receptor response, form central cellular components in the human innate immune response (Romani 2004; Zelante et al. 2007). During the commensal state and also during infection when *Candida* changes its morphology resulting in the formation of hyphae, the pathogen is directly attacked by the host innate immune response. Thus during evolution, the pathogen has established the means to interact and inactivate the innate immune responses at the levels of complement and macrophage attack (Romani 2004; Zipfel et al. 2007a).

1. Complement

The complement system plays an important role in the recognition and elimination of microbes (Walport 2001a, b). This system is rather efficient in opsonizing and damaging invading microbes immediately upon infection (Kraiczy and Wurzner 2006). The activated complement system is highly toxic and has devastating effects. Complement is activated by three pathways: the alternative, the classical and the lectin pathways. The *alternative pathway* is initiated by a spontaneous conformational change of the central complement component C3b. By exposing a highly active thioester, the activated protein can bind to any component in its vicinity. The initial reaction of this thioester exposed by C3 is in the range of milliseconds and is thus relatively short. Once activated, complement factor B is bound to C3b and activated by the serine protease factor D, to form the enzymatically active C3bBb complex. This C3 convertase generates more C3b molecules and initiates a rather powerful amplification cascade reaction (Fig. 11.1).

The *classical pathway* is initiated by antigen:antibody complexes. Upon complex formation, additional zymogens like C4 are bound and activated, which expose binding sites for other components of this cascade. Upon cleavage, e.g. by a serine protease, the enzymatically active C4b2a complex is generated. This enzyme represents the classical pathway C3 convertase. The *lectin pathway* is initiated when the mannan-binding lectin (MBL) binds to carbohydrates which are accessible on the surface of pathogens. After binding, MBL is activated by MBL-associated serine proteases (MASPs) and consequently activates C4 and leads to C1-independent formation of the classical pathway convertase C4b2a. The three complement pathways, which form two different types of C3 generating enzymes, i.e. C3 convertases (C3bBb, C4b2a), merge at the level of C3. Both enzymes initiate a powerful amplification reaction and generate more active C3, which – if left unrestricted – has the capacity to bind to any surface in the direct vicinity. Following binding, surface-bound amplification convertases are generated, which boost activation considerably. C3 cleavage generates the active component C3b and the anaphylactic and antimicrobial component C3a.

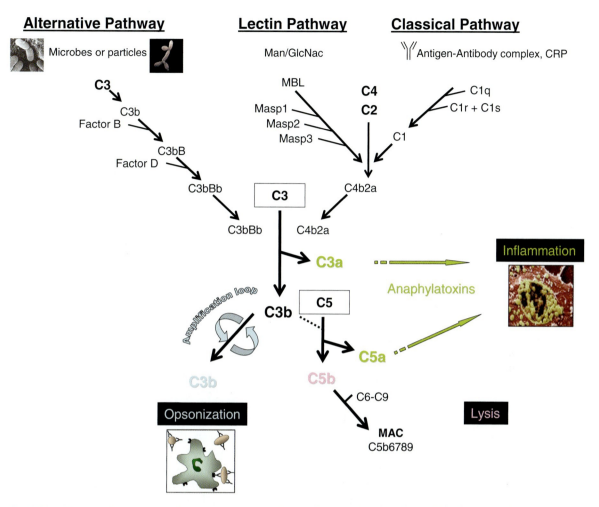

Fig. 11.1. The complement cascade: a pattern recognition system of innate immunity. The human complement system represents a central host defence system of innate immunity which is activated immediately upon contact of a microbe with host plasma or body fluids. This cascade-type system is triggered by three distinct pathways, the alternative, the lectin and the classical pathways. The *Alternative Pathway* is initiated by a spontaneous conformational change of the central complement component C3b which exposes its active thioester and, after binding the additional component *Factor B* and activation by the serine protease *Factor D*, forms an active C3bBb complex which displays enzymatic activity. This C3 convertase initiates a powerful amplification reaction and generate more C3b molecules. The sites of action of the host inhibitors that are utilized by human pathogenic fungi (e.g. *C. albicans*, *A. fumigatus*) *Factor H*, *FHL1* and *C4BP* are indicated. The *Lectin Pathway* is initiated when the mannan binding lectin (*MBL*) binds to carbohydrates which are accessible on the surface of pathogens. After binding, MBL is activated by MBL-associated serine proteases (MASPs) and consequently activates C4 and leads to the C1-independent formation of the classical pathway convertase C4b2a. The *Classical Pathway* is initiated by antigen:antibody complexes. Upon complex formation, additional zymogens like *C4* are activated and consequently form additional binding sites for other components of this cascade and for proteases. Upon cleavage, e.g. by a serine protease, active enzyme complexes are generated which form the classical pathway convertase C4b2a. Complement activation initiates several main effector functions and has devastating effects for a microbe. The activated and amplified cascade results in the coating of a microbial or foreign surface with C3 proteins, a process termed opsonization. Opsonization enhances recognition and phagocytosis of a microbe by host immune effector cells and macrophages. In addition, initiation of the terminal complement pathway results in the formation of the membrane attack complex (*MAC*), which generates a pore in the membrane of the target cells. During activation, potent anyphylactic components in the form of *C3a* and *C5a* are generated which initiate inflammatory responses

Activation of complement results in the deposition of a large number of C3b products on a surface. This process is termed opsonization and enhances phagocytosis of a C3b-labelled particle by macrophages which are equipped with specific C3 receptors termed complement receptors (CR1–CR4).

The second effector mechanism of the activated complement system is the generation of C5 convertases (C3bBbC3b or C4b2a3b) which cleave and activate C5 and form C5b as well as the anaphylactic and antimicrobial component C5a. C5b initiates the terminal complement pathway, by activating sequentially and non-enzymatically the terminal complement components C6, C7, C8 and C9, resulting in the formation of the membrane attack complex (MAC), also termed the terminal complement complex (TCC). The end-product of the MAC pathway is a pore which is composed of multiple assembled C9 proteins, which inserts into the membrane of the target and causes cell lysis and membrane disruption.

The initial phase of complement activation is indiscriminate and generates highly toxic effects on target cells or surfaces. Thus control is required on the surface of host cells (self cells). From the host's side, unrestricted activation is favoured on the surface of invaders, e.g. microbes that are consequently damaged and eliminated.

The important role of the complement system for the clearance of microbes becomes obvious by the complex mechanisms and evasion strategies utilized by pathogens for immune disguise and immune escape (Kraiczy and Wurzner 2006; Rooijakkers et al. 2006; Zipfel et al. 2007a). A wide spectrum of pathogenic microbes and multicellular organisms, including fungi and other pathogens such as Gram negative and Gram positive bacteria, protozoa, helminths as well as viruses (Lachmann 2002), have developed sophisticated means to control and inactivate host complement attack on the surface of the pathogen. This form of complement disguise is mediated by attacking the central host complement regulators on the surface of a pathogen and thus mimicking host surface characteristics for immune escape. The interaction of *Candida* species with the human complement system is an important aspect of the pathogenesis of *Candida* species and has been particularly studied for the human pathogenic yeast *C. albicans*.

Whole yeast cells activate the complement cascade; and depletion of the alternative complement system, e.g. by cobra venom factor in mice, results in a particular susceptibility for infection (Gelfand et al. 1978).

Complement activation results in opsonization with C3 activation products that are deposited on the surface of a microbe. These C3 activation products either trigger the formation of the terminal complement cascade, the membrane attack complex, or cause opsonization, i.e. deposition of the fungal surface with multiple C3b molecules. Attached to the surface, this C3 decoration induces and enhances phagocytosis by macrophages, which are equipped with specific C3 receptors. Thus deposition of C3b is critical for the induction of phagocytosis by macrophages and by activated professional antigen-presenting cells.

The opsonic activity of human serum is reduced by heat inactivation, which destroys the active complement components. Apparently the alternative pathway of complement, which is spontaneously activated on any kind of surface, is central for this opsonic effect. The alternative pathway of complement is the primary mechanism for C3 activation and C3b is the primary activation product that is deposited onto the surface of *Candida* cells (Kozel et al. 1987). In addition, antibodies such as mannan-specific IgG from normal human serum accelerate C3b binding to *Candida* via the alternative pathway (Zhang and Kozel 1998).

Human pathogenic fungi such as *C. albicans*, *Aspergillus fumigatus* and *Cryptococcus neoformans*, as well as dermatophytes *Trichophyton rubrum*, *T. mentagrophytes* and *Arthroderma benhamiae* activate the complement system of the host (Kozel et al. 1996). Although *C. albicans* activates all three pathways of the human complement system, an essential role is proposed for the alternative pathway of complement.

2. Binding of Host Complement Regulators

Human pathogenic fungi, like other pathogens, acquire host proteins on their surface, including complement regulators and components of the coagulation cascade. Binding of the central fluid phase complement regulators factor H, FHL-1 and C4-binding protein (C4BP) has been shown for *C. albicans* (Meri et al. 2002; Meri et al. 2004) and for *A. fumigatus* (Behnsen et al. 2007). Bound proteins are visualized (e.g. by fluorescence microscopy) and show a complete coating of the pathogen with the host immune effector proteins, resulting in immune disguise (Fig. 11.2). Bound

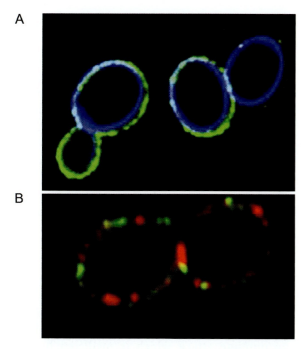

Fig. 11.2. The human pathogenic yeast *Candida albicans* binds human complement regulators factor H, FHL-1 and C4BP to its surface. **A** Human *Candida* cells bind the host immune regulators factor H and FHL-1 to its surface. *Candida* cells were incubated in human plasma and, after extensive washing, bound factor H and FHL-1 were identified with specific antiserum (*green fluorescence*) and visualized by immunofluorescence microscopy. The cell wall of the yeast is shown in *blue* due to calcofluor staining. Binding of the host alternative pathway regulator factor H (*green*) and the classical pathway regulator C4BP (*red*) occurs simultaneously on the surface of *C. albicans*

to the surface of these pathogens, the host regulators mediate regulatory activity and thus aid in the inactivation of the complement attack directly at the surface of the pathogen. Apparently several surface or cell wall proteins are expressed at the surface of the yeast cells which mediate binding. Some of these complement regulatory acquiring surface proteins have been cloned and characterized in more detail.

3. C3 Receptor

Receptors for C3 and the C3 degradation fragments C3d and iC3b have been identified on the surface of *C. albicans* (Heidenreich and Dierich 1985; Alaei et al. 1993). The specific binding of C3 fragments suggested that *Candida* has specific surface receptors for the cleavage products of C3. Immunological cross-reactivity and similar binding characteristics have been demonstrated for the host C3b fragments between these specific *C. albicans* complement receptors and the host C3 receptors CR2 and CR3. Consequently it was suggested that the *C. albicans* C3 receptors play a role in virulence.

4. Antifungal Activity of C3a and C5a

Complement activation during the early phase results in cleavage of the soluble plasma protein C3 into the fragments C3b and C3a. C3b is deposited on any kind of surface and both soluble as well as membrane-bound regulators influence the fate of newly deposited C3b molecules. C3b deposition on a microbial surface enhances and favours phagocytosis by macrophages or initiation of additional reactions, which results in the formation of the terminal pore-forming membrane attack complex (MAC). The other activation product, the soluble small C3a peptide, acts as an anaphylatoxin which, e.g. has chemotactic activity and facilitates infiltration by macrophages. In addition, recent work has shown that C3a has antimicrobial activity (Nordahl et al. 2004) and particulary antifungal activity (Sonesson et al. 2007). C3a as well as C3a-derived peptides, which bind to the *Candida* surface, have antifungal activity and induce membrane changes and the release of extracellular material. Apparently arginine residues of C3a are critical for the antifungal and membrane-breaking activity. C3 fragments are activated and bound from normal human serum onto *C. albicans*. The host complement system plays a central role for controlling the host immune response and finally clearance of *Candida*, as demonstrated in C5-deficient mice. The C5-deficient A/J mice showed high death rates upon infection within 24 h, while the corresponding C5-sufficient C56BL/6J mice lived longer (Mullick et al. 2004, 2006).

Complement activation generates small biological active fragments, e.g. the small activation products C3a and C5a. Both fragments display chemotaxis, interact with specific cellular receptors on the surface of host immune effector cells, display inflammatory activity and enhance the immune response (Walport 2001a, b; Nordahl et al. 2004). The anaphylatoxin C3a exerts antimicrobial effects against the yeast *Candida*. C3a-derived peptides attach to the surface of *Candida* where they induce membrane perturbations and also the release of extracellular material (Sonesson et al. 2007). Thus C3a as well as C3a-derived peptides may be useful for the development of peptide-based antifungal therapies.

III. Coagulation Cascade and Plasminogen Binding to Pathogenic Yeast

Host–pathogen interaction is a fascinating and quickly developing area of research. Pathogens – most likely in contrast to microbes – bind several soluble host proteins and utilize these host proteins for immune escape and survival within the host. These proteins include regulators of both the complement and the coagulation cascade. For *C. albicans*, binding of the the host complement regulators factor H, FHL-1 and C4bp was shown (Meri et al. 2002; Meri et al. 2004) and binding of plasminogen was also identified (Crowe et al. 2003). Similar results are reported for *A. fumigatus* (Behnsen et al. 2007).

The binding of host immune regulators not only results in complement control but also augments the invasive potential by generating proteolytic activity on the surface of the pathogen. For *C. albicans*, eight distinct plasminogen-binding proteins have been identified (Crowe et al. 2003). When bound to the yeast surface, the host serine protease plasminogen can be activated by mammalian plasminogen activators and the proteolytic active enzyme augments the invasive potential of the yeast.

The pathogenicity of *Candida* invasion is related to proteolysis and is associated with virulence (Fig. 11.3). *C. albicans* has a proteolytic environment at the surface and two forms are known: acquired proteases (e.g. in the form of plasminogen) and endogenous, pathogen encoded proteases such as aspartyl proteases. These proteolytic enzymes assist in the penetration of host barriers, immune evasion, degradation of host defence, tissue invasion and extracellular matrix degradation (Hube 1996).

C. albicans utilizes the host mammalian fibrinolytic system which mediates the degradation of both newly formed fibrin and also extracellular matrix components, e.g. during tissue remodelling. The coagulation cascade generates thrombin that cleaves fibrinogen, which results in the formation of fibrin monomers that spontaneously polymerize to proto-fibrin and form a fibrin array. Fibrinolysis, the solution or solubilization of such fibrin clots is crucial for wound healing, angiogenesis and vessel wall formation. Thus the acquisition of newly generated host fibrinogen is beneficial for the pathogen.

1. Acquisition of Host Coagulation Components in the Form of Plasminogen

Plasminogen activation, i.e. conversion to the active protease plasmin, is tightly regulated by tissue plasminogen activator (tPA) or urokinase palsminogen activator (uPA), both of which are synthesized by endothelial cells. Plasmin cleaves extracellular matrices and also is involved in the degradation of immunglobulins.

IV. Macrophage Response to Pathogenic Fungi

The course of infection of *Candida* species was covered by a recent review (Mavor et al. 2005). Following infection, *Candida* is exposed to the host immune system and in order to survive *Candida*, like any other microbe, must actively prevent the immune attack which is mediated by the host innate and adaptive immune system (Fig. 11.4). Here we focus on the initial events that are induced by the immediate-acting innate immune system. The later-acting adaptive response, which is mediated by T- and B-lymphocytes is also covered by recent reviews (Romani 2000; Fidel 2002).

1. Candida and the Innate Immune Response

The innate immune response is particularly mediated by the complement system and by macrophages which are directly triggered by a microbe. The host complement system plays a central role for controlling the host immune response and finally clearance of *Candida*, as demonstrated in C5-deficient mice. C5-deficient A/J mice showed high death rates within 24 h of infection, while the corresponding C5-sufficient C56BL/6J mice lived longer.

Toll-like receptors (TLRs) represent a major class of pattern-recognition receptors which recognize pathogen-associated molecular patterns (PAMPs) as monomers; and as dimers they induce activation of the host innate immune response (Fig. 11.5). TLRs play a central role in the recognition of fungal pathogens including *C. albicans*, *A. fumigatus* and *Cryptococcus neoformans*. Apparently TLRs (particularly TLR2, TLR4) play differential roles in the activation of immune effector

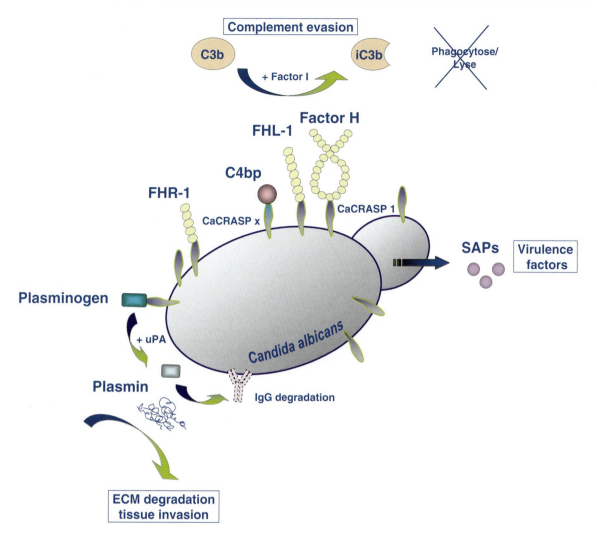

Fig. 11.3. Multiple Immune escape strategies are used by the pathogenic yeast *Candida albicans*. *C. albicans* escapes human complement attack by immune disguise, i.e. the pathogen binds soluble host immune regulators Factor H, FHL-1 and C4Bp via specific surface proteins termed CaCRASP (*C. albicans* complement regulator-acquiring surface proteins) Several distinct CRASP proteins are utilized on the cell surface of *C. albicans*, which in addition to host complement regulators also binds plasminogen. When bound to *C. albicans* CRASP-1, the host regulators factor H, FHL-1 and plasminogen are functionally active. *Factor H* aids in complement inhibition on the level of C3 and aids in the inactivation of the central component C3b. Also, attached plasminogen can be converted to the active protease plasmin, which displays proteolytic activity and most likely mediates tissue invasion and degradation of immunoglobulins. In addition, *C. albicans* expresses and secretes extracellular hydrolytic enzymes (*SAPs*; secreted aspartyl proteases) which also aid in virulence

cells and the response due to TLR stimulation triggers cytokine response and immune reactions (Netea et al. 2006).

This overview describes and details some current examples of immune escape strategies of the human pathogenic yeast *C. albicans* as well as other pathogenic fungi. The host immune defence acts on multiple levels and consequently the pathogen exploits several layers for immune disguise. The specific acquisition of host immune effector proteins establishes a counter-reaction at the surface of the pathogen in an immune-competent host and consequently aids in survival of the pathogen. Identifying the molecules at the surface of

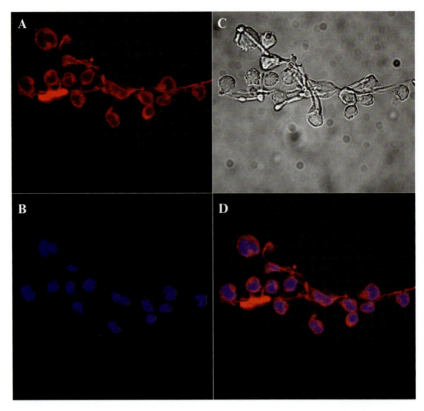

Fig. 11.4. *Candida albicans* interaction with human immune effector cells. *C. albicans* in the hyphal form was incubated with human macrophages. **A** Human macrophages were stained with wheat germ agglutinin (*red*) **B** their nuclei were indicated by DAPI staining (*blue*) **C** *Candida* hyphae were visualized by phase-contrast microscopy **D** An overlay of the three figures shows that human macrophages are in close contact and attack the yeast cells

the pathogen (e.g. *Candida* CRASP proteins) and characterizing the molecular interaction with the specific host immune effector proteins defines specific targets which can be used for interference with the host immune attack and in the development of specific antifungal compounds.

V. Conclusions

Complement forms the first defence line of innate immunity. In order to survive such a hostile complement attack, pathogens need to interfere with and inactivate this immediate-acting defence system. Pathogenic fungi like *C. albicans* and *A. fumigatus* are well equipped for this situation, as they camouflage their surface with host complement regulators and host proteolytic enzymes. The identification of several of such pathogenic surface proteins that bind the host complement regulators factor H and FHL1, as well as the fibrinolytic protease plasminogen, also allows a detailed characterization and understanding of these evasion mechanisms in molecular terms. Several of the fungal complement-binding surface proteins represent moonlighting proteins which were initially identified as cytoplasmic glycolytic enzymes. It will be of interest to define whether these proteins serve additional functions, likely acting as bridging molecules to facilitate the further steps of infection, i.e. adhesion to host cells and internalization. Thus these types of surface proteins are of interest in the study of pathogen–host interactions, but in addition they serve as interesting targets to interfere with the infection process.

Acknowledgement. The work of the authors is funded by the Deutsche Forschungsgemeinschaft (DFG) in the Priority Program 1160.

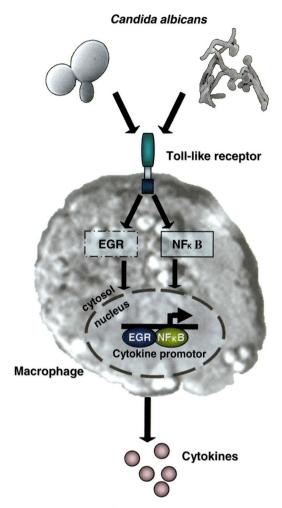

Fig. 11.5. *Candida* triggers macrophage response. *Candida* yeast forms or hyphal forms bind to and attach to host macrophages via specific receptors, including toll-like receptors. Attachment of the yeast or hyphae initiates a signal cascade which results in the induction of the transcription factors *EGR* (*e*arly *g*rowth *r*eponse genes) or *NFkB*. Following nuclear translocation, the activated transcription factor induces transcription of immune effector genes, e.g. cytokines and anti-inflammatory proteins

References

Alaei S, Larcher C et al (1993) Isolation and biochemical characterization of the iC3b receptor of *Candida albicans*. Infect Immun 61:1395–1399

Behnsen J, Narang P et al (2007) Environmental dimensionality controls the interaction of phagocytes with the pathogenic fungi *Aspergillus fumigatus* and *Candida albicans*. PLoS Pathog 3:e13

Calderone RA, Fonzi WA (2001) Virulence factors of *Candida albicans*. Trends Microbiol 9:327–335

Crowe JD, Sievwright IK et al (2003) *Candida albicans* binds human plasminogen: identification of eight plasminogen-binding proteins. Mol Microbiol 47:1637–1651

Fidel PL Jr (2002) Immunity to *Candida*. Oral Dis 8[Suppl 2]:69–75

Gelfand JA, Hurley DL et al (1978) Role of complement in host defense against experimental disseminated candidiasis. J Infect Dis 138:9–16

Heidenreich F, Dierich MP (1985) *Candida albicans* and *Candida stellatoidea*, in contrast to other *Candida* species, bind iC3b and C3d but not C3b. Infect Immun 50:598–600

Hube B (1996) *Candida albicans* secreted aspartyl proteinases. Curr Top Med Mycol 7:55–69

Kozel TR, Brown RR et al (1987) Activation and binding of C3 by *Candida albicans*. Infect Immun 55:1890–1894

Kozel TR, Weinhold LC et al (1996) Distinct characteristics of initiation of the classical and alternative complement pathways by *Candida albicans*. Infect Immun 64:3360–3368

Kraiczy P, Wurzner R (2006) Complement escape of human pathogenic bacteria by acquisition of complement regulators. Mol Immunol 43:31–44

Lachmann PJ (2002) Microbial subversion of the immune response. Proc Natl Acad Sci USA 99:8461–8462

Mavor AL, Thewes S et al (2005) Systemic fungal infections caused by *Candida* species: epidemiology, infection process and virulence attributes. Curr Drug Targets 6:863–874

Meri T, Hartmann A et al (2002) The yeast *Candida albicans* binds complement regulators factor H and FHL-1. Infect Immun 70:5185–5192

Meri T, Blom AM et al (2004) The hyphal and yeast forms of *Candida albicans* bind the complement regulator C4b-binding protein. Infect Immun 72:6633–6641

Mullick A, Elias M et al (2004) Dysregulated inflammatory response to *Candida albicans* in a C5-deficient mouse strain. Infect Immun 72:5868–5876

Mullick A, Leon Z et al (2006) Cardiac failure in C5-deficient A/J mice after *Candida albicans* infection. Infect Immun 74:4439–4451

Netea MG, Ferwerda G et al (2006) Recognition of fungal pathogens by toll-like receptors. Curr Pharm Des 12:4195–4201

Nordahl EA, Rydengard V et al (2004) Activation of the complement system generates antibacterial peptides. Proc Natl Acad Sci USA 101:16879–16884

Richardson MD (2005) Changing patterns and trends in systemic fungal infections. J Antimicrob Chemother 56[Suppl 1]:i5–i11

Romani L (2000) Innate and adaptive immunity in *Candida albicans* infections and saprophytism. J Leukoc Biol 68:175–179

Romani L (2004) Immunity to fungal infections. Nat Rev Immunol 4:1–23

Rooijakkers SH, Ruyken M et al (2006) Early expression of SCIN and CHIPS drives instant immune evasion by *Staphylococcus aureus*. Cell Microbiol 8:1282–1293

Sonesson A, Ringstad L et al (2007) Antifungal activity of C3a and C3a-derived peptides against *Candida*. Biochim Biophys Acta 1768:346–353

Walport MJ (2001a) Complement. First of two parts. N Engl J Med 344:1058–1066

Walport MJ (2001b) Complement. Second of two parts. N Engl J Med 344:1140–1144

Zelante T, Montagnoli C et al (2007) Receptors and pathways in innate antifungal immunity: the implication for tolerance and immunity to fungi. Adv Exp Med Biol 590:209–221

Zhang MX, Kozel TR (1998) Mannan-specific immunoglobulin G antibodies in normal human serum accelerate binding of C3 to *Candida albicans* via the alternative complement pathway. Infect Immun 66:4845–4850

Zipfel PF, Wurzner R et al (2007a) Complement evasion of pathogens: common strategies are shared by diverse organisms. Mol Immunol 44:3850–3857

Zipfel PF, Mihlan M, Skerka C (2007b) The alternative pathway of complement: a pattern recognition system. Adv Exp Med Biol. 598:80–92

12 Toll-Like Receptors and Fungal Recognition

Frank Ebel[1], Jürgen Heesemann[1]

CONTENTS

I. Introduction.......................... 243
II. The Innate Immune Response........... 243
III. The Family of Toll-Like Receptors 245
IV. The Role of Non-TLR Pattern Recognition Receptors in Antifungal Immune Responses.................... 247
V. Relevance of Distinct TLRs in the Immune Responses to Different Fungi............ 248
   A. Candida albicans.................... 249
   B. Aspergillus fumigatus 252
   C. Cryptococcus neoformans............ 254
   D. TLR-Mediated Recognition of Other Pathogenic Fungi................... 255
VI. Summary and Outlook................ 256
References........................... 257

I. Introduction

More than 100 000 fungal species exist and humans are commonly exposed to them. Despite this permanent encounter, only very few fungi are pathogenic to humans. Superficial skin infections caused by dermatophytes are common, but usually not severe and therefore not within the scope of this review. In contrast, some fungi are able to establish rare, but severe and life-threatening systemic mycoses. These fungal pathogens belong either to the so-called 'dimorphic fungi' (e.g. *Histoplasma capsulatum, Paracoccidioides brasiliensis, Blastomyces dermatitidis, Coccidioides immitis*) or are opportunistic fungal pathogens, like *Candida albicans, Aspergillus fumigatus* and *Cryptococcus neoformans*.

Infections with dimorphic fungi commonly start with the inhalation of fungal spores. The infectious morphotype is the yeast form, while hyphal growth is found under in vitro conditions at temperatures below 37 °C. Dimorphic fungi are endemic in the Americas and can cause systemic infections even in immunocompetent individuals. Opportunistic fungal pathogens can also cause systemic mycoses, but only in individuals with severe immunological deficiencies, like HIV patients, transplant recipients or patients with certain forms of cancer. The mortality rates associated with systemic mycoses are generally high, due to the still suboptimal diagnostic and/or therapeutic options. In addition, the number of patients at risk to acquire infections by opportunistic fungi is constantly rising as a result of medical progress, and therefore opportunistic fungal infections represent a severe problem in modern medicine.

II. The Innate Immune Response

Microbial infections are a common, serious threat to multicellular organisms and forced them to develop means to counteract this challenge. The sophisticated immune system found in vertebrates comprises two major branches: the adaptive and the innate immune system. In an adaptive immune response exposure to microorganisms leads to selection and clonal expansion of highly specific B- and T-cells, which help to eliminate the respective microbial pathogen and to establish a highly specific immunological memory. Although this process is extremely efficient, it requires several days in which the host depends on the alternative and evolutionary more ancient defense strategies of the innate immune system. This immune response relies on a limited number of germ line encoded, cell-bound or soluble receptors that detect microbes and induce their engulfment and elimination by a set of specialized phagocytic cells. These phagocytes, e.g. macrophages, neutrophils and dendritic cells, are not only responsible for the fast clearance of invading microbes, but are also required for the presentation of antigens, a feature which drives the process of clonal selection and expansion of B- and T-cells. This demonstrates the instructive role of the innate immune system

---

[1] Max-von-Pettenkofer-Institut, LMU München, Pettenkoferstrasse 9a, 80336 München, Germany; e-mail: ebel@mvp.uni-muenchen.de

for the adaptive immunity and underlines the interdependence of both branches of the immune system.

Those parts of the host which are exposed to the outside environment and are therefore especially susceptible to microbial infections, e.g. the gut or the lung, require an enhanced surveillance by the immune system. The specificity of the innate immune response has to be broad to deal with a wide variety of microbial challenges, but it relies only on a very limited number of receptors. Our knowledge on the molecular basis of innate immune recognition has increased substantially over the past decade. A very important step in this process was made by Medzhitov and Janeway (2000), who were the first to describe the basic principle of the innate immune recognition. According to this concept, germ line encoded 'pattern recognition receptors' (PRRs) recognize microbial patterns, which are commonly indicative for a certain subgroup of microorganisms. These 'pathogen-associated molecular patterns' (PAMPs) are structural components that are essential for the microbial survival and therefore conserved among many microorganisms (Table 12.1). An archetype of such a PAMP is lipopolysaccharide (LPS), which is an essential component of the outer membrane of Gram-negative bacteria and recognized by Toll-like receptor (TLR) 4 (Poltorak et al. 1998). As for LPS, such PAMPs are usually not pathogen-specific, but conserved between pathogenic and non-pathogenic microorganisms. It was therefore recently suggested to rename them 'microbe-associated molecular patterns' (MAMPs; Didierlaurent et al. 2002), but PAMP is still the commonly used term. Recognition of PAMPs by soluble or cell-bound PRRs may lead to different kinds of responses, depending for instance on the cell type involved. In general, enhanced phagocytosis and induction of an inflammatory response represent the major mechanisms that are triggered and that mediate the clearance of infection. However, activation of PRRs may also lead to the production of anti-inflammatory cytokines, like IL-10, which can result in an attenuation of an immune response. During infection signals derived from invading microorganisms are processed by several types of immune and non-immune cells, e.g. epithelial or endothelial cells, and this leads to a complex and orchestrated immune response that in its entirety is not easy to analyze.

The struggle of multicellular organisms with their microbial challengers started very early in evolution, and the horseshoe crab *Limulus polyphemus*, a so-called 'living fossil', represents a paradigm for an ancient immune system that already comprises elements and mechanisms characteristically found in the mammalian innate immune response (Iwanga and Lee 2005). Limulus recognizes LPS and β(1-3) glucan by soluble lectins that can activate an immune response cascade. The two PRRs, designated Factor C and Factor G, enable specific detection of Gram-negative bacteria and fungi. The horseshoe crab responds to the infection with a specific hemolyph clotting reaction. This highly specific and sensitive response is nowadays used in commercially available tests for detection and quantification of LPS and β(1-3) glucan (Muta 2006).

Table 12.1. Ligands recognized by human Toll-like receptors (TLRs). *LBP* lipopolysaccharide (LPS)-binding protein, *MD-2* myeloid differentiation protein 2, *PAMPs* pathogen-associated molecular patterns, * endogenous ligands

Toll-like receptor	Co-factors	Localization	PAMPs
TLR1	TLR2	Surface	Triacylated lipopeptides
TLR2		Surface	Peptidogycan, lipoproteins, lipopeptides, atypical LPS
TLR2	Dectin-1	Surface	Zymosan
TLR3		Endosome (?)	Double-stranded viral RNA
TLR4	MD-2, LBP	Surface	LPS, mannans, glucuronoxylmannan, glycoinositolphospholipids, viral fusion proteins, taxol, heat shock proteins*, fibrinogen*, fibronectin*
TLR5		Surface	Bacterial flagellin
TLR6	TLR2	Surface	Diacylated lipopeptides
TLR7		Endosome	Imidazoquinoline, single-stranded viral RNA
TLR8		Endosome (?)	Imidazoquinoline, single-stranded viral RNA
TLR9		Endosome	Non-methylated CpG DNA, chromatine immune complexes*
TLR10		Surface (?)	Unknown

III. The Family of Toll-Like Receptors

The Toll-like receptors (TLRs) represent the paradigmatic example of a PRR. The impulse for the identification of these receptors in humans came from work on a *Drosophila melanogaster* gene cassette that was originally shown to regulate the dorsoventral patterning of the fruit fly (Anderson et al. 1985). Later the *toll* gene, which is part of this cassette, turned out to be essentially required for protection against fungal infections (Lemaitre et al. 1996). Toll is a transmembrane protein comprising an extracellular domain with several leucin-rich repeats (LRRs) and an intracellular domain that shares significant homology with the intracellular domain of the human IL-1 receptor (Hultmark 1994). This Toll/IL-1 receptor (TIR) domain is required for the intracellular signalling of both receptors. The functional importance of Toll in the *Drosophila* immune response sparked the search for homologous human proteins (Rock et al. 1998). Eleven homologous sequences were identified in the human genome, of which ten encode for a functional protein. These proteins have been designated Toll-like receptors since they share all characteristic structural elements with *Drosophila* Toll.

Leucin-rich repeats are well characterized modules which are often engaged in protein-ligand interactions (Kobe and Deisenhofer 1995) and the LRRs present in the members of the TLR family are therefore likely (and in part proven) candidates for recognition of distinct PAMPs. The concave surface of the horseshoe-like structure of the LRR domains likely provides the interface for the interaction with the PAMP ligands. The recent elucidation of the structure of the ectodomain of human TLR3 furthermore revealed that the regular surface of the LRRs is in this case disrupted by two insertions that seem to be an essential element of the two putative PAMP binding sites of this receptor (Ben et al. 2006).

Ligation of PAMPs to the TLR ectodomains facilitates receptor dimerization, which represents a very early step in the signal transduction mechanism. The further signalling events require the myeloid differentiation marker protein MyD88 which also contains a TIR domain (Hultmark 1994) that interacts with the cytoplasmic TIR domain of the TLRs (Xu et al. 2000). The 'TIR domain-containing adaptor protein' TIRAP6 is involved in the MyD88-dependent signalling pathway of TLR2 and TLR4, whereas the 'TIR domain-containing adaptor inducing interferon-β' (TRIF) and the 'TRIF-related adaptor molecule' (TRAM) are involved in the MyD88-independent signalling of TLR3 and TLR4, leading to specific immunological responses upon each TLR stimulation (Fig. 12.1).

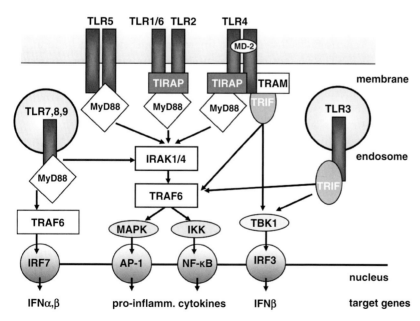

Fig. 12.1. TLR signalling pathways. *IRAK* Interleukin-1 receptor I-associated kinase, *IRF* IFN regulatory factor, *IFN* interferon, *MD-2* myeloid differentiation protein 2, *MyD88* myeloid differentiation marker 88, *TBK1* TANK-binding kinase 1, *TIR* Toll/IL-1 receptor, *TIRAP* TIR domain-containing adaptor protein, *TNF* tumor necrosis factor, *TRAF* TNF receptor-associated factor, *TRAM* TRIF-related adaptor molecule, TRIF TIR domain-containing adaptor inducing IFNβ

The TRIF-independent pathways converge at a complex comprising the 'TNF receptor-associated factor 6' (TRAF6) and the 'interleukin-1 receptor-associated kinase' (IRAK). The signalling downstream of the TRAF6/IRAK complex is highly conserved and culminates in the translocation of the nuclear factors NF-κB and AP-1 to the nucleus and the subsequent activation of certain sets of genes implicated in an inflammatory immune response (Akira and Takeda 2004). A very similar signalling cascade is triggered by the Toll receptor in Drosophila (Silverman and Maniatis 2001), with one interesting difference to the situation in mammals: TLRs are supposed to directly interact with their respective PAMPs, whereas Toll binds the Drosophila protein Spätzle, which is present in the Drosphila body fluid and serves as a kind of extracellular indicator protein. To be able to bind to Toll, Spätzle has to be processed by a serine protease that is activated during infection (Ligoxygakis et al. 2002). Thus, Toll does not function as a PRR according to the concept of Medzhitov and Janeway, but is merely a building block in the signalling cascade that originates from a still unknown Drosphila molecule that senses the microbial infection.

Of the ten functional human TLRs at least some seem to be able to recognize several, biochemically diverse PAMPs (Kawai and Akira 2005). This flexible interaction with different ligands might be facilitated by the engagement of different binding sites on one receptor (Ben et al. 2006) and/or the ability of certain TLRs to form functional heterodimers, e.g. TLR1/TLR2 and TLR2/TLR6 (Ozinsky et al. 2000). Since several PRRs are usually engaged in recognition of a certain microbial pathogen, the resulting cross-talk between these receptors and their signalling pathways may also contribute to the broad spectrum of possible outcomes.

To our current knowledge, TLR2 and TLR4 are the most promiscuous TLRs and both seem to be particularly important for the recognition of PAMPs that are localized on microbial surfaces. TLR2- and TLR4-signalling pathways have been shown to interact in a synergistic way to enhance and balance an inflammatory response (Sato et al. 2000). Depending on their specific function members of the TLR family are differentially expressed and localized within different types of cells. TLR3, TLR7 and TLR9, which recognize double-stranded RNA and so-called CpG DNA motifs (Hemmi et al. 2000; Alexopoulou et al. 2001; Lund et al. 2004), are exclusively found in endosomes, while TLR2 and TLR4, which either alone or in combination have been implicated in the immune response to several bacterial and fungal pathogens, are found on the cellular surface (Takeda and Akira 2005). However, depending on the cell type, both receptors may also reside either in part or exclusively in certain intracellular compartments (Hornef et al. 2002; Meng et al. 2004). During infection these TLRs may re-localize to phagosomes containing engulfed microbes and the phagosome is likely to represent the principal stage for the recognition of PAMPs by TLR2 and TLR4 (Underhill and Gantner 2004).

TLR-mediated recognition of microbes by macrophages usually leads to the release of pro-inflammatory molecules, which recruit and activate neutrophils and other macrophages (Takeda and Akira 2005). The pattern of the released cytokines will also drive the immune response towards an either T helper cell (Th) 1- or Th2-like type. Dendritic cells are of particular importance in this process, as they stand at the cross-roads of innate and adaptive immunity and their response decisively determines whether an anti-fungal immune response is protective or not (Romani et al. 2002; Mazzoni and Segal 2004). The TLR-induced modulation of phagocytosis by TLR signalling (Underhill and Gantner 2004; Luther et al. 2007) and the production of anti-microbial products, like reactive oxygen species or nitric oxide, provide further and more direct means for the clearance of microbial infections. In general, an optimal anti-fungal immune response combines oxidative and non-oxidative mechanisms, the latter consisting of an efficient phagosomal maturation, the release of anti-fungal effector molecules by degranulation and the sequestration of iron.

Macrophages and neutrophils are the main effector cells responsible for the elimination of fungal pathogens; and both have been shown to engulf single fungal cells and to produce reactive oxygen species for microbial killing. During fungal infections, neutrophils are of particular importance for the elimination of multicellular fungal organisms that escape phagocytosis, but are susceptible to effector molecules released by degranulation (Schaffner et al. 1982).

As outlined above, an immune response has to be regulated in a way that leads to the activation of those effector cells and mechanisms required for the elimination of the respective pathogen.

However, during infection, the inflammatory response has to be limited and controlled to avoid deleterious side effects and after elimination of the invading microorganism the immune response has to be brought back to homeostasis by anti-inflammatory mechanisms. Interestingly, certain TLRs have also been shown to participate in this process (Netea et al. 2006b) and other means exist which are capable of down-regulating a sustained TLR response, for which LPS tolerance is the most prominent example (Takeda and Akira 2005). In conclusion, TLRs can be engaged in the activation, modulation and repression of an inflammatory response.

IV. The Role of Non-TLR Pattern Recognition Receptors in Antifungal Immune Responses

The complement system comprises a complex, tightly regulated network of a multitude of soluble factors as well as a set of corresponding cell bound receptors. Complement has been shown to be an important element in the innate immune response to systemic fungal infections; and several components of the fungal cell wall have been implicated in complement activation, e.g. β glucan, galactomannan or glucuronoxylomannan (for a recent review, see Speth et al. 2004; also see Chapter 11 in this volume). Although fungi are resistant to the membrane attack complex, complement factors can enhance the phagocytic clearance by opsonization and additionally attract and activate phagocytes. Specific binding of factor H and FHL-1, two regulatory components of the complement cascade, to the surface of *Can. albicans* may therefore provide an important mean for immune evasion (Meri et al. 2002).

The major components of the fungal cell wall, e.g. β(1–3) glucan, chitin and different kinds of mannans, are not found in humans and therefore constitute potential PAMPs (Bowman and Free 2006). In recent decades, several soluble or cell-bound lectins have been described which recognize fungal pathogens (Brown 2006). The family of collectins, for example, comprises several prominent lectin PRRs, like the mannose-binding lectin (MBL), the surfactant proteins A and D and the long pentraxin-3 (PTX3). Collectins are calcium-dependent (C-type) trimeric lectins which bind to surface carbohydrates of pathogenic fungi and enhance their clearance (van de Wetering et al. 2004). The collectins surfactant protein A and D are present in the lumen of the lung and they bind to the surface of fungal pathogens. Opsonization by surfactant proteins can have different effects: (a) it may enhance the phagocytic uptake, as it has been reported for *A. fumigatus* conidia (Madan et al. 1997), (b) it may have no effect on phagocytosis (Walenkamp et al. 1999) or (c) it may even hamper phagocytosis by the formation of large fungal aggregates, as has been reported for *Can. albicans* and *Cry. neoformans* (van de Wetering et al. 2004). PTX3 is a long pentraxin that directly binds to the surface of *A. fumigatus* conidia and is essentially required for protection against *A. fumigatus* infections in mice (Garlanda et al. 2002).

Cell-bound mannose receptors have been implicated in the phagocytosis of fungi (Newman and Holly 2001). More recent data, however, indicate that the human mannose receptor is not a professional phagocytic receptor, but rather a binding molecule which mediates the endocytic uptake of soluble ligands and not the phagocytic internalization of particulate ligands, like zymosan (Le Cabec et al. 2005).

More recently three additional C-type lectins have been implicated in anti-fungal immunity: the dendritic cell-specific intercellular adhesion molecule 3-grabbing non-integrin (DC-SIGN), the SIGN-related 1 (SIGNR1) and dectin-1. DC-SIGN is a type II C-type lectin that recognizes *A. fumigatus* conidia and triggers their phagocytic uptake by dendritic cells and macrophages. This interaction could be blocked by *A. fumigatus* galactomannan or mannan preparations derived from other fungi (Serrano-Gomez et al. 2004). SIGNR1 has been shown to be engaged in the non-opsonic recognition of fungal particles by macrophages, although it was found to be only poorly phagocytic (Taylor et al. 2004). Specific binding to mannans has been established for DC-SIGN and SIGNR1 (Mitchell et al. 2000), while the third novel lectin, Dectin-1, has been identified as the major surface receptor for β(1–3) glucan on dendritic cells and macrophages (Brown et al. 2002). Several studies demonstrated a pivotal role for this non-TLR PRR in the recognition and internalization of different fungi (Brown 2006). Recent data suggest that apart from β(1–3) glucan, dectin-1 also recognizes PTX3 opsonized particles (Diniz et al. 2004). Binding of PTX3 to zymosan or *Paracoccidioides brasiliensis* yeast cells was shown to strongly enhance the dectin-1-dependent phagocytic uptake by mouse macrophages.

The importance of β(1–3) glucan as a particular fungal PAMP is underlined by recent data demonstrating a elaborate fungal immune evasion strategy. It has been known for some time that *Histoplasma capsulatum* strains or mutants lacking α(1–3) glucan have a reduced virulence (Klimpel et al. 1988; Marion et al. 2006). Rappleye and co-workers have now demonstrated that an outermost layer of α(1–3) glucan prevents recognition of the major cell wall constituent β(1–3) glucan by dectin-1 (Rappleye et al. 2007). This camouflage has been proposed as a determinant essentially defining the different pathogenic potentials of dimorphic fungal pathogens and opportunistic fungi. While a huge body of evidence supports the concept of dectin-1 being a crucial PRR for anti-fungal immunity, recent analysis of dectin-1 knock-out mice came to controversial results on the relevance of dectin-1 for certain fungal infections. Taylor and co-workers (2007) reported that dectin-1 deficient mice are susceptible to systemic *Can. albicans* infections accompanied by a reduced ability to recruit neutrophils and inflammatory monocytes to the site of infection. In contrast, Saijo et al. (2007) found that dectin-1 is not essential for clearance of *Can. albicans* infections, but is required for resistance against *Pneumocystis carinii*. The cytokine production during *Pneumocystis* infection was similar in dectin-1 knock-out and wild-type mice, but further analysis revealed that dectin-1-deficient macrophages showed a defective production of reactive oxygen species. Whether differences in the fungal strains or the infection protocols account for the contradictory results with respect to *Candida* infections remains to be determined. The earlier finding that deletion of CARD9, a downstream signalling molecule of dectin-1, leads to susceptibility to *Candida* infection (Gross et al. 2006) supports the concept of dectin-1 being required to mount an efficient anti-*Candida* immune response.

Recent research, starting with the results of Gantner et al. (2003), provided a large body of experimental evidence suggesting a tight functional interaction between dectin-1 and TLR2, which exemplifies a cross-talk between two structurally unrelated PRRs. Further collaborations between members of the TLR family and innate immune receptors have been described in the recent review by Mukhopadhyay et al. (2004). These data provide a first glimpse into the complex network of signalling receptors present on the surface of immune cells that drives and controls the immune response to microbial pathogens.

V. Relevance of Distinct TLRs in the Immune Responses to Different Fungi

The innate immune response is crucial for the control of fungal infections and the role of TLRs in anti-fungal immunity has been covered by several recent reviews (Levitz 2004; Netea et al. 2004; Roeder et al. 2004b; Romani 2004). Systemic fungal infections usually start either with the inhalation of fungal cells or from a persistent colonization of a mucosal or cutanous surface. During systemic spread, the fungus must be able to survive the attacks of immune cells and to breach epithelial or endothelial barriers. Survival in the infected tissue requires further adaptive abilities, e.g. the competition with the host for ferric iron or the evasion of immune defense mechanisms.

The pathogenic dimorphic fungi are exogenous pathogens that are usually taken up by inhalation and are capable of causing a systemic infection even in the immunocompetent host. The development of molecular biological tools for analysis of these pathogens and the availability of genomic information allowed the identification of virulence mechanisms employed by these pathogens to overcome the host pulmonary defenses (Rappleye and Goldman 2006). It appears likely that recognition by TLRs and other PRRs determines the outcome of infection. However, thus far only one report has been published on the interaction of a dimorphic fungus with TLRs. In this study, Viriyakosol and co-workers (2005) demonstrate that TLR2 plays an important role in the innate immune response to *Coccidioides posadasii*.

Our knowledge on the PRRs engaged in the defense of opportunistic fungal pathogens is much broader and the role of TLRs, especially TLR2 and TLR4, has recently been analyzed in much detail. Several studies have been published analyzing the role of TLRs in the immune response to *Can. albicans*, *A. fumigatus* and *Cry. neoformans*, and in less detail also to several other opportunistic fungi. It is generally accepted that a Th1 type immune response is required for clearance of fungal infections and that TLRs play a major role in modulating the anti-fungal immune response (Romani 2004). However, as outlined below, there is still a debate on the relative importance of certain TLRs (and non-TLR receptors) for the recognition of different opportunistic fungi. This problem is related to the fact that the relevant fungal TLR ligands have not been precisely defined yet.

A. *Candida albicans*

Candida is a commensal microorganism that is not typically found in the environment, but colonizes the human gastrointestinal tract and in particular its mucosal surfaces. It therefore represents the prototype of an opportunistic fungal pathogen causing endogenous infections. Severe candidiasis requires a local or systemic impairment of the immune surveillance which results in either a local and superficial or a severe and systemic infection. *Can. albicans* can switch from a yeast-like morphotype to hyphal growth, a process which has been discussed as a major virulence trait. During the past decades molecular genetic studies have confirmed this hypothesis and further demonstrated that genes governing cellular morphology are co-regulated with genes encoding conventional virulence factors such as proteases and adhesins (Kumamoto and Vinces 2005). The transcriptional regulatory networks of *Can. albicans* thus ensure that virulence factors are produced in hyphae, but not in the yeast form. Hyphae are able to exert a mechanical force, aiding the penetration of epithelial and endothelial barriers, which is a prerequisite for an efficient systemic spread of infection. The tightly regulated hyphal morphogenesis therefore represents an integral part of the overall virulence strategy of *Can. albicans*.

Although most of the research activities are still focussed on *Can. albicans* as a model organism for an opportunistic fungal pathogen, other *Candida* species are also well known to cause severe infections in immunocompromised individuals. These strains are of increasing importance due to the fact that the incidence of non-albicans infections is rising, at least in certain parts of the world, and that certain non-albicans species have an enhanced resistance to commonly used antimycotics, like fluconazol (Perfect 2004). Interestingly, *Can. glabrata* has only a limited ability to form hyphae, but is still pathogenic (Csank and Haynes 2000). This is of special interest, because changes in the fungal surface which normally accompany the infection process have severe consequences for recognition by PRRs and consequently for setting up an appropriate immune response, as is discussed in more detail below.

The finding that the TLR adaptor protein MyD88 is required for the response of macrophages to *Can. albicans* (Marr et al. 2003) and for the mouse resistance to *Candida* infections (Bellocchio et al. 2004; Villamon et al. 2004c) has already suggested that certain TLRs play an important role during systemic candidiasis. Due to the availability of mice defective in either TLR2 or TLR4, research was mainly focussed on these two receptors. In the initial study, Netea and co-workers (2002) found that the absence of TLR4-mediated signals rendered mice highly susceptible to disseminated candidiasis, whereas in a following study, mice lacking TLR2 turned out to be less affected (Netea et al. 2004). This suggested that signalling through either of these two TLRs alone may govern the immune response in different directions. However, both kinds of infected mice showed normal levels of TNFα, indicating that lack of a single TLR was not sufficient to completely abrogate a proinflammatory immune response. Instead, the susceptibility of TLR4-deficient mice was attributed to an impaired production of the chemokines 'keratinocyte-derived chemokine' (KC) and the 'macrophage inflammatory protein-2' (MIP-2) and consequently an impaired recruitment of neutrophils (Netea et al. 2002), whereas the resistance of TLR2-deficient mice correlated with a reduced production of the anti-inflammatory cytokine IL-10 and a decrease in the regulatory T-cell population (Netea et al. 2004a). These findings suggested a protective role for TLR4, whereas a TLR2-dominated anti-inflammatory signalling seemed to play a deleterious role during systemic candidiasis. However, in human mononuclear cells blocking of TLR2, but not of TLR4, led to a significant inhibition of TNFα production (Netea et al. 2002), implicating TLR2, but not TLR4 in the pro-inflammatory response to *Candida* infections in humans. These data demonstrate that results obtained in the mouse system cannot automatically be transferred to the situation in humans. However, the importance of TLR4 in anti-*Candida* immune response in humans has recently been underlined by the finding that TLR4 Asp299Gly/Thr399Ile polymorphisms are a risk factor for *Candida* bloodstream infections (Van der Graaf et al. 2006).

For a detailed understanding of the roles of certain PRRs in the immune response to certain microbial pathogens it is crucial to define the respective PAMPs that are recognized. In the case of fungal pathogens such PAMPs most likely reside within the cell wall, a complex and carbohydrate-rich structure (Bowman and Free 2006) that is unique to fungi and exposed to the surrounding environment. The outer layer of the *Candida* cell wall mainly consists of proteins which are glycosylated with either O- or N-linked carbohydrates, while β(1–3) glucan represents the major constituent of the underlying meshwork (Ruiz-Herrrera et al. 2006). Mannans have been demonstrated to induce the production of proinflammatory cytokines by TLR-4 signalling and might therefore represent a major fungal PAMP (Tada et al. 2002). Using a set of *Candida* mutant strains with defined defects in cell wall synthesis, Netea et al. (2006a) recently provided evidence that O-linked mannans are recognized by TLR4, whereas recognition of N-linked mannan is mediated by the mannose receptor. Mutants with defects in N-linked carbohydrate induced a severely impaired production of proinflammatory cytokines in human mononuclear cells and mouse macrophages. Similar, but less dramatic results were obtained with mutants harboring truncated O-linked mannans suggesting that the pro-inflammatory response to *Candida* can largely be attributed to TLR4- and mannose receptor-dependent signals. The residual mannan-independent proinflammatory cytokine production appears to be triggered by recognition of β(1–3) glucan through the dectin-1 receptor, a process which requires TLR2 for full efficiency (Netea et al. 2006a). This involvement of TLR2 in the dectin-1 mediated recognition of β(1–3) glucan provides an explanation for the reduced production of TNFα that was previously observed for human mononuclear cells treated with anti-TLR2 antibodies (Netea et al. 2002).

The study of Netea et al. (2006) is of special interest, as it identifies several ligands for PRRs and analyzes their relative contribution to an anti-*Candida* immune response. These data suggest that the use of mutants defective in defined structures or components of the cell wall provides a promising approach that is likely to gain detailed insights into the interplay between immune receptors and their PAMP counterparts in the cell wall.

The evidence for a important or even dominant role of mannan-receptor- and dectin-1-dependent signals for cytokine production underlines the importance of non-TLR receptors in the immune response to *Candida*, but these data also point to the complexity of the signalling network which in the response to every single microorganism consists of a cross-talk and interplay of different Toll-like and non-Toll-like receptors. As already mentioned above, two papers have been published very recently describing the phenotype of dectin-1 knock-out mice. One report came to the conclusion that dectin-1 has a fundamental function in antifungal immunity, evidenced by the susceptibility of such mice to systemic *Can. albicans* infections (Taylor et al. 2007), whereas the other revealed no obvious role for dectin-1 in an anti-*Candida* immune response, but in the host defense against *Pneumocystis carinii* infections (Saijo et al. 2007). The latter study furthermore provided evidence that the production of inflammatory cytokines in response to zymosan is largely due to a MyD88- (and likely TLR-) dependent signalling, whereas dectin-1-mediated signals are of less importance. These data should remind us that zymosan is not a homogenous β(1–3) glucan particle, as sometimes supposed, but a complex structure containing other components, including mannans, other glucans and chitins (Di Carlo et al. 1958).

While it is generally assumed that recognition via TLR4 favors the production of pro-inflammatory cytokines, the situation is less clear for TLR2. Like TLR4, this receptor is able to trigger the production of pro-inflammatory cytokines (Hirschfeld et al. 2001) and stimulation of an inflammatory response by TLR2 is especially prominent in response to Gram[+] bacteria (Takeuchi et al. 1999) and when TLR2 recognizes fungal PAMPs in concert with dectin-1 (Gantner et al. 2003). However under certain conditions TLR2 signalling can also provide a strong anti-inflammatory stimulus (Netea et al. 2004b). Recent data provided evidence that *Candida* yeasts and blastoconidia are recognized by TLR4, while hyphae are not (d'Ostiani et al. 2000; Van der Graaf et al. 2005); and it has been concluded that the resulting inability of hyphae to stimulate IL-12 or IFNγ production represents an immune evasion strategy (Netea et al. 2004b). Lack of TLR4-mediated signals during invasive hyphal growth may shift the balance towards a TLR2-mediated and more anti-inflammatory IL-10 dominated signalling.

Such TLR2-dominated signalling furthermore seems to stimulate the clonal expansion of CD4+ CD25+ regulatory T-cells (Treg; Netea et al. 2004), which are well known to down-regulate an inflammatory response. While this is a crucial step in bringing back a successful immune response to homeostasis, it can be deleterious during a still unsolved infection. This assumption is confirmed by the finding that depletion of Tregs improved the resistance of mice to systemic candidiasis (Netea et al. 2004). Overwhelming septic immune responses are typically found in systemic bacterial infections while they are less common in patients suffering from systemic mycoses. This may result from differences in the regulation of the immune responses to bacterial and fungal pathogens. It might be possible that, in comparison to bacterial infections, there is not such an urgent need for a strong anti-inflammatory signalling after clearance of fungal pathogens, whereas the limited pro-inflammatory signals induced by many fungal pathogens might be essentially required to establish a successful immune response.

From the data summarized so far a picture emerges for the immune response to systemic candidiasis that seems to be consistent, but there are still several points that need to be addressed in more detail. One question is whether or not surface exposed short O-linked mannans are present on the hyphal surface or whether they are shielded from TLR4 by certain, yet unknown surface structures. There are data suggesting that mannoproteins exist in the cell wall of Candida yeasts and hyphae, which contain O-linked mannans and might in principle be recognized by TLR4. Along this line are recent data demonstrating binding of the soluble mannan-binding lectin to Candida yeasts and hyphae in the infected tissue (Lillegard et al. 2006).

The data of Netea and co-workers provide strong evidence that TLR4 is required for protection to candidiasis, while TLR2 signalling is dispensable or under certain conditions even harmful for Candida-infected mice. However, the relative importance and function of TLR2 and TLR4 in the immune response to systemic Candida infections is still a matter of discussion. Villamón and co-workers (2004a) found that TLR2-deficient mice are more susceptible to candidiasis and Murciano et al. (2006) reported that TLR4 is dispensable in a mouse model of a hematogenously disseminated Candida infection. Consistent data obtained from in vitro experiments provided evidence that the production of proinflammatory cytokines to Candida yeasts and hyphae is impaired in TLR2-, but not in TLR4-deficient macrophages (Villamón et al. 2004a; Gil and Golzalbo 2006). Taken together, these data emphasize the importance of TLR2, but not TLR4, for the clearance of Candida infections. Additionally, it has been shown that the low molecular weight phospholipomannan of Can. albicans binds to macrophages and induces the production of TNFα in a TLR2-dependent manner (Jouault et al. 2003), again supporting the concept of TLR2 as a major factor involved in the inflammatory response to Candida.

In a study by Bellocchio and co-workers (2004) mice defective in either MyD88, TLR2, TLR4, TLR9 or IL-1 receptor were infected with yeasts or hyphae of Can. albicans. All mice infected intravenously with the highly pathogenic hyphal form succumbed to infection, whereas TLR4- and TLR9-deficient mice survived significantly longer. In contrast, only MyD88- and IL-1 receptor-deficient mice died after intravenous infection of the less virulent Can. albicans yeasts. These data suggest that neither TLR4, nor TLR2 is essentially required in this mouse model of systemic candidiasis and consistently, neutrophils from TLR2- and TLR4-deficient mice were not impaired in their antifungal activity to Candida yeasts and hyphae (Bellochio et al. 2004).

Mice having survived a primary infection with Candida yeasts were also re-infected with a normally lethal dose of Candida hyphae. While wild-type and TLR9-deficient mice were protected and survived this challenge, most TLR2- and TLR4-deficient mice died, suggesting a pivotal role for both receptors in the formation of an immunological memory (Bellochio et al. 2004). However, using a similar experimental setup, a more recent study came to the conclusion that TLR2 is dispensable for protection of mice immunized with a sub-lethal challenge of Can. albicans (Villamón et al. 2004b). The influence of anti-mycotica treatment on the immunological recognition of Can. albicans was analyzed in a study by Roeder et al. (2004a). Evidence was reported that anti-mycotica treatment has a strong impact on the relative importance of TLR2- and TLR4-signalling. This might result from differential surface exposition of certain cell wall components in anti-mycotica treated versus untreated Candida cells.

In summary, the data published so far on the relative importance of TLR2 and TLR4 in the mouse and human immune response to *Can. albicans* infections are not fully consistent. The reasons for the observed discrepancies can be multifaceted, but it is likely that differences in the strains used contribute to the controversial results. Subtle differences in the organization and regulation of *Candida* cell wall components might occur and have to be taken into account. Comparative studies using different strains might therefore provide new and important insights and could help to solve the current discrepancies.

B. *Aspergillus fumigatus*

Aspergillus fumigatus is a saprophytic fungus commonly found in the soil and on decaying organic matter. The ubiquitous presence of this mold in the environment results from the efficient distribution of its conidia through the air. These spores are taken up by inhalation and reach the alveoli of the lung due to their small size. Although humans inhale large numbers of *A. fumigatus* conidia every day, life-threatening, systemic *Aspergillus* infections are rare and restricted to severely immunocompromised patients, a fact which reflects the efficient surveillance of the lung by the innate immune system. Interestingly, recent findings nevertheless suggest that sub-clinical *A. fumigatus* infections may occur and precede hospitalization and immune suppression (Sarfati et al. 2006). Latent *Aspergillus* infections may therefore represent an underestimated risk factor for immunocompromised patients vulnerable to develop an invasive aspergillosis.

Alveolar macrophages are primarily responsible for the elimination of inhaled conidia, as they are able to track down and kill them by phagocytosis. Degradation of the ingested spores in the phagolysosome and the production of reactive oxygen species are means that have been implicated in the killing of engulfed conidia (Ibrahim-Granet et al. 2003; Philippe et al. 2003). In contrast to macrophages, neutrophils have to be recruited to the site of infection. They form a second line of defense and possess the unique ability to attack and kill not only conidia and germtubes, but also hyphae, which may escape phagocytosis due to their size. Neutrophils can recognize such elongated fungal elements by PRRs (Bellocchio et al. 2004b) and kill them by the degranulation and release of aggressive molecules, e.g. reactive oxygen species (Schaffer et al. 1982). Natural killer (NK) cells represent another type of immune cells that can be recruited to the *Aspergillus*-infected lung by the release of the 'chemokine ligand monocyte chemotactic protein-1', also designated 'chemokine (C-C motif) ligand-2' (MCP-1/CCL2). Recent data by Morrison et al. (2003) demonstrate that NK cells represent a previously unrecognized and critical element in the early host defense mechanism during invasive aspergillosis and contribute to the residual anti-fungal protection of neutropenic mice. Macrophages and dendritic cells can both engulf and kill fungal cells and can additionally communicate with the adaptive immune system. In this respect, dendritic cells, are of pivotal importance as they largely determine whether a protective pro-inflammatory Th1-like response is induced or not (Romani et al. 2002).

Like *Can. albicans*, *A. fumigatus* undergoes a morphological transition during infection, leading from small and metabolic inactive conidia to elongated, growing hyphae. Further intermediate morphotypes, like swollen conidia or germlings, also differ in their surface properties, and even hyphae expose certain molecules or structures only in restricted parts of their surface (Momany et al. 2004). This results in a complex pattern of biochemical distinct surface structures that changes substantially during infection, being mainly proteinaceous in resting conidia (Paris et al. 2003), but rich in carbohydrates in hyphae, germlings and swollen conidia (Bernard and Latge 2001). Consequently, the relative importance of certain PRRs may differ for the various morphotypes and the immune response to *A. fumigatus* therefore relies on a dynamic pattern of PRR-PAMP interactions.

While *Candida* has co-evolved with its human host over a long time, the situation is different for *A. fumigatus*. This mold is primarily an environmental and saprophytic organism. Severe systemic infections are rare and merely caused by the increased number of susceptible human hosts due to recent progress in modern medicine. It therefore appears unlikely that *Aspergillus* infections in humans had a substantial impact on the evolution of this fungus. It is, nevertheless, striking that from more than 200 *Aspergillus* species, only *A. fumigatus* is responsible for over 90% of all cases of invasive aspergillosis (Latge 1999).

This indicates that *A. fumigatus* indeed must have some distinct properties that facilitate a pathogenic life in the immunocompromised human host. The recent availability of genome sequences for *A. fumigatus* (Nierman et al. 2005), the non-pathogenic *A. oryzae* and *A. nidulans*, as well as fungal pathogens, like *Can. albicans* and *Cry. neoformans* enabled a search for characteristic traits of fungal virulence (Galagan et al. 2005). *A. fumigatus* seems to lack specific pathogenic elements, but is instead well equipped with numerous efflux pumps, an elaborate apparatus for environmental sensing, mechanisms and molecules to counteract reactive oxidants and an ability to retrieve and exploit a broad spectrum of nutrients, which, in combination enable this saprotrophic mold to survive in the unfriendly environment of the human host (Tekaia and Latge 2005).

A better understanding of the mechanisms that lead to progression of an *A. fumigatus* infection in the immunosuppressed host is clearly an urgent task. An important step in this direction is the recent attempts to unravel the interactions between the different morphotypes of this mold and the immune cells of the immunocompetent and immunocompromised host. In two early reports, Taramelli et al. (1996) and Grazziutti et al. (1997) found that *A. fumigatus* conidia and hyphae induce an inflammatory cytokine response in human and murine macrophages. A first attempt to identify the underlying PRR-PAMP interactions was performed by Wang et al. (2001). Using human adherent monocytes and blocking monoclonal antibodies, those authors showed that *A. fumigatus* hyphae trigger an inflammatory response through TLR4-, but not TLR2-signalling. This process was shown to be dependent on CD14, a protein that is known to be engaged in the TLR4-mediated recognition of LPS (Chow et al. 1999).

In the following years, several studies analyzed the response of macrophages derived from either TLR2- and/or TLR4-deficient mice to *A. fumigatus* hyphae and conidia. The published data are in part controversial and range from a specific requirement for TLR2, but not TLR4 (Mambula et al. 2002), a requirement for TLR4 and to a lesser extent for TLR2 (Meier et al. 2003), to a mainly TLR4-dependent response to conidia, but not to hyphae (Netea et al. 2003). The latter data suggested an immune evasion strategy of *A. fumigatus* hyphae similar to that proposed for *Candida* (Netea et al. 2004b).

While the relative importance of TLR2 and TLR4 has been analyzed in some detail, only one study has been published so far that attempted to define the relative importance of the other members of the human TLR family. From a set of human HEK293 cells transfected with any of the ten human TLR genes only those expressing TLR2 or TLR4 showed a significant response to *A. fumigatus* (Meier et al. 2003). Experiments with transfected cells are commonly performed as they provide a mean to study the relevance of single receptors. However, cells used for transfection are usually not part of an anti-fungal immune response and care has to be taken in the discussion of the resulting data. Transfected cells likely provide an appropriate model for many TLRs, but might be less suitable to study the relevance of intracellular TLRs, which recognize PAMPs that have to be released from the respective microorganisms. TLR9 recognizes non-methylated CpG-motifs which are commonly found in bacterial, but not in human DNA. TLR9 might theoretically be able to recognize also *Aspergillus* DNA, which seems to lack methylation of CpG motifs (Antequera et al. 1984). However recognition of *Aspergillus* CpG motifs by TLR9 would require fungal lysis in the phagosome of professional phagocytes, which is unlikely to occur in HEK293 cells. In the transfection experiments mentioned above, this problem could have been bypassed by the use of hyphal fragments, which may have released sufficient DNA, but additional experiments are clearly required to rule out a potential involvement of TLR9 in the recognition of *A. fumigatus*.

Evidence arguing against an important role of most TLRs came from experiments with mouse macrophages that were defective in both TLR2- and TLR4-signalling. These phagocytes were not able to respond to *A. fumigatus* using different read-out parameters (Meier et al. 2003) suggesting that other TLRs, apart from TLR2 and TLR4, are of only minor importance for recognition of *A. fumigatus* by isolated mouse macrophages.

According to a recent study, not only intact *A. fumigatus* cells, but also released *A. fumigatus* molecules have the capacity to activate innate immune cells through TLR2- and TLR4-signalling (Braedel et al. 2004). This finding opens up the interesting perspective to isolate and identify soluble *Aspergillus* PAMPs that are released from the fungal cell wall during growth.

As summarized above, the relative importance of TLR2 and TLR4 for the macrophage-mediated response to different *A. fumigatus* morphotypes is still a matter of discussion. The numerous reasons which may account for the sometimes controversial results obtained in the different studies have been discussed recently (Luther and Ebel 2006).

Infection experiments in mice defective in certain PRRs provide an important proof to test the relevance of certain receptors for the clearance of infection. So far two studies have been undertaken to analyze the role of TLR2 and TLR4 in a mouse model of systemic aspergillosis. Both sets of data indicate that immunocompetent TLR4- and TLR2-deficient mice showed a normal resistance to *A. fumigatus* infections, whereas an immunosuppressive treatment rendered them susceptible to infection (Bellocchio et al. 2004a; Balloy et al. 2005). This suggests that, in the immunocompetent host, resident phagocytes in the lung are able to clear the infection without any need for a recruitment of other effector cells, which likely requires a TLR-dependent signalling. After immunosuppression the residual capacities of the phagocytes has to be well orchestrated and coordinated to achieve protection and, in this context, TLR2 and TLR4 are obviously required. In line, a recent study with human patients revealed that certain polymorphisms in TLR1 and TLR6 are associated with a higher risk to develop an invasive aspergillosis after allogenic stem cell transplantation (Kesh et al. 2005). Since TLR1 and TLR6 are well known to form functional heterodimers with TLR2 (Ozinsky et al. 2000) these data also point to an important role of TLR2 in the anti-*Aspergillus* immune response.

Recently, the role of dectin-1 in the immune response to *A. fumigatus* was analyzed in some detail. Several studies came to the conclusion that recognition of β(1–3) glucan by dectin-1 is essentially required to stimulate a proinflammatory cytokine response and to trigger the recruitment of neutrophils to the site of infection. All studies revealed that the proinflammatory response to germinating conidia is much more pronounced than to resting conidia, which correlated well with differences in the surface display of β(1–3) glucan in both kinds of spores (Hohl et al. 2005; Steele et al. 2005; Gersuk et al. 2006). Consistent with the previous finding that dectin-1 can trigger phagocytosis of fungal particles (Herre et al. 2004) it was shown that dectin-1 is also engaged in phagocytosis of both *A. fumigatus* germlings and conidia (Gersuk et al. 2006; Luther et al. 2007).

Evidence for a collaborative induction of inflammatory responses by dectin-1 and TLR2 came originally from a study by Gantner et al. (2003) focussing on the response of macrophages to β(1–3) glucan-rich zymosan particles. Subsequent studies provided supportive evidence for a cross-talk between TLR2 and dectin-1, e.g. in the immune response to the bimorphic pathogenic fungus *Coccidioides posadasii* (Viriyakosol et al. 2005); and this has recently been confirmed for the immune response to *A. fumigatus* (Hohl et al. 2005; Gersuk et al. 2006). Evidence from other experimental systems already suggested that TLR signalling may also modulate phagocytosis (Underhill and Gantner 2004). For *A. fumigatus* TLR2-signalling, but not TLR4-signalling, was found to positively modulate the phagocytic uptake of conidia by macrophages (Luther et al. 2007), which suggests that the cross-talk between TLR2 and dectin-1 is not restricted to the production of proinflammatory cytokines.

C. *Cryptococcus neoformans*

Cryptococcus neoformans is the etiological agent of cryptococcosis, a disease that is a major cause for meningoencephalitis in immunocompromised individuals and is of particular importance in patients suffering from AIDS. The most striking feature of this opportunistic yeast is its capsule, which represents a major virulence trait evidenced by the fact that non-capsulated strains are non-pathogenic. The structure, assembly and regulation of this important anti-phagocytic factor have been recently reviewed by Bose et al. (2003). The capsule masks potential ligands in the cell wall and consequently, unopsonized *Cry. neoformans* are poorly phagocytosed. Glucuronoxylomannan (GXM) represents the major constituent of the capsule and the density of the GXM matrix was found to be much higher in capsules from cells isolated from infected tissue, than from yeasts harvested after in vitro growth (Gates et al. 2004). GXM is shedded during infection and detectable in blood, cerebrospinal fluid and the infected tissue. Furthermore it exhibits potent immunosuppressive properties (for a recent review, see Monari et al. 2006). Ligation of GXM to TLR4 on macrophages was shown to mediate a fast and long-lasting up-regulation of the Fas ligand, which enables GXM-loaded macrophages to induce apoptosis in Fas-expressing T-cells through activation of caspase-8 (Monari et al. 2005; Pericolini et al. 2006). Inhibition of neutrophil recruitment is another mean by which TLR4 signalling can impair the anti-*Cryptococcus* immune response. GXM has been shown to prevent the rolling and fixed binding of neutrophils to the endothelium, steps which are crucially required for the migration of neutrophils out of the vessel and into the tissue. This effect of GXM can be blocked using monoclonal antibodies directed to TLR4 or CD14 (Ellerbroek et al. 2004a), and O-acetylation of the GXM was shown to be essential for this interference with neutrophil migration (Ellerbroek et al. 2004b). In conclusion, these data suggest that 6-O-acetylated mannose might be a novel fungal TLR4 ligand.

In vitro experiments using transfected CHO cells demonstrated enhanced binding of GXM to cells expressing TLR2, TLR4 and/or CD14, whereas an activation of NF-kB was restricted to TLR4/CD14 expressing cells. Incubation of human peripheral blood monocytes (PBMCs) with GXM also led to the activation of NF-kB, but surprisingly not to a release of TNFα (Shoham et al. 2001), suggesting that GXM triggers several and interfering signalling events. Inhibition of an inflammatory response to LPS by GXM was already observed more than ten years ago by Vecchiarelli and co-workers (1996); and this effect was later attributed to the GXM-triggered secretion of the anti-inflammatory cytokine IL-10 (Vecchiarelli et al. 1996).

The data presently available emphasize an immunomodulatory role for released GXM, which provides a strong anti-inflammatory stimulus. The impact of TLR signalling during *Cryptococcus* infection has so far been addressed in two studies using murine models of cryptococcosis. In the first report, Yauch and co-workers (2004) found that, after intranasal infection, mice lacking TLR2 or CD14 were more susceptible than the corresponding wild-type mice, whereas no difference was observed after an intravenous infection route. Lack of TLR4 had no consequences for the outcome of infection, while an important role in the immune response to *Cryptococcus* was attributed to MyD88, an adaptor molecule essential for the signalling of IL1 receptor and most members of the TLR family (Yauch et al. 2004). An important role for MyD88 and TLR2 was confirmed in a study of Biondo et al. (2005), who also found that TLR4 is dispensable in this context. A more recent study came to the conclusion that TLR2 and TLR4 are only of limited importance for resistance of mice to *Cry. neoformans* infection (Nakamura et al. 2006) and this was supported by the finding that HEK293 cells expressing either TLR2/dectin-1 or TLR4/MD2/CD14 showed no activation of NF-kB in response to *Cryptococcus* (Nakamura et al. 2006). In summary, all published data suggest that TLR4-signalling is not protective during cryptococcosis, but that GXM-triggered TLR4-signalling might in fact hamper an efficient clearance of infection. At least two reports suggest a protective role for TLR2 and MyD88 in the anti-*Cryptococcus* immune response. Additional attempts are clearly required to define the role of TLR2 in the defeat of *Cry. neoformans* infections.

D. TLR-Mediated Recognition of Other Pathogenic Fungi

Macrophages respond to the infectious spherule form of the bimorphic pathogenic fungus *Coccidioides posadasii*, the causative agent of the coccidioidomycosis, in a TLR2-, MyD88- and dectin-1-dependent manner, but this inflammatory response appears to be independent of TLR4 (Viriyakosol et al. 2005).

Pseudallescheria boydii with its asexual form, *Scedosporium apiospermum*, is now recognized as an important emerging opportunistic pathogen causing invasive mycosis in immunocompromised patients (O'Bryan 2005). A linear 4-linked α-D-glucan from the conidial surface of *P. boydii* has been isolated and analyzed. This soluble α-D-glucan inhibits conidial phagocytosis and triggers the secretion of cytokines in a TLR2-, CD14- and MyD88-dependent manner (Bittencourt et al. 2006).

The lipophilic yeast *Malassezia furfur* is the causative agent of pityriasis versicolor (Crespo-Erchiga and Florencio 2006) a common disorder of the skin, which is characterized by scaly hypo- or hyperpigmented lesions. *M. furfur* was shown to induce expression of TLR2 and MyD88 in human keratinocytes and IL-8 expression and secretion was observed to depend on TLR2 (Baroni et al. 2006).

Pneumocystis jirovecii is a common cause of pneumonia in immunocompromised patients and substantially contributes to the morbidity and the mortality of patients suffering from AIDS or malignancies (Gigliotti and Wright 2005). β(1–3) Glucan-rich *Pneumocystis* cell wall fractions have been isolated and analyzed (Lebron et al. 2003). Pneumocystis β glucan trigger an activation of NF-kB and the production of TNFα in macrophages in a MyD88-dependent manner, whereas no evidence has been obtained for the involvement of TLR4 (Lebron et al. 2003).

Fungi are well known to produce a large spectrum of allergens (Horner et al. 1995) and TLRs are likely to play an important role in recognition of these molecules. A Th2-dominated immune response is required for the development of asthma and TLRs have the capacity to polarize the T helper cell bias of an adaptive immune response into opposing directions. One proposed explanation for the increased prevalence rates of allergic diseases in the developed countries is the so-called 'hygiene hypothesis' (Bach 2002), which hypothesized that infection in early childhood acquired through unhygienic contact helps to prevent the development of allergic disease, whereas the decreased exposure to microbes caused by modern public health practices is supposed to lead to deficiencies in an important source of immune education (Horner 2006a). Several recent epidemiologic surveys showing an inverse relationship between the frequency of infectious disease and the incidence of allergic diseases lend support to this hypothesis, but the underlying mechanisms are still poorly understood. So far, only a few attempts have been undertaken to define the role of TLR polymorphisms for the development of asthmatic diseases and the functional role of TLRs for the reaction of airways to inhaled allergens is still largely unexplored (Kauffman 2006). Nevertheless, pharmacologic interventions that target TLR-signalling are supposed to possess an important clinical potential for the prevention and treatment of allergic diseases.

VI. Summary and Outlook

A large body of evidence suggests that TLRs are important PRRs controlling innate immune responses to living fungal pathogens and non-infectious but immunostimulatory fungal materials. While the principal importance of TLRs for the innate immunity is generally accepted, there is a controversial discussion about the precise role of distinct TLRs in the immune response to different pathogenic fungi. There are many factors that might contribute to this discontenting situation, including: (a) the use of different fungal strains and morphotypes, (b) the use of different kinds of immune cells and (c) differences in the experimental design. Certainly the sometimes controversial results also reflect the difficulties in analyzing the complex interplay between an array of often poorly defined ligands on the one hand, and a network of signalling receptors on the other, some of them being not well characterized.

Research on the relevance of TLRs for fungal infections is severely hampered by the fact that our knowledge of the respective PAMPs is still in its infancy. However, some progress has been made recently and several fungal glycostructures have been implicated as ligands for either TLR2 or TLR4. From the data summarized in Table 12.2 it is striking that certain types of mannans seem to be of special importance. However, even subtle biochemical modifications might have a strong impact on the potential of the respective structure to trigger a TLR-dependent signalling and therefore even slight impurities or contaminations of the mannans under investigation have to be excluded. Since glycostructures isolated from a fungal cell wall are notorious for being heterogeneous and difficult to analyze in detail, a combined effort of (glyco-)biochemistry and immunology is required to identify the relevant structures.

The finding that TLRs control immune responses to fungal pathogens and determine their T helper cell bias raised the possibility to target such receptors for the manipulation of an immune

Table 12.2. Putative fungal pathogen-associated molecular patterns (PAMPs) recognized by Toll-like receptors (TLR)

PAMP	Fungal pathogen	TLR	Reference
Phospholipomannan	*Candida albicans*	TLR2	Jouault et al. (2003)
Linear α(1–4) glucan	*Pseudallescheria boydii*	TLR2	Bittencourt et al. (2006)
Mannan	*Can. albicans, Saccharomyces cerevisiae*	TLR4	Tada et al. (2002)
Glucuronoxylomannan	*Cryptococcus neoformans*	TLR4	Shoham et al. (2001)
O-Linked mannan	*Can. albicans*	TLR4	Netea et al. (2006a)

response. Pharmacologic interventions may be aimed at shifting the T helper cell bias towards either a protective Th1-like pro-inflammatory or an anti-inflammatory response. The application of anti-TLR2 monoclonal antibodies was shown to prevent a septic shock in mice challenged with lipopeptides (Meng et al. 2004); and humanized antibodies may provide excellent tools to block or reduce a hyper-activation of innate immune responses. Modulation of an antifungal immune response by TLR ligands has also been discussed, e.g. in the context of asthma (Horner 2006b). We are now facing novel perspectives for treatment of infections and allergic or autoimmune diseases, but we have to keep in mind that our knowledge of the underlying mechanisms is still very limited. The modulation of an immune response by the use of an array of defined TLR ligands is certainly an exiting perspective, but any manipulation of a TLR-mediated immune response may also have deleterious and life-threatening consequences for the patients (Ishii et al. 2006). Any attempt to modulate an existing immune response therefore needs to be accurately balanced and restricted to the safe side of its potential use.

References

Akira S, Takeda K (2004) Toll-like receptor signalling. Nat Rev Immunol 4:499–511

Alexopoulou L, Holt AC, Medzhitov R, Flavell RA (2001) Recognition of double-stranded RNA and activation of NF-kappaB by Toll-like receptor 3. Nature 413:732–738

Anderson, KV, Jürgens, G, Nüsslein-Volhard (1985) Establishment of dorsal-ventral polarity in the Drosophila embryo: genetic studies on the role of the Toll gene product. Cell 42:779–789

Antequera F, Tamame M, Villanueva JR, Santos T (1984) DNA methylation in the fungi. J Biol Chem 259:8033–8036

Bach JF (2002) The effect of infections on susceptibility to autoimmune and allergic diseases. N Engl J Med 347:911–920

Balloy V, Si-Tahar M, Takeuchi O, Philippe B, Nahori A, Tanguy M, Huerre M, Akira S, Latge JP, Chignard M (2005) Involvement of toll-like receptor 2 in experimental invasive pulmonary aspergillosis. Infect Immun 73:5420–5425

Baroni A, Orlando M, Donnarumma G, Farro P, Iovene MR, Tufano MA, Buommino E. (2006) Toll-like receptor 2 (TLR2) mediates intracellular signalling in human keratinocytes in response to *Malassezia furfur*. Arch Dermatol Res 297:280–288

Bell JK, Botos, I, Hall PR, Askins J, Shiloach J, Davies DR, Segal DM (2006) The molecular structure of the TLR3 extracellular domain. J Endotoxin Res 12:375–378

Bellocchio S, Montagnoli C, Bozza S, Gaziano R, Rossi G, Mambula SS, Vecchi A, Mantovani A, Levitz SM, Romani L (2004a) The contribution of the Toll-like/IL-1 receptor superfamily to innate and adaptive immunity to fungal pathogens in vivo. J Immunol 172:3059–3069

Bellocchio S, Moretti S, Perrucio K, Fallarino F, Bozza S, Montagnoli C, Mosci P, Lipford GB, Pitzurra L, Romani L (2004b). TLRs govern neutrophil activity in aspergillosis. J Immunol 173:7406–7415

Bernard M, Latge JP (2001) *Aspergillus fumigatus* cell wall: composition and biosynthesis. Med Mycol 39:9–17

Bose I, Reese AJ, Ory JJ, Janbon G, Doering TL (2003) A yeast under cover: the capsule of *Cryptococcus neoformans*. Eukaryot Cell 2:655–663

Biondo C, Midiri A, Messina, L, Tomasello F, Garufi G, Catania MR, Bombaci M, Beninati C, Teti G, Mancuso G (2005) MyD88 and TLR2, but not TLR4, are required for host defense against *Cryptococcus neoformans*. Eur J Immunol 35:870–878

Bittencourt VCB, Figueiredo RT, da Silva RB, Mourão-Sá DS, Fernandez PL, Sassaki GL, Mulloy B, Bozza MT, Barreto-Bergter E (2006) An α-glucan of *Pseudallescheria boydii* is involved in fungal phagocytosis and Toll-like receptor activation. J Biol Chem 32:22614–22623

Bowman SM, Free SJ (2006) The structure and synthesis of the fungal cell wall. Bioessays 28:799–808

Braedel S, Radsak M, Einsele H, Latge JP, Michan A, Loeffler J, Haddad Z, Grigoleit U, Schild H, Nebart H (2004) *Aspergillus fumigatus* antigens activate innate immune cells via toll-like receptors 2 and 4. Br J Haematol 125:392–399

Brown GD (2006) Dectin-1. A signalling non-TLR pattern-recognition receptor. Nat Rev Immunol 6:33–43

Brown GD, Taylor PR, Reid DM, Willment JA, Williams DL, Martinez-Pomares L, Wong SYC, Gordon S (2002) Dectin-1 is a major beta-glucan receptor on macrophages. J Exp Med 196:407–412

Chow JC, Young DW, Golenbock DT, Christ WJ, Gusovsky F (1999) Toll-like receptor-4 mediates lipopolysaccharide-induced signal transduction. J Biol Chem 274:10689–10692

Crespo-Erchiga V, Florencio VD (2006) Malassezia yeasts and pityriasis versicolor. Curr Opin Infect Dis 19:139–147

Csank C, Haynes K (2000) *Candida glabrata* displays pseudohyphal growth. FEMS Microbiol Lett 189:115–120

Di Carlo FJ, Fiore JV (1958) On the composition of zymosan. Science 127:756–757

Didierlaurent A, Sirard JC, Kraehenbuhl JP, Neutra MR (2002) How the gut senses its content. Cell Microbiol 4:61–72

Diniz SN, Nomizo R, Cisalpino PS, Teixeira MM, Brown GD, Mantovani A, Gordon S, Reis LF, Dias AA (2004) PTX3 function as a opsonin for the dectin-1-dependent internalization of zymosan by macrophages. J Leukoc Biol 75:649–656

Dobourdeau M, Athman R, Balloy V, Huerre M, Chignard M, Philpott DJ, Latge JP, Ibrahim-Granet O (2006) *Aspergillus fumigatus* induces innate immune responses in alveolar macrophages through the MAPK pathway independently of TLR2 and TLR4. J Immunol 177:3994–4001 d'Ostiani CF, Del Sero G, Bacci A, Montagnoli C, Spreca A, Mencacci A, Ricciardi-Castagnoli P, Romani L (2000) Dendritic cells discriminate between yeasts and hyphae of the fungus *Candida albicans*. Implications for initiation of T helper cell immunity in vitro and in vivo. J Exp Med 191:1661–1674

Ellerbroek PM, Ulfman LH, Hoepelman AI, Coenjaerts FE (2004a) Cryptococcal glucuronoxylomannan interferes with neutrophil rolling on the endothelium. Cell Microbiol 6:581–592

Ellerbroek PM, Lefeber DJ, van Veghel R, Scharringa J, Brouwer E, Gerwig GJ, Janbon G, Hoepelman AI, Coenjaerts FE (2004b) O-acetylation of cryptococcal capsular glucuronoxlomannan is essential for interference with neutrophil migration. J Immunol 173:7513–7520

Galagan JE, Calvo SE, Cuomo C, Ma LJ, Wortman JR, Batzoglou S, Lee SI, Basturkmen M, Spevak CC, Clutterbuck J et al (2005) Sequencing of *Aspergillus nidulans* and comparative analysis with *A. fumigatus* and *A. oryzae*. Nature 438:1105–1115

Gantner BN, Simmons RM, Canavera SJ, Akira S, Underhill DM (2003) Collaborative induction of inflammatory responses by dectin-1 and Toll-like receptor 2. J Exp Med 197:1107–1117

Garlanda C, Hirsch E, Bozza S, Salustri A, De Acetis M, Nota R, Maccagno A, Riva F, Bottazzi B, Peri G, Doni A, Vago L, Botto M, De Santis R, Carminati P, Siracusa G, Altruda F, Vecchi A, Romani L, Mantovani A (2002) Non-redundant role of the long pentraxin PTX3 in anti-fungal innate immune response. Nature 420:182–186

Gates MA, Thorkildson P, Kozel TR (2004) Molecular architecture of the *Cryptococcus neoformans* capsule. Mol Microbiol 52:13–24

Gersuk GM, Underhill DM, Zhu L, Marr KA (2006) Dectin-1 and TLRs permit macrophages to distinguish between different *Aspergillus fumigatus* cellular states. J Immunol 176:3717–3724

Gigliotti F, Wright TW (2005) Immunopathogenesis of *Pneumocystis carinii* pneumonia. Expert Rev Mol Med 7:1–16

Gil ML, Gozalbo D (2006) TLR2, but not TLR4, triggers cytokine production by murine cells in response to *Candida albicans* yeasts and hyphae. Microbes Infect 8:2299–2304

Grazziutti ML, Rex JH, Cowart RE, Anaissie EJ, Ford A, Savary CA. (1997) *Aspergillus fumigatus* conidia induce a Th1-type cytokine response. J Infect Dis 176:1579–1583

Gross O, Gewies A, Finger K, Schafer M, Sparwasser T, Peschel C, Forster I, Ruland J (2006) Card9 controls a non-TLR signalling pathway for innate anti-fungal immunity. Nature 442:651–656

Hemmi H, Takeuchi O, Kawai T, Kaisho T, Sato S, Sanjo H, Matsumoto M, Hoshino K, Wagner H, Takeda K, Akira S (2000) A Toll-like receptor recognizes bacterial DNA. Nature 408:740–745

Herre J, Marshall AS, Caron E, Edwards AD, Williams DL, Schweighoffer E, Tybulewicz V, Reis e Sousa C, Gordon S, Brown GD (2004) Dectin-1 uses novel mechanisms for yeast phagocytosis in macrophages. Blood 104:4038–4045

Hirschfeld M, Weiss JJ, Toshchakov V, Salkowski CA, Ward DC, Qureshi N, Michalek SM, Vogel SN (2001) Signalling by Toll-like receptor 2 and 4 agonists results in differential gene expression in murine macrophages. Infect Immun 69:1477–1482

Hohl TM, Van Epps HL, Rivera A, Morgan LA, Chen PL, Feldmesser M, Pamer EG (2005) *Aspergillus fumigatus* triggers inflammatory responses by stage-specific beta-glucan display. PLoS Pathog 1:e30

Hornef MW, Frisan T, Vandewalle A, Normark S, Richter-Dahlfors A (2002) Toll-like receptor 4 resides in the Golgi apparatus and colocalizes with internalized lipopolysaccharide in intestinal epithelial cells. J Exp Med 195:559–570

Horner AA (2006a) Toll-like receptor ligands and atopy: a coin with at least two sides. J Allergy Clin Immunol 117:1133–1140

Horner AA (2006b) Update on toll-like receptor ligands and allergy: implications for immunotherapy. Curr Allergy Asthma Rep 6:395–401

Horner WE, Helbing A, Salvaggio JE, Lehrer SB (1995) Fungal allergens. Clin Microbiol Rev 8:161–179

Hultmark D (1994) Macrophage differentiation marker MyD88 is a member of the Toll/IL-1 receptor family. Biochem Biophys Res Commun 199:144–146

Ibrahim-Granet O, Philippe B, Boleti H, Boisvieux-Ulrich E, Grenet D, Stern M, Latge JP (2003) Phagocytosis and intracellular fate of *Aspergillus fumigatus* conidia in alveolar macrophages. Infect Immun 71:891–903

Ishii KJ, Uematsu S, Akira S (2006) 'Toll' gates for future immunotherapy. Curr Pharm Des 12:4135–4142

Iwanaga, S, Lee, BL (2005) Recent advances in the innate immunity of invertebrate animals. J Biochem Mol Biol 38:128–150

Janeway C, Medzhitov R (2000) Innate immune recognition: mechanisms and pathways. Immunol Rev 173:89–97

Jouault T, Ibata-Ombetta S, Takeuchi O, Trinel PA, Sacchetti P, Lefebvre P, Akira S, Poulain D (2003) *Candida albicans* phospholipomannan is sensed through toll-like receptors. J Infect Dis 188:165–172

Kauffman HF (2006) Innate immune responses to environmental allergens. Clin Rev Allergy Immunol 30:129–140

Kawai T, Akira S (2005) Pathogen recognition with Toll-like receptors. Curr Opin Immunol 17:338–344

Kesh S, Mensah NY, Peterlongo P, Jaffe D, Hsu K, Van den Brink M, O'Reilly R, Pamer E, Satagopan J, Pananicolaou GA (2005) TLR1 and TLR6 polymorphisms are associated with susceptibility to invasive aspergillosis after allogenic stem cell transplantation. Ann N Y Acad Sci 1062:95–103

Klimpel KR, Goldman WE (1988) Cell walls from avirulent variants of *Histoplasma capsulatum* lack alpha-(1,3)-glucan. Infect Immun 56:2997–3000

Kobe B, Deisenhofer J (1995) A structural basis of the interactions between leucine-rich repeats and protein ligands. Nature 374:183–186

Kumamoton CA, Vinces MD (2005) Contributions of hyphae and hypha-co-regulated genes to *Candida albicans* virulence. Cell Microbiol 7:1546–1554

Latge JP (1999) *Aspergillus fumigatus* and aspergillosis. Clin Microbiol Rev 12:310–350

Lebron F, Vassallo R, Puri V, Limper AH (2003) *Pneumocystis carinii* cell wall beta-glucans initiate macrophage inflammatory responses through NF-kappaB activation. J Biol Chem 278:25001–25008

Le Cabec V, Emorine LJ, Toesca I, Cougoule C, Maridonneau-Parini I (2004) The human mannose receptor is not a professional phagocytic receptor. J Leukoc Biol 77:934–943

Lemaitre B, Nicolas E, Michaut L, Reichhart JM, Hoffmann JA (1996) The dorsoventral regulatory gene cassette spatzle/Toll/cactus controls the potent antifungal response in *Drosophila* adults. Cell 86:973–983

Levitz S (2004) Interactions of Toll-like receptors with fungi. Microbes Infect 6:1351–1355

Ligoxygakis P, Pelte N, Hoffmann JA, Reichhart J-M (2002) Activation of *Drosophila* Toll during fungal infection by a blood protease. Science 297:114–116

Lillegard JB, Sim RB, Thorkildson P, Gates MA, Kozel TR (2006) Recognition of *Candida albicans* by mannan-binding lectin in vitro and in vivo. J Infect Dis 193:1589–1597

Lund JM, Alexopoulou L, Sato A, Karow M, Adams NC, Gale NW, Iwasaki A, Flavell RA (2004) Recognition of single-stranded RNA viruses by Toll-like receptor 7. Proc Natl Acad Sci USA 101:5598–5603

Luther K, Ebel F (2006) Toll-like receptors: Recent advances, open questions and implications for aspergillosis control. Med Mycol 44:219–227

Luther K, Torosantucci A, Brakhage AA, Heesemann J, Ebel F (2007) Phagocytosis of *Aspergillus fumigatus* conidia by murine macrophages involves recognition by the dectin-1 beta-glucan receptor and Toll-like receptor 2. Cell Microbiol 9:368–381

Madan T, Eggleton P, Kishore U, Strong P, Aggrawal SS, Sarmu PU, Reid KB (1997) Binding of pulmonary surfactant proteins A and D to *Aspergillus fumigatus* conidia enhances phagocytosis and killing by human neutrophils and alveolar macrophages. Infect Immun 65:3171–3179

Mambula SS, Sau K, Henneke P, Golenbock DT, Levitz SM (2002) Toll-like receptor (TLR) signaling in response to *Aspergillus fumigatus*. J Biol Chem 277:39320–39326

Marion CL, Rappleye CA, Engle JT, Goldman WE (2006) An alpha-(1,4)-amylase is essential for alpha-(1,3)-glucan production and virulence in *Histoplasma capsulatum*. Mol Microbiol 62:970–983

Marr KA, Balajee SA, Hawn TR, Ozinski A, Pham U, Akira S, Aderem A, Liles WC (2003) Differential role of MyD88 in macrophage-mediated responses to opportunistic fungal pathogens. Infect Immun 71:5280–5286

Mazzoni A, Segal DM (2004) Controlling the Toll road to dendritic polarization. J Leukoc Biol 75:721–730

Meier A, Kirschning C, Nikolaus T, Wagner H, Heesemann J, Ebel F (2003) Toll-like receptor (TLR) 2 and TLR4 are essential for *Aspergillus*-induced activation of murine macrophages. Cell Microbiol 5:561–570

Meng G, Rutz M, Schiemann M, Metzger J, Grabiec A, Schwandner R, Luppa PB, Ebel F, Busch DH, Bauer S, Wagner H, Kirschning CJ (2004) Antagonistic antibody prevents toll-like receptor 2-driven lethal shock-like syndrome. J Clin Invest 113:1473–1481

Meri T, Hartmann A, Lenk D, Eck R, Würzner R, Hellwage J, Mei S, Zipfel PF (2002) The yeast *Candida albicans* binds complement regulators factor H and FHL-1. Infect Immun 70:5185–5192

Mitchell DA, Fadden AJ, Drickamer K (2001) A novel mechanism of carbohydrate recognition by the C-type lectins DC-SIGN and DC-SIGNR. Subunit organization and binding to multivalent ligands. J Biol Chem 276:28939–28945

Momany M, Lindsey R, Hill TW, Richardson EA, Momany C, Pedreira M, Guest GM, Fisher JF, Hessler RB, Roberts KA (2004) The *Aspergillus fumigatus* cell wall is organized in domains that are remodelled during polarity establishment. Microbiology 150:3261–3268

Monari C, Pericolini E, Bistoni F, Casadevall A, Kozel TR, Vecchiarelli A (2005) *Cryptococcus neoformans* capsular glucuronoxylomannan induces expression of fas ligand in macrophages. 174:3461–3468

Monari C, Bistoni F, Vecchiarelli A (2006) Glucuronoxylomannan exhibits potent immunosuppressive properties. FEMS Yeast Res 6:537–542

Morrison BE, Park SJ, Mooney JM, Mehrad B (2003) Chemokine-mediated recruitment of NK cells is a critical host defense mechanism in invasive aspergillosis. J Clin Invest 112:1862–1870

Mukhopadhyay S, Herre J, Brown GD, Gordon S (2004) The potential of Toll-like receptors to collaborate with other innate immune receptors. Immunology 112:521–530

Murciano C, Villamon E, Gozalbo D, Roig P, O'Connor JE, Gil ML. (2006) Toll-like receptor 4 defective mice carrying point or null mutations do not show increased susceptibility to *Candida albicans* in a model of hematogenously disseminated infection. Med Mycol 44:149–157

Muta T (2006) Molecular basis for invertebrate innate immune recognition of (1-3)-beta-glucan as a pathogen-associated molecular pattern. Curr Pharm Des 12:4155–4161

Nakamura K, Miyagi K, Koguchi Y, Kinjo Y, Uezu K, Kinjo T, Akamine M, Fujita J, KawamurA I, Mitsuyama M, Adachi Y, Ohno N, Takeda K, Akira S, Miyazato A, Kaku M, Kawakami K. (2006) Limited contribution of Toll-like receptors 2 and 4 to the host response to a fungal infectious pathogen, *Cryptococcus neoformans*. FEMS Immunol Med Microbiol 47:148–154

Netea MG, Van der Graaf CAA, Vonk AG, Verschueren I, Van der Meer JWM, Kullberg BJ (2002) The role of Toll-like receptor (TLR) 2 and TLR4 in the host defense against disseminated candidiasis. J Inf Dis 185:1483–1489

Netea MG, Warris A, Van der Meer JWN, Fenton MJ, Verver-Janssen TJ, Jacobs LE, Andresen T, Verweij PE, Kullberg BJ (2003) *Aspergillus fumigatus* evades immune recognition during germination through loss of Toll-like receptor-4-mediated signal transduction. J Infect Dis 188:320–326

Netea MG, Sutmuller R, Hermann C, Van der Graaf CAA, Van der Meer JWM, Van Krieken JH, Hartung T, Adema G, Kullberg BJ (2004a) Toll-like receptor 2 suppresses immunity against *Candida albicans* through induction of IL-10 and regulatory T cells. J Immunol 172:3712–3718

Netea MG, Van der Meer JWM, Kullberg B-J (2004b) Toll-like receptors as an escape mechanism from the host defense. Trends Microbiol 12:484–488

Netea MG, Van der Graaf CAA, Van der Meer JWM, Kullberg BJ (2004c) Recognition of fungal pathogens by

Toll-like receptors. Eur J Clin Microbiol Infect Dis 23:672–676

Netea MG, Gow NAR, Munro CA, Bates S, Collins C, Ferwerda G, Hobson RP, Bertram G, Hughes HB, Jansen T, Jacobs L, Buurman ET, Gijzen K, Williams DL, Torensma R, McKinnon A, MacCallum DM, Odds FC, Van der Meer JWM, Brown AJP, Kullberg BJ. (2006a) Immune sensing of *Candida albicans* requires cooperative recognition of mannans and glucans by lectin and Toll-like receptors. J Clin Invest 116:1642–1650

Netea MG, Van der Meer JWM, Kullberg B-J (2006b) Role of the dual interaction of fungal pathogens with pattern recognition receptors in the activation and modulation of host defence. Clin Microbiol Infect 12:404–409

Newman SL, Holly A (2001) *Candida albicans* is phagocytosed, killed, and processed for antigen presentation by human dendritic cells. 69:6813–6822

Nierman WC, Pain A, Anderson MJ, Wortman JR, Kim HS, Arroyo J, Berriman M, Abe K, Archer DB, Bermejo C et al (2005) Genomic sequence of the pathogenic and allergenic filamentous fungus *Aspergillus fumigatus*. Nature 438:1151–1156

O'Bryan (2005) Pseudoallescheriasis in the 21st century. Expert Rev Anti Infect Ther 3:765–773

Ozinsky A, Underhill DM, Fontenot JD, Hajjar AM, Smith KD, Wilson CB, Schroeder L, Aderem A (2000) The repertoire for pattern recognition of pathogens by the innate immune system is defined by cooperation between toll-like receptors. Proc Natl Acad Sci USA 97:13766–13771

Paris S, Debeaupuis JP, Crameri R, Carey M, Charles F, Prevost MC, Schmitt C, Philippe B, Latge JP (2003) Conidial hydrophobins of *Aspergillus fumigatus*. Appl Environ Microbiol 69:1581–1588

Pericolini E, Cenci E, Monari C, De Jesus M, Bistoni F, Casadevall A, Vecchiarelli A (2006) *Cryptococcus neoformans* capsular polysaccharide component glucuronoxylomannan induces apoptosis of human T-cells through activation of caspase-8. Cell Microbiol 8:267–275

Perfect JR (2004) Antifungal resistance: the clinical front. Oncology 18:15–22

Philippe B, Ibrahim-Granet O, Prevost MC, Gougerot-Pocidalo MA, Sanchez Perez M, van der Meeren A, Latge JP (2003) Killing of *Aspergillus fumigatus* by alveolar macrophages is mediated by reactive oxidant intermediates. Infect Immun 71:3034–3042

Poltorak A, He X, Smirnova I, Liu MY, van Huffel C, Du X, Birdwell D, Alejos E, Silva M, Galanos C, Freudenberg M, Ricciardi-Castagnoli P, Layton B, Beutler B (1998) Defective LPS signaling in C3H/HeJ and C57BL/10ScCr mice: mutations in Tlr4 gene. Science 282:2085–2088

Rappleye CA, Goldman WE (2006) Defining virulence genes in the dimorphic fungi. Annu Rev Microbiol 60:281–303

Rappleye CA, Eissenberg LG, Goldman WE (2007) *Histoplasma capsulatum* alpha-(1,3)-glucan blocks innate immune recognition by the beta-glucan receptor. Proc Natl Acad Sci USA 104:1366–1370

Rock FL, Hardiman G, Timans JC, Kastelein RA, Bazan JF (1998) A family of human receptors structurally related to Drosophila Toll. Proc Natl Acad Sci USA 95:588–593

Roeder A, Kirschning CJ, Schaller M, weindl G, Wagner H, Korting HC, Rupec RA (2004a) Induction of nuclear factor-kappa B and c-Jun/activator protein-1 via toll-like receptor 2 in macrophages by antimycotic-treated *Candida albicans*. J Infect Dis 190:1318–1326

Roeder A, Kirschning CJ, Rupec RA, Schaller M, Korting HC (2004b) Toll-like receptors and innate antifungal responses. Trends Microbiol 12:44–49

Romani L (2004) Immunity to fungal infections. Nat Rev Immunol 4:1–23

Romani L, Bistoni F, Puccetti P (2002) Fungi, dendritic cells and receptors: a host perspective of fungal virulence 10:508–514

Ruiz-Herrera J, Elorza MV, Valentin E, Sentandreu R (2006) Molecular organization of the cell wall of *Candida albicans* and its relation to pathogenicity. FEMS Yeast Res 6:14–29

Saijo S, Fujikado N, Furuta T, Chung SH, Kotaki H, Seki K, Sudo K, Akira S, Adachi Y, Ohno N, Kinjo T, Nakamura K, Iwakura Y (2007) Dectin-1 is required for host defense against *Pneumocystis carinii* but not against *Candida albicans*. Nat Immunol 8:39–46

Sarfati J, Monod M, Recoo P, Sulahian A, Pinel C, Candolfi E, Fontaine T, Debeaupuis JP, Tabouret M, Latgé JP (2006) Recombinant antigens as diagnostic markers for aspergillosis. Diagn Microbiol Infect Dis 55:279–291

Sato S, Nomura F, Kawai T, Takeuchi O, Mühlradt PF, Takeda K, Akira S (2000) Synergy and cross-tolerance between Toll-like receptor (TLR) 2- and TLR4-mediated signaling pathways. J Immunol 165:7096–7101

Schaffner A, Douglas H, Braude A (1982) Selective protection against conidia by mononuclear and against mycelia by polymorphnuclear phagocytes in resistance to *Aspergillus*. J Clin Invest 69:617–631

Serrano-Gomez D, Dominguez-Soto A, Ancochea J, Jimenez-Heffernan JA, Leal JA, Corbi AL (2004) Dendritic cell-specific intercellular adhesion molecule 3-grabbing nonintegrin mediates binding and internalization of *Aspergillus fumigatus* conidia by dendritic cells and macrophages. J Immunol 173:5635–5643

Shoham S, Huang C, Chen JM, Golenbock DT, Levitz SM. (2001) Toll-like receptor 4 mediates intracellular signaling without TNF-alpha release in response to *Cryptococcus neoformans* polysaccharide capsule. J Immunol 166:4620–4626

Silverman N, Maniatis T (2001) Nf-kappaB signaling pathways in mammalian and insect innate immunity. Genes Dev 15:2321–2342

Speth C, Rambach G, Lass-Flörl C, Dierich MP, Würzner R (2004) The role of complement in invasive fungal infections. Mycoses 47:93–103

Steele C, Rapaka RR, Metz A, Pop SM, Williams DL, Gordon S, Kolls JK, Brown GD (2005) The beta-glucan receptor dectin-1 recognizes specific morphologies of *Aspergillus fumigatus*. PLoS Pathog 1:e42

Tada H, Nemoto E, Shimauchi H, Watanabe T Mikami T, Matsumoto T, Ohno N, Tamura H, Shibata K, Akashi S, Miyake K, Sugawara S, Takada H (2002) *Saccharomyces cerevisiae*- and *Candida albicans*-derived mannan induced production of tumor necrosis factor alpha by human monocytes in a CD14- and

Toll-like receptor 4-dependent manner. Microbiol Immunol 46:503–512

Takeda K, Akira S (2005) Toll-like receptors in innate immunity. Int Immunol 17:1–14

Takeuchi O, Hoshino K, Kawai T, Sanjo H, Takada H, Ogawa T, Takeda K, Akira S (1999) Differential roles of TLR2 and TLR4 in recognition of gram-negative and gram-positive bacterial cell wall components. Immunity 11:443–451

Taramelli D, Malabarba MG, Sala G, Basilico N, Cocuzza G (1996) Production of cytokines by alveolar and peritoneal macrophages stimulated by *Aspergillus fumigatus* conidia or hyphae. J Med Vet Mycol 34:49–56

Taylor PR, Brown GD, Herre J, Williams DL, Willment JA, Gordon S (2004) The role of SIGNR1 and the βglucan receptor (dectin-1) in the nonopsonic recognition of yeast by specific macrophages. J Immunol 172:1157–1162

Taylor PR, Tsoni SV, Willment JA, Dennehy KM, Rosas M, Findon H, Haynes K, Steele C, Botto M, Gordon S, Brown GD (2007) Dectin-1 is required for beta-glucan recognition and control of fungal infection. Nat Immunol 8:31–38

Tekaia F, Latge JP (2005) *Aspergillus fumigatus*: saprophyte or pathogen? Curr Opin Microbiol 8:385–392

Underhill DM, Gantner B (2004) Integration of Toll-like receptor and phagocytic signaling for tailored immunity. Microbes Infect 6:1368–1373

Van der Graaf CAA, Netea MG, Verschueren I, Van der Meer JWM, Kullberg BJ (2005) Differential cytokine production and Toll-like receptor signaling pathways by *Candida albicans* blastoconidia and hyphae. Infect Immun 73:7458–7464

Van der Graaf CA, Netea MG, Morre SA, Den Heijer M, Verweiji PE, Van der Meer JW, Kullberg BJ (2006) Toll-like receptor 4 Asp 299Gly/Thr399Ile polymorphisms are a risk factor for Candida bloodstream infection. Eur Cytokine Netw 17:29–34

Van de Wetering JK, Van Golde LM, Batenburg JJ (2004) Collectins: players of the innate immune system. Eur J Biochem 271:1229–1249

Vecchiarelli A, Retini C, Pietrella D, Monari C, Tascini C, Beccari T, Kozel TR (1995) Downregulation by cryptococcal polysaccharide of tumor necrosis factor alpha and interleukin-1 beta secretion from human monocytes. Infect Immun 63:2919–2923

Vecchiarelli A, Retini C, Monari C, Tascini C, Bistoni F, Kozel TR (1996) Purified capsular polysaccharide of *Cryptococcus neoformans* induces interleukin-10 secretion by human monocytes. Infect Immun 64:2846–2849

Villamón E, Gozalbo D, Roig P, O'Connor JE, Ferrandiz ML, Fradelizi D, Gil ML. (2004a) Toll-like receptor 2 is dispensable for acquired host immune resistance to *Candida albicans* in a murine model of disseminated candidiasis. Microbes Infect 6:542–548

Villamón E, Gozalbo D, Roig P, O'Connor, JE Fradelizi D, Gil ML. (2004b) Toll-like receptor 2 is essential in murine defenses against *Candida albicans*. Microbes Infect 6:1–7

Viriyakosol S, Fierer J, Brown GD, Kirkland TN (2005) Innate immunity to the pathogenic fungus *Coccidioides posadasii* is dependent on Toll-like receptor 2 and dectin-1. Infect Immun 73:1553–1560.

Walenkamp AM, Verheul AF, Scharringa J, Hoepelman IM (1999) Pulmonary surfactant protein A binds to *Cryptococcus neoformans* without promoting phagocytosis. Eur J Clin Invest 29:83–92

Wang JE, Warris A, Ellingsen EA, Jorgensen PF, Flo TH, Espevik T, Solberg R, Verweij PE, Aasen AO (2001) Involvement of CD14 and toll-like receptors in activation of human monocytes by *Aspergillus fumigatus* hyphae. Infect Immun 69:2402–2406

Xu Y, Tao X, Shen B, Horng T, Medzhitov R, Manley JL, Tong L (2000) Structural basis for signal transduction by the Toll/interleukin-1 receptor domains. Nature 408:111–115

Yauch LE, Mansour MK, Shoham S, Rottman JB, Levitz SM. (2004) Involvement of CD14, toll-like receptors 2 and 4, and MyD88 in the host response to the fungal pathogen *Cryptococcus neoformans* in vivo. Infect Immun 72:5373–5382

13 Clinical Aspects of Dermatophyte Infections

Jochen Brasch[1], Uta-Christina Hipler[2]

CONTENTS

I. Introduction	263
II. Dermatophytes	263
III. Dermatophytoses	267
A. Pathophysiology	267
B. Clinical Manifestations	269
1. Tinea Pedis and Tinea Manuum	269
2. Onychomycosis	270
3. Tinea Capitis and Tinea Barbae	271
4. Tinea Corporis	272
5. Tinea Cruris	273
6. Deep Infections	274
7. Dermatophytids	274
IV. Epidemiology	274
V. Treatment	278
A. Topical Treatment	278
B. Systemic Treatment	278
C. Special Treatment Aspects of Distinct Forms of Tinea	279
D. Hygiene	279
VI. Conclusions	279
References	281

I. Introduction

Infections caused by dermatophytes are termed dermatophytoses, tinea or ringworm. They are among the most common and widespread endemic infectious diseases (Rippon 1988; Kates et al. 1990; Odom 1993; Gupta et al. 2003). In some geographic areas or environments more than 30% of the population is affected (Rippon 1985; Noguchi et al. 1995). Therefore, dermatophytoses pose a considerable worldwide health problem. In most cases dermatophytoses in humans remain superficial infections, restricted to the skin, nails, and hair (Ogawa et al.1998). These infections often lead to skin lesions, which are uncomfortable but not life-threatening. However, infections of deeper structures like subcutaneous soft tissue can occur under conducive conditions. The relationship between the pathogen and the human host depends very much on the species of dermatophyte involved and on the patient's immunocompetence, general health, and living conditions. Dermatophytoses provide a fascinating model for the interaction between highly specialized fungi and host defense. Because the skin is uniquely accessible to examination with the naked eye, a trained observer can examine this fungal battlefield without additional technical devices.

In the following chapter, after a short look at the dermatophytes as medically important fungi, some of the known mechanisms involved in dermatophyte infections are discussed in order to help better understand the development and pathogenesis of dermatophyte infections. The clinical aspects of the most common and characteristic types of dermatophytoses (or forms of tinea) are covered, followed by a review of their epidemiological aspects. Finally, general principles of treatment are described.

II. Dermatophytes

The term dermatophyte is defined primarily by functional characteristics and not by strict taxonomical criteria. A dermatophyte is a hyalohyphomycete that can degrade keratin and consequently cause communicable skin infections in humans and/or animals (generally mammals; Weitzman and Summerbell 1995; Weitzman and Padhye 1996). Keratin is the main constituent of the outermost layer of human skin, the stratum corneum, as well as hair and nails in humans, and hooves, fur, and feathers in animals. An intact stratum corneum is usually a sufficient protective barrier against microorganisms, but highly specialized keratinophilic dermatophytes can invade this outer shield of the skin and cause infections.

---

[1] Abteilung Dermatologie, Universität Kiel, Schittenhelmstraße 7, 24105 Kiel, Germany; e-mail: jbrasch@dermatology.uni-kiel.de
[2] Abteilung Dermatologie, Universität Jena, Erfurter Straße 35, 07743 Jena, Germany

The different species of dermatophytes are morphologically and physiologically very closely related (de Vroey 1985; Matsumoto and Ajello 1987). Comparisons of ribosomal RNA genes in fact indicate that they are monophyletic in origin and that radiation began only about 50×10^6 years ago. This means that the evolution of dermatophytes occurred synchronously with the evolution of their mammalian hosts (Harmsen et al. 1995). Clonal lineages can even be found in the medically relevant and host-associated dermatophytes, and such lineages are capable of maintaining populations and undergoing further evolutionary developments (Gräser et al. 2006).

Phylogenetic studies have suggested that dermatophytes developed from non-pathogenic, soil-colonizing fungi into species specialized for particular human or animal hosts as their ecological niche (Gräser et al. 1999). Better dermatophyte adaptation to the human host is generally accompanied by a reduced capability to produce spores like macroconidia and differentiated hyphal elements (Summerbell 2000). Interestingly, a strictly anthropophilic dermatophyte, *Trichophyton rubrum* (Fig. 13.1) is currently the most prevalent cause of dermatophyte infections of hairless skin (Sinski and Kelley 1991; Aly et al. 1994; Chinelli et al. 2003; Foster et al. 2004) and another strictly anthropophilic species, *T. tonsurans*, is the most common cause of dermatophyte scalp infections in large parts of the world (Babel et al. 1990; Foster et al. 2004).

The exact classification of dermatophytes is hampered by the fact that their pathogenic growth phase (which occurs in human skin) differs from the saprophytic phase obtained in vitro. As saprophytes, dermatophytes reproduce asexually in an anamorphic state by producing vegetative spores (conidia). In some species, however, teleomorphic states have been discovered that reproduce sexually (Takashio 1979). These teleomorphic states, or perfect forms, turned out to be members of the Subphylum Ascomycotina, Order Onygenales, Family Arthrodermataceae, Genus *Arthroderma* (Matsumoto and Ajello 87). The dualism of anamorphic and teleomorphic states has led, on the one hand, to a classification system based on anamorph states, and on the other hand, to a valid taxonomic classification based on teleomorphic states (Simpanya 2000). In recent years, genetic methods have generated a wealth of additional information pertaining to species differentiation of dermatophytes (De Hoog et al. 1998; Harmsen et al.1999; Summerbell et al. 1999). This on-going research, which has led to revisions of the classification, is continually evolving and being refined (Gräser 2001).This dynamic field of investigation is, however, beyond the scope of this clinically-focused chapter.

A nomenclature adequately reflecting the medically important features of separate species is required for clinical purposes. From a clinician's point of view, all strains of a distinct species should be characterized by species-related associations with particular diseases (e.g., infections of the nails or the scalp), host adaptation (e.g., adaptation to humans, or distinct species of animals like cats or mice), and epidemiology (i.e., geographic distribution). A strictly biological approach, generally emphasizing purely taxonomic aspects, does not always achieve this aim. The problem is that some genetically very closely related dermatophytes have long been viewed as distinct species because of their association with different clinical settings. Therefore, it remains an unsolved problem how to define dermatophyte species in a way that each recognized species is in congruence with both distinctive medically important characteristics and also a taxonomic system based on evolutionary data. The clinician must be able to correlate an infection with its causative agent. The identification of a specific pathogenic species should allow meaningful clinical conclusions on how the pathogen was likely acquired and what kind of infection

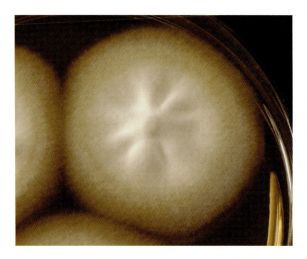

Fig. 13.1. *Trichophyton rubrum* on Sabouraud agar. This species is the most common dermatophyte worldwide. The thallus has a characteristic surface profile

it usually causes. In contrast, taxonomic perfection is less relevant in a clinical context. Dermatophytes isolated from skin infections are usually identified by culture techniques that only allow growth of their anamorphic state. All this explains why clinicians are accustomed to applying the classification system that is based on the anamorphic states and why clinicians continue to distinguish some morphologically divergent variants of now genetically unified species that are of interest with regard to differences in distribution, ecology or pathogenicity. The names of the anamorphic states are therefore used in this chapter on the clinical aspects of dermatophyte infections.

The anamorphic states of the dermatophytes belong to three genera: *Trichophyton*, *Microsporum* and *Epidermophyton*, which are easily distinguished by their morphological criteria (Weitzman and Summerbell 1995). All dermatophytes grow in filamentous form on standard mycological media like Sabouraud agar or Kimmig agar, developing dense mycelial thalli and vegetative spores (conidia). The genus *Trichophyton* (Figs. 13.1–13.3) characteristically produces small monocellular club-shaped microconidia, which are borne laterally on the hyphae. Thin- and smooth-walled multicellular macroconidia are often additionally found, although these can be quite rare. These macroconidia are club-shaped and usually not clustered. *Micosporum* species (Fig. 13.4) also have micro- and macroconidia, but the latter are rough-walled, with spindle-shaped walls thicker than those of *T. macroconidia*. The genus *Epidermophyton*

Fig. 13.3. *Trichophyton mentagrophytes* on Sabouraud agar. The variant shown has a granular surface and is often associated with animals

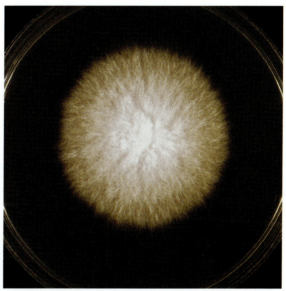

Fig. 13.4. *Microsporum canis* on Sabouraud agar. Cats are important hosts of this zoophilic dermatophyte. It is a common pathogen in scalp infection (tinea capitis) of children

does not produce microconidia, but forms multicellular macroconidia instead. These are club-shaped, thin- and smooth-walled, and arranged in bundles. The only clinically relevant species of this genus is *E. floccosum* (Fig. 13.5).

All dermatophytes grow within infected tissue by forming hyphae, allowing them to spread and penetrate adjacent host cells. In older hyphae transverse septa can lead to the separation of fungal cells that then develop thickened walls.

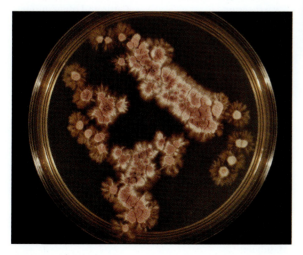

Fig. 13.2. *Trichophyton violaceum* on Sabouraud agar. The name of this anthropophilic species is explained by its strong red, or violaceous, pigmentation

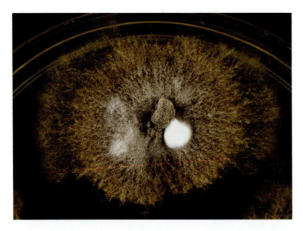

Fig. 13.5. *Epidermophyton floccosum* on Sabouraud agar. The name refers to the white mycelial balls that often occur on the surface of older thalli. This strictly anthropophilic dermatophyte is a common cause of crural infections (tinea cruris)

Finally the hyphae disintegrate into chains of single cells called arthrospores. Arthrospores are robust and resistant to environmental noxa like cold, dryness or UV radiation (Hashimoto and Blumenthal 1978). Even under unfavorable conditions arthrospores can remain viable and virulent for a long time, making them important for the distribution of dermatophytes. Dermatophytes infecting hair can form spores located either within the hair (endothrix) or covering the hair in a sleeve-like manner (ectothrix). These spores are also important propagules for fungal distribution.

According to their phylogenetic development, the main habitat of distinct dermatophyte species can be soil, animals, or humans, so that geophilic, zoophilic and anthropophilic species of dermatophytes must be distinguished (Summerbell 2000; Table 13.1). However, many dermatophytes not primarily adapted can cause human skin infections under conducive conditions. From a clinical point of view, this distinction is of interest because geophilic and zoophilic species tend to trigger more inflammatory lesions in humans than anthropophilic species (Blank et al. 1969; Grigoriu et al. 1984; Brasch 1990a–c) and because knowledge of the ecological group is necessary to identify and eradicate possible sources of infection. For example, cattle are the most common reservoir for *T. verrucosum*, whereas cats are common hosts of *M. canis*. Furthermore, some species or variants of dermatophytes are only endemic in certain geographical areas. Knowledge of geographical distribution is often helpful, although any species can

Table 13.1. Dermatophytes (anamorph species) according to Summerbell and Kane (1997), De Hoog et al. (2000), Brasch and Gräser (2005a, b)

Anthropophilic dermatophytes	Main geographic areas of distribution
Trichophyton concentricum	Pacific Islands, Southeast Asia, Central America
T. interdigitale	Worldwide
T. rubrum	Worldwide
former *T. megninii*	Portugal, Spain, Sardinia, Burundi
var. *kanei*	North America, Europe, Africa
var. *krajdenii*	North America, Europe
var. *raubitschekii*	Asia, Mediterranean, Africa, North America
T. schönleinii	Eurasia, North Africa
T. tonsurans	Worldwide
T. violaceum	Eastern Europe, Northern Africa, Central America
former *T. gourvilii*	West and Central Africa
former *T. soudanense*	Africa
former *T. yaoundei*	Central and Southeast Africa
Microsporum audouinii	Worldwide
M. ferrugineum	Asia, Eastern Europe, Africa
Epidermophyton floccosum	Worldwide
Zoophilic dermatophytes	Main host animals
Trichophyton erinacei	Hedgehogs
T. mentagrophytes	Rodents and other animals
T. simii	Monkeys
T. tonsurans i.e., former *T. equinum*	Horses
T. verrucosum	Cattle and other animals
Microsporum amazonicum	Rats
M. canis	Cats, dogs, and other animals
M. gallinae	Fowl, birds
M. nanum	Pigs
M. persicolor	Voles
M. praecox	Horses
Geophilic dermatophytes	
Trichophyton ajelloi	
T. eboreum	
T. flavescens	
T. gloriae	
T. phaseoliforme	
T. terrestre	
T. thuringiense	
T. vanbreuseghemii	
Microsporum cookei	
M. fulvum	
M. gypseum	
M. racemosum	

now be encountered at any location due to global travel and tourism (Brasch and Gräser 2005a, b).

Identification of dermatophyte species begins with the clinical assessment of an infection, taking epidemiological considerations, potential means of acquisition, localization of the infection, and the type of tinea into account. Examination under long-wave ultraviolet light (Wood light) is helpful in detecting certain dermatophyte species (e.g., *M. canis*), which can produce fluorescent metabolites. Infected superficial tissue (hair, nails, scrapings from flaky skin) is then collected from appropriate sites. In many cases, fungal elements can be microscopically detected in KOH-mounts of such specimens. Furthermore, cultures are grown from the collected material at room temperature or at 26–28 °C on mycological agars with supplementary antibiotics to prevent bacterial overgrowth and cycloheximide to suppress molds. The cultures are then identified by their macroscopic morphology, microscopic characteristics of fungal elements, and physiological characteristics (Brasch 1990a–c, 2004; Meinhof 1990). Genetic analyses and other innovative methods like MALDI-TOF are increasingly being used in specialized laboratories. The currently recognized distinct species of dermatophytes are listed in Table 13.1 (Summerbell and Kane 1997; De Hoog et al. 2000; Brasch and Gräser 2005a, b).

III. Dermatophytoses

A. Pathophysiology

Dermatophyte infections (tinea, ringworm) are transmitted by infectious particles, mostly arthrospores (or other spores or hyphal fragments) formed within infected tissue (Richardson 1990; Tsuboi et al. 1994; Rashid 2001). Transmission can occur directly from human or animal to human, or indirectly via contaminated objects such as hairbrushes, towels, pillows, furniture, floors, or objects in contact with animals like fences or blankets. Germination of adherent spores can be observed one day after inoculation, followed by penetration of hyphae after three days (Duek et al. 2004). However, it often takes two weeks or more until a lesion becomes clinically apparent.

The kind of infection that develops in an individual subject depends on the inoculation site, the fungal species involved, and on the host response, which is related to age, gender, immunocompetence, general health, and possibly to genetic disposition (Brasch 1990a–c; Faergemann et al. 2005). Dermatophyte infections generally remain restricted to keratinized tissue (Tsuboi et al. 1994; Ogawa et al. 1998). It can be said that zoophilic and geophilic species cause more acute and inflammatory infections than anthropophilic species, that immunodeficiency or otherwise compromised general health leads to more severe or chronic infections, and that invasion of non-viable tissues (nails, hair) tends to produce persistent infections. For example, an infection caused by *T. verrucosum* is usually much more inflammatory than an infection with *T. rubrum*, infections in AIDS patients are much more refractory than comparable infections in otherwise healthy individuals, and a dermatophyte infection of a nail has a much higher tendency to become chronic than an infection of facial skin.

Inoculation of infectious dermatophyte elements into the skin and adherence of such propagules to the stratum corneum are promoted by defects in the stratum corneum, occlusion, and moist skin. If the transmitted fungal propagules are left undisturbed with sufficient time for attachment (Zurita and Hay 1987) and growth (Aljabre et al.1993; Tsuboi et al. 1994), hyphae begin to spread radially within the stratum corneum. UV exposure may potentially promote the downward growth of dermatophytes (Brasch and Menz 1995). Dermatophytes are well equipped with a spectrum of enzymes, enabling them to penetrate the stratum corneum and digest keratins (Day et al. 1968; Cheung and Maniotis 1973; Apodaca and McKerrow 1989; Summerbell 2000; Viani et al. 2001; Jousson et al. 2004; Kaufman et al. 2005), lipids (Nobre and Viegas 1972; Das and Banerjee 1977; Hellgren and Vincent 1980) and other substrates (Hankin and Anagnostakis 1975; Hopsu-Havu and Tunnela 1976; Calvo et al. 1985; Brasch et al. 1991; Brasch and Zaldua 1994). Keratinolytic proteases appear relevant for their ability to utilize keratin (Ferreira-Nozawa et al. 2006). The epidermal barrier is markedly impaired by fungal invasion of the skin (Jensen et al. 2007). Other pathogenic factors of dermatophytes include xanthomegnin as a toxin (Gupta et al. 2000), mannans as immunosuppressive agents (Dahl 1994; Dahl and Grando 1994), hemagglutinins (Bouchara et al. 1987), and trigger factors for cooperative hemolytic reactions (Schaufuss et al. 2005). Antibiotic substances synthesized by dermatophytes (Youssef et al. 1978; Lappin-Scott et al. 1985) may help them to compete with bacteria. The inflammatory skin response triggered by the spreading dermatophytes is greatest at the advancing margin of the fungal invasion and can be recognized by marked erythema, scaling, and palpable infiltration of the skin, sometimes accompanied by pustules. The accentuated inflammatory rim of the skin lesions often has an arched or circular shape, explaining the use of the term "ringworm" to describe dermatophytoses.

This fungal invasion activates the host's defense mechanisms (Wagner and Sohnle 1995). The inflammatory response comprises innate immunity and, if previous sensitization had occurred, immunological mechanisms as well (Jones 1986; Martínez Roig and Torres Rodriguez 1987; Tagami et al. 1989, Calderon 1989; Brasch et al. 1993; Dahl 1994). Neutrophilic granulocytes and macrophages are attracted and activated by complement-dependent and complement-independent mechanisms (Sato 1983; Swan et al. 1983; Davies and Zaini 1984; Dahl and Carpenter 1986; Suite et al. 1987; Brasch et al. 1991; Kahlke et al. 1996; Campos et al. 2005). The significant increase in epidermal proliferation (Berk et al. 1976; Jensen et al. 2007) is thought to help slough the fungus from the skin surface. Serum factors like transferrin and fatty acids of the skin may help suppress infections by binding iron necessary for fungal growth (King et al. 1975) or by having antimycotic effects (Nathanson 1960; Carlisle et al. 1974; Das and Banerjee 1982; Garg and Müller 1993; Brasch and Friege 1994). Keratinocytes and infiltrating mononuclear cells are activated to release inflammatory cytokines like interferon-gamma, plus a broad panel of interleukins (Miyata et al. 1996; Koga et al. 2001; Nakamura et al. 2002; Shiraki et al. 2006a, b). Cytokines and interleukins are cellular mediators that can activate and regulate defense mechanisms. Epidermal keratinocytes can also express so-called defensins as antimicrobial peptides; human beta-defensin 2 has been detected in dermatophyte-infected skin (Kawai et al. 2006; Jensen et al. 2007). Langerhans cells play a decisive role in the initiation of epidermal cellular immune defense and accumulate at the site of infection (Emtestam et al. 1985; Brasch et al. 1993). Furthermore, lymphocytes invade the lesional skin and are activated (Brasch and Sterry 1992). These mechanisms contribute to a delayed, T cell-mediated immune response that develops during dermatophyte infections (Svejgaard 1986a, b) and is decisive for healing (Dahl 1985; Jones 1993; Dahl 1994). In contrast, a Th2 reponse appears to provide no protection (Leibovici et al. 1995). The terminal hair follicles of the scalp are the main habitat of dermatophytes infecting the scalp. They are connected to the sebaceous glands, forming so-called pilosebaceous units that decisively contribute to the endocrine functions of the skin by producing and metabolizing steroid hormones (Zouboulis 2000). Human steroid hormones encountered in this highly differentiated and distinct environment may have effects on dermatophytes dependening on the fungal species (Chattaway and Townsley 1962; Schär et al. 1986; Brasch and Gottkehaskamp1992; Brasch and Flader 1996; Brasch 1997; Brasch et al. 2002). Since steroid hormones are regulated differently in men and women, they may contribute to gender-related resistances to certain fungal infections (Stevens 1989). In most cases the combined effects of defense mechanisms lead to at least some temporary clearing in the central areas of infected skin, which may, however, be followed by new waves of invasion. This process can be mirrored in characteristic skin lesions (Fig. 13.6). Chronic infections are common with anthropophilic strains.

Antibodies against dermatophyte products were detected under suitable conditions in the blood and tissue of individuals with dermatophyte infections (Holden et al. 1981; Honbo et al. 1984; Svejgaard 1986a, b; Calderon et al. 1987; Lee et al. 1988) and blood cells and skin were shown to specifically respond to dermatophyte compounds (Espiritu et al. 1988; Koga et al. 1993). Such circulating cells and agents may explain hematogenously elicited sterile skin reactions observed in patients with inflammatory dermatophytoses in

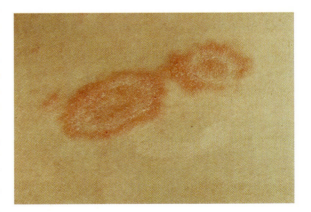

Fig. 13.6. Tinea lesions on glabrous skin with marked erythematous rings and concentric areas of clearing between them

non-infected skin, otherwise known as dermatophytids (Grappel et al. 1974).

B. Clinical Manifestations

As discussed above, dermatophyte infections begin with the adherence of contagious particles to the outmost layer or the epidermis (stratum corneum) or with the inoculation of infectious propagules into the epidermis. Hyphae then spread within the superficial epidermal layers and cause an inflammatory infection. This process can occur at any site on the skin surface, including nails and hair. The disease remains confined to the superficial skin in most cases and is called dermatophytosis, ringworm, or tinea. The latter name is usually supplemented by a Latin term designating the affected site, for example, tinea manuum for infection of the hand or tinea capitis for infection of the scalp. The disease is called tinea profunda when the dermis is invaded. In very rare cases extracutaneous dermatophyte infections have even been proven. Although the occurrence of systemically elicited sterile skin reactions triggered by dermatophyte infections at a distant site is named trichophytid or dermatophytid, the existence of such reactions is generally not acknowledged.

1. Tinea Pedis and Tinea Manuum

Tinea pedis is a dermatophyte infection of the skin of the feet, generally beginning in the toe webs (Fig. 13.7) and often affecting the soles. Tinea pedis is the most common fungal infection worldwide (Masri-Fridling 1996) and has an extremely high prevalence in shoe-wearing populations. Constricting footwear promotes perspiration of the feet and mechanical friction of the skin, followed by interdigital skin maceration, which offers ideal conditions for dermatophyte invasion. Certain professions (miners, athletes, soldiers) are particularly vulnerable to tinea pedis, otherwise known as athletes' foot, due to their high degree of physical activity and their wearing of tight boots or shoes. Individuals living and working in close conditions and sharing bathing facilities have a high risk of foot contact with dermatophyte-contaminated material. Under such conditions, tinea pedum is endemic, affecting nearly 100% of the individuals, even leading to considerable disability in some cases (Götz and Hantschke 1965; Allen and Taplin 1973).

Tinea pedis is most often caused by *T. rubrum*, *T. interdigitale*, variants of *T. mentagrophytes* (Brasch 2001) or other anthropophilic dermatophytes. These species generally lead to aphlegmasic and persistant, chronic lesions. However, geophilic species are also occasionally found due to exposure to contaminated soil and zoophilic species can trigger severe inflammatory infections.

Tinea pedis generally begins within the narrow interdigital space of the toe webs, often as a lesion between digits four and five (Fig. 13.7). It is characterized by itching, scaling, and maceration of the skin. It can spread to the soles and dorsal sides of the feet (Fig. 13.8). Infections of the plantar skin, characterized by dry scaling and often barely recognized as an infection by the host, are particularly likely to become chronic ("moccasin"-type tinea pedis). Scales containing contagious arthrospores are shed and thereby contribute to the further distribution of the disease. However, acute inflammatory courses can also be seen which are associated

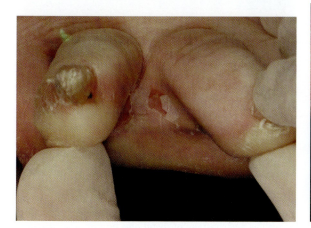

Fig. 13.7. Tinea pedis, interdigital form. Maceration and fissuring of the skin are typical features

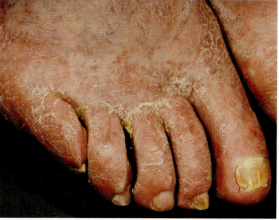

Fig. 13.8. Tinea pedis. The skin shows inflammation accompanied by erythema and scaling

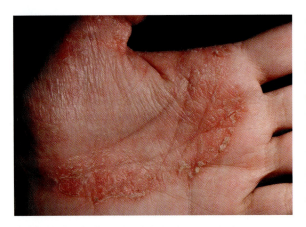

Fig. 13.9. Tinea manum. There is a sharply defined inflammatory response on the palmar skin spreading from the thenar area

with vesicles, pustules, and weeping, erosive lesions. Significant work absence may become necessary, with secondary bacterial infection a common complication in such cases (Leyden 1994). Longstanding tinea pedum is often associated with infection of the toenails (onychomycosis; Szepietowski et al. 2006) and is considered a major fungus reservoir that may lead to infection of other body areas (Daniel and Jellinek 2006). Tinea manuum is a dermatophytosis of the hand (Fig. 13.9). Tinea manuum is much less frequently encountered than tinea pedis (Blank and Mann 1975) and can often be attributed to autoinoculation of fungi from the feet to a hand. The synchronous infection of both feet and one hand ("two feet – one hand syndrome") is a rather common condition, with lesions on the feet normally preceding the infection of the hand (Daniel et al. 1997). Tinea manuum is most often due to *T. rubrum* (Arenas 1991), but zoophilic and geophilic species can also be found, particularly following exposure to animals or soil.

2. Onychomycosis

A fungal infection of the nails is called onychomycosis. When the fungus is a dermatophyte, this condition is named tinea unguium. Onychomycosis is an endemic and extremely common disease in developed countries (Roberts 1992; Perea et al. 2000; Effendy et al. 2005; Szepietowski et al. 2006). It has been estimated that up to 30% of dermatological patients in Germany, 15–20% of people between 40 and 60 years of age in the United States (Zaias 1985) and 8.4% of the entire population in Finland (Heikkilä and Stunn 1995) suffer from onychmycosis with dermatophytes as the main agents. The prevalence of onychomycosis in Europe, mostly due to dermatophytes, has been estimated at approximately 25% (Hay 2005). Dermatophytes can penetrate the nail plate and form mycelium and arthrospores within the nail (Rashid et al.1995; Scherer and Scherer 2004). Tinea unguium is typically seen in older patients and is comparatively rare in children (Philpot and Shuttleworth 1989). Predisposing factors include impaired blood circulation in the toes, endocrine disorders like diabetes mellitus, neurological abnormalities, a compromised immune system, and slow nail growth. Toenails are infected much more often than fingernails. Most cases of onychomycosis worldwide are caused by anthropophilic dermatophytes, predominantly *T. rubrum* (Gupta and Ryder 2004), although almost all other dermatophytes can also be found.

Depending on the route of fungal invasion and the part of the nail involved, different types of onychomycosis can be distinguished (Baran et al. 1998). The most frequent type is distal and lateral subungual onychomycosis, mainly due to *T. rubrum*. Fungal penetration begins with the inoculation of infectious material under the free distal or lateral part of the nail plate. From there the fungus grows proximally through the nail, leading to thickening, yellow discoloration, disintegration of the nail plate, and distal onycholysis (Fig. 13.10). Superficial white onychomycosis begins with invasion of the dorsal nail plate, resulting in a white and roughened nail surface. In

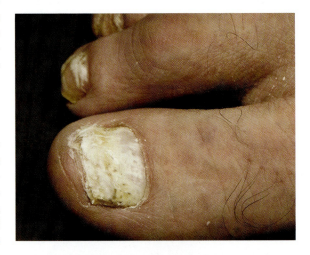

Fig. 13.10. Onychomycosis (tinea unguis). The distal two-thirds of this toenail show discoloration and destruction, as well as a early stage of distal onycholysis. The proximal part of the nail is less severely affected

addition to *T. rubrum*, *T. interdigitale* is often found. Proximal subungual onychomycosis is associated with fungal paronychia at the primary site of infection. In endonyx onychomycosis the nail plate shows lamellar splitting, due to fungal destruction of deep as well as superficial nail parts and a diffuse milky-white discoloration (Tosti et al. 1999). *T. soudanense* and *T. violaceum* are encountered. Finally, complete invasion of the nail results in total destruction of the nail, known as dystrophic onychomycosis. Onychomycosis is usually a chronic disease with little prospect of spontaneous resolution. As the nail plate is not accessible to cellular host defense, it provides a relatively safe habitat for fungi. Accordingly, inflammation is not a feature of onychomycosis.

3. Tinea Capitis and Tinea Barbae

Dermatophyte infections of the scalp and localized hair follicles are termed tinea capitis. Tinea capitis is a significant worldwide problem. It is usually caused by anthropophilic or zoophilic dermatophytes, depending on the country in which it is encountered (Elewski 1996; Aly et al. 2000; Gupta and Summerbell 2000). In tinea barbae the dermatophyte infection affects the bearded areas of the face and neck. The scalp and beard areas have a very high density of terminal hair follicles and associated sebaceous glands. The activity and proliferation of these pilosebaceous units is regulated by steroid hormones and undergoes considerable changes during puberty. The pilosebaceous units can, in fact, be seen as a cutaneous endocrine organ, producing hormones with in situ effects ("intracrinology"; Courchay et al.1996; Kintz et al. 1999; Labrie et al. 2000). This means that epithelial proliferation, including hair growth as well as the activity of sebaceous glands, is different in skin areas with hair follicles than in areas without. The particular ecological characteristics of the hair follicles are likely explanations for some distinctive features of tinea capitis and tinea barbae. Dermatophytes are susceptible to fatty acids and can bind and respond to certain steroid hormones (Capek and Simek 1971; Clemons et al. 1988; Brasch and Gottkehaskamp 1992; Brasch and Flader 1996; Hernández-Hernández et al. 1999; Brasch et al. 2002). It is not surprising therefore, that tinea capitis is an age-dependent infection. Tinea capitis is typically a pediatric disorder and represents one of the most common infectious diseases in children (Elewski 1996; Alvarez and Silverberg 2006).

Tinea capitis often resolves in puberty and is therefore much less frequently seen in adults. In contrast, tinea barbae, confined to adult males, is less often seen than tinea capitis.

Different species of dermatophytes are the predominant pathogens in tinea capitis in distinct geographical areas. In the Americas *T. tonsurans* is already the most common species and continues to increase in incidence (Babel et al. 1990; Foster et al. 2004), while *T. violaceum* is widespread in Africa, and *M. canis* is found most frequently in Northern Europe.

Fungal invasion of the hair causes disintegration and breakage, producing patchy areas of hair stumps or intrafollicular black debris (black dot ringworm; Fig. 13.11). Hyphal elements transform into arthrospores (Okuda et al. 1988, 1989) and, depending on the fungal species, spores accumulate within the hair (endothrix) or as sheaths on the

Fig. 13.11. Tinea capitis caused by anthropophilic *Trichophyton violaceum*. Patchy areas with scaling and hair stumps but without much inflammation are typical for this infection

outside of the hair shafts (ectothrix). *T. schoenleinii* can form endothrix mycelia. This species typically causes heavily crusted scalp lesions (kerion) covered with so-called scutula; this peculiar type of tinea capitis is termed favus. Zoophilic agents like *M. canis* are often acquired from animals like cats and dogs. Subsequent child to child transmission can lead to epidemic outbreaks of tinea capitis in schools. The zoophilic species often cause inflammatory infections with folliculitis and pustules and even regional lymphadenopathy (Fig. 13.12), whereas infections with anthropophilic species are normally mild with only slight erythema (Fig. 13.11). These infections can take chronic courses and persist into adulthood. Under suitable conditions anthropophilic dermatophytes can even exist in an asymptomatic state on the scalp of carriers (Sharma et al. 1988). It is noteworthy that the most common dermatophyte worldwide, *T. rubrum*, only very exceptionally causes scalp infection, similar to *E. floccosum*.

Involvement of the deeper parts of the hair follicles and the dermis is common in tinea barbae (De Lacerda et al. 1981; Bonifaz et al. 2003). Tinea barbae results in often strongly inflammatory and suppurative pustular lesions (Fig. 13.13) accompanied by painful swelling of the draining lymph nodes and even general malaise, especially if zoophilic dermatophytes are involved (Sabota et al. 1996). Hairs are loose and can easily be removed, and purulent material can be discharged. Occupational exposure to infected animals should always be considered

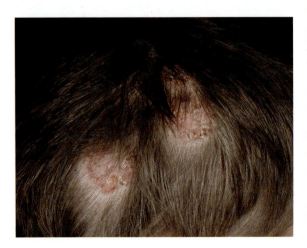

Fig. 13.12. Tinea capitis caused by zoophilic *Microsporum canis*. This is an inflammatory lesion accompanied by loss of hair, erythematous swelling, scaling and crusts

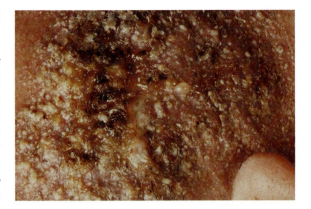

Fig. 13.13. Tinea barbae. The development of strongly inflammatory and suppurative pustular lesions is a typical feature often misdiagnosed as a bacterial infection

in tinea barbae (Rutecki et al. 2000), particularly cattle that may be carriers of *T. verrucosum* (Sabota et al. 1996). An old term for such purulent and inflammatory lesions is sycosis barbae. This type of infection is often initially misdiagnosed as a bacterial disease and is unsuccessfully treated with antibiotics (Roman et al. 2001). Such insufficiently treated lesions ultimately heal with scarring and permanent loss of hair.

4. Tinea Corporis

Dermatophytosis of the glabrous skin is called tinea corporis. Tinea corporis can be directly transmitted between humans. Direct physical contact, such as intimate contact or contact during sports, enables direct transmission to occur. Outbreaks of tinea corporis, referred to as tinea gladiatorum, have been observed among wrestlers and judokas, due to *T. verrucosum* (Frisk et al. 1966) and particularly *T. tonsurans* (Cohen and Schmidt 1992; Hradil et al. 1995; Brasch et al. 1999; Esteve et al. 2006; Shiraki et al. 2006a, b). However, most cases of tinea corporis are probably caused by self-infection through autoinnoculation of fungi from tinea unguium or tinea pedis, or by indirect transmission. *T. rubrum* is the main agent of tinea corporis and can cause chronic and persistent infections (Kemna and Elewski 1996). Reduced general health and impaired immunity are predisposing factors.

A typical lesion of tinea corporis originates from the centrifugal spreading of the fungus within the superficial epidermis, leading to the development of annular and sharply marginated erythematous plaques with raised and scaling borders, and incomplete central clearing in most cases (Fig. 13.14). Multiple lesions can give rise to extensive confluent patterns (Fig. 13.15). The degree of inflammation varies according to the fungal species and the host response. In severe cases, mostly due to geophilic or zoophilic dermatophytes, vesicles, pustules, and folliculitis are seen; anthropophilic species may trigger only mild erythema and scaling (Fig. 13.16).

5. Tinea Cruris

Tinea cruris is a dermatophyte infection of the groin, perineum and perianal region (Fig. 13.17). The prevalent pathogens are *T. rubrum* (Silva-Tavares et al. 2001) and *E. floccosum*. Predisposing factors are perspiration, occlusion and restricted hygiene (e.g., in certain communities, dormitories, etc.). It is noteworthy that tinea cruris is mainly found in men, particularly tinea cruris caused by *E. floccosum* (Dvorak and Otcenasek 1969; Blank and Mann 1975; Alteras and Feuerman 1983). This is possibly related to the susceptibility of the fungi to androgenic hormones (Chattaway and Townskey 1962; Schär et al. 1986; Brasch and Gottkehaskamp 1992; Brasch and Flader 1996; Brasch 1997; Brasch et al. 2002). Serum testosterone levels were found to be significantly lower in patients with Epidermophyton floccosum infections than in controls (Hashemi et al. 2004). Infections in females are comparatively rare, with *E. floccosum* identified in only one out of seven female prostitutes, already at high occupational risk of acquiring the infection

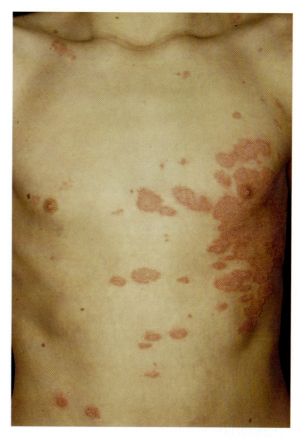

Fig. 13.15. Tinea corporis spreading on the trunk

through direct contact with their clients (Otero et al. 2002). Many patients with tinea cruris also suffer from tinea pedum or onychomycosis (Alteras 1968).

Clinical lesions in tinea cruris are similar to those in tinea corporis. Rather sharply delineated

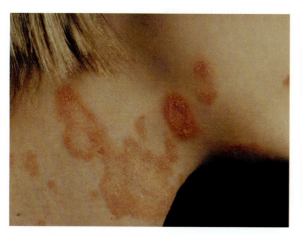

Fig. 13.14. Tinea corporis on the neck with confluent circinate lesions

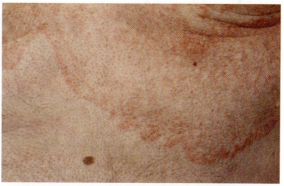

Fig. 13.16. Tinea corporis, a chronic type of infection caused by anthropophilic *Trichophyton rubrum* that triggers only discrete inflammation

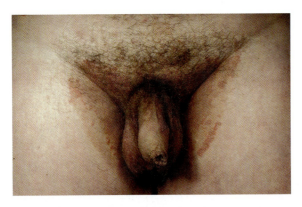

Fig. 13.17. Tinea cruris. This is a typical chronic infection with only minor inflammation and some hyperpigmentation caused by anthropophilic *Epidermophyton floccosum*

erythematous and scaling plaques extend with progressive and accentuated borders (Fig. 13.17). Acute inflammation can lead to vesicles and pustules, while chronic lesions often develop hyperpigmentation. Maceration can occur in the inguinal folds. The scrotum is often involved. Although patients may experience itching, the infection often remains unnoticed.

6. Deep Infections

Although dermatophytes are specialized on superficial infections, deep infections (tinea profunda) can occur preferentially with anthropophilic species and when the host is immunocompromised (Meinhof et al. 1976; Smith et al. 2001). These lesions cannot be clinically diagnosed with certainty and the fungus must be confirmed through histology and culture of biopsy material taken from the organ involved (Squeo et al. 1998; Fig. 13.18). Suppurative nodular eruptions can be caused by perforating granulomas (Majocchi's granuloma) within the deep reticular dermis (Padilha-Goncalves 1980; Kinbara et al. 1981; Chastain et al..2001). The terms dermatophyte pseudomycetoma (Ajello et al. 1980) or mycetoma (West and Kwon-Chung 1988) have been applied in cases showing discharge of whitish-yellow grains composed of hyphae from subcutaneous tissue. Even systemic dissemination of dermatophytes has been seen in very rare cases with lymph node involvement (Tejasvi et al. 2005) and infection of bone, CNS, and other organs (Hironaga et al. 1983; Hofmann 1994).

7. Dermatophytids

Dermatophytids are secondary inflammatory reactions of the skin at sites distant from the skin area infected with the dermatophyte. In particular, strongly inflammatory tinea lesions are viewed as elicitors of dermatophytids. Dermatophytid reactions may be triggered by an immune response carried by circulating antigens, T cells and antibodies. They can manifest as clinically heterogeneous efflorescences, including papules, vesicles, urticae, erythema nodosum-like eruptions, and erythema annulare (El-Mofty and Nada 1965; Veien et al. 1994; Iglesias et al. 1994; Gianni et al. 1996; Romano et al. 2006). Although there are many reports of such reactions in the literature the general concept of such "id"-reactions is not unanimously endorsed by all dermatologists and each case of a suspected dermatophytid must be critically confirmed (Kaaman and Torssander 1983).

IV. Epidemiology

Although dermatophytes are cosmopolitan fungi, certain species are more likely to be found in specific geographic areas and some species are more likely to be associated with particular forms of tinea. It has long been recognized, however, that such associations are not permanent and can change considerably in the course of time (Rippon 1985). Monitoring of these changing ecological and epidemiological patterns and surveying of factors influencing dermatophyte transmission are not only helpful for better understanding of the natural history of dermatophytes but also for the correct assessment of their current roles in diseases. Precise information regarding the present distribution of dermatophytes, their trans-

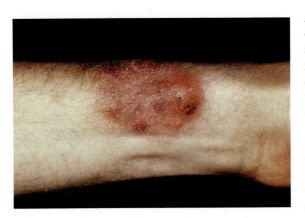

Fig. 13.18. Tinea profunda on the forearm. The erythematous nodular swelling indicates an infection of the dermis

mission and spreading and their relation to general disease patterns is essential for the planning of disease control measures. Epidemiological studies on dermatophytoses have therefore been conducted in many parts of the world. In the following, the up-to-date epidemiological findings pertaining to the most important species of dermatophytes and their related diseases are reviewed.

In **Italy** a study of the fungi responsible for skin mycoses showed that *M. canis* was the most common dermatophyte, followed by *T. rubrum*, *T. mentagrophytes* and *E. floccosum*. Tinea corporis was the most common mycosis, followed by tinea unguis, tinea capitis and tinea pedis. Men were chiefly carriers of tinea cruris and tinea pedis, women of tinea corporis, and children and adolescents of tinea capitis. Several examples are known of infection transmission via interhuman contact, via human–animal contact, and from soil (Filipello Marchisio et al. 1996).

In **Northern Greece** similar results were found in a study of dermatophytoses due to *T. rubrum* during 1981–1990. During this decade *T. rubrum* was the most frequent causative agent of dermatophyte infections in Northern Greece, especially in cases of tinea pedis, tinea cruris, tinea corporis, and tinea unguium, as well as dermatophytosis of the hands. In women tinea pedis and toenail infections prevailed, whereas men were particularly infected in the groin, hands, and face. Chronic follicular dermatophytosis of the lower legs was also present in women, while tinea corporis and fingernail infections showed no significant sex-related differences (Devliotou-Panagiotidou et al. 1992).

In **Central Poland** 7393 cases of dermatophytosis were studied in 1998, including 2204 (29.8%) cases of tinea glabrosa. Etiological factors in descending order were: *M. canis* (23.5%), *T. mentagrophytes* var. *granulosum* (21.6%), *T. rubrum* (17.8%), *T. tonsurans* (10.4%), *T. mentagrophytes* var. *quinckeanum* (6.0%), *M. gypseum* (5.3%), *T. violaceum* (3.7%), *T. mentagrophytes* var. *interdigitale* (2.3%), *M. equinum* (0.7%), *T. verrucosum* (0.4%), *Trichophyton* sp. (0.4%) and *M. cookei* (0.14%). At present tinea of the glabrous skin is the dominant clinical form of dermatophyte infections of skin and skin-appendages in Central Poland (Jeske et al. 1999).

In the general population of **Spain** the prevalence and risk factors of tinea unguium and tinea pedis were investigated in the year 2000. The prevalence of tinea unguium was 2.8% (4% for men, 1.7% for women), and the prevalence of tinea pedis was 2.9% (4.2% for men, 1.7% for women). The etiological agents of tinea unguium were identified as *T. rubrum* (82.1%), followed by *T. mentagrophytes* var. *interdigitale* (14.3%) and *T. tonsurans* (3.5%). *T. rubrum* (44.8%) and *T. mentagrophytes* (44.8%), followed by *E. floccosum* (7%) and *T. tonsurans* (3.4%), were the organisms isolated from patients with tinea pedis. The percentage of subjects suffering from both diseases was 1.1% (1.7% of men, 0.6% of women). In a multivariate logistic regression analysis, age [relative risk (RR) 1.03] and gender (RR 2.50) were independent risk factors for tinea unguium, while only gender (RR 2.65) was predictive for occurrence of tinea pedis. In both analyses, the presence of one of the two conditions was associated with a higher risk of other disease (RR >25; Perea et al. 2000).

In **Switzerland** the dermatophytes are important because 5–10% of dermatological consultations are related to mycotic infections. During an eight-year period (1993–2000) a study was conducted to obtain information about the prevailing species of dermatophytes in the southwest of Switzerland and their patterns of infection. A dermatophyte was detected in 4193 cultures out of a total of 33 725. *T. rubrum* was the most frequently isolated species, accounting for 62.5% of the strains, followed by *T. mentagrophytes* (24.5%) and *M. canis* (5.0%). The relative frequency of isolation of distinct dermatophyte species depends on the frequency of different types of tinea found in different countries, among other factors. The Swiss study reveals the importance of *T. rubrum*, the noteworthy frequency of *M. canis* in the native Swiss population, and the appearance of new species among immigrants (Monod et al. 2002).

In **Slovenia** dermatophyte infections were studied during the period 1995–2002. A total number of 42 494 samples were collected from 33 974 patients suspected of having dermatomycoses. 71.2% positive cultures could be identified. *M. canis* was the most frequenty isolated dermatophyte (46.8%), followed by *T. rubrum* (36.7%), *T. mentagrophytes* var. *interdigitale* (7.9%), and *T. mentagrophytes* var. *mentagrophytes* (4.9%). Less frequently isolated were *M. gypseum*, *T. verrucosum*, *E. floccosum*, *T. tonsurans*, and *T. violaceum*. The most common dermatophyte infections included tinea corporis, onychomycosis, tinea pedis, and tinea faciei. Zoophilic dermatophytes were most commonly recovered from children and adolescents with tinea capitis, tinea corporis, and tinea faciei. Anthropophilic species were identified mostly in adults with tinea pedis, onychomycosis, and tinea inguinalis. During the period studied, a decline in the rate of *M. canis* infection could be recorded, while infections produced by *T. rubrum* increased in frequency (Dolenc-Volj 2005).

In southern Iran the prevalence of dermatophytes was found to be 13.5%, and an incidence of 10.6 per 100 000 person-years was registered over a period of three years (1999–2001). *E. floccosum* was the most frequently isolated dermatophyte (31.4%), followed by *T. rubrum* (18.3%) and *M. gypseum* (4.1%). *E. floccosum* was the most commonly isolated dermatophyte in the age group 20–29 years (30.2%). Tinea corporis (31.4%) was the most common type of infection, followed by tinea cruris (20.7%), tinea manuum (15.4%), tinea capitis (12.4%), tinea pedis (10.6%), tinea faciei (7.1%), and tinea unguium (2.4%). The rates of all types of tinea were higher in males than in females. The anthrophilic species *E. floccosum* was the most common agent of tinea. The most prevalent fungal infection was tinea corporis caused by *E. floccosum* (Falahati et al. 2003).

In the same area of Iran a group of children aged ≤16 years suspected to have dermatophyte infections was examined over a period of three years (1999–2001). The incidence rate of dermatophytoses was 6.6 per 100 000 person-years. *T. violaceum* was the most frequent isolate (28.3%), followed by *M. canis* (15.1%), *E. floccosum* (15.1%), *T. rubrum* (13.2%), *T. mentagrophytes* (11.3%), *M. gypseum* (7.5%), and *T. verrucosum* (5.7%). Tinea capitis (39.6%) was the most common type of infection, followed by tinea corporis (30.2%), tinea faciei (18.9%), and tinea manuum (7.5%; Rastegar Lari et al. 2003).

In **Jordan** a similar spectrum of dermatophytes was found during 1997–1998. The frequencies of etiological agents isolated from patients were as follows: *T. mentagrophytes* var. *interdigitale* (32.7%), *T. rubrum* (28.6%), *E. floccosum* (20.1%), *M. canis* (11.1%), *T. schoenleinii* (4%), *T. verrucosum* (2%), *T. violaceum* (1%), and *M. gypseum* (0.5%). The most common superficial mycotic infection was tinea pedis (35.2%), followed by tinea capitis (23.1%), tinea unguium (21.6%), and tinea corporis (10.6%). Men were mainly affected by tinea cruris and tinea pedis, while women suffered from tinea pedis, tinea unguium, and tinea capitis (Abu-Elteen et al. 2004).

During the period 2003–2004 the first epidemiologic study was conducted in **Algeria**. A total of 1300 male subjects were clinically examined. Clinical diagnosis for tinea pedis and onychomycosis was suspected in 249 and 72 subjects, and confirmed in 197 and 60 cases, respectively. The yeast species *Candida parapsilosis* and the dermatophyte *T. rubrum* were shown to be the most common pathogens in both tinea pedis (*C. parapsilosis* 20.4%; *T. rubrum* 17%) and onychomycosis (*T. rubrum* 35%; *C. parapsilosis* 28.3% Djeridane et al. 2006).

In **Mexico** a total number of 2397 cases of dermatophytoses with superficial cutaneous lesions were reviewed between the years 1978 and 1990. The total numbers of cases were as follows: 726 tinea pedis (30.3%), 613 tinea unguium (25.6%), 441 tinea capitis (18.4%), 395 tinea corporis (16.5%), and 222 tinea cruris (9.3%). The most commonly isolated dermatophyte species was *T. rubrum* (45%), followed by *T. mentagrophytes* (23.7%), *T. tonsurans* (21%), *M. canis* (7.1%), and *E. floccosum* (2.5%). Less frequently *M. audouinii*, *M. gypseum*, *T. violaceum*, and *T. verrucosum* were isolated. Most of the cases were observed in the warmest months of the year (from March to September), and were equally distributed in both genders, except for tinea cruris which was more prevalent in men (3.5:1 ratio; Welsh et al. 2006). Table 13.2 gives an overview of the prevalence of dermatophyte species in different countries (Monod et al. 2002).

In **summary**, these recent epidemiological studies reveal the following facts. Some general trends have become apparent within the last decades. The spectrum of dermatophytes causing skin lesions has changed within the past 70 years. Before the Second World War, especially in Germany, *M. audouinii* and *E. floccosum* were most frequently found. Since the middle of the past century, *T. rubrum* has become the most frequently isolated dermatophyte, accounting for 80–90% of all strains isolated, followed by *T. mentagrophytes*. *T. rubrum* is the most common cause of tinea pedis, nail infections, tinea cruris, and tinea corporis worldwide. Although the incidence of tinea capitis is declining in developing nations, tinea pedis and onychomycosis are becoming more common. This development is typical for Central and Northern Europe and is connected with the increase in the incidence of tinea pedis. In contrast, zoophilic dermatophytes, such as *M. canis* and *T. verrucosum*, are now the most frequently isolated dermatophytes in Southern Europe and in the Middle East and North Africa. The increasing use of athletic

Table 13.2. Prevalence (%) of dermatophyte species in different countries (according to Monod et al. 2002, and references listed in that paper). In this table some old names of dermatophytes are listed that, according to current opinion, do not describe separate species (see Table 13.1): *Trichophyton soudanense* and *T. gourvilii* are now considered to be *T. violaceum*; *T. equinum* is now considered to be *T. tonsurans*. + Only very low numbers of isolates

	Switzerland (Lausanne)	Holland (Leiden)	Germany (Würzburg)	Finland (Oulu)	Poland (Gdansk)	Spain (Galicia)	Italy (Rome)	Greece (Crete)	Iran	USA	USA
	1993–2000	1972–1992	1976–1985	1982–1995	1984–1995	1951–1987	1985–1993	1992–1996	1986–1991	1993–1995	1985–1987
Epidermophyton floccosum	1.0	6.0	2.7	4.4	9.5	11.8	9.3	7.6	14.9	1.1	2.0
Microsporum canis	5.0	1.2	1.2	0.1	27.8	25.5	50.0	25.0	19.4	3.3	4.0
M. gypseum	0.2	0.2	+			5.2	2.3	0.3		0.4	0.6
M. nanum											+
M. ferrugineum		0.1	+							+	+
M. gallinae									0.2		
Trichophyton mentagrophytes	24.6	21.4	19.6	24.2	41.6	21.4	10.6	17.8	20.6	8.5	6.0
T. rubrum	62.1	64.2	73.9	67.5	15.5	24.6	27.0	44.4	16.5	41.3	54.8
T. soudanense	1.6					0.1				+	+
T. violaceum	1.7	0.5	+	0.3	0.3	1.2	0.6	3.1	8.7	0.2	0.08
T. verrucosum	1.3	2.5	1.0	3.2	1.0	3.1	+	1.8	11.5	0.3	0.2
T. tonsurans	0.1		0.2		4.3	3.9	0.16		1.3	44.9	31.3
T. gourvilii	+										
T. schoenleinii		0.3	+	0.06		2.5			5.5	+	+
T. terrestre			0.8								+
T. equinum						0.1					
T. erinacei									0.8		
T. concentricum										+	
Total number of strains analyzed in the study	4193	7111	8973	1543	1544	3351	2823	327	7712	26 815	14 696

shoes by both men and women and the popularity of communal bathing may be contributing factors. In contrast, the anthropophilic agents of scalp infections appear to have been eradicated in developing countries. The exception is *T. tonsurans*-related tinea capitis in North America. *M. canis* is a prevalent agent of tinea capitis in many parts of the world, and this could be related to a close association of humans with their pets. *T. violaceum* is endemic in certain parts of Eastern Europe, Africa, Asia, and South America, but not in North America (Aly 1994; Foster et al. 2004).

These epidemiological findings do not confirm the assumption that patients are genetically predisposed to *T. rubrum* infection in a dominant autosomal pattern (Seebacher 2003a–c).

V. Treatment

Once a dermatophyte is identified as the causal agent of a skin infection, adequate treatment is no longer problematic due to modern and effective antimycotic drugs. However, in many parts of the world, particularly Africa and Asia, such medications are either too expensive or are not available to patients in need. Dermatophytoses are therefore insufficiently treated with traditional methods in many populations, or simply remain untreated. In countries with highly developed medical systems, sufficient treatment is not impaired by a lack of resources. Here, difficulties in eradicating endemic dermatophyte infections and preventing their further spread are mainly due to the fact that wearing of shoes is unavoidable, living in close communities is common, the number of individuals with predisposing factors (old age, reduced immunocompetence, etc.) is increasing, and hygiene is often neglected.

All individuals, regardless of their circumstances, are equally affected by one relevant thereapeutic problem: dermatophytes can colonize nails and hair and often do so. In these habitats dermatophytes cannot be reached by defending host cells of innate or acquired immunity or by antimycotic serum factors or immunoglobulines. In addition, they can form resistant and dormant arthrospores in these niches. Tinea of nails and hair therefore requires particular treatment measures.

In the following some basic agents and principles of treatment are described. For more detailed information the current dermatological literature should be consulted (Gupta et al. 1998; Roberts et al. 2003; Kyle and Dahl 2004; Baran and Kaoukhov 2005; Borgers et al. 2005; Seebacher et al. 2006).

A. Topical Treatment

Superficial dermatophytoses not involving hair and nails, such as tinea corporis, tinea cruris, and tinea faciei, can be adequately treated with topical drugs (Kyle and Dahl 2004) if the lesions are not too extensive. Topical treatment can also be used for onychomycosis, if it is of superficial type or if no more than the distal half of the nail plate is infected. The base used for application of the antimycotic agent can be an ointment, cream, lotion, lacquer, or spray. It should be chosen according to the body site affected and the condition of the skin. Azoles (Hantschke and Reichenberger 1980; Gutierrez 1994), hydroxypyridones (Dittmar 1981; Hänel et al. 1988), allylamines (Balfour and Faulds 1992; Korting et al. 2001), and morpholines (Reinel and Clarke 1992; Zaug and Bergstraesser 1992) have all been proven effective. These agents must be applied once or twice daily for a sufficient period of time to an area exceeding the clinically visible lesion. The azoles are considered fungistatic, whereas hydroxypyridones and allylamines can be fungicidal, so that shorter treatment periods can be justified with the latter agents (Kyle and Dahl 2004).

B. Systemic Treatment

Systemic treatment is usually recommended for superficial tinea when it is widespread or recurrent, or when immunocompromised patients are affected. It should generally be combined with topical treatment. Tinea capitis and tinea profunda usually require systemic treatment. Onychomycosis also requires systemic treatment when more than the distal halves of a few nails are involved, and is usually combined with topical treatment. Well tested antimycotic substances for oral administration in dermatophytoses are griseofulvin, azoles, and allylamines (Reichenberger and Götz 1962; Niewerth and Korting 2000; Bell-Syer et al. 2003; Fleece et al. 2004; Borgers et al. 2005; Dasghaib et al. 2005). Griseofulvin is the oldest agent with a narrow spectrum comprising only dermatophytes. It interferes with the cellular microtubular system and has a fungistatic effect.

Azoles (ketoconazole, fluconazole, itraconazole) inhibit the fungal biosynthesis of ergosterol, which is an essential compound of the fungal cellular membranes. The effect is a fungistatic one. Griseofulvin and the oldest azole, ketoconazole, are gradually being replaced by the newer agents. Griseofulvin is considered less effective in most cases than terbinafine (Bell-Syer et al. 2002). Terbinafine, an allylamine, inhibits fungal ergosterol synthesis as well, but allylamines also initiate a toxic effect and can be fungicidal. The choice of a particular agent in a given situation depends on the type of tinea, potential contraindications for a particular drug and on interactions with the patient's other medications (Back et al. 1992).

Clinically relevant dermatophyte resistance to these substances is currently a rare exception. The ongoing development of new antimycotic agents aims to reduce interactions with other drugs, to eliminate side-effects and to increase fungicidal effects.

C. Special Treatment Aspects of Distinct Forms of Tinea

Tinea capitis is characterized by the formation of arthrospores around and/or within the hair. Cutting of the infected hair helps to rapidly eradicate this source of further fungal distribution. An additional topical treatment is generally recommended and helps stop infectiousness and shorten treatment time. Systemic treatment must normally be continued for several weeks. Tinea capitis due to *M. canis* may be less responsive to terbinafine in children (Baudraz-Rosselet et al.1996) because of reduced drug delivery via their not yet developed sebaceous glands (Fleece et al. 2004). Tinea capitis predominantly affects children. Modern drugs, however, have usually only been tested in adults and are often not yet officially approved for use in children in some countries (Seebacher 2006). These legal issues must also be considered in treatment.

Onychomycosis poses a considerable problem due to the arthrospores in the air-filled cavities of the keratotic material (Effendy and Strassman1999). Arthrospores in these cavities are inaccessible and often resistant to medications in their resting phase (Seebacher 2003a–c). Therefore, clipping of the nails and mechanical or chemical removal of infected nail parts is recommended. Antimycotics should by topically applied with use of solutions or lacquers that enhance drug penetration into the nails (Seebacher 2003a–c). Systemic treatment must be performed for several months, depending on the growth rate of he nails (Iorizzo et al. 2005; Gupta and Tu 2006). Accordingly, toenails require longer treament periods than fingernails. Different modes of intermittent drug application have been developed to reduce the total amount of drug necessary for treatment (Ginter and De Doncker 1998; Evans and Sigurgeirsson 1999). Despite these intensive therapeutic efforts, long-term cure of onychomycosis cannot always be achieved (Epstein 1998; Sigurgeirsson et al. 2002).

Tinea profunda always requires systemic therapy. In such cases potential immunosuppression of the patient must be considered and may necessitate a considerably prolonged treatment period.

D. Hygiene

Antifungal treatment of a patient must be accompanied by measures to prevent reinfection and further distribution of the disease. Since the infectious fungal propagules shed with scales, hair or other particles from tinea lesions remain infectious for a long time, care must be taken to eliminate such materials from the environment (Gupta et al. 2001; Tanaka et al. 2006). Clothing, shoes, home textiles, personal care items, and floors likely to be contaminated must be sufficiently cleaned and disinfected. The source of an infection, the reservoir of the dermatophyte, should be tracked down whenever possible. Depending on the species of dermatophyte in question, family members, other individuals within shared communities, pets, or other animals must be checked for tinea and treated, if necessary.

VI. Conclusions

Infections caused by dermatophytes are termed dermatophytoses, tinea, or ringworm. A dermatophyte is a hyalohyphomycete that can degrade keratin and can cause communicable skin infections in humans and/or animals. Dermatophytoses result from the interaction between these highly specialized fungi and the host defence. All over the world dermatophytoses are among the most common and widespread endemic infectious diseases. Dermatophytoses are typically superficial infections restricted to the skin and its appendages.

Dermatophytes probably developed from non-pathogenic, soil-colonizing fungi into species specialized for particular animal or human hosts as their ecological niche. Accordingly, geophilic, zoophilic, and anthropophilic species are recognized. Most of the dermatophyte species have no known sexual reproduction. Their taxonomy continues to be a matter of debate and recent genetic analyses led to some shifts in nomenclature. The anamorphic asexual states of the dermatophytes belong to three genera (*Trichophyton*, *Microsporum*, *Epidermophyton*) which can easily be distinguished by morphological criteria that are mainly based on the form of conidia. The type of dermatophytosis can give a first clue which pathogenic species may be involved. However, species identification needs a pure culture to assess morphologic and physiologic characteristics. In addition, genetic analyses are gaining increasing importance.

Transmission of dermatophytes (mostly by arthrospores as propagules) can occur directly from human or animal to human, or indirectly via contaminated objects. Usually it takes two weeks or longer until a lesion becomes clinically apparent. The kind of infection that develops depends on the inoculation site, the fungal species involved, and the host response, which is related to age, gender, immunocompetence, general health, and possibly to the genetic disposition of the host. If the transmitted fungal propagules are left undisturbed with sufficient time to attach to the skin of a new host, hyphae begin to spread radially within the stratum corneum. This triggers an inflammatory skin response which is most intense at the advancing margin of the fungal invasion and can clinically be recognized by marked erythema, scaling, and palpable infiltration of the skin, sometimes accompanied by pustules. The accentuated inflammatory rim of the skin lesions often has an arched or circular shape, explaining the term "ringworm" to describe dermatophytoses. Hair follicles, hair shafts, and nails can also be invaded, leading to different patterns of infection. Usually an immunologic host response develops over time.

Tinea pedis is a dermatophyte infection of the skin of the feet that generally starts in the toe webs and often reaches the soles. It is the most common fungal infection worldwide and has an extremely high prevalence in shoe-wearing populations. An infection of the nails by a dermatophyte is named tinea unguium. The nails of the feet are affected most frequently. It is an endemic and very common disease in developed countries. Dermatophyte infections of the scalp and its hair follicles are termed tinea capitis. Tinea capitis is a significant worldwide problem and represents one of the most common infectious diseases in children. Dermatophytosis of the glabrous skin is called tinea corporis. It is mainly caused by *Trichophyton rubrum*. Tinea cruris is a dermatophyte infection of the groin, perineum, and perianal regions. Although dermatophytes are specialists for superficial infections, infections of subcutaneous tissue (tinea profunda) can occur, especially when the host is immunocompromised.

Dermatophytes are cosmopolitan fungi but certain species are more likely to be found in specific geographic areas and some species are more likely to be associated with particular forms of tinea. *T. rubrum* is the most common cause of tinea pedis, nail infections, tinea cruris, and tinea corporis worldwide, followed by *T. mentagrophytes* in Northern Europe. In some countries, however, other species, like *Microsporum canis* or *T. tonsurans*, are isolated more often and some dermatophytes, like *T. violaceum* or *M. audouinii*, preferentially cause tinea capitis or tinea inguinalis (*Epidermophyton floccosum*).

A panel of modern antimycotic agents is available for the treatment of dermatophytoses. Superficial infections of small skin areas can be adequately treated by topical application of such substances in a suitable vehicle. However, widespread infections and infections that involve hair follicles or large parts of the nails require systemic therapy. For topical application hydroxypyridones, allylamines, morpholines, and azoles are preferred agents. For systemic treatment griseofulvin, azoles, and allylamines are approved drugs. Different treatment schemes and periods are recommended for the distinct agents and types of infection, and for the choice in an individual case possible side-effects and drug interactions need to be considered. A combined topical and systemic treatment is often advisable. Removal of infected tissue like nail material can be helpful to shorten therapy and hygienic measures are necessary to prevent reinfections.

Acknowledgement. We thank Mrs. Katherine Houghton for her meticulous proof-reading of the manuscript and for her grammatical corrections.

References

Abu-Elteen KH, Abdul Malek M (2004) Prevalence of dermatophytosis in the Zarqa district of Jordan. Mycopathologia 145:137–142

Ajello L, Kaplan W, Chandler FW (1980) Dermatophyte mycetomas: fact or fiction? Pan Am Health Org Sci Pub 396:135–140

Aljabre SHM, Richardson MD, Scott EM, Rashid A, Shankland GS (1993) Adherence of arthroconidia and germlings of anthropophilic and zoophilic varieties of *Trichophyton mentagrophytes* to human corneocytes as an early event in the pathogenesis of dermatophytosis. Clin Exp Dermatol 18:231–235

Allen AM, Taplin D (1973) Epidemic *Trichophyton mentagrophytes* infections in servicemen. JAMA 226:864–867

Alteras I (1968) Clinical, epidemiological and mycological aspects of tinea cruris. Mykosen 11:451–455

Alteras I, Feuerman EJ (1983) New data on the epidemiology of tinea cruris in Israel. Mycopathologia 83:115–116

Alvarez MS, Silverberg NB (2006) Tinea capitis. Cutis 78:189–196

Aly R (1994) Ecology and epidemiology of dermatophyte infections. J Am Acad Dermatol 31:S21–S25

Aly R, Hay RJ, Del Palacio A, Galimberti R (2000) Epidemiology of tinea capitis. Med Mycol 38[Suppl 1]:183–189

Apodaca G, McKerrow JH (1989) Purification and characterization of a 27,000-Mr extracellular proteinase from *Trichophyton rubrum*. Infect Immun 57:3072–3080

Arenas R (1991) Tinea manuum. Epidemiological and mycological data on 366 cases. Gac Med Mex 127:435–438

Babel DE, Rogers AL, Beneke ES (1990) Dermatophytes of the scalp: incidence, immune response, and epidemiology. Mycopathologia 109:69–73

Back DJ, Tjia JF, Abel SM (1992) Azoles, allylamines and drug metabolism. Br J Dermatol 126[Suppl 39]:14–18

Balfour JA, Faulds D (1992) Terbinafine. A review of its pharmacodynamic and pharmacokinetik properties, and therapeutic potential in superficial mycoses. Drugs 43:259–284

Baran R, Kaoukhov A (2005) Topical antifungal drugs for the treatment of onychomycosis: an overview of current strategies for monotherapy and combination therapy. J Eur Acad Dermatol Venereol 19:21–29

Baran R, Hay RJ, Tosti A, Haneke E (1998) A new classification of onychomycosis. Br J Dermatol 139:567–571

Baudraz-Rosselet F, Monod M, Joccoud S, Frenk E (1996) Efficacy of terbinafine treatment of tinea capitis varies according to the dermatophyte species. Br J Dermatol 35:114–116

Bell-Syer SEM, Hart R, Crawford F, Torgerson DJ, Russel I (2003) Oral treatment for fungal infections of the skin of the foot. Cochrane Database Syst Rev 2004/3:CD003584

Berk SH, Penneys NS, Weinstein GD (1976) Epidermal activity in annular dermatophytosis. Arch Dermatol 112:485–488

Blank F, Mann SJ (1975) *Trichopyton rubrum* infections according to age, anatomical distribution and sex. Br J Dermatol 92:171–174

Blank H, Taplin D, Zaias N (1969) Cutaneous *Trichophyton mentagrophytes* infections in Vietnam. Arch Dermatol 99:135–144

Bonifaz A, Ramirez-Tamayo T, Saul A (2003) Tinea barbae (tinea sycosis): experience with nine cases. J Dermatol 30:898–903

Borgers M, Degreef H, Cauwenbergh G (2005) Fungal infections of the skin: infection process and antimycotic therapy. Curr Drug Targets 6:849–862

Bouchara JP, Robert R, Chabasse D, Senet JM (1987) Evidence for the lectin nature of some dermatophyte haemagglutinins. Ann Inst Pasteur Microbiol 138:729–726

Brasch J (1990a) Erreger und Pathogenese von Dermatophytosen. Hautarzt 41:9–15

Brasch J (1990b) Nachweis und Identifizierung von Pilzen bei Dermatomykosen. Dtsch Med Wochenschr 115:1280–1283

Brasch J (1990c) Kutane Entzündungsreaktion am Beispiel mykotischer Infektionen. Hautarzt 41[Suppl X]:13–15

Brasch J (1997) Hormones, fungi and skin. Mycoses 40[Suppl 1]:11–16

Brasch J (2001) *Trichophyton mentagrophytes* var. *nodulare* causing tinea pedis. Mycoses 44:426–431

Brasch J (2004) Standard and recently developed methods for the differentiation of dermatophytes. Hautarzt 55:136–142

Brasch J, Flader S (1996) Human androgenic steroids affect growth of dermatophytes in vitro. Mycoses 39:387–392

Brasch J, Friege B (1994) Dicarboxylic acids affect the growth of dermatophytes in vitro. Acta Derm Venereol 74:347–350

Brasch J, Gottkehaskamp D (1992) The effect of selected human steroid hormones upon the growth of dermatophytes with different adaptation to man. Mycopathologia 120:87–92

Brasch J, Gräser Y (2005a) Die Variante Raubitschekii von *Trichophyton rubrum* hat Deutschland erreicht. Hautarzt 56:473–477

Brasch J, Gräser Y (2005b) *Trichophyton eboreum* sp. nov. isolated from human skin. J Clin Microbiol 43:5230–5337

Brasch J, Menz A (1995) UV-Susceptibility and negative phototropism of dermatophytes. Mycoses 38:197–203

Brasch J, Sterry W (1992) Immunophenotypical characterization of inflammatory cellular infiltrates in tinea. Acta Derm Venereol 72:345–347

Brasch J, Zaldua M (1994) Enzyme patterns of dermatophytes. Mycoses 37:11–16

Brasch J, Martins BS, Christophers E (1991) Enzyme release by *Trichophyton rubrum* depends on nutritional conditions. Mycoses 34:365–368

Brasch J, Schröder JM, Christophers E (1991) Chemotaktische Wirkung von Dermatophyten-Extrakten auf neutrophile Granulozyten. In: Hornstein OP, Meinhof W (eds) Fortschritte der Mykologie. Perimed, Erlangen, pp 71–76

Brasch J, Martens H, Sterry W (1993) Langerhans cell accumulation in chronic tinea pedis and pityriasis versicolor. Clin Exp Dermatol 18:329–332

Brasch J, Rüther T, Harmsen D (1999) *Trichophyton tonsurans* var. *sulfureum* subvar. *perforans* bei Tinea gladiatorum. Hautarzt 50:363–367

Brasch J, Flader S, Roggentin P, Wudy S, Homoki J, Shackleton CHL, Sippell W (2002) Metabolismus von Dehydroepiandrosteron durch *Epidermophyton floccosum*. Mycoses 45[Suppl 1]:37–40

Calderon RA (1989) Immunoregulation of dermatophytosis. Crit Rev Microbiol 16:339–368

Calderon RA, Hay RJ, Shennan GI (1987) Circulating antigens and antibodies in human and mouse dermatophyotosis: use of monoclonal antibody reactive to phosphorylcholine-like epitopes. J Gen Microbiol 133:2699–2705

Calvo MA, Brughera T, Cabanes FJ, Calvo RM, Trape J, Abarca L (1985) Extracellular enzymatic activities of dermatophytes. Mycopathologia 92:19–22

Campos MR, Russo M, Gomes E, Almeida SR (2005) Stimulation, inhibition and death of macrophages infected with *Trichophyton rubrum*. Microbes Infect 8:372–379

Capek A, Simek A (1971) Antimicrobial agents. IX. Effect of steroids on dermatophytes. Folia Microbiol (Praha) 16:299–302

Carlisle DH, Inouye JC, King RD, Jones HE (1974) Significance of serum fungal inhibitory factor in dermatophytosis. J Invest Dermatol 63:239–241

Chastain MA, Reed RJ, Pankey GA (2001) Deep dermatophytosis: report of 2 cases and review of the literature. Cutis 67:457–462

Chattaway FW, Townsley JD (1962) The effect of certain steroids upon the growth of *Trichophyton rubrum*. J Gen Microbiol 28:437–441

Cheung SSC, Maniotis J (1973) A genetic study of an extracellular elastin-hydrolysing protease in the ringworm fungus *Arthroderma benhamiae*. J Gen Microbiol 74:299–304

Chinelli PA, Sofiatti Ade A, Nunes RS, Martins JE (2003) Dermatophyte agents in the city of Sao Paulo, from 1992 to 2002. Rev Inst Med Trop Sao Paulo 45:259–263

Clemons KV, Schär G, Stover EP, Feldman D, Stevens DA (1988) Dermatophyte-hormone relationships; characterization of progesterone-binding specificity and growth inhibition in the genera *Trichophyton* and *Microsporum*. J Clin Microbiol 26:2110–2115

Cohen BA, Schmidt C (1992) Tinea gladiatorum. New Engl J Med 327:820

Courchay G, Boyera N, Bernard BA, Mahe Y (1996) Messenger RNA expression of steroidgenesis enzyme subtypes in the human pilosebaceous unit. Skin Pharmacol 9:169–176

Dahl MV (1985) Resistance factors in dermatophyte infections. Aust J Dermatol 26:98–101

Dahl MV (1994) Dermatophytosis and the immune response. J Am Acad Dermatol 31:S34–S41

Dahl MV, Carpenter R (1986) Polymorphonuclear leukocytes, complement, and *Trichophyton rubrum*. J Invest Dermatol 86:138–141

Dahl MV, Grando SA (1994) Chronic dermatophytosis: what is special about *Trichophyton rubrum*? Adv Dermatol 9:97–109

Daniel CR, Jellinek NJ (2006) The pedal fungus reservoir. Arch Dermatol 142:1344–1345

Daniel CR, Gupta AK, Daniel MP, Daniel CM (1997) Two feet – one hand syndrome: a retrospective multicenter survey. Int J Dermatol 36:658–660

Das SK, Banarjee AB (1977) Lipolytic enzymes of *Trichophyton rubrum*. Sabouraudia 15:313–323

Das SK, Banerjee AB (1982) Effect of undecanoic acid on the production of exocellular lipolytic and keratinolytic enzymes by undecanoic acid-sensitive and -resistant strains of *Trichophyton rubrum*. Sabouraudia 20:179–184

Dasghaib L, Azizzadeh M, Jafari P (2005) Therapeutic options for the treatment of tinea capitis: griseofulvin versus fluconazole. J Dermatol Treat 16:43–46

Davies RR, Zaini F (1984) *Trichophyton rubrum* and the chemotaxis of polymorphous leucocytes. Sabouraudia 22:65–71

Day WC, Tonic P, Stratman SL, Leeman U, Harmon SR (1968) Isolation and properties of an extracellular protease of *Trichophyton granulosum*. Biochem Biophys Acta 167:597–606

De Hoog GS, Bowman B, Gräser Y, Haase G, El Fari M, Gerrits van den Ende AH, Melzer-Krick B, Untereiner WA (1998) Molecular phylogeny and taxonomy of medically important fungi. Med Mycol 26 [Suppl 1]:52–56

De Hoog GS, Guarro J, Gené J, Figueras MJ (2000) Atlas of clinical fungi, 2nd edn. Centraalbureau voor Schimmelcultures, Amsterdam

De Lacerda MH, Caldeira JB, Dlfino JP, Nunes FP, Goncalves H, Lobo C (1981) Sycosis of the beard (tinea barbae). Analysis of 42 cases. Med Cutan Ibero Lat Am 9:161–178

Devliotou-Panagiotidou D, Koussidou-Eremondi T, Karakatsanis G, Minas A, Chrysomallis F, Badillet G (1992) Dermatophytosis due to *Trichophyton rubrum* in northern Greece during the decades 1981–1990. Mycoses 35:375–380

De Vroey C (1985) Epidemiology of ringworm (dermatophytosis). Semin Dermatol 4:185–200

Dittmar W (1981) Offene, außereuropäische Studien zur Wirksamkeit und Verträglichkeit von Ciclopiroxolamin bei Dermatomykosen. Arzneim Forsch 31[Suppl II]:1381–1385

Djeridane A, Djeridane Y, Ammar-Khodja A (2006) Epidemiological and aetiological study on tinea pedis and onychomycosis in Algeria. Mycoses 49:190–196

Dolenc-Voljč M (2005) Dermatophyte infections in the Ljubljana region, Slovenia. Mycoses 48:181–186

Duek L, Kaufman G, Ulman Y, Berdicevsky I (2004) The pathogenesis of dermatophyte infections in human skin sections. J Infect 48:175–180

Dvorak J, Otcenasek M (1969) Zur Ätiologie und Epidemiologie der Tinea cruris. Hautarzt 20:362–363

Effendy I, Strassman K (1999) Longitudinal studies on survivability and antifungal susceptibility of dermatophytes causing toenail mycosis. Mycoses 42:172–173

Effendy I, Lecha M, Feuilhade de Chauvin M, Di Chiacchio N, Baran R (2005) Epidemiology and clinical classification of onychomycosis. J Eur Acad Dermatol Venereol 19 [Suppl 1]:8–12

Elewski B (1996) Tinea capitis. Dermatol Clin 14:23–31

El-Mofty AM, Nada MM (1965) Erythema annulare centrifugum als Dermatophytid. Hautarzt 16:123–125

Emtestam L, Kaaman T, Hovmark A, Asbrink E (1985) An immunohistochemical staining of epidermal Langerhans' cells in tinea cruris. Acta Derm Venereol 65: 240–243

Epstein E (1998) How often does oral treatment of toenail onychomycosis produce a disease-free nail? An analysis of published data. Arch Dermatol 134:551–554

Espiritu BR, Szpindor-Watson A, Zeitz HJ, Thomas LL (1988) IgE-mediated sensitivity to *Trichophyton rubrum* in a patient with chronic dermatophytosis and Cushing's syndrome. J Allergy Clin Immunol 81:847–851

Esteve E, Rousseau D, Defo D, Poisson DM (2006) Outbreak of cutaneous dermatophytosis in the Judo French Programme in Orleans: September 2004 – June 2005. Ann Dermatol Venereol 123:525–529

Evans EGV, Sigurgeirsson B (1999) Double blind, randomised study comparing continuous terbinafine with intermittent itraconazole in the treatment of toenail onychomycosis. Br Med J 318:1031–1035

Faergemann J, Correia O, Nowicki R, Ro BI (2005) Genetic predisposition – understanding underlying mechanisms of onychomycosis. J Eur Acad Dermatol Venereol 19[Suppl 1]:17–19

Falahati M, Akhlaghi L, Rastegar Lari A, Alaghehbandam R (2003) Epidemiology of dermatophytosis in an area south of Teheran, Iran. Mycopathologia 156:279–287

Ferreira-Nozawa MS, Silveira HCS, Ono CJ, Fachin AL, Rossi A, Martinez-Rossi NM (2006) The pH signaling transcription factor PaC mediates the growth of *Trichophyton rubrum* on human nail in vitro. Med Mycol 44:641–645

Filipello Marchisio V, Preve L, Tullio V (1996) Fungi responsible for skin mycoses in Turin (Italy). Mycoses 39:141–150

Fleece D, Gaughan JP, Aronoff SC (2004) Griseofulvin versus terbinafin in treatment of tinea capitis: a meta-analysis of randomized, clinical trials. Pediatrics 114:1312–1315

Foster FW, Ghannoum MA, Elewski BE (2004) Epidemiologic surveillance of cutaneous fungal infection in the United States from 1999 to 2002. J Am Acad Dermatol 50:748–752

Frisk A, Hilborn H, Melén B (1966) Epidemic occurrence of trichophytosis among wrestlers. Acta Derm Venereol 46:453–456

Garg AP, Müller J (1993) Fungitoxicity of fatty acids against dermatophytes. Mycoses 36:51–63

Gianni C, Betti R, Crosti C (1996) Psoriasiform id reaction in tinea corporis. Mycoses 39:307–308

Ginter G, De Doncker P (1998) An intermittent itraconazole 1-week dosing regimen for the treatment of toenail onychomycosis in dermatological practice. Mycoses 41:235–238

Götz H, Hantschke D (1965) Einblicke in die Epidemiologie der Dermatomykosen im Kohlenbergbau. Hautarzt 16:543–548

Gräser Y (2001) Konsequenzen molekularbiologischer Typisierungsmethoden für die Taxonomie der Dermatophyten. Hygiene Mikrobiol 3:102–107

Gräser Y, El Fari M, Vigalys R, Kuijpers AF, De Hoog GS, Presber W, Tietz H (1999) Phylogeny and taxonomy of the family Arthrodermataceae (dermatophytes) using sequence analysis of the ribosomal ITS region. Med Mycol 37:105–114

Gräser Y, De Hoog S, Summerbell RC (2006) Dermatophytes: recognizing species of clonal fungi. Med Mycol 44:199–209

Grappel SF, Bishop CT, Blank F (1974) Immunology of dermatophytes and dermatophytosis. Bacteriol Rev 38:222–250

Grigoriu D, Delacrétaz J, Borelli D (1984) Lehrbuch der medizinischen Mykologie. Huber, Bern

Gupta AK, Summerbell RC (2000) Tinea capitis. Med Mycol 38:255–287

Gupta AK, Tu LQ (2006) Therapies for onychomycosis: a review. Dermatol Clin 24:375–379

Gupta AK, Einarson TR, Summerbell RC, Sher NH (1998) An overview of topical antifungal therapy in dermatomycoses. A North American perspective. Drugs 55:645–674

Gupta AK, Ahmad I, Borst I, Summerbell RC (2000) Detection of xanthomegnin in epidermal materials infected with *Trichophyton rubrum*. J Invest Dermatol 115:901–905

Gupta AK, Ahmad I, Summerbell RC (2001) Comparative efficacies of commonly used disinfectants and antifungal pharmaceutical spray preparations against dermatophytic fungi. Med Mycol 39:321–328

Gupta AK, Chaudhry M, Elewski B (2003) Tinea corporis, tinea cruris, tinea nigra, and piedra. Dermatol Clin 21:395–400

Gupta AK, Ryder JE, Summerbell RC (2004) Onychomycosis: classification and diagnosis. J Drugs Dermatol 3:51–56

Gutierrez EQ (1994) Multizentrische Phase-III-Studie über Wirksamkeit und Sicherheit von 2%iger Sertaconazol-Creme im Vergleich zu 2%iger Miconazol-Creme bei Patienten mit dermatologischen Pilzerkrankungen. Ärztl Forsch 3:9–43

Hänel H, Raether W, Dittmar W (1988) Evaluation of fungicidal action in vitro and in a skin model considering the influence of penetration kinetics of various standard antimycotics. Ann N Y Acad Sci 544:329–337

Hankin L, Anagnostakis SL (1975) The use of solid media for detection of enzyme production by fungi. Mycologia 67:597–607

Hantschke D, Reichenberger M (1980) Doppelblinde, randomisierte vergleichende in vivo Untersuchungen zwischen den Antimykotika Clotrimazol, Tolnaftat und Naftifin. Mykosen 23:657–668

Harmsen D, Schwinn A, Weig M, Brocker EB, Heesemann J (1995) Phylogeny and dating of some pathogenic keratinophilic fungi using small ribosomal subunit RNA. J Med Vet Mycol 33:299–303

Harmsen D, Schwinn A, Brocker EB, Frosch M (1999) Molecular differentiation of dermatophyte fungi. Mycoses 42:67–70

Hashemi SJ, Sarasgani MR, Zomorodian K (2004) A comparative survey of serum androgenic hormones levels between male patients with dermatophytosis and normal subjects. Jpn J Infect Dis 57:60–62

Hashimoto T, Blumenthal HJ (1978) Survival and resistance of *Trichophyton mentagrophytes* arthrospores. Appl Environ Microbiol 35:274–277

Hay R (2005) Literature review. Onychomycosis. J Eur Acad Dermatol Venereol 19[Suppl 1]:1–7

Heikkilä H, Stunn S (1995) The prevalence of onychomycosis in Finland. Br J Dermatol 133:699–703

Hellgren L, Vincent J (1980) Lipolytic activity of some dermatophytes. J Med Microbiol 13:155–157

Hernández-Hernández F, López-Martinez R, Camacho-Arroyo I. Mendoza-Rodríguez CA, Cerbón MA (1999) Detection and expression of corticosteroid binding protein gene in human pathogenic fungi. Mycopathologia 143:127–130

Hironaga M, Okazaki N, Saito K, Watanabe S (1983) Trichophyton granulomas. Unique systemic dissemination to lymph nodes, testes, vertebrae, and brain. Arch Dermatol 119:482–490

Hofmann H (1994) Soft tissue and bone infections caused by dermatophytes. Hautarzt 45:48

Holden CA, Hay RJ, MacDonald DM (1981) The antigenicity of *Trichophyton rubrum*: in situ studies by an immunoperoxidase technique in light and electron microscopy. Acta Derm Venereol 61:207–211

Honbo S, Jones HE, Artis WMA (1984) Chronic dermatophyte infection: evaluation of the Ig class-specific antibody response reactive with polysaccharide and peptide antigens derived from *Trichophyton mentagrophytes*. J Invest Dermatol 82:287–290

Hopsu-Havu VK, Tunnela E (1976) Production of elastase, urease and sulphatase by *Epidermophyton floccosum* (Harz) Langeron et Milochevitch (1930). Mykosen 20:91–96

Hradil E, Hersle K, Nordin P, Faergemann J (1995) An epidemic of tinea corporis caused by *Trichophyton tonsurans* among wrestlers in Sweden. Acta Derm Venereol 75:305–306

Iglesias ME, Espana A, Idoate MA, Quintanilla E (1994) Generalized skin reaction following tinea pedis (dermatophytids). J Dermatol 21:31–34

Iorizzo M, Piraccini BM, Rech G, Tosti A (2005) Treatment of onychomycosis with oral antifungal agents. Expert Opin Drug Deliv 2:435–440

Jensen JM, Pfeiffer S, Akaki T, Schroeder JM, Kleine M, Neumann C, Proksch E, Brasch J (2007) Barrier function, epidermal differentiation and human β-defensin 2 expression in tinea corporis. J Invest Dermatol 127:720–727

Jeske J, Lupa S, Seneczko F, Glowacka A, Ochecka-Szymanska A (1999) Epidemiology of dermatophytes of humans in Central Poland. Part V. Tinea corporis. Mycoses 42:661–663

Jones HE (1986) Cell-mediated immunity in the immunopathogenesis of dermatophytosis. Acta Derm Venereol Suppl (Stockh) 121:73–83

Jones HE (1993) Immune response and host resistance of humans to dermatophyte infection. J Am Acad Dermatol 28:S12–S18

Jousson O, Lechenne B, Bontems O, Mignon B, Reichard U, Barblan J, Quadroni M, Monod M (2004) Secreted subtilisin gene family in Trichophyton rubrum. Gene 339:79–88

Kaaman T, Torssander J (1983) Dermatophytid – a misdiagnosed entity? Acta Derm Venereol 63:404–408

Kahlke B, Brasch J, Christophers E, Schröder JM (1996) Dermatophytes contain a novel lipid-like leukocyte activator (LILA). J Invest Dermatol 107:108–112

Kates SG, Nordstrom KM, McGinley KJ, Leyden JJ (1990) Microbial ecology of interdigital infections of toe web spaces. J Am Acad Dermatol 22:578–582

Kaufman G, Berdicevsky I, Woodfolk JA, Horwitz BA (2005) Markers for host-induced gene expression in *Trichophyton dermatophytosis*. Infect Immun 73:6584–6590

Kawai M, Yamazaki M, Tsuboi R, Miyajima H, Ogawa H, Tsuboi R (2006) Human beta-defensin-2, an antimicrobial peptide, is elevated in scales collected from tinea pedis patients. Int J Dermatol 45:1389–1390

Kemna ME, Elewski BE (1996) A U.S. epidemiologic survey of superficial fungal diseases. J Am Acad Dermatol 35:539–542

Kinbara T, Hayakawa Y, Taniguchi S, Takiguchi T (1981) Multiple subcutaneous *Trichophyon rubrum* abscesses – a case report and review of the Japanese literature. Mykosen 24:588–593

King RD, Khan HA, Foye JC, Greenberg JH, Jones HE (1975) Transferrin, iron, and dermatophytes. I. Serum dermatophyte inhibitory component definitively identified as unsaturated transferring. J Lab Clin Med 86:204–212

Kintz P, Cirimele V, Ludes B (1999) Physiological concentrations of DHEA in human hair. J Anal Toxicol 23:424–428

Koga T, Duan H, Urabe K, Furue M (2001) Immunohistochemical detection of interferon-gamma-producing cells in dermatophytosis. Eur J Dermatol 11:105–107

Korting HC, Tietz H-J, Bräutigam M, Mayser P, Rapatz G, Pauls C (2001) One week terbinafine 1% cream (Lamisil) once daily is effective in the treatment of interdigital tinea pedis: a vehicle controlled study. Med Mycology 39:335–340

Kyle AA, Dahl MV (2004) Topical therapy for fungal infections. Am J Clin Dermatol 5:443–451

Labrie F, Luu-The V, Labrie C, Pelletier G, El-Aöly M (2000) Intracrinology and the skin. Horm Res 54:218–229

Lappin-Scott HM, Rogers ME, Adlard MW, Holt G, Noble WC (1985) High-performance liquid chromatographic identification of betalactam antibiotics produced by dermatophytes. J Appl Bacteriol 59:437–441

Lee KH, Lee JB, Lee MG, Song DH (1988) Detection of circulating antibodies to purified keratinolytic proteinase in sera from guinea pigs infected with *Microsporum canis* by enzyme-linked immunosorbent assay. Arch Dermatol Res 280:45–49

Leibovici V, Evron R, Axelrod O, Westerman M, Shalit M, Barak V, Frankenburg S (1995) Imbalance of immune responses in patients with chronic and widespread fungal skin infection. Clin Experimental Dermatol 20:390–394

Leyden JL (1994) Tinea pedis pathophysiology and treatment. J Am Acad Dermatol 31:S31–S33

Martínez Roig A, Torres Rodriguez JM (1987) The immune response in childhood dermatophytoses. Mykosen 30:574–580

Masri-Fridling GD (1996) Dermatophytes of the feet. Dermatol Clin 1996:33–40

Matsumoto T, Ajello L (1987) Current taxonomic concepts pertaining to the dermatophytes and related fungi. Int J Dermatol 26:491–499

Meinhof W (1990) Isolierung und Identifizierung von Dermatophyten. Zentralbl Bakteriol 73:229–245

Meinhof W, Hornstein OP, Scheiffarth F (1976) Multiple subkutane *Trichophyton rubrum*-Abszesse. Pathomorphose einer generalisierten superfiziellen Tinea bei gestörter Infektabwehr. Hautarzt 27:318–327

Miyata T, Fujimura T, Masuzawa M, Katsuoka K, Nishiyama S (1996) Local expression of IFN-gamma mRNA in skin lesions of patients with dermatophytosis. J Dermatol Sci 13:167–172

Monod M, Jaccoud S, Zaugg C, Lechénne B, Baudraz F, Panizzon R (2002) Survey of dermatophyte infections in the Lausanne area (Switzerland). Dermatology 205:201–203

Nakamura Y, Kano R, Hasegawa A, Watanabe S (2002) Interleukin-8 and tumor necrosis factor alpha production in human epidermal keratinocytes induced by *Trichophyton mentagrophytes*. Clin Diagn Lab Immunol 9:935–937

Nathanson RB (1960) The fungistatic action of oleic, linoleic, and linolenic acids on *Trichophyton rubrum* in vitro. J Invest Dermatol 35:261–263

Niewerth M, Korting HC (2000) The use of systemic antimycotics in dermatotherapy. Eur J Dermatol 10:155–160

Nobre G, Viegas MP (1972) Lipolytic activity of dermatophytes. Mycopathol Mycol Appl 46:319–323

Noguchi H, Hiruma M, Kawada A, Ishibashi A, Kono S (1995) Tinea pedis in members of the Japanese Self-Defence Forces: relationships of its prevalence and its severity with length of military service and width of interdigital spaces. Mycoses 38:494–499

Odom R (1993) Pathophysiology of dermatophyte infections. J Am Acad Dermatol 28:S2–S7

Ogawa H, Summerbell RC, Clemons KV, Koga T, Ran YP, Rashid A, Sohnle PG, Stevens DA, Tsuboi R (1998) Dermatophytes and host tissue defence in cutaneous mycoses. Med Mycol 36[Suppl 1]:166–173

Okuda C, Ito M, Sato Y (1988) Fungus invasion into human hair tissue in black dot ringworm: light and electron microscopic study. J Invest Dermatol 90:729–733

Okuda C, Ito M, Sato Y, Oka K (1989) Fungus invasion of human hair tissue in tinea capitis caused by *Microsporum canis*: light and electron microscopic study. Arch Dermatol Res 281:238–246

Otero L, Palacio V, Vazquez F (2002) Tinea cruris in female prostitutes. Mycopathologia 153:29–31

Padilha-Goncalves A (1980) Granulomatous dermatophytia. Pan Am Health Org Sci Pub 396:141–147

Perea S, Ramos MJ, Garau M, Gonzalez A, Noriega AR, Palacio A del (2000) Prevalence and risk factors of tinea unguium and tinea pedis in the general population in Spain. J Clin Microbiol 38:3226–3230

Philpot CM, Shuttleworth D (1989) Dermatophyte onychomycosis in children. Clin Exp Dermatol 14:203–205

Rashid A (2001) Arthroconidia as vectors of dermatophytosis. Cutis 67[Suppl]:23

Rashid A, Scott E, Richardson MD (1995) Early events in the invasion of the human nail plate by *Trichophyton mentagrophytes*. Br J Dermatol 133:932–940

Rastegar Lari A, Akhlaghi L, Falahati M, Alaghehbandam R (2003) Characteristics of dermatophytosis among children in an area south of Teheran, Iran. Mycoses 48:32–37

Reichenberger M, Götz H (1962) Zur Therapie der Tinea unguium mit Griseofulvin. In: Götz H (ed) Die Griseofulvinbehandlung der Dermatomykosen. Springer, Heidelberg, pp 68–71

Reinel D, Clarke C (1992) Comparative efficacy and safety of amorolfine nail lacquer 5% in onychomycosis, once-weekly versus twice-weekly. Clin Exp Dermatol 17[Suppl 1]:44–49

Richardson MD (1990) Diagnosis and pathogenesis of dermatophyte infections. Br J Clin Practice 44[Suppl 71]:98–102

Rippon JW (1985) The changing epidemiology and emerging patterns of dermatophyte species. In: McGinnis MR (ed) Current topics in medical mycology. Springer, Heidelberg, pp 208–234

Rippon JW (1988) Medical mycology, 3rd edn. Saunders, Philadelphia

Roberts DT (1992) Prevalence of dermatophyte onychomycosis in the United Kingdom: results of an omnibus survey. Br J Dermatol 126[Suppl 39]:23–27

Roberts DT, Taylor WD, Boyle J (2003) Guidelines for treatment of onychomycosis. Br J Dermatol 148:402–410

Roman C, Massai L, Gianni C, Crosti C (2001) Case reports. Six cases of infection due to *Trichophyton verrucosum*. Mycoses 44:334–337

Romano C, Rubegni P, Ghilardi A, Fimiani M (2006) A case of a bullous tinea pedis with dermatophyid reaction caused by *Trichophyton violaceum*. Mycoses 49:249–250

Rutecki GW, Wurtz R, Thomson RB (2000) From animal to man: tinea barbae. Curr Infect Dis Rep 2:433–437

Sabota J, Brodell R, Rutecki GW, Hoppes WL (1996) Severe tinea barbae due to *Trichophyton verrucosum* infection in dairy farmers. Clin Infect Dis 23:1308–1310

Sato A (1983) Ultrastructure of the epidermis in tinea cruris. J Dermatol 10:511–518

Schär G, Stover EP, Clemons KV, Feldman D, Stevens DA (1986) Progesterone binding and inhibition of growth in *Trichophyton mentagrophytes*. Infect Immun 52:763–767

Schaufuss P, Brasch J, Steller U (2005) Dermatophytes can trigger cooperative (CAMP-like) haemolytic reactions. Brit J Dermatol 153:584–590

Scherer WP, Scherer MD (2004) Scanning electron microscope imaging of onychomycosis. J Am Podiatr Med Assoc 94:356–362

Seebacher C (2003a) The change of dermatophyte spectrum in dermatomycosis. Mycoses 46:42–46

Seebacher C (2003b) Action mechanisms of modern antifungal agents and resulting problems in the management of onychomycosis. Mycoses 46:506–510

Seebacher C (2003c) Onychomykose. J Dtsch Dermatol Ges 1:74–78

Seebacher C, Abeck D, Brasch J, Cornely O, Daeschlein G, Effendy I, Ginter-Hanselmayer G, Haake N, Hamm G, Hipler UC, Hof H, Korting HC, Kramer A, Mayser P, Ruhnke M, Schlacke KH, Tietz HJ (2006) Tinea capitis. J Dtsch Dermtol Ges 12:1085–1092

Sharma V, Hall JC, Knapp JF, Sarai S, Galloway D, Babel DE (1988) Scalp colonization by *Trichophyton tonsurans*

in an urban pediatric clinic. Arch Dermatol 124: 1511–1513
Shiraki Y, Hiruma M, Hirose N, Sugita T, Ikeda S (2006a) A nationwide survey of *Trichophyton tonsurans* infection among combat sport club members in Japan using a questionnaire form and the hairbrush method. J Am Acad Dermatol 54:622–626
Shiraki Y, Ishibashi Y, Hiruma M, Nishikawa A, Ikeda S (2006b) Cytokine secretion profiles of human keratinocytes during *Trichophyton tonsurans* and *Arthroderma benhamiae* infections. J Med Microbiol 55:1175–1185
Sigurgeirsson B, Olafsson JH, Steinsson JP, Paul C, Billstein S, Evans EG (2002) Long-term effectiveness of treatment with terbinafine vs itraconazole in onychomycosis: a 5-year blinded prospective follow-up study. Arch Dermatol 138:353–357
Silva-Tavares H, Alchorne MM, Fischman O (2001) Tinea cruris epidemiology (Sao Paulo, Brazil). Mycopathologia 149:147–149
Simpanya MF (2000) Dermatophytes: their taxonomy, ecology and pathogenicity. In: Kushwaha RKS, Guarro J (eds) Biology of dermatophytes and other keratinophilic fungi. Revista Iberoamericana de Micología, Bilbao, pp 1–12
Sinski JT, Kelley LM (1991) A survey of dermatophytes from human patients in the United States from 1985 to 1987. Mycopathologia 114:117–126
Smith KJ, Welsh M, Skelton H (2001) Trichophyton rubrum showing deep dermal invasion directly from the epidermis in immunosuppressed patients. Br J Dermatol 145:344–348
Squeo RF, Beer R, Silvers D, Weitzman I, Grossman M (1998) Invasive *Trichophyton rubrum* resembling blastomycosis infection in the immunocompromised host. J Am Acad Dermatol 39:379–380
Stevens DA (1989) The interface of mycology and endocrinology. J Med Vet Mycol 27:133–140
Suite M, Moore MK, Hay RJ (1987) Leucocyte chemotaxis to antigens of dermatophytes causing scalp ringworm. Clin Exp Dermatol 12:171–174
Summerbell RC (2000) Form and function in the evolution of dermatophytes. In: Kushwaha RKS, Guarro J (eds) Biology of dermatophytes and other keratinophilic fungi. Revista Iberoamericana de Micología, Bilbao, pp 30–43
Summerbell RC, Kane J (1997) The genera *Trichophyton* and *Epidermophyton*. In: Kane J, Summerbell R, Sigler L, Krajden S, Land G (eds) Laboratory handbook of dermatophytes. Star, Belmont, pp 131–191
Summerbell RC, Haugland RA, Li A, Gupta AK (1999) rRNA gene internal transcribed spacer 1 and 2 sequences of asexual, anthropophilic dermatophytes related to *Trichophyton rubrum*. J Clin Microbiol 17:4005–4011
Svejgaard E (1986a) Humoral antibody responses in the immunopathogenesis of dermatophytosis. Acta Derm Venereol [Suppl 121]:85–91
Svejgaard E (1986b) Immunologic properties of a fraction of *Trichophyton rubrum* with affinity to concanavalin A. J Med Vet Mycol 24:271–280

Swan JW, Dahl MV, Coppo PA, Hammerschmidt DE (1983) Complement activation by *Trichophyton rubrum*. J Invest Dermatol 80:156–158
Szepietowski JC, Reich A, Garlowska E, Kulig M, Baran E (2006) Factors influencing coexistence of toenail onychomycosis with tinea pedis and other dermatomycoses. Arch Dermatol 142:1279–1284
Tagami H, Kudoh K, Takematsu H (1989) Inflammation and immunity in dermatophytosis. Dermatologica 179[Suppl 1]:1–8
Takashio M (1979) Taxonomy of the dermatophytes based on their sexual states. Mycologia 71:968–976
Tanaka K, Katoh T, Irimajiri J, Taniguchi H, Yokozeki H (2006) Preventive effects of various types of footwear and cleaning methods on dermatophyte adhesion. J Dermatol 33:528–536
Tejasvi T, Sharma VK, Sethurman G, Singh MK, Xess I (2005) Invasive dermatophytosis with lymph node involvement in an immunocompetent patient. Clin Exp Dermatol 30:506–508
Tosti A, Baran R, Piraccini BM, Fanti PA (1999) "Endonyx" onychomycosis: a new modality of nail invasion by dermatophytes. Acta Derm Venereol 79:52–53
Tsuboi R, Ogawa H, Bramono K, Richardson MD, Shankland GS, Crozier WJ, Sei Y, Ninomiya J, Nakabayashi A, Takaiuchi I, Payne CD, Ray TL (1994) Pathogenesis of superficial mycoses. J Med Vet Mycol 32[Suppl 1]: 91–104
Veien NK, Hattel T, Laurberg G (1994) Plantar *Trichophyton rubrum* infections may cause dermatophytids on the hands. Acta Derm Venereol 74:403–404
Viani FC, Dos Santos JI, Paula CR, Larson CE, Gambale W (2001) Production of extracellular enzymes by *Microsporum canis* and their role in its virulence. Med Mycol 39:463–468
Wagner DK, Sohnle PG (1995) Cutaneous defenses against dermatophytes and yeasts. Clin Microbiol Rev 8:317–335
Weitzman I, Padhye AA (1996) Dermatophytes: gross and microscopic. Dermatol Clin 14:9–22
Weitzman I, Summerbell RC (1995) The dermatophytes. Clin Microbiol Rev 8:240–259
Welsh O, Welsh E, Ocampo-Candiani J, Gomez M, Vera-Cabrera L (2006) Dermatophytes in Monterrey, Mexico. Mycoses 49:119–123
West BC, Kwon-Chung KJ (1980) Mycetoma caused by *Microsporum audouinii*: first reported case. Am J Clin Pathol 73:447–454
Youssef N, Wyborg CHE, Holt G (1978) Antibiotic production by dermatophyte fungi. J Gen Microbiol 105: 105–111
Zaias N (1985) Onychomycosis. Dermatol Clin 3:445
Zaug M, Bergstraesser M (1992) Amorolfine in the treatment of onychomycoses and dermatomycoses (an overview). Clin Exp Dermatol 17[Suppl 1]:61–70
Zouboulis CC (2000) Human skin: an independent peripheral endocrine organ. Horm Res 54:230–242
Zurita J, Hay RJ (1987) Adherence of dermatophyte microconidia and arthroconidia to human keratinocytes in vitro. J Invest Dermatol 89:529–534

Biosystematic Index

A
A. aleyrodis, 36
Acanthamoeba castellanii, 79
Aedes aegypti, 11–13
Allantomyces, 7, 12
Amblyomma americanum, 41
Amoebidium, 3, 5, 7, 8, 12, 15
Amoebidium parasiticum, 7, 12, 15
Anopheles gambiae, 53
Arthroderma, 264
Arthrodermataceae, 264
Arundinula abyssicola, 5
Arundinula, 7, 12
Ascomycotina, 264
Ashbya gossypii, 96, 97, 100
Aspergillus, 21–29, 38, 63, 64, 67, 77, 78, 176
    A. flavus, 64
    A. fumigatus, 64, 67, 69–72, 75–78, 86, 155, 156, 188, 189, 194, 235, 236, 238, 240, 243, 252
    A. nidulans, 49, 75, 253
    A. niger, 39, 75, 76
    A. oryzae, 75, 253
    A. terreus, 64, 75
Astreptonema, 7, 12
Austrosmittium biforme, 15, 16

B
B. amorpha, 51
B. (Cordyceps) bassiana, 34, 43
B. cylindrosporum, 40
B. nivea, 40, 51
B. tabaci, 41
B. tabaci-argentifolii, 50
Barbatospora, 8
Beauveria bassiana, 34–40, 42–53
Beauveria brongniartii, 36, 40, 41
Beauveria sulfurescens, 48
Bemisia argentifolii, 39, 41, 44
Bemisia tabaci, 60 (In ref. only-Chapter 3)
Bipolaris, 28
Blaberus giganteus, 37, 52
Blastomyces dermatitidis, 193, 243
Blastoschizomyces capitatus, 229
Brassica napus, 41

C
C. albicans genome, 194, 196
Caenorhabditis elegans, 78, 79
Candida spp., 63, 65, 72, 73, 83, 85, 86, 225–227, 230
C. albicans, 47, 63, 65, 66, 68, 69, 71–74, 77, 78, 83–91, 95–110, 155, 188, 194–196, 198, 200–206, 208–217, 233, 237, 239, 240, 243, 249, 256
    commensal organism, 194
    pathogenesis, 188, 194, 212
C. dubliniensis, 99
C. glabrata, 83, 155, 188, 194, 249
C. kefyr, 73
C. krusei, 73, 83, 155
C.lusitaniae, 73, 83
C.parapsilosis, 83
C.tropicalis, 73, 83, 155
growth, 84, 86, 87, 89, 91
morphogenesis, 89–91
morphogenetic transition, 84
Capniomyces, 8
Chilo partellus, 48
Coccidiodes immitis, 243
Coccidioides posadasii, 248, 254, 255
Coccoides imitis, 193
Conidiobolus coronatus, 38, 49
Cordyceps, 34, 42, 43
Cordyceps heteropoda, 40, 42
Cosmopolites sordidus, 43
Cryptococcus, 63, 66, 155
    C. neoformans, 63, 66, 69, 71, 72, 74, 75, 78, 84–87, 91, 188, 190, 194, 227, 243, 254, 256
    C. neoformans varians *gattii*, 74
    C. neoformans varians *neoformans*, 74
Culex pipiens, 9
Culex quinquefasciatus, 53
Cunninghamella bertholletiae, 26
Curvularia, 28

D
Daphnia, 5
Dermacentor variabilis, 53
Dermatophytes, 71
Dictyostelium discoideum, 79
Distribution, 118, 127, 129, 140
Drechslera, 28
Drosophila melanogaster, 79, 84, 245, 246

E
E. coli, 38
E. coronata, 36
E. muscae, 39
Eccrinidus, 12
Eccrinoides, 12
EFG1, 104–109
Ejectosporus, 7

Enteromyces, 5
    E. callianassae, 6, 7, 13
    E. sexuale, 5
Entomophaga aulicae, 48
Entomophaga maimaiga, 48
Entomophthorales, 33, 42, 51
Ephemerellomyces, 7
Epidermophyton, 193, 265, 266, 273, 274, 277, 280
Epidermophyton floccosum, 266, 273, 274, 277, 280
Erynia, 36
Erynia neoaphidis, 47

F
Filobasidiaceae, 116
Furculomyces boomerangus, 15, 16
Furculomyces, 8
Fusarium, 21, 22, 26, 27, 29, 155

G
Galleria mellonella, 37, 41, 48, 79
Genistelloides, 8
Genistellospora, 8
Genistellospora homothallica, 9, 11
Glotzia, 12
β(1–3) Glucan, 244, 247, 248, 250, 254, 256
Graminella, 7
Graminelloides, 8

H
Harpella, 8
Harpella melusinae, 7, 9, 10
Hebeloma cylindrosporum, 91
HGC1, 97, 104
Histoplasma capsulatum, 188, 193, 243, 248
Hoplia philanthus, 49

I
Ixodes scapularis, 53

L
L. hesperus, 41
Lagenidium giganteum, 36
Lecanicillium muscarium, 51
Legeriomyces, 12
Limulus polyphemus, 244
Lipid dependence, 115, 116, 119, 121
Liposcelis bostrychophila, 39
Lymantria dispar, 37, 48, 52

M
M. anisopliae, 34–42, 44–53
M. anisopliae var. *anisopliae*, 39, 50, 51
M. domestica, 39
M. flavoviride, 34, 35, 38, 43
Magnaporthe grisea, 85
Malassezia, 115–117
    M. dermatis, 116, 117, 121, 137
    M. equi, 116–118
    M. furfur (Robin), 115–122, 124, 125, 127–131, 133–135, 137–139
    M. globosa, 116–118, 120, 128, 129, 135, 137, 139
    M. japonica, 116, 117, 137
    M. nana, 116–118
    M. obtusa, 116–118, 129
    M. pachydermatis (Weidman), 115–119, 121, 122, 125, 128, 129, 131–133, 136, 139, 140
    M. restricta, 116–118, 129, 137
    M. sloofiae, 116, 118, 131
    M. sympodialis, 116–118, 120, 121, 128, 129, 131, 134, 137–139
    M. yamatoensis, 116, 117
Malassezia folliculitis, 133, 136
Malassezia sepsis, 136
Manduca sexta, 34, 37, 38, 48, 52
Mecostibus pinivorus, 43
Metarhizium, 40, 48
Methanosarcina thermophila, 85
Microsporum, 193, 265, 266, 272, 277, 280
Microsporum canis, 265, 272, 280
Molecular differentiation, 117
Mucor, 155

N
Neurospora crassa, 39, 40, 85
Nezara viridula, 44
Nomuraea rileyi, 36, 43, 44

O
Onygenales, 264
Oospora destructor, 41
Orchesellaria, 5
Orphella, 6
Oxidus gracilis, 7

P
P. neoaphidis, 51
Paecilomyces, 40, 41
    P. farinosus, 40
    P. fumosoroseus, 35, 39–44, 49, 50
Palavascia, 7
Palavascia sphaeromae, 7
Paracoccidioides brasiliensis, 193, 243, 247
Paramoebidium, 6, 8, 12, 13
Pennella, 8
Pichia pastoris, 76
Plasmodium falciparum, 87
Plutella xylostella, 49
Pneumocystis, 248, 256
    P. carinii, 248, 250
    P. jirovecii, 255
Popilla japonica, 37, 52
Pseudallescheria boydii, 21, 22, 27–29, 255, 256
Pteromaktron, 6

R
Rhipicephalus sanguineus, 53
Rhizopus, 26, 155
    R. oryzae, 26
Rhodnius prolixus, 48
Rhopalosiphum padi, 47
RIM101, 105, 106

S
S. exigua, 43, 47
S. frugiperda, 41

Saccharomyces, 63
    S. cerevisiae, 40, 68, 77, 78, 86, 91, 95–97, 102, 104–109, 165, 167, 170–173
Saksenaea vasiformis, 26
Scedosporium apiospermum, 255
Schizaphis graminum, 41
Simuliomyces, 8, 12
Sitobion avenae, 47
Smittium, 6–9, 12–16
    S. culisetae, 8, 11–16
    S. morbosum, 8
Stachylina, 6, 7
Stipella, 8
Streptomyces avermitilis, 39, 50
Streptomyces coelicolor, 39, 50
Synechocystis, 87

T
T. geodes, 40, 41
T. niveum, 40, 41
Taeniella, 7, 12
Taeniella carcini, 6, 7, 13
Tenebrio molitor, 38
Trichoderma harzianum, 38

Trichophyton, 193, 264–266, 271, 273, 275, 277, 280
Trichosporon spp., 155, 228, 229, 231
Trichophyton mentagrophytes, 265
Trichophyton rubrum, 264, 273, 280
Trichophyton schoenleinii, 272, 276, 277
Trichophyton tonsurans, 264, 266, 271, 272, 275–278, 280
Trichophyton verrucosum, 266, 267, 272, 275–277
Trichophyton violaceum, 265, 266, 271, 275–278, 280

V
Verticillium, 36, 37, 48, 50, 51
    V. albo-atrum, 37
    V. coccosporum, 37
    V. dahliae, 37, 50
    V. hemipterigenum, 42
    V. lecanii, 36, 37, 47, 48, 50, 51
    V. fungicola, 37

X
Xenopus, 42

Z
Zootermopsis angusticollis, 52
Zygopolaris, 6, 12

Subject Index

A
Actin Cytoskeleton, 96, 97, 99, 103, 105, 106
Adaptive immune response, 132, 133
Adaptive immune system, 233, 238, 252
Adenylate cyclase, 86–91, 103, 105, 107, 108
    soluble, 86, 87
AEDS, 129–132, 136, 137
Agent, 83, 91
    antifungal, 91
AIDS, 254, 256
AIDS-defining illness, 226
Alimentary tract, 226, 227
Allergen, 256
Allergic *Aspergillus* sinusitis, 21, 22, 24, 25, 28
Allylamines, 278–280
Alternative pathway, 234–237
Aminopeptidase, 44
Amoebidiales, 3–7, 12, 17
Amphipoda, 5
Amphotericin B, 25, 27, 28
Anabaena, 87
Anamorphic state, 264, 265
Anidulafungin, 226, 227
Anthropophilic, 264–275, 278, 280
Antibodies, 167, 174, 175, 268, 274
Antifungal chemotherapy, 230
Antifungal drugs, 156, 171
Antigens, 172–174
Antimycotic agents, 279, 280
Antioxidant defense, 171
Apoptosis, 255
Appressorium, 37, 47
APSES proteins
    EFG1 and EFH1, 207–208
    expression of metabolic genes, 207
    FLO8, 208
    morphogenesis, 207, 208
    overexpression, 208
Aquaporin, 90
Arthroconidia, 229
Arthrospores, 266, 267, 269–271, 278–280
Aryl hydrocarbon receptor, 124, 133
Asellariales, 3, 5, 7, 8, 13, 17
Aspergillosis, 70, 71, 76
a_w, 34
Azelaic acid, 121, 130, 132, 135
Azoles, 196, 278–280
    azole resistance, 197, 199
    fluconazole, 198, 201
    ketoconazol, 197
    sterol metabolism, 197

B
Basidiomycetes, 116, 120
Bassianin, 36, 37, 40
Bbchit1, 39, 50
BBP, 37
B. cylindrospora, 51
Beauveria, 35, 48, 52
Beauvericin, 40–42
Beauverolide L, 41
Benomyl, 199, 200
Bicarbonate, 85–87, 91
Biofilm formation, 99, 102, 108
    amino acid biosynthesis, 216
    Bcr1, 217
    cell surface contact, 216, 217
    farnesol, 216
    Gcn4p, 216
    hyphal formation, 216
    methionine and cysteine biosynthetic genes, 217
Biofilms, 99, 101–103, 106, 108, 171–172
Blastoconidia, 226, 229, 230
    clear circulating, 226
Blastospore, 34, 40, 43, 44, 47
Brain abscess, 22, 26, 28
Bronchopulmonary aspergillosis, 21, 22, 24, 25

C
Calcium oxalate, 24
cAMP
    Cdc35, 208–210
    Efg1, 209, 210
    Ras1p, 209
    yeast-to-hyphal transition, 209
Candida, 226, 227, 230
    antigens, 226
    blastoconidia, peripheral blood monocytes phagocytose, 226
    infections, invasive, 226
    osteomyelitis, 227
    species, 226, 227, 230
    species, fluconazole-resistant, 226
Candidiasis, 73, 76, 78, 83, 87, 169, 173–175, 226, 230, 249, 251
    disseminated, 83, 89
    esophageal, 226
    gastrointestinal, 226
    hepatic, 229
    hepatosplenic, 227
    mucocutaneous, 226
    oral, 226

Candidiasis (*continued*)
    renal, 227
    single-organ, 227
    superficial, 83
    tissue-proven, 227
    vaginal, 226
Capsule, 63, 64, 66, 69, 74, 254
Carbon dioxide, 84–91
    physiological, 85–88, 90, 91
    sensing, 84, 86, 87, 91
Carbonic anhydrase, 85–87, 91
CARD9, 248
Caspofungin, 25
Catalytic triad, 72
Catheters, 227, 230
    venous, 227, 229, 230
CD14, 253, 255
CDEP-1, 37
cDNA microarrays, 34, 37, 46, 52
CDR1/2 regulon, 200
Cell-mediated immunity (CMI), 227
Cell separation, 95, 97, 100
Cells, immune, 226, 227
Cellular immune systems, 47
Cellular response and cytokine production, 130
Cell wall, 116, 119, 121, 131, 168, 172–173
Cell wall biogenesis
    cell wall transcriptome, 207
    Efg1p, 207
Cell wall proteome, 172, 173
Ceratopogonidae, 5, 10
Chironomidae, 5, 10, 15, 16
Chitinase(s), 36, 38–39, 44, 46, 47, 49, 50
Chitin deacetylase, 39
Chitinolytic enzymes, 38, 46
Chlamydomonas reinhardtii, 90
Chromosomal genes, 49, 50
Chronic granulomatous disease, 22, 23
Chronic mucocutaneous candidiasis, 226
Chymoelastase, 45, 46
Chymoleastases, 36
Ciclopiroxolamine, iron chelator, 201
Classical pathway, 234, 235, 237
Clinical isolates resistant to antimycotics, 198–200
Coleoptera, 5
Collagenase, 36
Collembola, 5, 7
Colony morphology, 99, 101
Commensalism, 8
Complement, 233–240, 247
Complement system, 130
Conidia, 34, 35, 37, 39, 40, 43–46, 48, 50, 53, 156, 175, 264, 265, 280
Crustacea, 5
Cryptococcal meningitis, 228
Cryptococcosis, 227
    disseminated, 228
Cryptococcus neoformans, 227
CSF, 228
CT scan, 228
Culicidae, 5, 7, 9, 10, 15
Culturability, 7–8
Culture media, 35
Cutaneous cells, 131
Cutaneous immune response, 133
Cutaneous lesions, 229
Cuticle degradation, 36, 44
Cuticle-degrading enzymes, 35, 52
Cuticular penetration, 35
Cyclase, 103, 105, 107, 108
Cyclic AMP, 87, 89, 103–104
Cyclic peptide toxin, 40, 41
Cyclins, 96, 97
Cyclosporin A, 41
Cyclotetrapeptide, 41
Cytokines, 268
Cytokinesis, 95, 96, 98
Cytotoxic chemotherapy, 226

D
Database, 163, 164, 168, 177
Decapoda, 5
Defense mechanisms, 268
Defensins, 268
Deferoxamine, 25, 26
Dematiaceous molds, 21, 22, 28
Depigmention, 135
Depsipeptides, 41
Dermatitidis, 230
    model Wangiella, 230
Dermatophytes, 263–280
Dermatophytids, 269, 274
Dermatophytoses, 263, 267, 268, 275, 276, 278–280
Desmosterol, 12, 16
Destruxins (DTX), 35, 40, 41
Detection, 160–161
Diagnosis, 173
Dimorphic fungi, 243, 248
Dipeptidyl peptidase (Bb DPP), 48
Dipicolinic acid, 41
Diplopoda, 5, 7
Diptera, 5, 6, 8
Diseases, in animals, 139, 140
Disseminated candidiasis, 227
    acute, 227, 230
    chronic, 227
Disseminated infection, tissue-proven, 229
Dixidae, 5
DNA-microarrays, 107
DNA probes, 51, 52
DNA repair, 35, 39
Drug response element, 200
DTX. *See* Destruxins
DTX A, 41

E
Eccrinales, 3–8, 12, 13, 17
Echinocandines, 196
    β-1,3-glucan synthase, 198
    Caspofungin, 197
Echinocandins, 226, 227
    caspofungin, 227
Efrapeptins, 40, 41
Eicosanoids, 66, 69, 71
Electrospray ionisation, 161, 162

Endocytosis, 103, 105, 106
Entomopathogenic fungi, 33–45, 47–53
Enzymes, 267
EPF. *See* Entomopathogenic fungi
Ephemeroptera, 5
Epidemiological studies, 275, 276
Epidemiology, 264, 274
Ergosterol, 279
EST analysis, 34, 37, 49, 50, 52
Extracellular enzymes, 175
Extracellular proteins, 175

F
Factor H, 235–240
Farnesol, 103, 106
Fatty acid, 64–69, 71, 116, 119–121, 136
    lipid, 66–71, 76
Favus, 272
5-FC, 228
Filamentation, 84–90
Filaments, 115, 119, 120, 132, 140
Flocculins, 108
Fluconazole, 226–230
Flucytosine, 228
Fluorescent dyes, 160
Fluorochromes, 122, 135
Fluphenazine, 199, 200
Folate, 77
    para-aminobenzoic acid (PABA) 77
Folliculitis, 230
Fractionation, 165–167
Fungal cell wall
    alpha-glucan, 248
    beta glucan, 247
    galactomannan, 247
    glucoronoylomannan (GXM), 254, 255
    mannan, 247, 256
Fungal mycosis, 156
Fungus ball, 24, 25

G
2D gel electrophoresis, 158–163
General stress response, 202
    cross-protection, 202
Genome sequence(s), 156, 217
    candida albicans, 188, 189
        fungal genome sequencing projects, 188
        saccharomyces cerevisiae, 188
Genome-wide microarrays, 196, 205
Genome-wide transcriptomics, 188
Genomic sequences, 188, 217
Geographical origin, 50, 51
Geophilic, 266, 267, 269, 270, 273, 280
Germination, 33, 34, 39, 42–44, 47, 49, 53
Gluconeogenesis, 67, 77, 78
Glycolysis, 77, 78
Glyoxylate cycle, 67–71, 78
    isocitrate lyase, 67–71
    malate synthase, 67, 68
GPI-anchored proteins, 107, 108
Granulocytopenic, 228, 229
Griseofulvin, 278–280
GTPase, 96, 103, 105

H
Harpellales, 3–8, 10, 12, 13, 17
Heat shock, 34
Hematophagous insects, 53
Hemocoel, 35, 43, 47
Hemocytes, 43, 47, 48
Hemolymph, 35, 36, 38, 41, 43, 47, 48
Hexadepsipeptides, 41
Hirsutide, 40, 41
HIV infection, 226–228
Host complement regulators, 236, 238–240
Host-pathogen interaction
    antioxidant genes, 214
    arginine biosynthetic pathway, 213
    cell surface genes, 216
    gluconeogenesis/glyoxylate pathways, 213
    glycolysis, 213
    growth inhibition, 214
    human blood, 214
    in vitro adhesion assay, 216
    macrophages, 212–214
    mammalian epithelia, 215
    methionine and arginine pathways, 212–213
    neutrophils, 194, 212–213
    nitrogen and carbohydrate starvation, 214
    oxidative stress responses, 213
    polymorphonuclear leukocytes, 214
    starvation conditions, 213
Hosts, 3, 5–8, 10–3, 15–17, 227, 230
Host specificity, 6, 7
    defenses, 226, 227
    mucosal, 226
    passing mucosal, 227
Human pathogenic yeast, 233, 236, 237, 239
Hydrophobin(s), 36, 43, 44
Hydroxypyridones, 278, 280
Hygiene, 273, 278, 279
Hyperpigmentation, 121, 132, 135, 139
Hyphae, 84, 87–91
    hyaline, 229
Hyphal growth, 95, 97, 99, 102–109
Hyphomycetes, 33

I
Image analysis, 161
Immune evasion, 233, 238, 247, 248, 250, 253
Immune homeostasis, 247, 251
Immune response, 233, 234, 237, 238, 268, 274
Immune system, 176
Immunity, 42, 52, 268, 272, 278
Immunocompromised patient, 252, 255
Immunological memory, 243, 251
Immunoproteomics, 173–175
Immunosuppressive, 155
Induction of resistance, 200
Infections, 155, 156, 166, 170, 171, 173, 174, 176
    cryptococcal, 227
Infectious disease, 187
Innate immune system, 233, 234, 238, 243, 252
Insecta, 5
Integrative data analyses, 177
Invasive aspergillosis, 156, 175, 252, 254
Invasive *aspergillus* sinusitis, 22, 23, 25

Invasive candidiasis, 226, 227
Invasive fungal infections, 230
Invasive pulmonary aspergillosis, 22–26
Invasive pulmonary mucormycosis, 22, 26, 27
IPG dry strips, 159, 160
IPG strips, 159, 160, 166
IPM, 53
Iron, 156
Isoelectric focusing, 159–160
Isoelectric point, 157, 159, 166, 168
Isopoda, 5
Isotopic tag, 165
Isozymes, 44, 45
Itraconazole, 195, 196, 198, 226, 230
ITS, 50

K
Keratin, 263, 267, 268, 279
Keratinocytes, 124, 131–133, 135, 139
Keto-malassezin, 124, 125, 136

L
LC-MS/MS, 164, 165, 167, 173
Lecciones, 227
Lectin pathway, 234, 235
Lesions, 230
Lethal, 8–9, 16, 17
Leucinostins, 40, 41
Leucin-rich repeats (LRR), 245
Linear and cyclic peptide toxins, 41–42
Linear peptide toxins, 40, 41
Lipases, 39, 120, 140
Lipids, 230
    amphotericin, 227, 228
    discontinuing parenteral, 230
    formulations, 227
Lipid-supplemented total parenteral nutrition, 230
Lipohilic yeasts, 115, 230
Lipoxygenases, 39
Liquid chromatography, 163–164
Lysine, 77

M
Macroconidia, 264, 265
Macrophages, 64, 66, 68–70, 73, 74, 77, 78, 164, 176–177, 226
    phagocytosis, 64, 69, 70, 75, 78
Maintenance therapy, 228
Majocchi's granuloma, 274
Malassezin, 124, 125, 133, 136
MAP-kinase cascade, 104–106, 109
Marine, 5–7, 13
Mass spectrometry, 161, 163–165
Mating
    opaque cells, 211
    positive regulator, 211, 212
Matrix-assisted laser desorption/ionisation, 161, 162
Melanin, 156, 172
    production, 121
Melanization, 38, 47
Melanocytes, 121, 124, 125, 132, 133, 135
Membrane proteins, 165–166, 168
Meningoencephalitis, 227
    cryptococcal, 228

Mental status, altered, 228
Mesomycetozoea, 3, 4
Metabolism, 63, 64, 67, 69–71, 76–78, 85, 91
Metalloprotease, 38, 45, 49
Methylcitrate cycle, 76
    methylcitrate synthase, 76
M. Furfur
    folliculitis, 230
    fungemia, 230
    yeasts, 230
Micafungin, 226, 227
Microarray, 156, 168, 188, 192, 193, 195–201, 203–208, 210, 214, 217
Microconidia, 265
Microenvironment, 44, 45
Mitochondria, 165, 169
Mitochondrial genes, 50
Mitochondrial genome, 50, 51
Mitogen-activated protein kinase, 87–90
Morphogenesis, 96, 97, 102–109
    CPH1, 206–207
    EFG1, 206–210
    hyphal induction, 206–208, 216
    yeast to hyphae transition, 206–207, 210
Morpholines, 278, 280
Mortality, 229, 230
Mucormycosis, 21–23, 25–27, 29
Mucosal
    candidiasis, 226
    experimental, 226
    immunity, 226
    refractory, 226
MyD88, 245, 249–251, 255, 256

N
Natural killer (NK) cells, 252
NER. *See* Nucleotide excision repair
Neutropenia, 22–26
Neutropenic patients, 226, 227, 229
Neutrophils, 226, 229
Nitric oxide, 202, 203
Non-peptide toxins, 40, 41
Nuclear factor (NF-kB), 246
Nuclear migration, 99, 100
Nucleotide analoga, 196
    5-flucytosine, 197
    purine and pyrimidine biosynthesis, 198
Nucleotide excision repair, 39, 40

O
Onychomycosis, 270, 271, 273, 275, 276, 278, 279
Oosporin, 40, 41
Opportunistic fungal pathogens, 188, 193–194
Opportunistic fungi, 243, 248, 249
Opportunistic infections, *Candida*, 187
Opportunistic pathogen, 156
Opsonisation, 235, 236, 247, 254
Oropharyngeal candidiasis, 226
Osmoadaptation, 170
Ovarian cysts, 9–11, 13
Oxalic acid, 40, 41, 48
Oxidative stress, 169, 170, 174, 176

P

Pathogen associated molecular pattern (PAMP)
    mannan, 250, 256
    phospholipomannan, 256
    ß1–3 glucan, 247, 248, 250
Pathogenesis
    dense biological network, 217
    individual virulence, 217
    metabolic pathways, 217
    new diagnostics, therapeutics and vaccines, 217
Pathogenicity, 33, 35, 42, 44, 47–53, 95, 107
Pathophysiology, 267
Pattern recognition receptor
    complement, 247
    DC-SIGN, 247
    dectin-1, 247, 248
    mannose receptor, 247
    mannose-binding lectin, 247
    pentraxin-3 (PTX3), 247
    SIGNR1, 247
    surfactant protein, 247
    Toll-like receptor (TLR), 247, 248
PEG, 34
Penetration peg, 35
Peptide mass fingerprinting, 163
Persist-uncorrected proof, 229
Phagocyte
    alveolar macrophage, 252
    dendritic cell, 243
    macrophage, 243, 247, 248, 252, 254
    monocyte, 248, 252
    neutrophil, 243, 252
Phagocytosis, 244, 246, 247, 252, 254, 255
Pharmacodynamics, 198
Phenoloxidase, 38, 48
Phenotypic degeneration, 35
Phialidic conidiogenesis, 116
Phospholipase, 64–67, 69, 71
    phospholipase A, 64, 65, 67
    phospholipase activity, 65–67
    phospholipase B, 64–67, 69
    phospholipase C, 64–67
    phospholipase D, 64–66
Phospholipids, 64–67, 69, 71, 91
    glycerophospholipids, 64
    lysophospholipids, 65, 67
    phosphoglycerols, 64
    sphingolipids, 64, 66
Phosphoproteins, 166–167
Phosphorylation, 166–167
pH stress, 204
Pigment, 119, 121, 122, 125, 127, 134–136, 140, 141
Pigmentation, 39
Pityriacitrin, 122, 125, 136
Pityrialactone, 122, 136
Pityriarubins, 123, 125, 126, 130, 136
Pityriasis versicolor, 115, 120, 122–125, 132, 134–135, 140
Plasminogen, 238–240
Plecoptera, 5
Polarisome, 97, 99, 105
Polarity, 95–97, 99, 106
Polyenes, 196
    amphotericin B, 197
    disruption of plasma membrane, 198
Polymorphism, 205, 206
Posaconazole, 25, 27, 28
Post-translational modifications, 157, 163, 166
Pr1, 36–38, 44–46, 49
Pr1A, 36–38, 41
Pr1B, 36–38, 41
Pr2, 45, 46
Precipitation, 159, 168, 175
Propensity, high, 230
Prophenol oxidase, 36
Protease, 34, 36–38, 44, 45, 47–49, 52, 64, 71–76
    aspartic protease, 72–74, 76, 78
    cysteine protease, 72
    endoprotease, 71, 72, 76
    exoproteases, 71, 72, 76
    metalloprotease, 72, 73, 76
    protease activity, 73, 74
    proteolytic activity, 73–75
    sedolisins, 76
    serine protease, 72, 75, 76
Protein, 161–163, 171
Protein kinase A, 87–91
Proteomics, 164–166, 168–172
Protoplast fusion, 48
Protoplasts, 38, 48
PRR polymorphism, 254
Pseudohyphae, 226
Psoriasis, 134, 137, 139
Psychodidae, 5

R

RAPD, 38, 49, 51, 52
RAPD-PCR, 49, 51
Reference map, 169
Regulatory T cell (Treg), 249, 251
Regulons, 199, 210
Resistance mechanisms to antimycotics, 195–202
RFLP, 37, 38, 50, 52
Rhesus, 90
Rhinocerebral mucormycosis, 22, 23, 26–27
Ribosomal RNA, 264
Ringworm, 263, 267–269, 271, 279, 280
ROS, 123, 130
RT-PCR, 37

S

Sample, 158–159, 165–166, 168
Sample preparation, 158–159, 168
Saprophyte, 156
SDS-polyacrylamide gel electrophoresis, 158
Secondary metabolism, 122, 135, 140
Secreted proteins, 175
Seizures, 228
Serum, 84, 86–90
Signal transduction pathway, 84, 86–88
Simuliidae, 5, 6, 8–10, 15
Single-stranded conformational polymorphism (SSCP), 50, 52
Skin, 230
Small rRNA gene, 51

Social insect, 42, 52
SOD. *See* Superoxide dismutases
Solid-state fermentation, 35
Somatic hybrids, 48
Species, 263–277, 279, 280
Species differentiation, 120
Spitzenkörper, 96, 97, 99, 100, 106
Spleen, 226, 227
Sporangiospore, 5–7, 12–15, 17
Spore(s)
    extrusion, 13, 14
    germination, 39, 44, 53
    hydration, 44
SSF. *See* Solid-state fermentation
Staining, 160
Standards, 167
Steroid hormones, 268, 271
Stratum corneum, 263, 267, 269, 280
Submerged conidia, 34, 43, 44
Subtilisin-like protease, 36–38
Subtilisin(s), 36–38, 41, 49, 50
Superoxide dismutases, 40
Switch, phenotypic, 84, 91
Systemic infection, 243, 248, 249, 252
Systemic mycoses, 243, 251
Systems biology, 177

T
Tandem-mass spectrometry, 163
Taxonomy, 280
Teleomorph, 116
Teleomorphic state, 264
Tenellin, 40
T helper cell
    Th1, 246, 257
    Th2, 246, 256
Therapeutic interventions, 229
Therapy, 156, 228, 279, 280
    administered corticosteroid, 226
    inhalational corticosteroid, 226
    topical, 226
Tinea, 263, 265–276, 278–280
    T. barbae, 271, 272
    T. capitis, 265, 269, 271, 272, 275, 276, 278–280
    T. corporis, 272, 273, 275, 276, 278, 280
    T. cruris, 266, 273–276, 278, 280
    T. gladiatorum, 272
    T. manuum, 269, 270, 276
    T. pedis, 269, 270, 272, 275, 276, 280
    T. profunda, 269, 274, 278–280
    T. unguium, 270, 272, 275, 276, 280
Tipulidae, 5
Tissues
    gut-associated lymphoid, 226
    soft, 230
Toll-like receptors, 234, 238, 241
Topical treatment, 278, 279
Toxins, 40–42, 52, 53
TPN, lipid-supplemented, 230

Transcriptional repressors
    carbon metabolism, 210
    filamentous growth programme, 210
    NRE, 211
    Nrg1, 210
    Rfg1, 210
    Ssn6, 210
    Tup1, 210
    yeast to hyphal switch, 210
Transcriptomics
    of fungal pathogens, 193–194
    gene expression, 195, 217
Trehalase, 78
Trehalose, 70, 77, 78
Trichomycetes, 3–9, 12, 15–18
Trichoptera, 5
Trichosporon, 228, 229
    genus, 228
    infections, 229
        acute disseminated, 229
        disseminated, 229
    pneumonitis, 229
    spp., 228, 229
Tripeptidyl peptidase (Bb TPP), 48
Trypsin, 38, 45, 46, 48, 49
Tryptophan, 119, 122, 125–127, 130, 133–135, 140
Tyrosinase inhibitor, 125, 135

U
Uridine, 77

V
Vaccines, 174
Virulence, 33, 36–39, 41, 42, 44, 47–50, 52, 95, 102, 103, 105–110, 175–176
Virulence factors, 233
Voriconazole, 25, 226, 229, 230

W
Wangiella infections, 230
Water
    activity, 34
    stress, 34
White-opaque switching
    Efg1, 211
    homozygosis at the mating type locus, 211
    mating type locus, 211, 212
Wiskott–Aldrich syndrome protein, 106
Wood light, 267

Y
Yeast, 83–86, 89
    dematiaceous, 230

Z
Zoophilic, 265–267, 269–273, 275, 276, 280
Zygomycota, 3, 5, 6
Zygospores, 5, 6, 9, 17
Zymosan, 244, 247, 250, 254